Petrodiesel Fuels

Petrodiesel Fuels

Science, Technology, Health, and Environment

Edited by
Ozcan Konur

CRC Press is an imprint of the
Taylor & Francis Group, an **informa** business

First edition published 2021
by CRC Press
6000 Broken Sound Parkway NW, Suite 300, Boca Raton, FL 33487-2742

and by CRC Press
2 Park Square, Milton Park, Abingdon, Oxon, OX14 4RN

CRC Press is an imprint of Taylor & Francis Group, LLC

ISBN: 978-0-367-45616-0 (hbk)
ISBN: 978-0-367-70888-7 (pbk)
ISBN: 978-0-367-45625-2 (ebk)

Typeset in Times
by SPi Global, India

Contents

Preface

Crude oils have been primary sources of energy and fuels, such as petrodiesel. However, significant public concerns about the sustainability, price fluctuations, and adverse environmental impact of crude oils have emerged since the 1970s. Thus, biooils and biooil-based biodiesel fuels have emerged as alternatives to crude oils and crude oil-based petrodiesel fuels, respectively, in recent decades. Nowadays, although petrodiesel fuels are still used extensively, biodiesel fuels are being used increasingly as petrodiesel–biodiesel blends in the transportation and power sectors. Therefore, there has been great public interest in the development of environment and human-friendly and sustainable petrodiesel and biodiesel fuels. However, it is necessary to reduce the total cost of biodiesel production by reducing the feedstock cost through the improvement of biomass and lipid productivity. It is also necessary to mitigate the adverse impact of petrodiesel fuels on the environment and human health.

Although there have been over 1,500 reviews and book chapters in this field, there has been no review of the research as a representative sample of all the population studies done in the field of both petrodiesel and biodiesel fuels. Thus, this third volume of the handbook on petrodiesel fuels provides a representative sample of all the population studies in this field.

The major research fronts are determined from the sample and population paper-based scientometric studies. Table 1.1 presents information on the major and secondary research fronts emanating from these scientometric studies. There are four secondary research fronts in this third volume on petrodiesel fuels: crude oils, petrodiesel fuels in general, petrodiesel fuel emissions, and the health impact of these emissions.

As seen from Table 1.1 there is a substantial correlation between the distribution of research fronts in this handbook and the 100-most-cited sample papers. Papers on petrodiesel fuels are under-represented significantly in both this handbook and the 100-most-cited sample papers, compared to the population papers: 33, 34 v. 67.4%.

Table 1.2 extends Table 1.1 and provides more information on the content of this handbook, including chapter numbers, paper references, primary and secondary research fronts, and finally the titles of the papers presented.

The data presented in the tables and figures show that a small number of authors, institutions, funding bodies, journals, keywords, research fronts, subject categories, and countries have shaped the research in this field.

The findings show the importance of the progression of efficient incentive structures for the development of the research in this field as in other fields. It further seems that although research funding is a significant element of these incentive structures, it might not be the sole solution for increasing the incentives for research. On the other hand, it seems there is more to do to reduce the significant gender deficit in this field, as in other fields of science and technology.

The information provided on nanotechnology applications suggests that there is ample scope for the expansion of advanced applications in this research field.

Following substantial public concerns about the adverse impact of the emissions of petrodiesel fuels on the environment and human health, the research has intensified in the areas related to the reduction of these adverse effects. Thus, bioremediation of spills from crude oils and petrodiesel fuels at sea and in soils, as well as the desulfurization of petrodiesel fuels, have emerged as publicly important research areas.

Similarly, emissions from diesel fuel exhausts, due to their adverse effects on both human health and the environment, have been researched more in recent years. These emissions cover particulate emissions, aerosol emissions, and NO_x emissions.

The research on the adverse health impact of petrodiesel fuel exhaust emissions has primarily progressed along the lines of respiratory illnesses, cancer, and other illnesses, such as cardiovascular illnesses, brain illnesses, and reproductive system illnesses through human, animal, and *in vitro* studies.

It is clear that these illnesses caused by petrodiesel fuel exhaust emissions have been one of the most significant reasons to develop alternative biodiesel fuels.

Thus, this handbook is a valuable source for stakeholders primarily in the research fields of Energy Fuels, Chemical Engineering, Environmental Sciences, Biotechnology and Applied Microbiology, Physical Chemistry, Petroleum Engineering, Environmental Engineering, Multidisciplinary Chemistry, Thermodynamics, Analytical Chemistry, Mechanical Engineering, Agricultural Engineering, Marine Freshwater Biology, Green Sustainable Science Technology, Applied Chemistry, Multidisciplinary Geosciences, Microbiology, Multidisciplinary Materials Science, Mechanics, Toxicology, Multidisciplinary Sciences, Biochemistry and Molecular Biology, Water Resources, Plant Sciences, Multidisciplinary Engineering, Transportation Science Technology, Geochemistry and Geophysics, Food Science Technology, Ecology, Public Environmental Occupational Health, Meteorology and Atmospheric Sciences, Electrochemistry, and Biochemical Research Methods.

This handbook is also particularly relevant in the context of biomedical sciences for Public and Environmental Occupational Health, Pharmacology, Immunology, Respiratory System, Allergy, Genetics Heredity, Oncology, Experimental Medical Research, Critical Care Medicine, General Internal Medicine, Cardiovascular Systems, Physiology, Medicinal Chemistry, and Endocrinology and Metabolism.

Ozcan Konur

Acknowledgements

This handbook was a multi-stakeholder project from its inception to its publication. CRC Press and Taylor & Francis Group were the major stakeholders in financing and executing it. Marc Gutierrez was the executive editor. Eighty-three authors have kindly contributed chapters despite the relatively low level of incentives, compared to journals. As stated in many chapters, a small number of highly cited scholars have shaped the research on biodiesel and petrodiesel fuels. The contribution of all these stakeholders is greatly acknowledged.

Acknowledgments

[illegible]

Editor's Biography

Ozcan Konur, as both a materials scientist and social scientist by training, has focused on the bibliometric evaluation of research in the innovative high-priority research areas of algal materials and nanomaterials for energy and fuel as well as for biomedicine at the level of researchers, journals, institutions, countries, and research areas, including the social implications of the research conducted in these areas.

He has also researched extensively in the development of social policies for disadvantaged people on the basis of disability, age, religious beliefs, race, gender, and sexuality at the interface of science and policy.

He has edited a book titled *Bioenergy and Biofuels* (CRC Press, January 2018) and a handbook titled *Handbook of Algal Science, Technology, and Medicine* (Elsevier, April 2020).

Contributors

F. Buttignol
Paul Scherrer Institute, Switzerland

I. M. Rizwanul Fattah
University of Technology Sydney, Australia

Merv Fingas
Spill Science, Canada

Jianbing Gao
University of Leeds, UK

Bryan M. Hedgpeth
ExxonMobil Biomedical Sciences Inc., USA

Adrian Ilinca
University of Quebec, Canada

Mohamad Issa
Quebec Maritime Institute, Canada

Moon Sik Jeong
Hanyang University, South Korea

Ozcan Konur
Formerly, Ankara Yildirim Beyazit University, Turkey

O. Kröcher
Paul Scherrer Institute; Federal Institute of Technology Lausanne, Switzerland

Kun Sang Lee
Hanyang University, South Korea

Ye Liu
University of Leeds, UK

T. M. Indra Mahlia
University of Technology Sydney, Australia

A. Marberger
Hug Engineering AG, Switzerland

Kelly M. McFarlin
ExxonMobil Biomedical Sciences Inc., USA

M. Mofijur
University of Technology Sydney, Australia

R. J. G. Nuguid
Paul Scherrer Institute, Switzerland

Hwai Chyuan Ong
University of Technology Sydney, Australia

Shyam Pandey
University of Petroleum and Energy Studies, India

Roger C. Prince
Stonybrook Apiary, USA

Amit Kumar Sharma
University of Petroleum and Energy Studies, India

Part IX

Crude Oils

40 Crude Oils

A Scientometric Review of the Research

Ozcan Konur

CONTENTS

40.1 INTRODUCTION

Crude oil-based fuels such as diesel fuels have been primary sources of energy and fuels (Chisti, 2007, 2008; Konur, 2012g, 2015; Lapuerta et al., 2008; Marchetti et al., 2007; Srivastava and Prasad, 2000; van Gerpen, 2005). However, significant public concerns about the sustainability, price fluctuations, and adverse environmental impact of crude oils have emerged since the 1970s (Ahmadun et al., 2009; Atlas, 1981; Babich and Moulijn, 2003; Kilian, 2009; Moldowan et al., 1985; Perron, 1989;

Peters, 1986; Peterson et al., 2003; Sadorsky, 1999; van Hamme et al., 2003; Zivot and Andrews, 2002). Thus, studies on the science and technology of crude oils have arisen as a distinct research field (Ahmadun et al., 2009; Atlas, 1981; Babich and Moulijn, 2003; Moldowan et al., 1985; Peters, 1986; Peterson et al., 2003; van Hamme et al., 2003). Although biooil-based biodiesel fuels have begun to be used extensively in recent decades, crude oil-based petrodiesel fuels are still used (Konur, 2021a–g).

However, for the efficient progression of the research in this field, it is necessary to develop efficient incentive structures for the primary stakeholders and to inform these stakeholders about the research (Konur, 2000, 2002a–c, 2006a–b, 2007a–b; North, 1991a–b).

Scientometric analysis offers ways to evaluate the research in a respective field (Garfield, 1955, 1972; Konur, 2011, 2012a–n, 2015, 2016a–f, 2017a–f, 2018a–b, 2019a–b). However, there is no scientometric study of this field.

This chapter presents such a study using two datasets. The first dataset includes the 100-most-cited papers (n = 100 sample papers) whilst the second set includes population papers (n = over 53,000 population papers) published between 1980 and 2019.

The data on the indices, document types, authors, institutions, funding bodies, source titles, 'Web of Science' subject categories, keywords, research fronts, and citation impact are presented and discussed.

40.2 MATERIALS AND METHODOLOGY

The search for the literature was carried out in the 'Web of Science' database in January 2020. It contains the 'Science Citation Index-Expanded' (SCI-E), the 'Social Sciences Citation Index' (SSCI), the 'Book Citation Index-Science' (BCI-S), the 'Conference Proceedings Citation Index-Science' (CPCI-S), the 'Emerging Sources Citation Index' (ESCI), the 'Book Citation Index-Social Sciences and Humanities' (BCI-SSH), the 'Conference Proceedings Citation Index-Social Sciences and Humanities' (CPCI-SSH), and the 'Arts and Humanities Citation Index' (A&HCI).

The keywords for the search of the literature were collated from the screening of the abstract pages for the first 1,000 highly cited papers. This keyword set is provided in the Appendix. The papers on econometric studies relating to the price fluctuations of crude oils were excluded (Kilian, 2009; Perron, 1989; Sadorsky, 1999; Zivot and Andrews, 2002).

Two datasets are used for this study. The highly cited 100 papers comprise the first dataset (sample dataset, n = 100 papers) whilst all the papers form the second dataset (population dataset, n = over 53,000 papers).

The data on the indices, document types, publication years, institutions, funding bodies, source titles, countries, 'Web of Science' subject categories, citation impact, keywords, and research fronts are collated from these datasets. The key findings are provided in the relevant tables and figure, supplemented with explanatory notes in the text. The findings are discussed, a number of conclusions are drawn, and a number of recommendations for further study are made.

40.3 RESULTS

40.3.1 Indices and Documents

There are over 71,600 papers related to crude oils in the 'Web of Science' as of January 2020. This original population dataset was refined for the document type (article, review, book chapter, book, editorial material, note, and letter) and language (English), resulting in over 53,000 papers comprising over 74% of the original population dataset.

The primary index is the SCI-E for both the sample and population papers. About 92% of the population papers are indexed by the SCI-E database. Additionally 5.4, 3.3, and 2.1% of these papers are indexed by the CPCI-S, ESCI, and BCI-S databases, respectively. The papers on the social and humanitarian aspects of this field are relatively negligible with 4.0 and 0.2% of the population papers indexed by the SSCI and A&HCI, respectively.

Brief information on the document types for both datasets is provided in Table 40.1. The key finding is that article types of documents are the primary documents for the population papers whilst reviews form 35% of the sample papers. Articles are under-represented by –26.7% whilst reviews are over-represented by 32.4% in the sample papers.

40.3.2 Authors

Brief information about the most-prolific 23 authors with at least two sample papers each is provided in Table 40.2. Around 370 and 91,500 authors contribute to the sample and population papers, respectively.

The most-prolific authors are 'Ronald M. Atlas', 'Terry C. Hazen', 'Ian M. Head', and 'D. Martin Jones' with three sample papers each, all working on 'oil spill bioremediation'. The other researchers have two papers each. The top four authors have the most impact.

TABLE 40.1
Document Types

	Document Type	Sample Dataset (%)	Population Dataset (%)	Difference (%)
1	Article	65	91.7	–26.7
2	Review	35	2.6	32.4
3	Book chapter	0	2.2	–2.2
4	Proceeding paper	6	5.4	0.6
5	Editorial material	0	3.7	–3.7
6	Letter	0	0.8	–0.8
7	Book	0	0.2	–0.2
8	Note	0	0.9	–0.9

TABLE 40.2
Authors

	Author	Sample Papers (%)	Population Papers (%)	Surplus (%)	Institution	Country	Research Front
1	Atlas, Ronald M.	3	0.1	2.9	Univ. Louisville	USA	Oil spill bioremediation
2	Hazen, Terry C.	3	0.1	2.9	Univ. Calif. Berkeley	USA	Oil spill bioremediation
3	Head, Ian M.	3	0.1	2.9	Univ. Newcastle	UK	Oil spill bioremediation
4	Jones, D. Martin	3	0.1	2.9	Univ. Newcastle	UK	Oil spill bioremediation
5	Camilli, Richard	2	0.1	1.9	Woods Hole Oceanog. Inst.	USA	Oil spill bioremediation
6	Corma, Avelino*	2	0.1	1.9	Univ. Polytech. Valencia	Spain	Crude oil refining
7	Golyshin, Peter N.	2	0.1	1.9	GBF	Germany	Hydrocarbon bioremediation
8	Larter, Steve R.	2	0.1	1.9	Univ. Newcastle	UK	Oil spill bioremediation
9	Marshall, Alan G.	2	0.1	1.9	Florida State Univ.	USA	Crude oil characterization
10	Mcintyre, Cameron P.	2	0.1	1.9	Woods Hole Oceanog. Inst.	USA	Oil spill bioremediation
11	Moldowan, J. Michael	2	0.1	1.9	Chevron	USA	Crude oil geology
12	Morrow, Norman R.	2	0.1	1.9	New Mexico Inst. Mining Technol.	USA	Crude oil recovery
13	Mullins, Oliver C.	2	0.1	1.9	Schlumberger	USA	Crude oil characterization
14	Peters, Kenneth E.	2	0.1	1.9	Chevron	USA	Crude oil geology
15	Rabus, Ralf	2	0.1	1.9	Albert Ludwigs Univ.	Germany	Hydrocarbon bioremediation
16	Reddy, Christopher M.	2	0.1	1.9	Woods Hole Oceanog. Inst.	USA	Oil spill bioremediation
17	Rodgers, Ryan P.	2	0.1	1.9	Florida State Univ.	USA	Crude oil characterization
18	Sylva, Sean P.	2	0.1	1.9	Woods Hole Oceanog. Inst.	USA	Oil spill bioremediation
19	Timmis, Kenneth N.	2	0.1	1.9	Helmholtz Assoc.	Germany	Hydrocarbon bioremediation
20	Van Mooy, Benjamin A.S.	2	0.1	1.9	Woods Hole Oceanog. Inst.	USA	Oil spill bioremediation
21	Wang, Zihendi	2	0.1	1.9	Env. Canada	Canada	Oil spill bioremediation
22	Widdel, Friedrich	2	0.1	1.9	Max Planck Inst.	Germany	Hydrocarbon bioremediation
23	Yakimov, Michail M.	2	0.1	1.9	Ital. Res. Ctr.	Italy	Hydrocarbon bioremediation

* Highly cited researcher in 2019.

The most-prolific institution for these top authors is 'Woods Hole Oceanographic Institute' of the USA with five, followed by the 'University of Newcastle' of the UK with three authors. The other prolific institutions with two authors are 'Chevron' and 'Florida State University' of the USA. In total, 15 institutions house these top authors.

It is notable that only one of these top researchers are listed in the 'Highly Cited Researchers' (HCR) in 2019 (Clarivate Analytics, 2019; Docampo and Cram, 2019).

The most-prolific country for these top authors is the USA with 13 authors. The other prolific countries are Germany and the UK with four and three authors, respectively. In total, six countries contribute to these top papers.

There are six key research fronts for these top researchers. The top front is 'crude oil spill bioremediation' with 11 authors. The other prolific research fronts are 'hydrocarbon bioremediation', 'crude oil characterization', and 'crude oil geology' with five, three, and two authors, respectively.

It is further notable that there is a significant gender deficit among these top authors as all the researchers are male (Lariviere et al., 2013; Xie and Shauman, 1998).

40.3.3 Publication Years

Information about the publication years for both datasets is provided in Figure 40.1. This figure shows that 14, 33, 40, and 13% of the sample papers and 10.3, 13.0, 21.1, and 55.1% of the population papers were published in the 1980s, 1990s, 2000s, and 2010s, respectively.

Similarly, the most-prolific publication years for the sample dataset are 1998, 2000, and 2003 with seven papers each. On the other hand, the most-prolific publication years for the population dataset are 2015, 2016, 2017, 2018, and 2019 with at

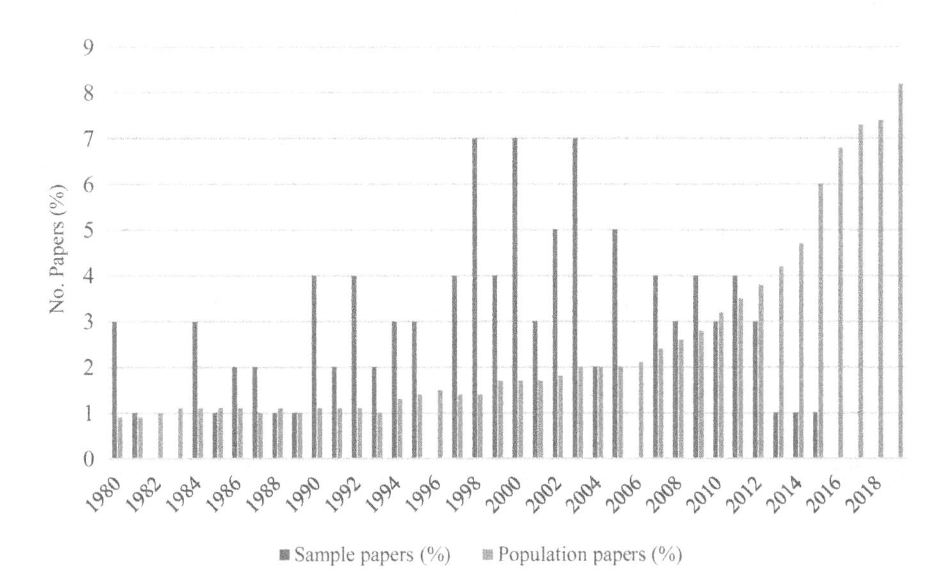

FIGURE 40.1 Research output between 1980 and 2019.

least 6% of the population papers each. It is notable that there is a sharply rising trend for the population papers, particularly in the 2010s.

40.3.4 Institutions

Brief information on the top 14 institutions with at least 3% of the sample papers each is provided in Table 40.3. In total, around 170 and 16,400 institutions contribute to the sample and population papers, respectively.

These top institutions publish 53.0 and 7.1% of the sample and population papers, respectively. The top institution is 'Woods Hole Oceanographic Institution' of the USA with seven sample papers and a 6.8% publication surplus. The other top institutions are 'British Petroleum', 'Chevron', 'Helmholtz Association', 'Newcastle University', the US 'Department of Interior', the US 'Geological Survey', and the 'University of California Berkeley' with four sample papers each.

The most-prolific country for these top institutions is the USA with 10 institutions. The other countries are France, Germany, and the UK.

The institutions with the most impact are 'Woods Hole Oceanographic Institution', 'Newcastle University', the 'University of California Berkeley', 'British Petroleum', and 'Chevron' with at least a 3.5% publication surplus each. On the other hand, the institutions with the least impact are 'Florida State University', the 'University of Wyoming', the 'French Institute of Petroleum', and the French 'Scientific Research National Center' with at least a 1.7% publication surplus each.

TABLE 40.3
Institutions

	Institution	Country	No. of Sample Papers (%)	No. of Population Papers (%)	Difference (%)
1	Woods Hole Oceanogr. Inst.	USA	7	0.2	6.8
2	British Petroleum	UK	4	0.5	3.5
3	Chevron	USA	4	0.5	3.5
4	Helmholtz Assoc.	Germany	4	0.8	3.2
5	Newcastle Univ.	UK	4	0.3	3.7
6	Dept. Interior	USA	4	0.9	3.1
7	Geol. Survey	USA	4	0.7	3.3
8	Univ. Calif. Berkeley	USA	4	0.3	3.7
9	Sci. Res. Natl. Ctr.	France	3	1.3	1.7
10	Florida State Univ.	USA	3	0.3	2.7
11	French Inst. Petrol.	France	3	0.6	2.4
12	Lawrence Berkeley Natl. Lab.	USA	3	0.2	2.8
13	Univ. Louisville	USA	3	0.2	2.8
14	Univ. Wyoming	USA	3	0.3	2.7

It is notable that some institutions with a heavy presence in the population papers are under-represented in the sample papers: the 'China University of Petroleum', the 'University of Alberta', the 'Chinese Academy of Sciences', the 'Russian Academy of Sciences', the 'China National Petroleum Corporation', and the 'University of Calgary' with at least a 1% presence in the population papers each.

40.3.5 Funding Bodies

Brief information about the top seven funding bodies with at least 3% of the sample papers each is provided in Table 40.4. It is significant that only 74 and 38% of the sample and population papers declare any funding, respectively.

The top funding body is the 'National Natural Science Foundation of China', funding 26.0% and 7.4% of the sample and population papers, respectively, with an 18.6% publication surplus. The other top funding bodies are the 'China Postdoctoral Science Foundation', the 'Fundamental Research Funds for the Central Universities', and the 'National Key R D Program of China' with six, four, and four sample papers, respectively.

It is notable that some top funding agencies for the population studies do not enter this top funding body list. Some of them are the 'Fundamental Research Funds for the Central Universities' of China (1.6%), the 'National Science Foundation' of the USA (0.9%), CAPES of Brazil (0.7%), the 'European Union' (0.5%), 'Petrobras' of Brazil (0.5%), the 'National Institutes of Health' of the USA (0.4%), the 'Natural Science Foundation of Shandong Province' of China (0.4%), the 'Chinese Academy of Sciences' (0.4%), and the 'Research Council of Norway' (0.4%).

It is notable that the most-prolific country for these top funding bodies is China with five.

TABLE 40.4
Funding Bodies

	Institution	Country	No. of Sample Papers (%)	No. of Population Papers (%)	Difference (%)
1	National Natural Science Foundation of China	China	26	7.4	18.6
2	China Postdoctoral Science Foundation	China	6	0.5	5.5
3	Fundamental Research Funds for the Central Universities	China	4	1.3	2.7
4	National Key R D Program of China	China	4	0.1	3.9
5	National Council for Scientific and Technological Development	Brazil	3	1.1	1.9
6	National Key Research and Development Program of China	China	3	0.2	2.8
7	Natural Sciences and Engineering Research Council of Canada	Canada	3	2.0	1.0

40.3.6 Source Titles

Brief information about the top seven source titles with at least three sample papers each is provided in Table 40.5. In total, 56 and 5,150 source titles publish the sample and population papers, respectively. On the other hand, these top seven journals publish 36.0 and 3.5% of the sample and population papers, respectively.

The top journal is 'Applied and Environmental Microbiology', publishing eight sample papers with a 7.6% publication surplus. This top journal is closely followed by the 'AAPG Bulletin American Association of Petroleum Geologists' and 'Nature' with seven sample papers and a more than 6.3% publication surplus each.

Although these journals are indexed by seven subject categories, the top categories are 'Biotechnology and Applied Microbiology', 'Multidisciplinary Sciences', 'Engineering Environmental', and 'Environmental Sciences'.

The journals with the most impact are 'Applied and Environmental Microbiology', 'Nature', and the 'AAPG Bulletin American Association of Petroleum Geologists' with at least a 6.3% publication surplus each. On the other hand, the journals with the least impact are the 'Journal of Hazardous Materials', 'Current Opinion in Biotechnology', and 'Science' with at least a 2.3% publication surplus each.

It is notable that some journals are relatively under-represented in the sample papers. Some of them are 'Energy Fuels', 'Fuel', the 'Oil Gas Journal', 'Petroleum Science and Technology', the 'Journal of Petroleum Science and Engineering', and the 'Marine Pollution Bulletin' with at least a 1.7% presence in the population papers each.

TABLE 40.5
Source Titles

Source Title	WOS Subject Category	No. of Sample Papers (%)	No. of Population Papers (%)	Difference (%)
1 Applied and Environmental Microbiology	Biot. Appl. Microb., Microbiol.	8	0.4	7.6
2 AAPG Bulletin American Association of Petroleum Geologists	Geosci. Mult.	7	0.7	6.3
3 Nature	Mult. Sci.	7	0.2	6.8
4 Environmental Science Technology	Eng. Env., Env. Sci.	5	1.3	3.7
5 Current Opinion in Biotechnology	Bioch. Mol. Biol., Biot. Appl. Microb.	3	0.1	2.9
6 Journal of Hazardous Materials	Eng. Env., Env. Sci.	3	0.7	2.3
7 Science	Mult. Sci.	3	0.1	2.9
		36	3.5	32.5

40.3.7 Countries

Brief information about the top ten countries with at least three sample papers each is provided in Table 40.6. In total, 26 and over 270 countries contribute to the sample and population papers, respectively.

The top country is the USA, publishing 43.0 and 22.1% of the sample and population papers, respectively. The other prolific countries are the UK, Germany, and Canada publishing 16, 11, and 10 sample papers, respectively.

On the other hand, China publishes 3.0 and 16.4% of the sample and population papers, respectively. The European and Asian countries represented in this table publish altogether 42 and 10% of the sample papers and 14.3 and 22.1% of the population papers, respectively.

It is notable that the publication surplus for the USA and these European and Asian countries is 20.9, 27.7, and –12.1%, respectively.

It is also notable that some countries are relatively under-represented in the sample papers. Some of them are Russia, Iran, Brazil, Japan, Spain, South Korea, and Italy with at least a 1.6% presence in the population papers each.

40.3.8 Web of Science Subject Categories

Brief information about the top 15 'Web of Science' subject categories with at least three sample papers each is provided in Table 40.7. The sample and population papers are indexed by 34 and 221 subject categories, respectively.

For the sample papers, the top subjects are 'Biotechnology Applied Microbiology', 'Microbiology', and 'Environmental Sciences' with 18, 17, and 15 papers, respectively. The other prolific subjects are 'Energy Fuels', 'Multidisciplinary Sciences', 'Engineering Chemical', and 'Chemistry Physical' with at least ten papers each.

TABLE 40.6
Countries

	Country	No. of Sample Papers (%)	No. of Population Papers (%)	Difference (%)
1	USA	43	22.1	20.9
2	UK	16	5.7	10.3
3	Germany	11	2.7	8.3
4	Canada	10	10.6	–0.6
5	France	6	3.3	2.7
6	Netherlands	6	1.3	4.7
7	Australia	4	2.6	1.4
8	India	3	3.1	–0.1
9	Norway	3	1.3	1.7
10	China	3	16.4	–13.4
	Europe-5	42	14.3	27.7
	Asia-3	10	22.1	–12.1

TABLE 40.7
'Web of Science' Subject Categories

	Subject	No. of Sample Papers (%)	No. of Population Papers (%)	Difference (%)
1	Biotechnology Applied Microbiology	18	4.4	13.6
2	Microbiology	17	3.2	13.8
3	Environmental Sciences	15	18.8	–3.8
4	Energy Fuels	12	26.9	–14.9
5	Multidisciplinary Sciences	12	2.1	9.9
6	Engineering Chemical	11	23.9	–12.9
7	Chemistry Physical	10	5.7	4.3
8	Chemistry Multidisciplinary	9	3.9	5.1
9	Engineering Environmental	9	6.4	2.6
10	Geosciences Multidisciplinary	9	7.1	1.9
11	Materials Science Multidisciplinary	8	2.8	5.2
12	Engineering Petroleum	5	17.1	–12.1
13	Biochemical Research Methods	4	0.8	3.2
14	Geochemistry Geophysics	4	2.9	1.1
15	Marine Freshwater Biology	4	3.5	0.5

It is notable that the publication surplus is most significant for 'Microbiology', 'Biotechnology Applied Microbiology', and 'Multidisciplinary Sciences' with 13.8, 13.6, and 9.9% surpluses, respectively. On the other hand, the subjects with least impact are 'Energy Fuels', 'Engineering Chemical', and 'Engineering Petroleum' with –14.9, –12.9, and –12.1% publication deficits, respectively.

It is notable that some subject categories with a heavy presence in the population papers are under-represented in the sample papers: 'Chemistry Analytical', 'Water Resources', 'Toxicology', 'Thermodynamics', 'Chemistry Applied', 'Engineering Mechanical', and 'Oceanography' with at least a 1.4% presence in the population papers each.

40.3.9 Citation Impact

These sample papers receive about 47,000 citations as of January 2020. Thus, the average number of citations per paper is 470.

40.3.10 Keywords

Although a number of keywords are listed in the Appendix for the datasets related to this field, some of them are more significant for the sample papers. The most-prolific keyword is 'oil*' with 71 occurrences. The other prolific keywords are 'petrol*', 'hydrocarbon', and '*degrad*' with 38, 36, and 31 citations, respectively. Further keywords are 'petroleum source rock*', 'desulfurization', 'bacteria', 'pah*', '*remed*', 'contam*', 'asphaltene', and 'corrosion' with 6, 7, 9, 7, 14, 12, 5, and 5 occurrences, respectively.

40.3.11 Research Fronts

Brief information about the key research fronts is provided in Table 40.8. There are seven research fronts for these sample papers: 'hydrocarbon contamination and bioremediation' (Leahy and Colwell, 1990; Haritash and Kaushik, 2009) 'crude oil spill bioremediation' (Atlas, 1981; Peterson et al., 2003), 'crude oil properties and characterization' (Marshall and Rodgers, 2004; Zhang et al., 2013), 'crude oil recovery' (Ahmadun et al., 2009; Vermeiren and Gilson, 2009), 'crude oil geology' (Peters, 1986; Scholle and Arthur, 1980), 'crude oil refining' (Babich and Moulijn, 2003; Otsuki et al., 2000), and 'crude oil pipelines' (Finsgar and Jackson, 2014; Nesic, 2007).

The most-prolific research front is 'hydrocarbon contamination and bioremediation' with 35 sample papers. The other prolific research fronts are 'crude oil spill bioremediation', 'crude oil properties and characterization', and 'crude oil recovery' with 20, 16, and 15 sample papers, respectively.

40.4 DISCUSSION

The size of the research on crude oils has increased to over 53,000 papers as of January 2020. It is expected that the number of the population papers in this field will exceed 150,000 papers by the end of the 2020s.

The research has developed more in the technological aspects of this field, rather than the social and humanitarian pathways, as evidenced by the negligible number of population papers in the indices of the SSCI and A&HCI. However, the econometric studies on the price fluctuations in crude oils have also been a significant research field (Kilian, 2009; Perron, 1989; Sadorsky, 1999; Zivot and Andrews, 2002).

The article types of documents are the primary documents for both datasets and reviews are over-represented by 32.4% in the sample papers (Table 40.1). Thus, the contribution of reviews by 35% of the sample papers in this field is highly exceptional (cf. Konur, 2011, 2012a–n, 2015, 2016a–f, 2017a–f, 2018a–b, 2019a–b).

Twenty-three authors from 15 institutions have at least two sample papers each (Table 40.2). Thirteen of these authors are from the USA and the remaining ones are from Germany and the UK.

TABLE 40.8
Research Fronts

	Research Front	No. of Sample Papers (%)
1	Hydrocarbon contamination and bioremediation	35
2	Crude oil spill bioremediation	20
3	Crude oil properties and characterization	16
4	Crude oil recovery	15
5	Crude oil geology	13
6	Crude oil refining	10
7	Crude oil pipelines	2

These authors focus on 'crude oil spill bioremediation' whilst the other prolific research fronts are 'hydrocarbon bioremediation', 'crude oil characterization', and 'crude oil geology'.

It is significant that there is ample 'gender deficit' among these top authors as all of them are male (Lariviere et al., 2013; Xie and Shauman, 1998).

About 14, 33, 40, and 13% of the sample papers and 10.3, 13.0, 21.1, and 55.1% of the population papers are published in the 1980s, 1990s, 2000s, and 2010s, respectively (Figure 40.1). This finding suggests that the population papers have built on the sample papers primarily published in the 1990s and 2000s. Following this rising trend, particularly in the 2010s, it is expected that the number of papers will reach 150,000 by the end of the 2020s, tripling the current size.

The engagement of the institutions in this field at the global scale is significant as over 170 and 6,400 institutions contribute to the sample and population papers, respectively.

Fourteen top institutions publish 53.0 and 7.1% of the sample and population papers, respectively (Table 40.3). The top institution is the 'Woods Hole Oceanographic Institution' of the USA with seven sample papers and a 6.8% publication surplus. The other top institutions are 'British Petroleum', 'Chevron', the 'Helmholtz Association', 'Newcastle University', the US 'Department of Interior', the US 'Geological Survey', and the 'University of California Berkeley' with four sample papers each. As in the case of the top authors, the most-prolific countries for these top institutions are the USA, the UK, and Germany. It is notable that some institutions with a heavy presence in the population papers are under-represented in the sample papers.

It is significant that only 74 and 38% of the sample and population papers declare any funding, respectively. The most-prolific country for these top funding bodies is China with five (Table 40.4). It is notable that some top funding agencies for the population studies do not enter this top funding body list.

Five Chinese funding bodies dominate this top funding table. This finding is in line with studies showing the heavy research funding in China and the NSFC is the primary funding agency (Wang et al., 2012).

The sample and population papers are published by 56 and over 1,050 journals, respectively. It is significant that the top seven journals publish 36.0 and 3.5% of the sample and population papers, respectively (Table 40.5).

The top journals, 'Applied and Environmental Microbiology', the 'AAPG Bulletin American Association of Petroleum Geologists', and 'Nature' publish together 22 sample papers with 20.2% publication surpluses.

The top subject categories for these top journals are 'Biotechnology and Applied Microbiology', 'Multidisciplinary Sciences', Engineering Environmental', and 'Environmental Sciences'. It is notable that some journals are relatively under-represented in the sample papers.

In total, 26 and over 270 countries contribute to the sample and population papers, respectively. The top country is the USA publishing 43.0 and 22.2% of the sample and population papers, respectively, with a 20.8% publication surplus (Table 40.6). This finding is in line with studies arguing that the USA is not losing ground in science and technology (Leydesdorff and Wagner, 2009).

The other prolific countries are the UK, Germany, and Canada publishing 16, 11, and 10 sample papers, respectively. These findings are in line with studies showing that European countries have a superior publication performance in science and technology (Youtie et al., 2008).

The European and Asian countries represented in this table publish altogether 42 and 10% of the sample papers and 14.3 and 22.1% of the population papers, respectively.

It is notable that the publication surplus for the USA and these European and Asian countries is 20.9, 27.7, and –12.1%, respectively. It is further notable that China has a significant publication deficit (–13.4%). This finding is in contrast with China's efforts to be a leading nation in science and technology (Zhou and Leydesdorff, 2006), but it is in line with the findings of Guan and Ma (2007) and Youtie et al. (2008) relating to China's performance in nanotechnology.

Similarly, Russia, Iran, Brazil, Japan, Spain, South Korea, and Italy have no place in this top country table, although they make significant contributions to the population papers (Bordons et al., 2015; Glanzel et al., 2006; Leydesdorff and Zhou, 2005; Moin et al., 2005; Oleinik, 2012).

The sample and population papers are indexed by 34 and 221 subject categories, respectively. For the sample papers, the top subjects are 'Biotechnology Applied Microbiology', 'Microbiology', and 'Environmental Sciences' with 18, 17, and 15 papers, respectively (Table 40.7). The other prolific subjects are 'Energy Fuels', 'Multidisciplinary Sciences', 'Engineering Chemical', and 'Chemistry Physical' with at least ten papers each.

It is notable that the publication surplus is most significant for 'Microbiology', 'Biotechnology Applied Microbiology', and 'Multidisciplinary Sciences' with 13.8, 13.6, and 9.9% surpluses, respectively. On the other hand, the subjects with least impact are 'Energy Fuels', 'Engineering Chemical', and 'Engineering Petroleum' with –14.9, –12.9, and –12.1% publication deficits, respectively. It is notable that some subject categories with a heavy presence in the population papers are under-represented in the sample papers.

These sample papers received about 47,000 citations as of January 2020. Thus, the average number of citations per paper is 470. Hence, the citation impact of the top 100 papers in this field has been significant.

Although a number of keywords are listed in the Appendix for the datasets related to this field, some of them are more significant for the sample papers. The most-prolific keyword is 'oil*' with 71 occurrences. The other prolific keywords are 'petrol*', 'hydrocarbon', '*degrad*' with 38, 36, and 31 citations, respectively. Further keywords are 'petroleum source rock*', 'desulfurization', 'bacteria', 'pah*', '*remed*', 'contam*', 'asphaltene', and 'corrosion' with 6, 7, 9, 7, 14, 12, 5, and 5 occurrences, respectively. As expected, these keywords provide valuable information about the pathways of the research in this field.

Seven research fronts emerge from the examination of the sample papers: 'hydrocarbon contamination and bioremediation', 'crude oil spill bioremediation', 'crude oil properties and characterization', 'crude oil recovery', 'crude oil geology', 'crude oil refining', and 'crude oil pipelines' (Table 40.8).

The most-prolific research front is 'hydrocarbon contamination and bioremediation' with 35 sample papers. The other prolific research fronts are 'crude oil spill bioremediation', 'crude oil properties and characterization', and 'crude oil recovery' with 20, 16, and 15 sample papers, respectively.

The key emphasis in these research fronts is the exploration of the structure–processing–property relationships of crude oils (Cheng and Ma, 2011; Konur and Matthews, 1989; Rogers and Hopfinger, 1994; Scherf and List, 2002).

40.5 CONCLUSION

This chapter has mapped the research on crude oils using a scientometric method.

The size of over 53,000 population papers shows the public importance of this interdisciplinary research field. However, it is significant that the research has developed more in the technological aspects, rather than the social and humanitarian pathways.

Articles and reviews dominate the sample papers, primarily published in the 1990s and 2000s. The population papers, primarily published in the 2010s, build on these sample papers.

The data presented in the tables and figure show that a small number of authors, institutions, funding bodies, journals, keywords, research fronts, subject categories, and countries have shaped the research in this field.

It is notable that the authors, institutions, and funding bodies from the USA, the UK, and Germany dominate the research in this field. Furthermore, China, Russia, Iran, Brazil, Japan, Spain, South Korea, and Italy are under-represented significantly in the sample papers.

These findings show the importance of the development of efficient incentive structures for the development of the research in this field as in other fields. It seems that the USA and European countries (such as the UK and Germany) have efficient incentive structures for the development of the research in this field, contrary to China, Russia, Iran, Brazil, Japan, Spain, South Korea, and Italy.

It further seems that although research funding is a significant element of these incentive structures, it might not be the sole solution for increasing incentives for research in this field, as is the case in China, Russia, Iran, Brazil, Japan, Spain, South Korea, and Italy.

On the other hand, it seems there is more to do to reduce the significant gender deficit in this field as in other fields of science and technology (Lariviere et al., 2013; Xie and Shauman, 1998).

The data on the research fronts, keywords, source titles, and subject categories provide valuable evidence for the interdisciplinary (Lariviere and Gingras, 2010; Morillo et al., 2001) nature of the research in this field.

There is ample justification for the broad search strategy employed in this study due to the interdisciplinary nature of this research field as evidenced by the top subject categories. The search strategy employed in this study is in line with those employed in related and other research fields (Konur, 2011, 2012a–n, 2015, 2016a–f, 2017a–f, 2018a–b, 2019a–b).

Seven research fronts emerge from the examination of the sample papers: 'hydrocarbon contamination and bioremediation', 'crude oil spill bioremediation', 'crude oil properties and characterization', 'crude oil recovery', 'crude oil geology', 'crude oil refining', and 'crude oil pipelines' (Table 40.8).

It is recommended that further scientometric studies are carried out for each of these research fronts, building on the pioneering studies in these fields.

ACKNOWLEDGMENTS

The contribution of the highly cited researchers in the field of crude oils is greatly acknowledged.

40.A APPENDIX

40.A.1 Crude Oils Keyword Set

TI = ("heavy oil*" or "oil sand*" or (oil* and shale) or "crude oil*" or "oil-recovery" or "oil-water emuls*" or "water-in-oil emuls*" or "petroleum production" or "oil-spill*" or "oil plume*" or "oil slick*" or "heavy petroleum" or "petroleum hydrocarbon*" or ((petroleum or oil*) and (refin* or upgrading or *cracking)) or "petroleum resid*" or "petroleum oil*" or "petroleum fraction*" or "petroleum-reservoir*" or "petroleum resin*" or "oil fraction*" or "light petroleum" or "black oil*" or "oil and gas" or "fuel oil*" or "petroleum crude*" or "crude petroleum" or "oil*field*" or "oil hydrocarbon*" or "oil-reservoir*" or "oil formation" or (refin* and "water network*") or "petroleum formation" or "petroleum spill*" or ((refinery or oil* or petrol*) and (effluent* or wastewater* or sludge* or pipeline* or corrosion or "water treatment")) or ((petroleum or oil) and industry) or "petroleum compound*" or "oil seepage" or "petroleum seepage" or "oil dispers*" or "petroleum mixture*" or "petroleum microbiol*" or "Deepwater-Horizon" or "Exxon-Valdez" or "Amoco-cadiz" or "oil toxicity" or "petroleum toxicity" or "petroleum platform*" or "oil platform*" or "petroleum worker*" or "oil worker*" or "petroleum distil*" or "petroleum feedstock*" or "petroleum microbial*" or "petroleum recovery" or ((petroleum or oil) and (generation or migration)) or "petroleum installation*" or "petroleum-exploration" or "oil exploration" or "petroleum source r*" or "oil source r*" or "petroleum field*" or "petroleum geology" or "petroleum geochem*" or "oil deposit*" or "petroleum deposit*" or "petroleum system*" or "oil discovery" or "Petroleum origin*" or "oil origin*" or "petroleum crude*" or "petroleum matur*" or "Petroleum inclusion*" or "petroleum basin*" or "oil basin*" or "oil accumulation*" or "petroleum accumulation*" or "petroleum expulsion" or "petroleum potential" or "petroleum biomarker*" or Petroleomic* or "petroleum product*" or "petroleum fluid*" or "petroleum composition*" or ((*desulfur* or *desulphur*) and (oil* or petrol* or refin*)) or "petroleum analysis" or "petroleum component*" or asphaltene or "oil well*" or "petroleum well*" or "naphthenic acid*" or ((petroleum or oil* or hydrocarbon* or pahs or pah or naphthalen*) and (contamin* or bioremediat* or *degrad* or weathering or remediat*))) or SO = ("spill science*")

NOT (TI = ("Petroleum coke" or "petroleum gas" or "bio-crude" or "bio-oil*" or petrolog* or petrogenesis or biomass* or vegetable* or *alga* or *diesel or "bio-refin*" or "hydrocarbon gas" or mart* or meteor* or aryl or "essential oil*" or olive or soy* or coal or coke or polyethylene or ether or palm or sewage or *seed* or biofuel* or *stellar or methanol or nigella* or "oils and fats" or eucasanoid*) or WC = (econ* or food* or astro*))

REFERENCES

Ahmadun, F. R., A. Pendashteh and L. C. Abdullah, et al. 2009. Review of technologies for oil and gas produced water treatment. *Journal of Hazardous Materials* 170:530–551.

Atlas, R. M. 1981. Microbial degradation of petroleum hydrocarbons: An environmental perspective. *Microbiological Reviews* 45:180–209.

Babich, I. V. and J. A. Moulijn. 2003. Science and technology of novel processes for deep desulfurization of oil refinery streams: A review. *Fuel* 82:607–631.

Bordons, M., B. Gonzalez-Albo, J. Aparicio and L. Moreno. 2015. The influence of R&D intensity of countries on the impact of international collaborative research: Evidence from Spain. *Scientometrics* 102:1385–1400.

Cheng, Y. Q. and E. Ma. 2011. Atomic-level structure and structure–property relationship in metallic glasses. *Progress in Materials Science* 56:379–473.

Chisti, Y. 2007. Biodiesel from microalgae. *Biotechnology Advances* 25:294–306.

Chisti, Y. 2008. Biodiesel from microalgae beats bioethanol. *Trends in Biotechnology* 26:126–131.

Clarivate Analytics. 2019. Highly cited researchers: 2019 Recipients. Philadelphia, PA: Clarivate Analytics. https://recognition.webofsciencegroup.com/awards/highly-cited/2019/ (accessed January 3, 2020).

Docampo, D. and L. Cram. 2019. Highly cited researchers: A moving target. *Scientometrics* 118:1011–1025.

Finsgar, M. and J. Jackson. 2014. Application of corrosion inhibitors for steels in acidic media for the oil and gas industry: A review. *Corrosion Science* 86:17–41.

Garfield, E. 1955. Citation indexes for science. *Science* 122:108–111.

Garfield, E. 1972. Citation analysis as a tool in journal evaluation. *Science* 178:471–479.

Glanzel, W., J. Leta and B. Thijs. 2006. Science in Brazil. Part 1: A macro-level comparative study. *Scientometrics* 67:67–86.

Guan, J. C. and N. Ma. 2007. China's emerging presence in nanoscience and nanotechnology: A comparative bibliometric study of several nanoscience 'giants'. *Research Policy* 36:880–886.

Haritash, A. K. and C. P. Kaushik. 2009. Biodegradation aspects of polycyclic aromatic hydrocarbons (PAHs): A review. *Journal of Hazardous Materials* 169:1–15.

Kilian, L. 2009. Not all oil price shocks are alike: Disentangling demand and supply shocks in the crude oil market. *American Economic Review* 99:1053–1069.

Konur, O. 2000. Creating enforceable civil rights for disabled students in higher education: An institutional theory perspective. *Disability & Society* 15:1041–1063.

Konur, O. 2002a. Access to nursing education by disabled students: Rights and duties of nursing programs. *Nurse Education Today* 22:364–374.

Konur, O. 2002b. Assessment of disabled students in higher education: Current public policy issues. *Assessment and Evaluation in Higher Education* 27:131–152.

Konur, O. 2002c. Access to employment by disabled people in the UK: Is the Disability Discrimination Act working? *International Journal of Discrimination and the Law* 5:247–279.

Konur, O. 2006a. Participation of children with dyslexia in compulsory education: Current public policy issues. *Dyslexia* 12:51–67.

Konur, O. 2006b. Teaching disabled students in higher education. *Teaching in Higher Education* 11:351–363.

Konur, O. 2007a. A judicial outcome analysis of the Disability Discrimination Act: A windfall for the employers? *Disability & Society* 22:187–204.

Konur, O. 2007b. Computer-assisted teaching and assessment of disabled students in higher education: The interface between academic standards and disability rights. *Journal of Computer Assisted Learning* 23:207–219.

Konur, O. 2011. The scientometric evaluation of the research on the algae and bio-energy. *Applied Energy* 88:3532–3540.

Konur, O. 2012a. Evaluation of the research on the social sciences in Turkey: A scientometric approach. *Energy Education Science and Technology Part B: Social and Educational Studies* 4:1893–1908.

Konur, O. 2012b. Prof. Dr. Ayhan Demirbas' scientometric biography. *Energy Education Science and Technology Part A: Energy Science and Research* 28:727–738.

Konur, O. 2012c. The evaluation of the biogas research: A scientometric approach. *Energy Education Science and Technology Part A: Energy Science and Research* 29:1277–1292.

Konur, O. 2012d. The evaluation of the educational research: A scientometric approach. *Energy Education Science and Technology Part B: Social and Educational Studies* 4:1935–1948.

Konur, O. 2012e. The evaluation of the global energy and fuels research: A scientometric approach. *Energy Education Science and Technology Part A: Energy Science and Research* 30:613–628.

Konur, O. 2012f. The evaluation of the research on the Arts and Humanities in Turkey: A scientometric approach. *Energy Education Science and Technology Part B: Social and Educational Studies* 4:1603–1618.

Konur, O. 2012g. The evaluation of the research on the biodiesel: A scientometric approach. *Energy Education Science and Technology Part A: Energy Science and Research* 28:1003–1014.

Konur, O. 2012h. The evaluation of the research on the bioethanol: A scientometric approach. *Energy Education Science and Technology Part A: Energy Science and Research* 28:1051–1064.

Konur, O. 2012i. The evaluation of the research on the biofuels: A scientometric approach. *Energy Education Science and Technology Part A: Energy Science and Research* 28:903–916.

Konur, O. 2012j. The evaluation of the research on the biohydrogen: A scientometric approach. *Energy Education Science and Technology Part A: Energy Science and Research* 29:323–338.

Konur, O. 2012k. The evaluation of the research on the microbial fuel cells: A scientometric approach. *Energy Education Science and Technology Part A: Energy Science and Research* 29:309–322.

Konur, O. 2012l. The scientometric evaluation of the research on the production of bioenergy from biomass. *Biomass and Bioenergy* 47:504–515.

Konur, O. 2012m. The scientometric evaluation of the research on the deaf students in higher education. *Energy Education Science and Technology Part B: Social and Educational Studies* 4:1573–1588.

Konur, O. 2012n. The scientometric evaluation of the research on the students with ADHD in higher education. *Energy Education Science and Technology Part B: Social and Educational Studies* 4:1547–1562.

Konur, O. 2015. Current state of research on algal biodiesel. In *Marine Bioenergy: Trends and Developments*, ed. S. K. Kim and C. G. Lee, 487–512. Boca Raton, FL: CRC Press.

Konur, O. 2016a. Scientometric overview in nanobiodrugs. In *Nanoarchitectonics for Smart Delivery and Drug Targeting*, ed. A. M. Holban and A. M. Grumezescu 405–428. Amsterdam: Elsevier.

Konur, O. 2016b. Scientometric overview regarding nanoemulsions used in the food industry. In *Emulsions: Nanotechnology in the Agri-Food Industry*, ed. A. M. Grumezescu, 689–711. Amsterdam: Elsevier.

Konur, O. 2016c. Scientometric overview regarding the nanobiomaterials in antimicrobial therapy. In *Nanobiomaterials in Antimicrobial Therapy*, ed. A. M. Grumezescu, 511–535. Amsterdam: Elsevier.

Konur, O. 2016d. Scientometric overview regarding the nanobiomaterials in dentistry. In *Nanobiomaterials in Dentistry*, ed. A. M. Grumezescu, 425–453. Amsterdam: Elsevier.

Konur, O. 2016e. Scientometric overview regarding the surface chemistry of nanobiomaterials. In *Surface Chemistry of Nanobiomaterials*, ed. A. M. Grumezescu, 463–486. Amsterdam: Elsevier.

Konur, O. 2016f. The scientometric overview in cancer targeting. In *Nanoarchitectonics for Smart Delivery and Drug Targeting*, ed. A. M. Holban and A. Grumezescu, 871–895. Amsterdam: Elsevier.

Konur, O. 2017a. Recent citation classics in antimicrobial nanobiomaterials. In *Nanostructures for Antimicrobial Therapy*, ed. A. Ficai and A. M. Grumezescu, 669–685. Amsterdam: Elsevier.

Konur, O. 2017b. Scientometric overview in nanopesticides. In *New Pesticides and Soil Sensors*, A. M. Grumezescu, ed. 719–744. Amsterdam: Elsevier.

Konur, O. 2017c. Scientometric overview regarding oral cancer nanomedicine. In *Nanostructures for Oral Medicine*, ed. E. Andronescu, A. M. Grumezescu, 939–962. Amsterdam: Elsevier;

Konur, O. 2017d. Scientometric overview regarding water nanopurification. In *Water Purification*, ed. A. M. Grumezescu, 693–716. Amsterdam: Elsevier.

Konur, O. 2017e. Scientometric overview in food nanopreservation. In *Food Preservation*, ed. A. M. Grumezescu, 703–729. Amsterdam: Elsevier.

Konur, O. 2017f. The top citation classics in alginates for biomedicine. In *Seaweed Polysaccharides: Isolation, Biological and Biomedical Applications*, ed. J. Venkatesan, S. Anil, S. K. Kim, 223–249. Amsterdam: Elsevier.

Konur, O. 2018a. Scientometric evaluation of the global research in spine: An update on the pioneering study by Wei et al. *European Spine Journal* 27:525–529.

Konur, O. 2018b. Bioenergy and biofuels science and technology: scientometric overview and citation classics. In *Bioenergy and Biofuels*, ed. O. Konur, 3–63. Boca Raton: CRC Press.

Konur, O. 2019a. Cyanobacterial bioenergy and biofuels science and technology: A scientometric overview. In *Cyanobacteria: From Basic Science to Applications*, ed. A. K. Mishra, D. N. Tiwari and A. N. Rai, 419–442. Amsterdam: Elsevier.

Konur, O. 2019b. Nanotechnology applications in food: A scientometric overview. In *Nanoscience for Sustainable Agriculture*, ed., R. N. Pudake, N. Chauhan and C. Kole, 683–711. Cham: Springer.

Konur, O., ed. 2021a. *Handbook of Biodiesel and Petrodiesel Fuels: Science, Technology, Health, and Environment*. Boca Raton, FL: CRC Press.

Konur, O., ed. 2021b. *Handbook of Biodiesel and Petrodiesel Fuels: Science, Technology, Health, and Environment. Volume 1. Biodiesel Fuels: Science, Technology, Health, and Environment*. Boca Raton, FL: CRC Press.

Konur, O., ed. 2021c. *Handbook of Biodiesel and Petrodiesel Fuels: Science, Technology, Health, and Environment. Volume 2. Biodiesel Fuels based on the Edible and Nonedible Feedstocks, Wastes, and Algae: Science, Technology, Health, and Environment*. Boca Raton, FL: CRC Press.

Konur, O., ed. 2021d. *Handbook of Biodiesel and Petrodiesel Fuels: Science, Technology, Health, and Environment. Volume 3. Petrodiesel Fuels: Science, Technology, Health, and Environment*. Boca Raton, FL: CRC Press.

Konur, O. 2021e. Biodiesel and petrodiesel fuels: Science, technology, health, and environment. In *Handbook of Biodiesel and Petrodiesel Fuels: Science, Technology, Health, and Environment. Volume 1. Biodiesel Fuels: Science, Technology, Health, and Environment*, ed. O. Konur. Boca Raton, FL: CRC Press.

Konur, O. 2021f. Biodiesel and petrodiesel fuels: A scientometric review of the research. In *Handbook of Biodiesel and Petrodiesel Fuels: Science, Technology, Health, and Environment. Volume 1. Biodiesel Fuels: Science, Technology, Health, and Environment*, ed. O. Konur. Boca Raton, FL: CRC Press.

Konur, O. 2021g. Biodiesel and petrodiesel fuels: A review of the research. In *Handbook of Biodiesel and Petrodiesel Fuels: Science, Technology, Health, and Environment. Volume 1. Biodiesel Fuels: Science, Technology, Health, and Environment*, ed. O. Konur. Boca Raton, FL: CRC Press.

Konur, O. 2021h Nanotechnology applications in the diesel fuels and the related research fields: A review of the research. In *Handbook of Biodiesel and Petrodiesel Fuels: Science, Technology, Health, and Environment. Volume 1. Biodiesel Fuels: Science, Technology, Health, and Environment*, ed. O. Konur. Boca Raton, FL: CRC Press.

Konur, O. 2021i. Biooils: A scientometric review of the research. In *Handbook of Biodiesel and Petrodiesel Fuels: Science, Technology, Health, and Environment. Volume 1. Biodiesel Fuels: Science, Technology, Health, and Environment*, ed. O. Konur. Boca Raton, FL: CRC Press.

Konur, O. 2021j. Characterization and properties of biooils: A review of the research. In *Handbook of Biodiesel and Petrodiesel Fuels: Science, Technology, Health, and Environment. Volume 1. Biodiesel Fuels: Science, Technology, Health, and Environment*, ed. O. Konur. Boca Raton, FL: CRC Press.

Konur, O. 2021k. Biomass pyrolysis and pyrolysis oils: A review of the research. In *Handbook of Biodiesel and Petrodiesel Fuels: Science, Technology, Health, and Environment. Volume 1. Biodiesel Fuels: Science, Technology, Health, and Environment*, ed. O. Konur. Boca Raton, FL: CRC Press.

Konur, O. 2021l. Biodiesel fuels: A scientometric review of the research. In *Handbook of Biodiesel and Petrodiesel Fuels: Science, Technology, Health, and Environment. Volume 1. Biodiesel Fuels: Science, Technology, Health, and Environment*, ed. O. Konur. Boca Raton, FL: CRC Press.

Konur, O. 2021m. Glycerol: A scientometric review of the research. In *Handbook of Biodiesel and Petrodiesel Fuels: Science, Technology, Health, and Environment. Volume 1. Biodiesel Fuels: Science, Technology, Health, and Environment*, ed. O. Konur. Boca Raton, FL: CRC Press.

Konur, O. 2021n. Propanediol production from glycerol: A review of the research. In *Handbook of Biodiesel and Petrodiesel Fuels: Science, Technology, Health, and Environment. Volume 1. Biodiesel Fuels: Science, Technology, Health, and Environment*, ed. O. Konur. Boca Raton, FL: CRC Press.

Konur, O. 2021o. Edible oil-based biodiesel fuels: A scientometric review of the research. In *Handbook of Biodiesel and Petrodiesel Fuels: Science, Technology, Health, and Environment. Volume 2. Biodiesel Fuels based on the Edible and Nonedible Feedstocks, Wastes, and Algae: Science, Technology, Health, and Environment*, ed. O. Konur. Boca Raton, FL: CRC Press.

Konur, O. 2021p. Palm oil-based biodiesel fuels: A review of the research. In *Handbook of Biodiesel and Petrodiesel Fuels: Science, Technology, Health, and Environment. Volume*

2. Biodiesel Fuels based on the Edible and Nonedible Feedstocks, Wastes, and Algae, ed. O. Konur. Boca Raton, FL: CRC Press.

Konur, O. 2021q. Rapeseed oil-based biodiesel fuels: A review of the research. In *Handbook of Biodiesel and Petrodiesel Fuels: Science, Technology, Health, and Environment. Volume 2. Biodiesel Fuels based on the Edible and Nonedible Feedstocks, Wastes, and Algae*, ed. O. Konur. Boca Raton, FL: CRC Press.

Konur, O. 2021r. Nonedible oil-based biodiesel fuels: A scientometric review of the research. In *Handbook of Biodiesel and Petrodiesel Fuels: Science, Technology, Health, and Environment. Volume 2. Biodiesel Fuels based on the Edible and Nonedible Feedstocks, Wastes, and Algae: Science, Technology, Health, and Environment*, ed. O. Konur. Boca Raton, FL: CRC Press.

Konur, O. 2021s. Waste oil-based biodiesel fuels: A scientometric review of the research. In *Handbook of Biodiesel and Petrodiesel Fuels: Science, Technology, Health, and Environment. Volume 2. Biodiesel Fuels based on the Edible and Nonedible Feedstocks, Wastes, and Algae: Science, Technology, Health, and Environment*, ed. O. Konur. Boca Raton, FL: CRC Press.

Konur, O. 2021t. Algal biodiesel fuels: A scientometric review of the research. In *Handbook of Biodiesel and Petrodiesel Fuels: Science, Technology, Health, and Environment. Volume 2. Biodiesel Fuels based on the Edible and Nonedible Feedstocks, Wastes, and Algae: Science, Technology, Health, and Environment*, ed. O. Konur. Boca Raton, FL: CRC Press.

Konur, O. 2021u. Algal biomass production for biodiesel production: A review of the research. In *Handbook of Biodiesel and Petrodiesel Fuels: Science, Technology, Health, and Environment. Volume 2. Biodiesel Fuels based on the Edible and Nonedible Feedstocks, Wastes, and Algae*, ed. O. Konur. Boca Raton, FL: CRC Press.

Konur, O. 2021v. Algal biomass production in wastewaters for biodiesel production: A review of the research. In *Handbook of Biodiesel and Petrodiesel Fuels: Science, Technology, Health, and Environment. Volume 2. Biodiesel Fuels based on the Edible and Nonedible Feedstocks, Wastes, and Algae*, ed. O. Konur. Boca Raton, FL: CRC Press.

Konur, O. 2021w. Algal lipid production for biodiesel production: A review of the research. In *Handbook of Biodiesel and Petrodiesel Fuels: Science, Technology, Health, and Environment. Volume 2. Biodiesel Fuels based on the Edible and Nonedible Feedstocks, Wastes, and Algae*, ed. O. Konur. Boca Raton, FL: CRC Press.

Konur, O. 2021x. Crude oils: A scientometric review of the research. In *Handbook of Biodiesel and Petrodiesel Fuels: Science, Technology, Health, and Environment. Volume 3. Petrodiesel Fuels: Science, Technology, Health, and Environment*, ed. O. Konur. Boca Raton, FL: CRC Press.

Konur, O. 2021y. Petrodiesel fuels: A scientometric review of the research. In *Handbook of Biodiesel and Petrodiesel Fuels: Science, Technology, Health, and Environment. Volume 3. Petrodiesel Fuels: Science, Technology, Health, and Environment*, ed. O. Konur. Boca Raton, FL: CRC Press.

Konur, O. 2021z. Bioremediation of petroleum hydrocarbons in the contaminated soils: A review of the research. In *Handbook of Biodiesel and Petrodiesel Fuels: Science, Technology, Health, and Environment. Volume 3. Petrodiesel Fuels: Science, Technology, Health, and Environment*, ed. O. Konur. Boca Raton, FL: CRC Press.

Konur, O. 2021ab. Desulfurization of diesel fuels: A review of the research. In *Handbook of Biodiesel and Petrodiesel Fuels: Science, Technology, Health, and Environment. Volume 3. Petrodiesel Fuels: Science, Technology, Health, and Environment*, ed. O. Konur. Boca Raton, FL: CRC Press.

Konur, O. 2021ac. Diesel fuel exhaust emissions: A scientometric review of the research. In *Handbook of Biodiesel and Petrodiesel Fuels: Science, Technology, Health, and*

Environment. Volume 3. Petrodiesel Fuels: Science, Technology, Health, and Environment, ed. O. Konur. Boca Raton, FL: CRC Press.

Konur, O. 2021ad. The adverse health and safety impact of diesel fuels: A scientometric review of the research. In *Handbook of Biodiesel and Petrodiesel Fuels: Science, Technology, Health, and Environment. Volume 3. Petrodiesel Fuels: Science, Technology, Health, and Environment*, ed. O. Konur. Boca Raton, FL: CRC Press.

Konur, O. 2021ae. Respiratory illnesses caused by the diesel fuel exhaust emissions: A review of the research. In *Handbook of Biodiesel and Petrodiesel Fuels: Science, Technology, Health, and Environment. Volume 3. Petrodiesel Fuels: Science, Technology, Health, and Environment*, ed. O. Konur. Boca Raton, FL: CRC Press.

Konur, O. 2021af. Cancer caused by the diesel fuel exhaust emissions: A review of the research. In *Handbook of Biodiesel and Petrodiesel Fuels: Science, Technology, Health, and Environment. Volume 3. Petrodiesel Fuels: Science, Technology, Health, and Environment*, ed. O. Konur. Boca Raton, FL: CRC Press.

Konur, O. 2021ag. Cardiovascular and other illnesses caused by the diesel fuel exhaust emissions: A review of the research. In *Handbook of Biodiesel and Petrodiesel Fuels: Science, Technology, Health, and Environment. Volume 3. Petrodiesel Fuels: Science, Technology, Health, and Environment*, ed. O. Konur. Boca Raton, FL: CRC Press.

Konur, O. and F. L. Matthews. 1989. Effect of the properties of the constituents on the fatigue performance of composites: A review. *Composites* 20:317–328.

Lapuerta, M., O. Armas and J. Rodriguez-Fernandez. 2008. Effect of biodiesel fuels on diesel engine emissions. *Progress in Energy and Combustion Science* 34:198–223.

Lariviere, V. and Y. Gingras. 2010. On the relationship between interdisciplinarity and scientific impact. *Journal of the American Society for Information Science and Technology* 61:126–131.

Lariviere, V., C. Ni, Y. Gingras, B. Cronin and C. R. Sugimoto. 2013. Bibliometrics: Global gender disparities in science. *Nature News* 504:211–213.

Leahy, J. G. and R. R. Colwell. 1990. Microbial degradation of hydrocarbons in the environment. *Microbiological Reviews* 54:305–315.

Leydesdorff, L. and Wagner, C. 2009. Is the United States losing ground in science? A global perspective on the world science system. *Scientometrics* 78:23–36.

Leydesdorff, L. and P. Zhou. 2005. Are the contributions of China and Korea upsetting the world system of science? *Scientometrics* 63:617–630.

Marchetti, J. M., V. U. Miguel and A. F. Errazu. 2007. Possible methods for biodiesel production. *Renewable and Sustainable Energy Reviews* 11:1300–1311.

Marshall, A. G. and R. P. Rodgers. 2004. Petroleomics: The next grand challenge for chemical analysis. *Accounts of Chemical Research* 37:53–59.

Moin, M., M. Mahmoudi and N. Rezaei. 2005. Scientific output of Iran at the threshold of the 21st century. *Scientometrics* 62:239–248.

Moldowan, J. M., W. K. Seifert and E. J. Gallegos. 1985. Relationship between petroleum composition and depositional environment of petroleum source rocks. *AAPG Bulletin-American Association of Petroleum Geologists* 69:1255–1268.

Morillo, F., M. Bordons and I. Gomez. 2001. An approach to interdisciplinarity through bibliometric indicators. *Scientometrics* 51:203–222.

Nesic, S. 2007. Key issues related to modelling of internal corrosion of oil and gas pipelines: A review. *Corrosion Science* 49:4308–4308.

North, D. C. 1991a. *Institutions, Institutional Change and Economic Performance*. Cambridge, MA: Cambridge University Press.

North, D. C. 1991b. Institutions. *Journal of Economic Perspectives* 5:97–112.

Oleinik, A. 2012. Publication patterns in Russia and the West compared. *Scientometrics* 93:533–551.

Otsuki, S., T. Nonaka and N. Takashima, et al. 2000. Oxidative desulfurization of light gas oil and vacuum gas oil by oxidation and solvent extraction. *Energy & Fuels* 14: 1232–1239.

Perron, P. 1989. The great crash, the oil price shock, and the unit root hypothesis. *Econometrica: Journal of the Econometric Society* 57:1361–1401.

Peters, K. E. 1986. Guidelines for evaluating petroleum source rock using programmed pyrolysis. *AAPG Bulletin-American Association of Petroleum Geologists* 70:318–329.

Peterson, C. H., S. D. Rice and J. W. Short, et al. 2003. Long term ecosystem response to the Exxon Valdez oil spill. *Science* 302:2082–2086.

Rogers, D. and A. J. Hopfinger. 1994. Application of genetic function approximation to quantitative structure-activity relationships and quantitative structure-property relationships. *Journal of Chemical Information and Computer Sciences* 34:854–866.

Sadorsky, P. 1999. Oil price shocks and stock market activity. *Energy Economics* 21:449–469.

Scherf, U. and E. J. List. 2002. Semiconducting polyfluorenes—towards reliable structure–property relationships. *Advanced Materials* 14:477–487.

Scholle, P. A. and M. A. Arthur. 1980. Carbon isotope fluctuations in cretaceous pelagic limestones: Potential stratigraphic and petroleum exploration tool. *AAPG Bulletin-American Association of Petroleum Geologists* 64:67–87.

Srivastava, A. and R. Prasad. 2000. Triglycerides-based diesel fuels. *Renewable and Sustainable Energy Reviews* 4:111–133.

Van Gerpen, J. 2005. Biodiesel processing and production. *Fuel Processing Technology* 86:1097–1107.

Van Hamme, J. D., A. Singh and O. P. Ward. 2003. Recent advances in petroleum microbiology. *Microbiology and Molecular Biology Reviews* 67:503–549.

Vermeiren, W. and J. P. Gilson. 2009. Impact of zeolites on the petroleum and petrochemical industry. *Topics in Catalysis* 52:1131–1161.

Wang, X., D. Liu, K. Ding and X. Wang. 2012. Science funding and research output: A study on 10 countries. *Scientometrics* 91:591–599.

Xie, Y. and K. A. Shauman. 1998. Sex differences in research productivity: New evidence about an old puzzle. *American Sociological Review* 63:847–870.

Youtie, J, P. Shapira and A. L. Porter. 2008. Nanotechnology publications and citations by leading countries and blocs. *Journal of Nanoparticle Research* 10:981–986.

Zhang, W. B., Z. Shi and F. Zhang, et al. 2013. Superhydrophobic and superoleophilic PVDF membranes for effective separation of water-in-oil emulsions with high flux. *Advanced Materials* 25:2071–2076.

Zhou, P. and L. Leydesdorff. 2006. The emergence of China as a leading nation in science. *Research Policy* 35:83–104.

Zivot, E. and D. W. K. Andrews. 2002. Further evidence on the great crash, the oil price shock, and the unit-root hypothesis. *Journal of Business & Economic Statistics* 20:25–44.

41 Introduction to Oil Spill Behavior

Merv Fingas

CONTENTS

41.1 INTRODUCTION

Almost immediately after oil is spilled on either land or water, it undergoes multiple transformations in the physical-chemical structure of the resulting slick. The mass and physical-chemical structure of the spilled oil, however, is also strongly influenced by the initial oil properties as well as environmental forces. The processes that cause changes to the mass and nature of the oil define what is called 'oil weathering'. For most surface spills, the dominant weathering processes are evaporation of oil into the atmosphere and water-into-oil emulsification (Boehm et al., 2008; Diez et al., 2007; National Research Council, 2003; Short et al., 2007).

A second group of processes is related to the movement of oil in the environment. Spill modeling combines the knowledge of initial oil properties, release information, and forecasted environmental conditions to predict future locations as well as the state of the oil (Lehr et al., 2014). Weathering and movement processes can overlap, with weathering strongly influencing how oil is moved in the environment and vice versa. All processes depend very much on the type of oil spilled and the weather conditions during the spill.

41.2 AN OVERVIEW OF WEATHERING

The specific behavior processes that occur after an oil spill determine how the oil should be cleaned up and its effect on the environment. For example, if spilled oil evaporates rapidly, less oil is required to be cleaned up, though the hydrocarbons in the oil enter the atmosphere. An oil slick could be carried by surface currents or winds to the vicinity of a bird colony or to a shore where seals or sea lions are breeding and severely affect wildlife and their habitat. On the other hand, a slick could be carried out to sea where it has less immediate effect on the environment.

The fate and effects of a particular spill are determined by the behavior processes which are in turn almost entirely determined by the type of oil and the environmental conditions at the time of the spill. Spill responders need to know the ultimate fate of the oil in order to take measures to minimize the overall impact of the spill.

Oil spilled on water undergoes a series of changes in physical and chemical properties which in combination are termed 'weathering'. Weathering processes occur at very different rates, but begin immediately after oil is spilled into the environment. Weathering rates are not consistent throughout the duration of an oil spill and are usually highest immediately after it. Both weathering processes and the rates at which they occur depend very much on the type of oil and the environmental conditions. Most weathering processes are highly temperature-dependent, and will often slow to insignificant rates as temperatures approach zero degrees Celsius.

The processes included in weathering are evaporation, emulsification, natural dispersion, dissolution, photooxidation, sedimentation, interaction with mineral fines, biodegradation, and the formation of tar balls. These processes are listed in order of importance in terms of their effect on the slick's properties and its fate.

41.3 EVAPORATION

Evaporation is usually the most important weathering process (Fingas, 2015a). It has the greatest effect of all weathering processes on the amount of oil remaining on water or land after a spill, especially for that of a light oil. Over a period of several days, a light fuel such as gasoline evaporates completely at typical ambient temperatures, whereas only a small percentage of a heavier Bunker C oil evaporates. The more volatile the components that an oil or fuel contains, the greater the extent and rate of its evaporation. Many components of heavier oils will not evaporate at all, even over long periods of time and at high temperatures.

Oil and petroleum products evaporate in a slightly different manner than water and the process is much less dependent on wind speed and surface area. Oil evaporation can be considerably slowed down, however, by the formation of a 'crust' or 'skin' on top of the oil. This happens primarily on land or in calm areas where the oil layer does not get mixed. The skin or crust is formed when the smaller compounds in the oil are removed, leaving the larger compounds, such as waxes and resins, at the surface. This crust then seals off the remainder of the oil and slows evaporation. Stranded oil from old spills has been reexamined over many years and it has been found that when this crust has formed, there has been no significant evaporation in the oil underneath. When this crust has not formed, oil could be weathered to the hardness of wood over the same amount of years.

The rate of evaporation is very rapid immediately after a spill and then slows considerably. About 80% of the evaporation that takes place, occurs in the first two days after a spill. The evaporation of most oils follows a logarithmic curve with time. Some oils such as diesel fuel, however, evaporate as the square root of time, at least for the first few days. This means that the evaporation rate slows very rapidly in both cases, after a few days.

The properties of oil can change significantly with the extent of evaporation. If about 40% (by weight) of an oil evaporates, its viscosity could increase by as much as a thousand-fold. Its density could rise by as much as 10% and its flash point by as much as 400%. In many spills, evaporation is the key process for changing slick properties, such as viscosity and density. However, for oils susceptible to emulsification, this latter process may dominate.

The basis for most of the earlier oil evaporative work is the extensive studies on the evaporation of water (Brutsaert, 1982; Jones, 1992). In fact, some of the model equations still employ portions of these equations. The pioneering work in the development of water evaporation equations was carried out by Sutton (1934). Sutton proposed the following water evaporation equation:

$$E = Ms\ C_s\ U^{7/9}\ d^{1/9}\ Sc^{-r} \tag{41.1}$$

where E is the mean evaporation rate per unit area, Ms is the mass transfer coefficient, C_s is the concentration of the evaporating fluid (mass/volume), U is the wind speed, d is the area of the square or circular pool, Sc is the 'Schmidt number', and r is an empirical exponent which is assigned values from 0 to 2/3. The Schmidt number (Sc) is a dimensionless number, defined as the ratio of air momentum diffusivity (viscosity) to mass diffusivity, and is used to characterize flows in which there are simultaneous momentum and mass diffusion convection processes.

The most frequently used work in older spill modeling is that of Stiver and Mackay (1984). This is based on some of the earlier work by Mackay and Matsugu (1973). The formulation was initiated with assumptions about the evaporation of a liquid. If a liquid is spilled, the rate of evaporation is given by:

$$N = \mathrm{KAP}/(\mathrm{RT}) \tag{41.2}$$

where N is the evaporative molar flux (mol/s), K is the mass transfer coefficient under the prevailing wind (ms^{-1}), A is the area (m^2), and P is the vapor pressure of the bulk liquid.

Equation (41.2) could not be directly related to oil data and was replaced with a new equation developed using laboratory empirical data:

$$Fv = (T/K1)\ln(1+K1\theta/T)\exp(K2-K3/T) \tag{41.3}$$

where F_v is the volume fraction evaporated and $K_{1,2,3}$ are empirical constants.

A value for K_1 was obtained from the slope of the F_v vs. log θ curve from pan or bubble evaporation experiments. For θ greater than 10^4, K_1 was found to be approximately 2.3T divided by the slope. The expression $\exp(K_2 - K_3/T)$ was then calculated, and K_2 and K_3 determined individually from evaporation curves at two different temperatures.

Fingas (2012, 2015a) studied oil evaporation and found that oil did not evaporate in the same manner as water. Instead of being air-boundary-layer regulated, it was found that oil was regulated by diffusion through the oil mass. This allowed for a simplified oil prediction of the type:

$$\text{Percentage oil evaporated} = c\ln(t) \tag{41.4}$$

where c is an empirical constant at a given temperature and t is the time in minutes.

This was further expanded to provide a relationship with temperature for a variety of oils to yield:

$$\text{Percentage evaporated} = \left[C + 0.045(T - 15)\right]\ln(t) \tag{41.5}$$

where C is a constant for each type of oil or petroleum product, T is the temperature in Celsius, and t is the time in minutes.

A large number of experiments were performed on oils to directly measure their evaporation curves. Examples of the empirical equations that result are given Table 41.1.

It was found that oils and fuels evaporated as two distinct types: those that evaporated as a logarithm of time and those that evaporated as a square root of time (Fingas, 2015a). Most oils typically evaporated as a logarithm (natural) with time. Diesel fuel and similar oils, such as jet fuel and kerosene, evaporate as a square root of time. The reasons for this are that diesel fuel and similar oils have a narrower range of compounds, which evaporate at similar rates, yielding rates, which sum as a square root.

For those oils such as diesel fuel, that evaporate as a square root with time, the equation is:

$$\text{Percentage evaporated} = \left[C + 0.01(T - 15)\right]\sqrt{t} \tag{41.6}$$

where the parameters are as in Equation (41.5) above. Again, some examples of experimental constants are given in Table 41.1.

The question then becomes, what does one do if you do not have the empirical constant, which takes days to measure? A procedure to use only distillation data provides a simple estimation method (Fingas, 2015a):

For oils that follow a logarithmic equation:

$$\text{Percentage evaporated} = \left[0.165(\%D) + .045(T - 15)\right]\ln(t) \tag{41.7}$$

For oils that follow a square root equation:

$$\text{Percentage evaporated} = \left[0.0254(\%D) + .01(T - 15)\right]\sqrt{t} \tag{41.8}$$

where $\%D$ is the percentage (by weight) distilled at 180°C.

TABLE 41.1
Example Equations for Predicting Evaporation

Oil Type	Equation	Oil Type	Equation
Alaska North Slope (2002)	%Ev = (2.86 + .045T)ln(t)	Jet A1	%Ev = (.59 + .013T)/t
Alberta Sweet Mixed Blend	%Ev = (3.24 + .054T)ln(t)	Komineft, Russian	%Ev = (2.73 + .045T)ln(t)
Arabian Medium	%Ev = (1.89 + .045T)ln(t)	Lago, Angola	%Ev = (1.13 + .045T)ln(t)
Arabian Heavy	%Ev = (1.31 + .045T)ln(t)	Lago Treco, Venezuela	%Ev = (1.12 + .045T)ln(t)
Arabian Light	%Ev = (3.41 + .045T)ln(t)	Maui, New Zealand	%Ev = (−0.14 + .013T)/t
ASMB - Standard #5	%Ev = (3.35 + .045T)ln(t)	Maya, Mexico	%Ev = (1.38 + .045T)ln(t)
Aviation Gasoline 100 LL	ln(%Ev) = (0.5 + .045T)ln(t)	Ninian, United Kingdom	%Ev = (2.65 + .045T)ln(t)
Barrow Island, Australia	%Ev = (4.67 + .045T)ln(t)	Norman Wells, Canada	%Ev = (3.11 + .045T)ln(t)
Belridge Crude, CA, USA	%Ev = (.01 + .013T)/t	North Slope - Southern Pipeline	%Ev = (2.47 + .045T)ln(t)
Boscan, Venezuela	%Ev = (−0.15 + .013T)/t	Nugini, New Guinea	%Ev = (1.64 + .045T)ln(t)
Brent, United Kingdom	%Ev = (3.39 + .048T)ln(t)	Odoptu, Russian	%Ev = (4.27 + .045T)ln(t)
Bunker C - Light (IFO~250)	%Ev = (.0035 + .0026T)/t	Prudhoe Bay (new stock)	%Ev = (2.37 + .045T)ln(t)
Bunker C - long term	%Ev = (−.21 + .045T)ln(t)	Rangely, CO, USA	%Ev = (1.89 + .045T)ln(t)
California API 11	%Ev = (−0.13 + .013T)/t	Sahara Blend, Algeria	%Ev = (0.001 + .013T)/t
California API 15	%Ev = (−0.14 + .013T)/t	Sahara Blend (long Term)	%Ev = (1.09 + .045T)ln(t)
Cano Limon, Colombia	%Ev = (1.71 + .045T)ln(t)	Sakalin, Russia	%Ev = (4.16 + .045T)ln(t)
Carpenteria, CA, USA	%Ev = (1.68 + .045T)ln(t)	Scotia Light	%Ev = (6.87 + .045T)ln(t)
Chavyo, Russia	%Ev = (3.52 + .045T)ln(t)	Sockeye (2001)	%Ev = (1.52 + .045T)ln(t)
Cold Lake Bitumen, AB Canada	%Ev = (−0.16 + .013T)/t	South Louisiana	%Ev = (2.39 + .045T)ln(t)
Cook Inlet - Granite Point	%Ev = (4.54 + .045T)ln(t)	South Louisiana (2001)	%Ev = (2.74 + .045T)ln(t)
Cusiana, Colombia	%Ev = (3.39 + .045T)ln(t)	Statfjord, Norway	%Ev = (2.67 + .06T)ln(t)
Delta West Block 97, USA	%Ev = (6.57 + .045T)ln(t)	Sumatran heavy, Indonesia	%Ev = (−0.11 + .013T)/t
Diesel (2002)	%Ev = (0.02 + .013T)/t	Sumatran Light	%Ev = (0.96 + .045T)ln(t)
Diesel Fuel (2002)	%Ev = (5.91 + .045T)ln(t)	Taching, China	%Ev = (−0.11 + .013T)/t
Ekofisk, Norway	%Ev = (4.92 + .045T)ln(t)	Takula, Angola	%Ev = (1.95 + .045T)ln(t)
Empire Crude, LA, USA	%Ev = (2.21 + .045T)ln(t)	Tapis, Malaysia	%Ev = (3.04 + .045T)ln(t)
Federated, AB, Canada	%Ev = (3.47 + .045T)ln(t)	Tchatamba Crude, Gabon	%Ev = (3.8 + .045T)ln(t)
Gasoline	%Ev = (13.2 + .21T)ln(t)	Thevenard Island, Australia	%Ev = (5.74 + .045T)ln(t)
Genesis, GOM, USA	%Ev = (2.12 + .045T)ln(t)	Troll, Norway	%Ev = (2.26 + .045T)ln(t)
Gulfaks, Norway	%Ev = (2.29 + .034T)ln(t)	Udang, Indonesia	%Ev = (−0.14 + .013T)/t
Hibernia, NL, Canada	%Ev = (2.18 + .045T)ln(t)	Vasconia, Colombia	%Ev = (0.84 + .045T)ln(t)
Hondo, CA, USA	%Ev = (1.49 + .045T)ln(t)	West Texas Intermediate	%Ev = (2.77 + .045T)ln(t)
Hout, Kuwait	%Ev = (2.29 + .045T)ln(t)	West Texas Intermediate	%Ev = (3.08 + .045T)ln(t)

The equations noted above were all measured at a slick thickness of 1.5 mm which is typical of at-sea values. The thickness effect has been studied and a correction factor for it can be applied as:

$$\text{Corrected equation factor}\left(C'\right) = \text{equation factor}\left(C\right) + 1 - 0.78 * \sqrt{t} \qquad (41.9)$$

where C' is the thickness corrected factor for application in Equation (41.5), C is the empirical equation factor, and t is the thickness of the slick in mm. It should be noted that this thickness adjustment is typically not needed for many situations in which oils spread quickly below 1.5 mm, such as for many light and medium crude oils.

41.4 WATER UPTAKE AND EMULSIFICATION

Emulsification is the process by which one liquid is dispersed into another one in the form of small droplets (Fingas, 2015b). Water droplets can remain in an oil mass in a stable form and the resulting material is completely different from unemulsified oil. These water-in-oil emulsions are sometimes called 'mousse' or 'chocolate mousse'.

The mechanism of emulsion formation is not yet fully understood, but it probably starts with sea energy forcing the entry of small water droplets, about 10–25 μm (or 0.010–0.025 mm) in size, into the oil. If the oil is only slightly viscous, these small droplets will not leave the oil quickly. On the other hand, if the oil is too viscous, droplets will not enter the oil to any significant extent. Once in the oil, the droplets slowly gravitate to the bottom of the oil layer. Asphaltenes and resins in the oil will interact with the water droplets to stabilize them. Depending on the quantity and type of asphaltenes and resins, as well as the sea surface energy, an emulsion may be formed. The conditions required for emulsions of any stability to form may only be reached after a period of evaporation. Evaporation increases the viscosity to the critical value and increases the resin and asphaltene percentage in the oil.

Water can be present in oil in five ways. First, some oils contain about 1% water as soluble water. This water does not significantly change the physical or chemical properties of the oil. The second way is called entrainment, whereby water droplets are simply held in the oil by its viscosity to form an unstable mixture. These are formed when water droplets are incorporated into oil by the sea's wave action and there are not enough asphaltenes and resins in the oil to stabilize the water droplets. The third way is that of unstable emulsions or those oils that simply do not form water-in-oil types. Unstable emulsions break down into water and oil within minutes or a few hours at most, once the sea energy diminishes. The properties and appearance of the unstable emulsion are almost the same as those of the starting oil, although the water droplets may be large enough to be seen with the naked eye.

Mesostable emulsions represent the fourth way water can be present in oil. These are formed when the small droplets of water are stabilized to a certain extent by a combination of the viscosity of the oil and the interfacial action of asphaltenes and resins. These emulsions generally break down into oil and water or sometimes into water, oil, and a stable residue within a few days. Mesostable emulsions are viscous liquids that are reddish-brown in color.

The fifth way that water exists in oil is in the form of stable emulsions (Fingas, 2015b–c). These form in a way similar to mesostable emulsions except that the oil contains a sufficient amount of resins and asphaltenes to stabilize the water droplets. The viscosity of stable emulsions is 800–1,000 times higher than that of the starting oil; the emulsion will remain stable for weeks and even months after formation. Stable emulsions are reddish-brown in color and appear to be nearly solid. Because of their high viscosity and near solidity, these emulsions do not spread and tend to remain in lumps or mats on the sea or shore.

The formation of emulsions is an important event in an oil spill. First, and most importantly, emulsification substantially increases the actual volume of the spill. Emulsions that contain about 70% water triple the volume of the oil spill. Even more significantly, the viscosity of the oil increases by as much as 1,000 times, depending on the type of emulsion formed. For example, oil that has the starting viscosity of motor oil can triple in volume and become almost solid through the process of emulsification (Fingas, 2015b).

These increases in volume and viscosity make clean-up operations more difficult. Emulsified oil is difficult or impossible to disperse, or to recover with skimmers. Emulsions can be broken down with special chemicals in order to recover the oil with skimmers. It is thought that emulsions break down into oil and water by further weathering, oxidation, mixing with unemulsified oil, and freeze–thaw action. Meso- or semi-stable emulsions are relatively easy to break down, whereas stable emulsions may take months or years to break down naturally, if they ever do break down.

Emulsion formation also changes the fate of the oil. It has been noted that when oil forms stable or mesostable emulsions, evaporation slows considerably. Biodegradation also appears to slow down. The dissolution of soluble components from oil may also cease once emulsification has occurred.

The various types form 'areas' when the natural logarithm of viscosity is plotted against the asphaltene times of the resin content (Fingas, 2015b–c). As there are other factors involved such as the asphaltene/resin ratio, the total saturate content, and the density, there is some overlap between regions. A simplified version of this is shown in Figure 41.1. Correlation of water-in-oil types shows that the most important factors are the starting oil viscosity and the asphaltene and resin contents. Even a simple graphical presentation of these three oil properties shows that the resulting water-in-oil type can be predicted with relative accuracy. Correlation of these factors alone shows distinct regions where the four kinds of water-in-oil types exist.

The data show that the water-in-oil types are physically stabilized by the oil viscosity and chemically stabilized by both asphaltenes and resins. For greater stability, resin content should exceed the asphaltene content slightly. Excess resin content (A/R < than about 0.6) apparently destabilizes the emulsion. A high asphaltene content (typically > 10%) increases the viscosity of the oil such that a stable emulsion will not form. Viscous oils will only uptake water as entrained water and will slowly lose much of this water over a period ranging up to months. Oils with low viscosity or without significant amounts of asphaltenes and resins will not form any water-in-oil types and will retain less than about 6% water. Oils of very high viscosity (typically > 10,000 MPAs) will not form any of these water-in-oil types. These screening procedures might be used to differentiate some of the more obvious types such as entrained water-in-oil types, and high and low viscosity oils that do not form (Fingas, 2015c).

Entrained water-in-oil types show a unique character, that is they show a starting oil density of greater than 0.96 g/mL, but less than 1.0 g/mL. Further, the starting oils have a viscosity greater than 2,300 MPas and less than 200,000 MPAs. The screening criteria can be applied to all the oils, and if they meet this requirement, they will entrain water. In doing the screening for entrained types, it is noted that only three oils out of more than 200 were characterized incorrectly using the entrained screening (> 2,300 and < 200,000 MPAs viscosity).

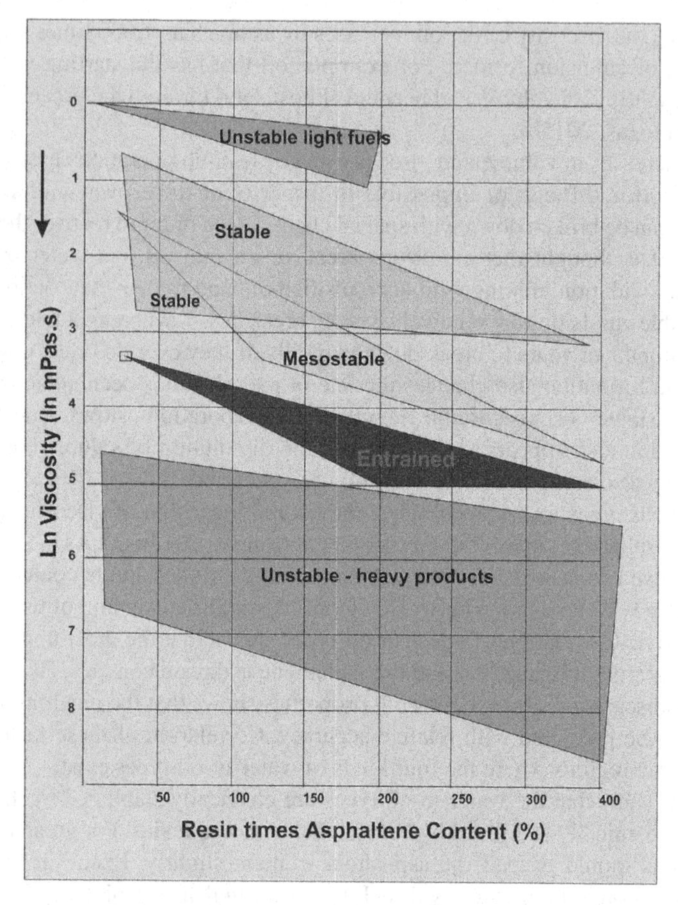

FIGURE 41.1 A simplified drawing of the regions of formation for various water-in-oil types.

Note: In reality, there is more overlap between regions.

A second obvious cut-off screen are those low-viscosity oils that do not form a water-in-oil type. The cut off of viscosity < 50 MPAs can be applied. Such oils as gasoline, diesel fuel, and very light crudes will never form a water-in-oil type. This screening results in a low error rate of only about five oils out of the more than 200 examined.

The third screening can be applied to those oils that have a very high viscosity and do not form any water-in-oil type. This type is easily screened using only the criterion of viscosity, that of viscosity > 200,000 MPAs. This results in a very high accuracy rate and a very low error rate, but there are few candidates for this oil in this class of compounds.

This third screening leaves only the mesostable and stable types. Several attempts to separate these were unsuccessful. Modeling can be more successful in separating these two groups.

The modeling of emulsion formation is somewhat complex but can be carried out by several different means. There are three approaches now available:

1. Empirical (compare requested oil to the database of oils)
   - Benefits: highly accurate;
   - Disadvantages: requires a large database (~ 430 oils at different weathering percentages), needs another method for oil input.
2. Fuzzy logic (use Yetilmezsoy model (Yetilmezsoy et al., 2012))
   - Benefits: fairly accurate;
   - Disadvantages: difficult having to deal with a 'black box' model, may have difficulty in obtaining and integrating this into typical spill models.
3. Calculate using regression model (use Fingas (2015b) regression model)
   - Benefits: somewhat accurate, have control over inputs and outputs;
   - Disadvantages: less accurate than above two approaches.

41.5 NATURAL DISPERSION

Natural dispersion occurs when fine droplets of oil are created and transferred into the water column by wave action or turbulence. Small oil droplets (less than 20 μm or 0.020 mm) are somewhat stable in water and will remain so for short periods of time. Large droplets tend to rise, and larger droplets (more than 50 μm) will not stay in the water column for more than a few seconds. Depending on oil conditions and the amount of sea energy available, natural dispersion can be insignificant or it can remove the bulk of the oil. In 1993, the oil from a stricken ship, the *Braer*, dispersed and sedimented almost entirely as a result of high seas off Scotland at the time of the spill and the dispersible nature of the oil cargo (Lunel, 1995). It should be noted that natural dispersion might just be an intermediary in the oil-sediment reaction. The *Braer* spill is more likely a good example of oil sedimentation because of oil–particle interaction than natural dispersion.

Natural dispersion is dependent on both the oil properties and the amount of sea energy (Delvigne and Sweeney, 1988). Heavy oils such as Bunker C or a heavy crude will not disperse naturally to any significant extent, whereas diesel fuel and even light crudes can disperse significantly if the saturate content is high and the asphaltene and resin contents are low. In addition, significant wave action is needed to disperse oil. In 40 years of monitoring spills on the oceans, those spills where oil has dispersed naturally have all occurred in very energetic seas – up to Beaufort 7.

The long-term fate of dispersed oil is not known, although it may degrade to some extent as it consists primarily of saturate components. Some of the dispersed oil may rise and form another surface slick or it may become associated with sediment and be precipitated to the bottom.

Delvigne and associates (Delvigne and Sweeney, 1989) studied the natural dispersion of oil in a series of wave tanks and a series of smaller apparatuses. The turbulence energy was related to the creation of naturally dispersed droplets. Of prime importance to oil spill modeling, the following empirical equation was created after simplification:

$$Q(\mathrm{d}) = Co\left(34.4\,H^2\mathrm{rms}\right)^{0.57} d^{0.7}\,\Delta d\,S\,\mathrm{cov}\left(0.032\left(U-5\right)/Tw\right) \qquad (41.10)$$

where:

$Q_{(d)}$ is the entrained mass rate of droplet sizes in the interval around d per unit surface area and per unit time – given in kg/m^2s;

C_o is a constant for a given oil, a light oil is about 1,000–1,800, a medium oil about 500 to 1,000, and a heavier oil < 500;

H_{rms} is the r.m.s. value of the wave height (m);

d, Δd is the droplet size and range of droplet size (interval) in m;

S_{cov} is the fraction of the sea covered by oil (0 to 1);

U_i is the wind speed at the initiation of breaking waves (~ 5 m/s);

T_w is the wave period (s).

Several issues have been noted about the formation developed above (Fingas, 2013):

1. The natural dispersion predicted was measured as a temporary phenomenon – that is the instantaneous droplets in the water column. Their persistence was not measured. The equation was designed to yield only the temporary transport in the water. Delvigne actually proposed a model for resurfacing of these droplets in the water column. Later workers incorrectly assumed that this amount was 'permanently' dispersed.
2. The equation over-predicts natural dispersion in most cases, especially in cases of low sea states.
3. The given Q is a rate with no input factor, thus mass on the ocean is not input or defined; further, it was stated that this is independent of the oil on the surface, also stated to be independent of oil thickness, an obvious difficulty.
4. The oil constants appear to be arbitrary and the rules that they are 1/viscosity does not work for cSt.
5. The wave periods in the wave tank experiments were very short.
6. The initiation wind should be greater than 5 m/s – perhaps as high as 1020 m/s.

These can be partially solved in terms of the Delvigne tank experiments (Fingas, 2013). First, the experiments were mostly conducted with a thickness of 0.6 mm, which constitutes an oil coverage of 0.6 kg/m^2. Dividing Equation (41.10) times this factor and making other substitutions:

$$F(\mathrm{d}) = 6.3\times10^{-4} / \rho^{1.5}\left(34.4\, H^2\mathrm{rms}\right)^{0.57}\left(0.032\left(U-15\right)/Tw\right) \quad (41.11)$$

where: $F_{(d)}$ is the fraction of the entrained mass rate of droplet sizes in the interval from 10 to 30 µm – given in reciprocal hours (/hour);

ρ is the viscosity of the oil in MPAs;

H_{rms} is the r.m.s. value of the wave height (m);

U is the wind speed in m/s;

T_w is the wave period, s.

41.6 OTHER PROCESSES

41.6.1 Dissolution

Through the process of dissolution, some of the soluble components of the oil are lost to the water under the slick (Danchuk and Willson, 2008). These include some of the lower molecular weight aromatics. As only a small amount, usually much less than a fraction of a percent of the oil, actually enters the water column, dissolution does not measurably change the mass balance of the oil. The significance of dissolution is that the soluble aromatic compounds are particularly toxic to fish and other aquatic life. If a spill of oil containing a large amount of soluble aromatic components occurs in shallow water and creates a highly localized concentration of compounds, then significant numbers of aquatic organisms can be killed.

Gasoline, diesel fuel, and light crude oils are the most likely to cause aquatic toxicity. Highly weathered oil is unlikely to dissolve into the water. On open water, the concentrations of hydrocarbons in the water column are unlikely to kill aquatic organisms.

Dissolution occurs immediately after the spill occurs and the rate of dissolution decreases rapidly after the spill as soluble substances are quickly depleted. Some of the soluble compounds also evaporate rapidly.

There are a variety of estimation techniques for dissolution; however, there is no good continuous model at this time.

41.6.2 Photooxidation

Photooxidation can change the composition of an oil (Aeppli et al., 2014). This occurs when the sun's action on an oil slick causes oxygenation and thus forms new products that sometimes may be resins. The resins may be somewhat soluble and dissolve into the water or they may cause water-in-oil emulsions to form. It is not well understood how photooxidation specifically affects oils, although certain oils are susceptible to the process, while others are not. For most oils, photooxidation is not an important process in terms of changing their immediate fate or mass balance after a spill.

41.6.3 Sedimentation, Adhesion to Surfaces, and Oil–Fines Interaction

Sedimentation is the process by which oil is deposited on the bottom of the sea or other water body. While the process itself is not well understood, certain facts about it are. Most sedimentation noted in the past has occurred when oil droplets reached a higher density than water after interacting with mineral matter in the water column. This interaction sometimes occurs on the shoreline or very close to the shore. Once oil is on the bottom, it is usually covered by other sediment and degrades very slowly. In a few well-studied spills, a significant amount (about 10%) of the oil was sedimented on the sea floor (Chanton et al., 2015). Such amounts can be very harmful to biota, which inevitably meet the oil on the sea bottom. Because of the difficulty of studying sedimentation, data are limited.

Oil is very adhesive, especially when it is moderately weathered, and binds to shorelines or other mineral material with which it comes in contact. A significant amount of oil can be left in the environment after a spill in the form of residual amounts adhering to shorelines and man-made structures such as piers and artificial shorelines. As this oil usually contains a high percentage of aromatics and asphaltenes with high molecular weight, it does not degrade significantly and can remain in the environment for decades.

Oil slicks and oil on shorelines sometimes interact with mineral fines suspended in the water column and the oil is thereby transferred to the water column (Sun et al., 2014). Particles of mineral with oil attached may be heavier than water and sink to the bottom as sediment, or the oil may detach and refloat. Oil–fines interaction does not generally play a significant role in the fate of most oil spills in their early stages, but can have an impact on the rejuvenation of an oiled shoreline over the long term.

41.6.4 Biodegradation

A large number of microorganisms are capable of degrading petroleum hydrocarbons. Many species of bacteria, fungi, and yeasts metabolize petroleum hydrocarbons as a food energy source (Xue et al., 2015). Bacteria and other degrading organisms are most abundant on land in areas where there have been petroleum seeps, although these microorganisms are found everywhere in the environment. As each species can utilize only a few related compounds at most, however, broad-spectrum degradation does not occur. Hydrocarbons metabolized by microorganisms are generally converted to an oxidized compound, which may be further degraded, may be soluble, or may accumulate in the remaining oil. The aquatic toxicity of the biodegradation products is sometimes greater than that of the parent compounds.

The rate of biodegradation depends primarily on the nature of the hydrocarbons and then on the temperature. Generally, rates of degradation tend to increase as the temperature rises. Some groupings of bacteria, however, function better at lower temperatures and others function better at higher temperatures. Indigenous bacteria and other microorganisms are often the best adapted and most effective at degrading oil as they are acclimatized to the temperatures and other conditions of the area. Adding 'super-bugs' to the oil does not necessarily improve the degradation rate.

The rate of biodegradation is greatest on saturates, particularly those containing approximately 12–20 carbons. Resins and asphaltenes, which have a high molecular weight, biodegrade very slowly, if at all. This explains the durability of roof shingles containing tar and roads made of asphalt, as both tar and asphalt consist primarily of high-molecular weight aromatics and asphaltenes. On the other hand, diesel fuel is a highly degradable product, as it is largely composed of degradable saturates and lower aromatics. Light crudes are also degradable to a degree. While gasoline contains degradable components, it also contains some compounds that are toxic to some microorganisms. These compounds generally evaporate more rapidly, but in almost all cases, most of the gasoline will evaporate before it can degrade. Heavy crudes contain little material that is readily degradable and Bunker C contains almost none.

The rate of biodegradation is also highly dependent on the availability of oxygen. On land, oils such as diesel can degrade rapidly at the surface, but very slowly if at all only a few centimeters below the surface, depending on oxygen availability. In water, oxygen levels can be so low that degradation is limited. It is estimated that it would take all the dissolved oxygen in approximately 400,000 L of seawater to completely degrade 1 L of oil. The rate of degradation also depends on the availability of nutrients such as nitrogen and phosphorus, which are most likely to be available on shorelines or on land. Finally, the rate of biodegradation also depends on the availability of the oil to the bacteria or microorganism. Oil degrades significantly at the oil–water interface at sea and, on land, mostly at the interface between soil and the oil.

Biodegradation can be a very slow process for some oils. It may take weeks for 50% of a diesel fuel to biodegrade under optimal conditions and years for 10% of a crude oil to biodegrade under similar conditions. For this reason, biodegradation under natural conditions is not considered an important weathering process in the short term.

41.6.5 Sinking and Over-Washing

If oil is denser than the surface water, it may sometimes sink. Some rare types of heavy crudes, Bitumen and Bunker C, can reach these densities and sink. When this occurs, the oil may sink to a denser layer of water rather than to the bottom. Less dense layers of water may override these denser layers of water. This occurs for example when waves in the seas are not high and warmer fresh water from land overrides dense seawater. Fresh water has a density of about 1.00 g/mL and the seawater a density of about 1.03 g/mL. Oil with a density greater than 1.00 but less than 1.03 would sink through the layer of fresh water and ride on the layer of salt water. The layer of fresh water usually varies in depth from about 1 to 10 m. If the sea energy increases, the oil may actually reappear on the surface, as the density of the water increases from 1.00 to about 1.03.

It is important to note that sinking of any form, whether to the bottom or to the top of a layer of dense seawater, is rare. When oil does sink, it complicates clean-up operations as the oil can be recovered only with underwater suction devices or special dredges.

Over-washing is another phenomenon that occurs quite frequently and can hamper clean-up efforts. At moderate sea states, a dense slick can be over-washed with water. When this occurs, the oil can disappear from view, especially if the spill is being observed from an oblique angle, as would occur if someone were observing a slick from a ship. Over-washing causes confusion about the fate of an oil spill as it can give the impression that the oil has sunk and then resurfaced.

A summary of the modeling formulations for sinking and submergence are given in the literature (Fingas et al., 2006).

41.6.6 Formation of Tar Balls

Tar balls are agglomerations of thick oil less than about 10 cm in diameter. Larger accumulations of the same material ranging from about 8 cm to 1 m in diameter are called tar mats. Tar mats are pancake-shaped, rather than round. Their formation is still not completely understood, but it is known that they are formed from the residuals of heavy crudes and Bunker C. After these oils weather at sea and slicks are broken up, the residuals remain in tar balls or tar mats. The reformation of droplets into tar balls and tar mats has also been observed, with the binding force being simply adhesion.

The formation of tar balls is the ultimate fate of many oils. These tar balls are then deposited on shorelines around the world. The oil may come from spills, but it is also residual oil from natural oil seeps or from deliberate operational releases such as from ships. Tar balls are regularly recovered by machine or by hand from recreational beaches.

REFERENCES

Aeppli, C., R. K. Nelson, and J. R. Radovic, et al. 2014. Recalcitrance and degradation of petroleum biomarkers upon abiotic and biotic natural weathering of Deepwater Horizon oil. *Environmental Science & Technology* 48: 6726–6734.

Boehm, P. D., D. S. Page, and J. S. Brown, et al. 2008. Distribution and weathering of crude oil residues on shorelines 18 years after the Exxon Valdez spill. *Environmental Science & Technology* 42: 9210–9216.

Brutsaert, W. 1982. *Evaporation into the Atmosphere: Theory, History and Applications.* Dordrecht: Springer.

Chanton, J., T. Zhao, and B. E. Rosenheim, et al. 2015. Using natural abundance radiocarbon to trace the flux of petrocarbon to the sea floor following the Deepwater Horizon oil spill. *Environmental Science & Technology* 49: 847–854.

Danchuk, S. and C. S. Willson, 2008. Numerical modeling of oil spills in the inland waterways of the lower Mississippi River delta. *International Oil Spill Conference Proceedings* 2008: 887–891.

Delvigne, G. A. L. and C. E. Sweeney. 1988. Natural dispersion of oil. *Oil and Chemical Pollution* 4: 281–310.

Delvigne, G. A. L. and C. E. Sweeney. 1989. *Natural Dispersion of Oil.* Delft: Delft Hydraulic Laboratory.

Diez, S., E. Jover, J. M. Bayona, and J. Albaiges. 2007. Prestige oil spill. III. Fate of a heavy oil in the marine environment. *Environmental Science & Technology* 41: 3075–3082.

Fingas, M. 2012. Studies on the evaporation regulation mechanisms of crude oil and petroleum products. *Advances in Chemical Engineering and Science* 2: 246–256.

Fingas, M. 2013. A review of natural dispersion models. In *Proceedings of the Arctic and Marine Oilspill Program (AMOP) Technical Seminar*, 207–233. Ottawa: Environment Canada.

Fingas, M. 2015a. Oil and petroleum evaporation. In *Handbook of Oil Spill Science and Technology*, ed. M. Fingas, 207–223. New York, NY: John Wiley and Sons.

Fingas, M. 2015b. Water-in-oil emulsion formation, *Handbook of Oil Spill Science and Technology*, ed. M. Fingas, 225–270. New York, NY: John Wiley and Sons.

Fingas, M. 2015c. A simplified procedure to predict water-in-oil emulsions. In *Proceedings of the Arctic and Marine Oilspill Program (AMOP) Technical Seminar*, 207–255. Ottawa: Environment Canada.

Fingas, M. F., B. Hollebone, and B. Fieldhouse. 2006. The density behaviour of heavy oils in water. In *Proceedings of the Arctic and Marine Oilspill Program (AMOP) Technical Seminar*, 57–77. Ottawa: Environment Canada.

Jones, F. E. 1992. *Evaporation of Water*. Chelsea, MI: Lewis Publishers.

Lehr, W., D. Simecek-Beatty, A. Aliseda, and M. Boufadel. 2014. Review of recent studies on dispersed oil droplet distribution. In *Proceedings of the Arctic and Marine Oilspill Program (AMOP) Technical Seminar*, 1–8. Ottawa: Environment Canada.

Lunel, T. 1995. The Braer spill: Oil fate governed by dispersion. *International Oil Spill Conference Proceedings*, 1995: 955–956.

Mackay, D. and R. S. Matsugu. 1973. Evaporation rates of liquid hydrocarbon spills on land and water. *Canadian Journal of Chemical Engineering* 51: 434–439.

National Research Council. 2003. *Oil in the Sea III, Inputs, Fates and Effects*. Washington, DC: National Academies Press.

Short, J. W., G. V. Irvine, and D. H. Mann, et al. 2007. Slightly weathered Exxon Valdez oil persists in Gulf of Alaska beach sediments after 16 years. *Environmental Science & Technology* 41: 1245–1250.

Stiver, W. and D. Mackay. 1984. Evaporation rate of spills of hydrocarbons and petroleum mixtures. *Environmental Science & Technology* 18: 834–840.

Sun, J., A. Khelifa, C. Zhao, D. Zhao, and Z. Wang. 2014. Laboratory investigation of oil-suspended particulate matter aggregation under different mixing conditions. *Science of the Total Environment* 473–474: 742–749.

Sutton, O. G. 1934. Wind structure and evaporation in a turbulent atmosphere. *Proceedings of the Royal Society of London. Series A* 146: 701–722.

Xue, J., Y. Yu, Y. Bai, L. Wang, and Y. Wu. 2015. Marine oil-degrading microorganisms and biodegradation process of petroleum hydrocarbon in marine environments: A review. *Current Microbiology* 71: 220–228.

Yetilmezsoy, K, M. Fingas, and B. Fieldhouse. 2012. Modeling water-in-oil emulsion formation using fuzzy logic. *Journal of Multiple-Valued Logic & Soft Computing* 18: 329–353.

42 Introduction to Oil Spills and their Clean-up

Merv Fingas

CONTENTS

42.1 SOURCES OF OIL SPILLS

The exploration, production, and consumption of oil and petroleum products are increasing worldwide and the threat of oil pollution increases accordingly. The movement of petroleum from the oil fields to the consumer involves as many as 10–15 transfers between many different modes of transportation including tankers, pipelines, railcars, and tank trucks. Oil is stored at transfer points, terminals, and refineries along the route. Accidents can happen at any of these exploration, production, and transportation steps or storage facilities.

Obviously, an important part of protecting the environment is ensuring that there are as few spills as possible. Both government and industry are working to reduce the risk of oil spills, with the introduction of strict new legislation and stringent operating codes. Industry has invoked many operating and maintenance procedures to reduce accidents that could lead to spills. In fact, the rate of spillage has decreased in the past 20 years. This is especially true for tanker accidents at sea. Intensive training programs have been developed to reduce the potential for human error. Despite these efforts, spill experts estimate that 30–50% of oil spills are either directly or indirectly caused by human error, with 20–40% of all spills caused by equipment failure or malfunction (Fingas, 2012).

42.1.1 Oil Spill Statistics

Oil spills are a frequent occurrence, particularly because of the extensive use of oil and petroleum products in our daily lives (Fingas, 2012). About 450,000 tons of oil and petroleum products are used in Canada every day. The USA uses about ten times this amount and, worldwide, about 20 million tons are used per day. In the USA, about half of the approximately 4 million tons of oil and petroleum products used per day is imported, primarily from Canada, Saudi Arabia, and Africa. About 40% of the daily demand in the USA is for automotive gasoline, and about 15% is for diesel fuel used in transportation. About 40% of the energy used in the USA comes from petroleum, 25% from natural gas, and 20% from coal. In both Canada and the USA, much of the refined oil goes into powering transportation.

Spill statistics are collected by a number of agencies in any country. In the USA, the U.S. Coast Guard maintains a database of spills in navigable waters, while state agencies keep statistics on spills on land, which are sometimes gathered into national statistics. The US Bureau of Safety and Environmental Enforcement (BSEE) maintains records of spills from offshore exploration and production activities. The International Tanker Owners Pollution Federation Limited (ITOPF) has maintained a worldwide database of spills from tankers, combined carriers, and barges since 1974.

It can sometimes be misleading to compare oil spill statistics, however, because different methods are used to collect the data. In general, statistics on oil spills are difficult to obtain, and any dataset should be viewed with caution. The spill volume or amount is the most difficult to determine or estimate. For example, in the case of a vessel accident, the volume in a given compartment may be known before the accident, but the remaining oil may have been transferred to other ships immediately after the accident. Some spill accident databases do not include the amounts burned, if or when that occurs, whereas others include all the oil lost by whatever means. Sometimes the actual character or physical properties of the oil lost are not known, which leads to different estimates of the amount lost. Spill statistics compiled in the past are less reliable than more recent data because few agencies or individuals collected spill statistics before about 1975. More recently, techniques for collecting statistics are continually improving.

Reporting procedures vary in different jurisdictions and organizations, such as government or private companies. Minimum spill amounts that must be reported vary according to different regulations, depending on the spill source and location. For example, in the USA, reporting thresholds for pipelines are spills > 800 L to land, but any quantity to water that generates a sheen.

The number of spills reported also depends on the minimum size or volume of the spill. In both Canada and the USA, most oil spills reported are more than 4,000 L (about 1,000 gallons). In Canada, there are about 12 such oil spills every day, of which only about one is spilled into navigable waters. These 12 spills amount to about 40 tons of oil or petroleum product. In the USA, there are about 15 spills per day into navigable waters and an estimated 85 spills on land or into freshwater.

Despite the large number of spills, only a small percentage of oil used in the world is actually spilled. Oil spills in the year 2010 in Canada and the USA (basically North America) are summarized in Table 42.1 in terms of the volume of oil spilled and the actual number of spills (Fingas, 2012; Schmidt-Etkin, 2011). Although there are differences between the two countries, there are many similarities, and this makes the dataset more viable as a predictor for many developed countries.

There are more spills from barges into navigable waters in the USA proportionately than in Canada because more oil is transported by barge. In fact, the largest volume of oil spilled to water in the USA comes from barges, while the largest number of spills comes from vessels other than tankers, bulk carriers, or freighters.

In Canada and the USA, most spills take place on land, and pipeline spills account for the highest volume. In terms of the actual number of spills, most happen at petroleum production facilities, wells, production collection facilities, and battery sites.

TABLE 42.1
North American Spill Statistics 2010

	Percentage of Total Spills	
Source	**Volume**	**Numbers**
I Land spills (85% volume, 90% numbers)		
Pipelines	40	20
Wells, production, and storage facilities	25	25
Storage refineries	12	25
Retail and delivery	5	10
Trucks	6	11
Rail	7	4
Other	5	5
II On-water spills (15% volume, 10% numbers)		
Non-tank vessels	25	30
Tank barges	15	10
Terminals/refineries	25	30
Tankers	20	20
Platforms and pipelines	15	10
III Types of products spilled		
Crude oil	35	
Diesel heating	20	
Bunkers	15	
Marine	10	
Gasoline	8	
Condensates	3	
Waste and residuals	3	
Other oils	6	

On water, the greatest volume of oil spilled comes from marine or refinery terminals, although the largest number is from the same source as in the USA – vessels other than tankers, bulk carriers, or freighters.

An important question concerns the amount of spillage that is likely to occur in the future. There are several indications of the trend. There are government databases showing trends on oil production (NEB, 2019; EIA, 2019). The existing statistics and the projected future production rates of oil are shown in Table 42.2. Furthermore, there are studies on trends (Etkin, 2014).

There are significant trends that should be noted:

1. In Canadian production, the amount of bitumen being marketed is rapidly increasing. In addition to being sold as a diluted product called 'Dilbit', bitumen is also being upgraded into a synthetic crude. Spills of Dilbit have caused concern because some of the unique spill properties that it has shown, namely that once weathered for a period of time, it may sink in fresh water.
2. In North Dakota, the Bakken oil field is currently producing oil and the expansion of this field is quite rapid. This oil is also of concern when spilled as it is very flammable and has caused considerable damage in spills such as the Lac Megantic spill in Quebec.
3. There is a pipeline shortage to transport the above two products and these are increasingly being transported by rail, which also has increased the risk of spills from this source. Spills of oil from trains are believed to pose a higher risk than from pipelines (Frittelli et al., 2014).
4. Pipelines themselves are being built at a very rapid pace and many have been modified to carry products to the North American south rather than carrying other products north. The spills from pipelines have been decreasing in size (Etkin, 2014). It should be noted that the number of pipeline spills might increase as a result of the increasing number of pipelines.
5. The volume and number of spills from tanker vessels has been constantly decreasing over the past 20 years. Tanker spills contribute very little to the spillage in many countries.

TABLE 42.2
Canadian and US Oil Production, 2019 (Thousands of Cubic Meters per year)

	Canada	USA
Total crude	270,000	1,011,000
Conventional crude	105,100	
Upgraded bitumen	153,400	
Condensates	100	
Heavy conventional	100	
Bitumen	4,200	
Estimated Dilbit	7,300	
Estimated Bakken crude		75,000

42.1.2 Cost of Spill Clean-up and Prevention

There are many deterrents to oil spills which include government fines and the high cost of clean-up. In the USA, clean-up costs average about \$300 per liter spilled, not counting the costs of environmental damage (Etkin, 1999, 2003). If one counts the natural resource damages and the socio-economic impacts, this cost can rise to \$900 per liter. It is important to note that clean-up costs vary significantly for each spill and depend on where the oil is spilled, the volume spilled, what is impacted, and the methods used to respond to the oil spill. Removal of oil from shorelines is usually the most expensive clean-up process. Costs to clean up a shoreline can vary between about \$5/m^2 to as high as \$120/m^2 depending on the situation.

42.1.3 Oil Inputs into the Sea

Spills are only one source of oil released to the sea (National Research Council, 2003). The sources of such reeases for the period 1990–1999 are shown in Figure 42.1. About half of the oil in the sea derives from natural sources, including the many natural seeps or discharges from oil-bearing strata on the ocean floor. About 38% of the oil reaching the sea is the runoff of oil and fuel from land-based sources, mostly from urban areas. Significant amounts of lubricating oil finds its way into wastewater, which is often discharged directly into the sea. About 13% of oil reaching the sea comes from the transportation sector, which includes tankers, freighters, barges, and other vessels.

42.1.4 Review of the Top Worldwide Spills

Spill response knowledge is very empirical and past spill responses are mined for lessons to be learned on oil spill behavior and the effectiveness of countermeasures to provide information on how to deal with future spills in similar circumstances. Table 42.3 lists some of the top spills.

42.2 SUMMARY OF SOURCES

Spills occur from a wide variety of sources and as a result of a large number of causes. The largest source of spills on land is from pipeline spills, and secondly from

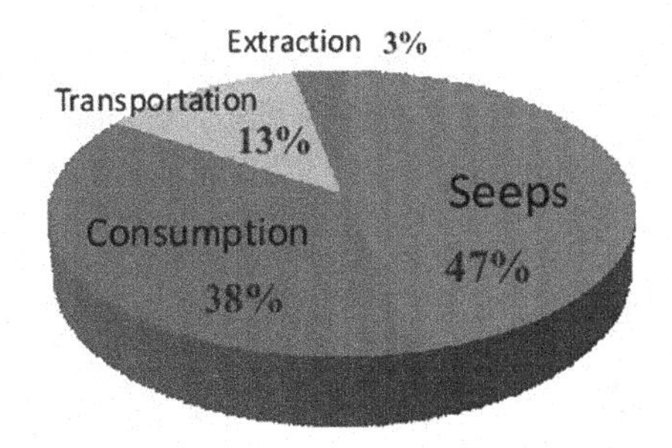

FIGURE 42.1 Sources of oil into the sea.

TABLE 42.3
Some Large Spills and The Lessons Learned

Spill Name	Size (Thousand Tons)	Year	Location	Oil Type	Nature of Spill	Some Lessons Learned
Gulf War Spill	800	1991	Arabian Gulf	Heavy crude	Oil released from storage and ships as an act of war	Recovery of sandy shorelines; mass recovery of near-shore oil
Deepwater Horizon	500	2010	Gulf of Mexico	Light crude	Oil well blowout	Behavior and fate of deep releases; submerged oil mat behavior
Amoco Cadiz	223	1978	Brittany, off France	Medium crude	Tanker grounding	Persistence of oil on rocky shorelines; mass recovery on foreshore areas
Prestige	63	2002	North of Spain	Heavy fuel oil	Tanker breakup	Dealing with very heavy oil spill; need for safe havens for stricken ships
Sea Empress	72	1996	South-west Britain	Light crude	Tanker grounding	Oil coming ashore after at-sea countermeasures; extensive wildlife effects, but could be worse
Exxon Valdez	37	1989	Prince William Sound, Alaska	Medium crude	Tanker grounding	Dealing with rocky oiled shorelines; fate and behavior in northern environments; shoreline clean-up methods
Ixtoc I	470	1979	Gulf of Mexico West	Heavy crude	Oil well blowout	Long travel distances of heavy oil; emulsification occurs at depth
Enbridge Pipeline	4	2010	Kalamazoo Michigan	Dilbit (diluted bitumen)	Pipeline rupture	Dilbit sinks after several days of weathering/interaction with particles; lessons on how to deal with sunken oil

oil production facilities. The volume of spills is approximately related to the large volume of oil handled by these two sources. Non-tank vessels and terminals/refineries are the largest sources of spills on water, again corresponding to the frequency of activities and high volumes handled. Trends have been analyzed and show that there is an emerging pattern of increased risk of spills from pipelines and railroads; however, the overall ranking of spill sources is likely to remain similar in the future to that in the past. The most frequently spilled products are crude oils, followed by light refined products such as diesel and No. 2 fuel oils.

42.3 CONSEQUENCES

The consequences will be discussed for spills on land and biological resources and habitats.

42.3.1 Consequences of Spills on Land

Oil spilled on land does not spread quickly, unlike on water, and the effects remain localized. Most types of oil will penetrate the soil and contaminate organisms there (Fingas, 2012). A full coating of fresh crude oil or diesel fuel will kill most plants and small trees on contact. Because of the usually limited area of impact, however, the effects of oil on land environments are not as great a concern as for marine environments.

Soil is defined as the loose unconsolidated material located near the surface, while rock is the hard-consolidated material, i.e. bedrock, usually found beneath the soil. Most soils consist of small fragments or grains that form openings or pores when compacted together. If these pores are sufficiently large and interconnected, the soil is said to be permeable and oil or water can pass through it. Sand and gravel are the most permeable types of soil. Materials such as clay, silt, or shale are termed impervious as they have extremely small, poorly interconnected pores and allow only limited passage of fluids. Soils also vary in terms of long-term retentivity. Loam tends to retain the most water or oil due to its high organic content.

As most soils are a heterogeneous mixture of these different types of sediments, the degree of spreading and penetration of oil can vary considerably in a given location. The types of soil are often arranged in layers, with loam on top and less permeable materials such as clay or bedrock underneath. If rock is fractured and contains fissures, oil can readily pass through it.

The oil's ability to permeate soils and its adhesion properties also vary significantly. Viscous oils, such as heavy fuel oil, often form a tarry mass when spilled and move slowly, particularly when the ambient temperature is low. Non-viscous products, such as gasoline, move in a manner similar to water, in both summer and winter. For such light products, most spreading occurs immediately after the spill.

Crude oils have intermediate adhesion properties. In an area with typical agricultural loam, spilled crude oil usually saturates the upper 10 to 20 cm of soil and rarely penetrates more than 60 cm. Generally, the oil only penetrates to this depth if it has formed pools in dry depressions. If the depressions contain water, the oil may not penetrate at all.

The fate of oil once on soil varies, and generally clean-up is slow and difficult. After clean-up, there is usually residual contamination which may linger for years. On agricultural land, highly weathered residual oil may not be a health problem and often crops can be grown in the soil the following year after clean-up and tilling/nutrient addition.

42.3.2 Oil Toxicity and Exposure

Before discussing the actual effects of an oil spill on various elements of the environment such as birds and fish, the types of effects will be discussed (Fingas, 2012; Shigenaka, 2011). Toxic effects are classified as acute or chronic, which refers to the time of effect of the contaminant on an organism. Acute means toxic effects occur within a short period of exposure in relation to the life span of the organism, and are generally the result of exposure to high contaminant concentrations. For example, acute toxicity to fish could be an effect observed within four days of exposure. Chronic means occurring during a relatively long period, usually 10% or more of the life span of the organism, and is generally the result of exposure to low contaminant concentrations. It might take a significant portion of the life span for a chronic toxic effect to be observable, although it could have been induced by exposure to a substance that was normally acutely toxic. Chronic toxicity refers to long-term effects that are usually related to changes in such things as metabolism, growth, reproduction, behavior, or ability to survive.

The effects of exposure to a toxic substance can be lethal or sub-lethal. A lethal effect is typically described in terms of the concentration of the toxicant that causes death to 50% of a test population of the species within a specified period of exposure time (typically four days). This is referred to as the LC_{50}. For example, tests of the effects of various crude oils on *Daphnia magna*, the water flea, show that 5–40 mg/L of the oil for a period of 24 hours is lethally toxic. The units of milligrams/liter are approximately equivalent to parts per million (ppm). Sub-lethal means detrimental to the test organism, but below the level that directly causes death within the test period. A sub-lethal effect is typically described in terms of the concentration of the toxicant that causes an adverse effect or response to 50% of a test population of the species within a specified period of exposure time. This is referred to as the EC_{50}. For example, it has been found that a concentration of 2 ppm of crude oil in water causes disorientation in *Daphnia magna* when exposed for 48 hours; this could be stated as $EC_{50} = 2$ ppm.

Oil can affect animals in many ways, including changing their reproductive and feeding behavior and causing tainting and loss of habitat. Oiling of more highly developed animals such as birds may result in behavioral changes, such as failure to take care of their nests, resulting in the loss of eggs. Even a light oiling can cause some species of birds to stop laying eggs altogether.

Feeding behavior might also change. Seals sometimes react to oiling by not eating, which compounds the negative effects of the oil. The loss of an organism's habitat due to oiling can be as harmful as direct oiling because alternative habitats may not be available, and the animal can perish from exposure or starvation.

Finally, tainting becomes of concern with fish and shellfish after an oil spill. Tainting occurs when the organism takes in enough hydrocarbons to cause an off-flavor or off-odor in seafood. These organisms are unsuitable for human consumption until this taint disappears, which takes from days to up to a year after exposure, depending on the species. After an oil spill, seafood samples in the area are often tested using both chemical methods and sensory testing. The area is sometimes closed to commercial fishing as a precaution, to prevent direct contamination of fishing gear and catch as well as to maintain public confidence. Loss of confidence in seafood safety and quality can influence seafood markets long after any actual risk to seafood from a spill has subsided, resulting in serious economic consequences.

Oil can enter organisms by several exposure routes: physical exposure, ingestion, absorption, and through the food chain. Animals or birds can come into direct contact with oil on the surface of water, on shorelines, or on land. The effects from fouling of bird feathers or sea otter fur can lead to mortality due to loss of the ability to thermoregulate in cold waters. Ingestion occurs when an organism directly consumes oil, usually by accident as in the case of birds when oil is ingested as they preen or groom their feathers.

Absorption of volatile components of oil is a common method of exposure, especially for plants and sessile (immobile) organisms, although it also occurs in birds and mammals. Fresh crude oil has a relative abundance of volatile compounds such as benzene and toluene that are readily absorbed through the skin or plant membrane and which are toxic to the organism. After a spill, organisms can also be exposed to oil that passes through several organisms via the food chain. Bioaccumulation rarely occurs because the components of oil are generally metabolized and excreted by fish and mammals or eliminated by shellfish when the exposure to the oil decreases.

The effects of oil on the flora and fauna of a region are influenced by many factors, including the relative abundance (thus how much of the population may be affected), its likelihood of coming into contact with the spilled oil, sensitivity to oil exposure (which varies by season, life stage, physiology, etc.), and how quickly a species population would be able to recover.

42.4 OIL SPILL COUNTERMEASURES

42.4.1 Containment on Water

Containment of an oil spill refers to the process of confining the oil, either to prevent it from spreading to a particular area, to divert it to another area where it can be recovered or treated, or to concentrate the oil so it can be recovered or burned (Fingas, 2012). Containment booms are the basic and most frequently used piece of equipment for containing an oil spill on water. Booms are generally the first equipment mobilized at a spill and are often used as long as the oil persists on the water surface. While many pipeline spills occur on land, most of the oil ends up in water bodies and, as such, booms are frequently used.

A boom is a floating mechanical barrier designed to stop or divert the movement of oil on water. Booms resemble a vertical curtain with portions extending above and below the water line. They are constructed in sections, usually 15 or 30 m long, with connectors installed on each end so that sections of the boom can be attached to each other, towed, or anchored.

Booms are used to enclose floating oil and prevent it from spreading, to protect biologically sensitive areas, to divert oil to areas where it can be recovered or treated, and to concentrate oil and maintain an adequate thickness so that skimmers can be used or other clean-up techniques, such as *in situ* burning, can be applied. Booms are used primarily to contain oil, although they are also used to deflect oil. When used for containment, they are often arranged in a U configuration. The U-shape is created by the current pushing against the center of the boom. The critical requirement is that the current in the apex of the U does not exceed 0.35 m/s or 0.7 knots, which is referred to as the critical velocity. This is the speed of the current flowing perpendicular to the boom, above which oil will be lost by entrainment into the water and under the boom. If used in areas where the currents are likely to exceed 0.35 m/s, such as in rivers and estuaries, booms must be deployed at an angle to the current. The oil can then be deflected out of the strong currents to areas where it can be collected or to less sensitive areas.

If strong currents prevent the best positioning of the boom in relation to the current, several can be deployed in a cascading pattern to progressively move oil toward one side of the watercourse. This technique is effective in wide rivers or where strong currents may cause a single boom to fail. When booms are used for deflection, the forces of the current on the boom are usually so powerful that stronger booms are required which must be anchored along their entire length.

A boom's performance and its ability to contain oil are affected by water currents, waves, and winds. Either alone or in combination, these forces often lead to boom failure and loss of oil. The most critical factor is the current speed relative to the boom. Failures will occur when this exceeds 0.35 m/s (0.7 knots).

42.4.2 Oil Recovery on Water

Recovery is the next step after containment in an oil spill clean-up operation. Even in the case of land spills, the oil most often flows to a water body from where it is recovered. An important objective of containment is to concentrate oil into thick layers to facilitate recovery. In fact, the containment and recovery phases of an oil spill clean-up operation are often carried out at the same time. As soon as booms are deployed at the site of spill, equipment and personnel are mobilized to take advantage of the increased oil thickness, favorable weather, and less weathered oil. After oil spreads or becomes highly weathered, recovery becomes less viable and is sometimes impossible.

Skimmers are mechanical devices designed to remove floating oil from a water surface. They vary greatly in size, application, and capacity, as well as in recovery efficiency. They are classified according to the area where they are used, for example, inshore, offshore, in shallow water, or in rivers, and by the viscosity of the oil they are intended to recover. Most function best when the oil slick is relatively thick. The oil must, therefore, be collected in booms or against a shoreline or floating ice before skimmers can be used effectively. The skimmer is placed wherever the oil is most concentrated in order to recover as much oil as possible. Weather conditions at a spill site have a major effect on the efficiency of skimmers. Depending on the skimmer type, most will not work effectively in waves greater than 1 m in height or in currents > 0.5 m/s. Most do not operate effectively in water with ice or debris such as branches, seaweed, and floating waste. Some have screens around the intake to prevent debris or ice from entering, conveyors or similar devices to remove or deflect debris, and cutters to deal with seaweed. Very viscous oils, tar balls, or oiled debris can clog the intake or entrance and make it impossible to pump oil from the skimmer's recovery system. Figure 42.2 shows a boom and a skimmer recovering oil.

Skimmers are also classified according to their basic operating principles: oleophilic surface skimmers, weir skimmers, suction skimmers or vacuum devices, elevating skimmers, and submersion skimmers. Over time, all skimming systems become less effective because of the oil's spread into thinner slicks and weathering into a more viscous liquid.

Sorbents are materials that recover oil through either absorption or adsorption. They play an important role in oil spill clean-up and are used in the following ways: to clean up the final traces of oil spills on water or land; as a backup to other containment means, such as sorbent booms; as a primary recovery means for very small spills; and as a passive means of clean-up. An example of such a passive clean-up is when sorbent booms are anchored off lightly oiled shorelines to absorb any remaining oil released from the shore and prevent further contamination or re-oiling of the shoreline.

Sorbents can be natural or synthetic materials. Natural sorbents are divided into organic materials, such as peat moss or wood products, and inorganic materials, such as vermiculite or clay. Sorbents are available in a loose form, which includes granules, powder, chunks, and cubes, often contained in bags, nets, or socks. Sorbents are also available in the form of pads, rolls, blankets, and pillows. The use of synthetic sorbents in oil spill recovery has increased in the last few years. These sorbents are

FIGURE 42.2 A boom and skimmer being used together to collect and recover oil.

often used to wipe other oil spill recovery equipment, such as skimmers and booms, after a clean-up operation. Sheets of sorbent are often used for this purpose.

Sorbent booms are deployed on the water when the oil slick is relatively thin, i.e. for the final polishing of an oil spill, to remove small traces of oil or sheen, or as a backup to other booms. Sorbent booms can be placed off a shoreline to recover oil that is mobilized by wave or tidal action; this strategy is often used for marshes where other response options are likely to cause additional harm. They are not absorbent enough to be used as a primary countermeasure technique for any significant amount of oil.

42.4.3 Treating Agents

Treating the oil with specially formulated chemicals is another option for dealing with oil spills (Fingas, 2012). An assortment of chemical spill-treating agents is available to assist in the cleaning up oil. Approval must be obtained from appropriate authorities before these chemical agents can be used. In addition, these agents are not always effective and the treated oil may still have toxic effects on aquatic and other wildlife. Dispersants are chemical spill-treating agents that promote the formation of small droplets of oil that are dispersed throughout the top layer of the water column by wave action and currents. These small oil droplets are less likely to resurface into slicks; instead, they are somewhat subject to dilution, transport by currents, and natural weathering processes that include dissolution, microbial degradation, and sedimentation. The use of dispersants remains a controversial issue, and special permission is required in most jurisdictions. Generally, in freshwater or land applications, their use is banned.

Dispersants have a relatively narrow window of effectiveness because oil becomes less dispersible as it weathers, mostly due to an increase in viscosity. There must be sufficient mixing energy from waves or currents, and the wind must not be too strong to prevent the dispersant droplets from landing on the oil slick. There are many restrictions on when and where dispersants may be used as an oil spill countermeasure, including water depth (usually > 10 m), distance from shore (often > 5 km), season, no spraying over birds, sea turtles, or marine mammals, and duration of application. The use of dispersants is considered as part of a trade-off assessment of the impacts of the dispersed oil versus the impacts of the untreated oil slick on sensitive resources.

42.4.4 *In situ* Burning

In situ burning is an oil spill clean-up technique that involves controlled burning of the oil at or near the spill site (Fingas, 2018). The major advantage of this technique is its potential for removing large amounts of oil over an extensive area in less or about the same time as other techniques but with the distinct advantage of being a final solution. The technique has been used at actual spill sites for some time, especially on land (including wetlands) and in ice-covered waters where the oil is contained by the ice. During the 2010 Deepwater Horizon oil spill in the Gulf of

Mexico, it was used extensively and removed 15% of oil from the water surface (Fingas, 2018).

42.4.5 Shoreline Clean-up

Oil spilled on water, be it at sea or inland, is seldom completely contained and recovered and some of it eventually reaches the shoreline or margin of a water body (Fingas, 2012). It is more difficult and time consuming to clean up shoreline areas than it is to carry out containment and recovery operations at sea. Physically removing oil from some types of shoreline can also result in more ecological and physical damage than if oil removal is left to natural processes. The decision to initiate clean-up and restoration activities on oil-contaminated shorelines should be based on a careful evaluation of socio-economic, aesthetic, and ecological factors. These include the behavior of oil in shoreline regions, the types of shoreline and their sensitivity to oil spills, the assessment process, shoreline protection measures, and recommended clean-up methods. Similarly, some of the shoreline types are related to land types, and the same clean-up applies.

The type of shoreline or similar land surface is crucial in determining the fate and effects of an oil spill as well as the clean-up methods to be used. In fact, the shoreline's basic structure and the size of material present are the most important factors in terms of oil spill clean-up. There are many types of shorelines, all of which are classified in terms of sensitivity to oil spills and ease of clean-up. These are in order of increasing sensitivity: exposed bedrock and man-made solid structures, sand beaches, mixed sand and gravel beaches, gravel beaches, riprap, exposed tidal flats, sheltered rocky shores, peat, sheltered tidal flats, marshes, mangroves, swamps, and low-lying tundra. These types occur on both seashore and freshwater shores.

Many methods are available for removing oil from shorelines or land forms. All of them are costly and take a long time to carry out. The selection of the appropriate clean-up technique is based on the type of substrate, the depth of oil penetration or burial in the sediments, the amount and type of oil and its present form/condition, the ability of the shoreline to support foot and vehicular traffic, the environmental, human, and cultural sensitivity of the shoreline, and the prevailing ocean and weather conditions.

Some recommended shoreline clean-up methods are natural recovery, manual removal, flooding or washing, use of vacuums, mechanical removal, tilling and aeration, sediment reworking, sorbents, and chemical cleaning agents. Surface-washing agents are intended to be applied to shorelines or surfaces to increase the amount of oil released from the surface, or to reduce the pressure and temperature of flushing. Only those agents that cause the released oil to float are used on the shoreline, so that the released oil can be recovered. Other agents are essentially industrial cleaners that may be appropriate for cleaning equipment when removed from the water, but not on shorelines.

Numerous guides have been developed to assist responders in selecting the most appropriate clean-up method depending on the type and degree of oiling, the habitat type, and environmental considerations.

Sometimes the best response to an oil spill on a shoreline may be to leave the oil and monitor the natural recovery of the affected area. This would be the case if more damage would be caused by clean-up than by leaving the environment to recover on its own. This option is suitable for small spills in sensitive environments and on a beach that will recover quickly on its own, such as on exposed shorelines and with non-persistent oils such as diesel fuel on impermeable beaches. This is not an appropriate response if important ecological or human resources are threatened by the long-term persistence of the oil.

42.4.6 Clean-up of Oil Spills on Land

When dealing with oil spills on land, clean-up operations should begin as soon as possible. It is important to prevent the oil from spreading by containing it and to prevent further contamination by removing the source of the spill. It is also important to prevent the oil from penetrating the surface and possibly contaminating the groundwater.

Berms or dikes can be built to contain oil spills and prevent oil from spreading horizontally. Caution must be taken, however, that the oil does not back up behind the berm and permeate the soil. Berms can be built with soil from the area, sand bags, or construction materials. They are removed after clean-up to restore the area's natural drainage patterns. Sorbents can also be used to recover some of the oil and to prevent further spreading. The contaminated area can sometimes be flooded with water to slow penetration and possibly float oil to the surface, although care must be taken not to increase spreading and to ensure that water-soluble components of the oil are not carried down into the soil with the water. Shallow trenches can be dug as a method of containment, which is particularly effective if the water table is high and oil will not permeate the soil. Oil can either be recovered directly from the trenches or burned in the trenches. After the clean-up, trenches are filled in to restore natural water levels and drainage patterns.

Suction hoses, pumps, vacuum trucks, and certain skimmers and sorbents are generally effective in removing excess oil from the surface, especially from ditches or low areas. The use of sorbents can complicate clean-up operations, however, as contaminated sorbents must be disposed of appropriately. Sorbents are best used to remove the final traces of oil from a water surface. Any removal of surface soil or vegetation also entails replanting.

Mechanical recovery equipment, such as bulldozers, scrapers, and front-end loaders, can cause severe and long-lasting damage to sensitive environments. These devices can be used in a limited capacity to clean oil from urban areas, roadsides, and possibly on agricultural land. The unselective removal of a large amount of soil leads to the problem of disposing of the contaminated material. Contaminated soil must be treated, washed, or contained before it can safely be disposed of in a landfill site. This could cost thousands of dollars per ton.

42.5 CONCLUSIONS AND OUTLOOK

Oil spills will occur as long as we continue to explore for, produce, and consume oil. Major spills trigger a renewed focus on oil spill prevention and improved response capabilities. For example, after the 2010 Deepwater Horizon oil spill in the northern Gulf of Mexico, industry and government spent much effort on blow out prevention and control, and on subsea dispersant injection systems. In 2014, there was a shift to focus on inland spill prevention and response to address concerns with the evolving risks from pipeline and rail transportation of both oil sands and Bakken crudes. However, over time, the lessons learned appear to fade, funding drops off, and prevention and readiness capabilities decrease, until the next major spill. Many spills are preventable and require a combination of systems, equipment, and trained people. Systems and equipment must be continuously updated and tested. People must be retrained and exercised using the systems and equipment in simulated emergencies. There have been incremental improvements in spill response technologies, but no major advances. Spills will likely continue to be cleaned mostly by manual and mechanical methods.

REFERENCES

EIA. 2019. *US Field Production of Crude Oil and Petroleum Products*. Washington, DC: Energy Information Administration.

Etkin, D. 1999. Estimating cleanup costs from oil spills. *International Oil Spill Conference Proceedings* 1999: 35–39.

Etkin, D. S. 2003. Estimation of shoreline response cost factors. *International Oil Spill Conference Proceedings* 2003: 1243–1253

Etkin, D. S. 2014. Risk of crude and bitumen pipeline spills in the United States: Analyses of historical data and case studies (1968–2012). In *Proceedings of the Arctic and Marine Oilspill Program (AMOP) Technical Seminar*, 297–316. Ottawa, ON: Environment Canada.

Fingas, M. 2012. *The Basics of Oil Spill Cleanup*. 3rd edn. Boca Raton, FL: CRC Press.

Fingas, M. 2018. *In-situ Burning as an Oil Spill Countermeasure*. Boca Raton, FL: CRC Press.

Frittelli, J., P. W. Parfomak, and J. L. Ramseur, et al. 2014. *U.S. Rail Transportation of Crude Oil: Background and Issues for Congress*. Washington, DC: Congressional Research Service.

National Research Council. 2003. *Oil in the Sea III: Inputs, Fates, and Effects*. Washington, DC: National Academies Press.

NEB. 2019. *Estimated Production of Canadian Crude Oil and Equivalent*. Calgary, Alberta: National Energy Board of Canada.

Schmidt-Etkin, D. 2011. Spill occurrences: a world overview. In *Oil Spill Science and Technology*, ed. M. Fingas, 7–48. New York, NY: Gulf Publishing Company.

Shigenaka, G. 2011. Effects of oil in the environment. In *Oil Spill Science and Technology*, ed. M. Fingas, 985–1024. New York, NY: Gulf Publishing Company.

43 Crude Oils and their Fate in the Environment

Bryan M. Hedgpeth
Kelly M. McFarlin
Roger C. Prince

CONTENTS

43.1 INTRODUCTION

Crude oils provide about a third of the world's energy, with current consumption (EIA, 2019) averaging about 100 million barrels (bbl) per day (MBD) (Table 43.1). Despite competition from renewable fuels, this consumption is likely to increase for at least the next decade, and current estimates of proven and probable reserves suggest this could continue for at least the next century unless political directives interfere (EIA, 2019), albeit with the potential for periodic oil crunches (Hacquard et al., 2019). Production is concentrated in the Middle East (Saudi Arabia, UAE, Kuwait, Iraq, Iran, etc.), the USA, and Russia, which together produce approximately half of the total, but important reserves are found around the world. The USA was both the largest producer (19.6 MBD) and the largest consumer (20.5 MBD) of crude oil in 2018, and production exceeded consumption in late 2019 (EIA, 2019). China is currently the largest importer of crude oil. The logistics of our global consumption are staggering – some 18 billion ton-miles of oil were moved by sea in 2018 (UNCTAD, 2020).

43.2 ORIGINS OF CRUDE OIL

Crude oils are generated from biomass, principally oil-rich aquatic algae, deposited in sediments long ago (Tissot and Welte, 1984). Of course, most biomass has always been metabolized by aerobic microbes before it reaches the sediment, but if it descends rapidly, perhaps associated with marine snow or mineral particles; it can become entombed with mineral sediment. The burial of these agglomerates tends to preserve oil-prone molecules because sediments become anaerobic within centimeters of the surface of the seafloor. In particular, oil-prone sediments can contain > 10% organic material (Walters, 2017).

Once organic biomass becomes buried on the seafloor it is naturally heated as more sedimentation occurs, and a process known as diagenesis ensues. Within a few hundred meters the temperature, microbial metabolism, and pressure convert the biomass to kerogen and the minerals to a consolidated source rock. Kerogen is a complex organic material with little free hydrocarbon, although decarboxylation reactions lower the O/C ratio without much change in the H/C ratio. Additional burial increases the temperature, and the source rocks enter the zone of catagenesis. Here kerogen begins to decompose to bitumen, a dense and highly viscous hydrocarbon, and then with increasing thermal stress to liberate the hydrocarbons of crude oil. These are

TABLE 43.1
Approximate Conversion Values for Crude Oils

Barrels (bbl)	Gallons (US)	Gallons (UK)	Liters	m^3	Tonnes	Energy (MJ)
1	42	35	159	0.159	0.14	6,678

Source: BP (2020a).

typically expelled from the matrix and migrate; carrying some dissolved polar molecules (Tissot and Welte, 1984; Walters, 2017).

At still higher temperatures, residual bitumen undergoes thermal cracking, releasing methane and small (C_2–C_5) alkanes, and then just methane. Finally, the now much altered rocks enter the zone of metagenesis and are no longer capable of generating anything except small quantities of methane (Tissot and Welte, 1984).

These thermal processes depend on both time and temperature, but it is generally agreed that diagenesis occurs at temperatures of < ~80°C, catagenesis from ~80 to ~200°C, and metagenesis > ~200°C. The oil window of maximal oil generation is between ~140 and ~175°C (Tissot and Welte, 1984). Obviously, the age of the source rock imposes an upper limit on the time available for these changes. Young source rocks, such as those of the late Miocene (10–14 million years ago, Ma) Monterey formation in California that are the source of the oil in the Midway-Sunset Field, were buried quite quickly by the active tectonics of the region and have had only a relatively short period to generate oil (Hall and Link, 1990). Some non-commercial amounts of crude oil are being generated today where geothermal sources are cooking recent biomass, such as in the Kamchatka volcanic region in the Russian Far East and the Guaymas Basin in Mexico's Gulf of California (Simoneit et al., 2009).

However, most oils are far older. A large portion was generated from Jurassic (~200–145 Ma) and Cretaceous (145.5–66 Ma) formations (Tissot and Welte, 1984). These source rocks have experienced a range of geothermal histories, with some generating oil within a few million years after deposition while others have only recently entered the oil window. Older source rocks are also productive, and though rare, there are some commercial oils produced from Precambrian (> 541 Ma) sources, such as the Tahe field in Xinjang, China (Liu et al., 2017).

In order to become an exploitable conventional resource, crude oil expelled from its source rock must migrate, often over hundreds of kilometers, through faults and permeable strata and accumulate in a reservoir. Oil accumulates in highly porous rocks, such as sandstones and carbonates, when there is a trapping structure and overlaying sealing rocks such as salt or shales. A conventional reservoir is likely to be a blend of oils from one source at different levels of maturity or from multiple sources arriving at the reservoir at different times (Tissot and Welte, 1984; Walters, 2017). In contrast, unconventional oils are produced from rocks with low permeability and require long horizontal wells that are hydraulically fractured to obtain sufficient volumes to be economic (Ma and Holditch, 2015).

Not all oil is trapped underground. Sometimes there is no trap and the petroleum reaches the surface. Elsewhere a reservoir may fill with oil and gas and overflow to the surface, or the sealing rock may fail. Some seeps are terrestrial, for example those that gave rise to the eternal fires near Baku in Azerbaijan, Manggarmas in Indonesia, and Clarington in USA (Etiope et al., 2013). Others are marine: far more oil gets into the sea every year from seeps than from catastrophic tanker accidents, even from catastrophic and prolonged blowouts (National Research Council, 2003). For centuries, oil accumulations were discovered at their surface seepage, including the first wells in Pennsylvania and Azerbaijan.

43.3 PRODUCTION OF CRUDE OIL

When initially discovered, most reservoirs are under high pressure, and reservoir engineers take great care to exploit this pressure safely. The first spectacular failure in controlling high reservoir pressure was the Spindletop gusher on January 10, 1901 in Beaumont, Texas – the released oil reached a height of about 50 m and produced some 100,000 bbl a day for nine days until it was capped (House, 1946). Although the lake of oil was ignited by a spark from a sightseeing train, later production started the Texas oil industry. The largest blowout in the USA was likely the 1910 Lakeview Gusher in Kern County, California, from the Midway-Sunset field. The well blew out and produced an estimated 8.25 million bbl of oil over 544 days of uncontrolled flow (Sims and Frailing, 1950). More recently the uncontrolled blowout of the Macondo well in the Gulf of Mexico during drilling by Deepwater Horizon led to the death of 11 men and the release of 3.19 million bbl of oil over 86 days (Barbier, 2015).

Reservoir engineers strive to maintain reservoir pressure during production to maximize oil flow (Manrique et al., 2010). Pressure originating from an aquifer in contact with oil or gas is relatively easily maintained by injecting water into the aquifer as the petroleum is produced. The water may be recycled connate fluid or seawater, a common enhancement to oil production in the North Sea (Awan et al., 2008). Gaseous hydrocarbons can be separated from the oil and injected back into a reservoir's gas cap to maintain pressure. In some cases, CO_2 is injected for this purpose (Muggeridge et al., 2014). Unconventional wells produce oil by solution gas drive, where the pressure drop due to oil production lowers the reservoir pressure so that the gas dissolved in single-phase oils expands to force oil into the well bore.

When primary pressure is insufficient to push oil to the surface, pumps provide artificial lift. Electrically driven submersible and gas lift pumps are used on high flow-rate wells, which are common offshore. Mechanical pumps, such as sucker rod pumps, are used for low flow-rate stripper wells that produce < 10 barrels/day. This is the production method in many mature fields in the USA, obvious from their arrays of nodding-donkey sucker rod pumps. Such wells typically produce oil and water, sometimes as an emulsion, and the water is often a strong brine. The careless disposal of such brines, and spills, have caused significant environmental problems (Whittemore, 2007).

Much has been made of the concept of peak oil, where production and consumption outweigh new discoveries, causing reserves to fall and production to decline (Bardi, 2019). Technology has certainly postponed this event, and few now predict an imminent irreversible peak in production. Obviously, the amount of oil underground must be finite, but the track record for predicting potential reserves is poor. One may estimate the amount of buried biomass (as reduced carbon) by calculating the amount of oxygen produced by photosynthesis. The only significant known mechanism for the accumulation of oxygen in our atmosphere is photosynthetic water scission, with the burial of the hydrogen as reduced carbon (i.e. biomass). There are about 1×10^{18} tons of oxygen in the atmosphere, so there must be approximately this much-reduced carbon in the lithosphere. Total oil production to date has been about 1.9×10^{11} tons.

43.4 TYPES OF CRUDE OIL

Crude oils produced from individual wells are combined and blended in gathering systems and sold as commodities with specified physical and chemical criteria. These are then transported via rail, pipelines, or tankers to refineries. About 200 different crude oils are produced at large scale (BP, 2020b; ECCC, 2020; ExxonMobil, 2020; Total Oil, 2020). They are described using the following variety of characteristics.

43.4.1 Heavy vs. Light

Almost all oils in commerce float because they are principally hydrocarbons: molecules made exclusively of carbon and hydrogen, which predominantly have a density of < 1 (Murray et al., 2020). However, almost all oils also contain some fraction of polar molecules, the asphaltenes and resins, that contain heteroatoms in addition to carbon and hydrogen (Tissot and Welte, 1984). These polar molecules are denser than hydrocarbons and raise the density of the oil. They are also more difficult to refine, making a crude oil rich in polar material but less valuable to a refiner. Oil density is routinely measured as a simple proxy for the presence of these undesirable molecules. The petroleum industry measures density in degrees of American Petroleum Institute (API) gravity, calculated by: API gravity = (141.5/specific gravity) – 131.5.

Thus a liquid with a density of 1 (water) would have an API gravity of 10. By convention, light oils have an API gravity > 31, medium 22–31, heavy oils 10–23, and extra heavy < 10, but this varies by region. For example, four common grades of Saudi Arabian oil, known as Arab Extra Light, Light, Medium, and Heavy, have API gravities of 40, 33, 31, and 28, respectively (McKinsey, 2020). The majority of oil produced in the USA has an API gravity of 40 or above, although most California oils are heavier, with an API gravity < 30. Bitumens, such as those found in Venezuela and Western Canada, can have API of < 10 and viscosities of > 750,000 cP. These oils need partial upgrading or dilution with light solvents before they can be transported (Gray, 2019).

In general, lighter oils are more valuable because they contain a higher proportion of molecules in the gasoline–diesel–jet fuel range. Condensates, with API gravities > 50 are particularly valuable. It is worth reiterating, however, that heavy oils are still less dense than water unless their API < 10. Only a few commercial crude oils and bitumens fall into that category.

43.4.2 Mature vs. Immature

As noted above, immature oils were generated at relatively low temperatures, and tend to be heavy and somewhat viscous (Walters, 2017). Examples include those from the Wilmington field in California, such as THUMS oil from Long Beach, with an API of 16 and a viscosity of 20–70 cP (for comparison, the viscosity of water is 0.89 cP). Such oils contain less hydrocarbon and more resins and asphaltenes (see below) than more mature oils. Mature reservoirs, such as those in the Murban field offshore from Abu Dhabi, have oil with an API close to 40 and a viscosity of only 4 cP.

43.4.3 Sweet vs. Sour

Sulfur plays an important role in the initial preservation of the biomass that will eventually become kerogen (Tissot and Welte, 1984), and some sulfur ends up in oil as organic sulfur (principally thiophenic). It seems that early drillers estimated this sulfur content by tasting the oil, apparently detecting a sour taste when organic sulfur levels were high; a more quantitative measure is used today, and an oil is termed sour if it contains > 0.5% sulfur. Gas drillers use a subtly different definition for sour gas, where the sourness comes from H_2S and volatile thiols and mercaptans. H_2S is acutely toxic, and natural gas is considered 'sour' if it contains > 4 ppm H_2S or > 8 ppm mercaptan sulfur. Many reservoirs contain both gas and crude oil; for example the Tengiz reservoir in Kazakhstan near the Caspian Sea contains a relatively sweet (0.51% S) oil with an API gravity of 43 but 15% H_2S (Dehghani et al., 2008). The H_2S was once processed into elemental sulfur on site, but today the sour gas is injected back into the reservoir to maintain pressure for oil production. Since sulfur has to be removed from most refinery products to meet fuel specifications, sour oils are less valuable than sweet ones.

43.4.4 Biodegraded Oils

Most oils in commerce have not been subject to microbial degradation before they are recovered, but others have clearly been degraded by microbes that have taken advantage of this rich source of carbon and energy. Since microbes have distinct preferences for degrading some hydrocarbons before others, biodegradation is easily detected by the preferential absence of molecules known to be the most readily consumed, such as the *n*-alkanes (Prince and Walters, 2016). Such biodegradation can be relatively minor; for example some oils from the Gullfaks reservoirs in the North Sea have lost only a few percent of their initial composition (Fingas, 2011), and are still light (API = 39.6). But biodegradation can also be very substantial, for example the bitumens in the Western Sedimentary Basin of Alberta, Canada (Athabasca, Cold Lake, and Peace River) have lost almost all their readily biodegradable components and the residuum has an API = 8–12 (Zhou et al., 2008). Biodegradation does not seem to occur at reservoir temperatures much above 80°C, or in reservoirs that were heated to such temperatures before the oil arrived and then cooled by uplift – a phenomenon known as paleo-pasteurization (Wilhelms et al., 2001). Higher temperatures typically begin at depths of about 2000 m.

Biodegradation removes the most readily processed molecules in crude oil, so biodegraded oils are typically less valuable than pristine ones. In addition, of course, biodegradation also reduces the total volume of producible oil, since the microbes have consumed it; this can have a substantial influence on the economics of a given reservoir.

An important concern with biodegradation is that it can begin after a reservoir is in production, especially if that production involves saltwater injection to maintain reservoir pressure. Seawater contains significant amounts of sulfate (typically 2.7 g/L; 28 mM) that can serve as the electron acceptor for anaerobic microbial growth. Such metabolism releases H_2S that can dissolve in the oil and be released during production; this souring of a reservoir can be a very expensive problem (Johnson et al., 2017).

43.4.5 Corrosive Oils

Although hydrocarbons are relatively inert, the molecules containing heteroatoms can be corrosive, and some oils are more corrosive than others are. Corrosivity is measured by the total acid number (TAN) – the amount of KOH (in mg) needed to neutralize the acids in 1 g of oil, in mgs, and typically ranges from 0 to 2 for conventional oils, and higher for bitumens. The corrosivity is typically attributed to naphthenic acids (Figure 43.1), a heterogeneous group of cyclic saturated carboxylic acids (Headley et al., 2016), mainly formed during biodegradation, that become more of a problem after they become concentrated in the heavy oil fractions after distillation of the more volatile hydrocarbons.

43.4.6 Benchmark Crude Oils

The diversity of oils in commerce has led to the designation of three crude oils as principal benchmark crudes: West Texas Intermediate (WTI), Brent Blend, and Dubai. WTI is the US benchmark, with an API of 40 and 0.4% sulfur; its price reflects delivery at Cushing, Oklahoma, a regional hub. Brent Blend is a mixture from several fields in the Brent and Ninian systems of the North Sea and is an international benchmark. It has an API of 40 and 0.35% sulfur; and its price reflects delivery at Sullom Voe in the Shetland Islands. Thus WTI and Brent are similar light sweet crudes, and their prices generally track each other. In contrast, Dubai crude oil serves as a benchmark for medium sour crudes, particularly those sold to Asia; it comes from the Fateh reservoir offshore from Dubai, and has an API of 31 and 2% sulfur. Other oils typically trade at prices benchmarked to these standards – some more valuable, especially lighter crudes and some less so.

43.5 COMPOSITION OF CRUDE OILS

The composition of produced oil depends not only on the source kerogen and its thermal history, but also on its migration and any secondary alterations. Oils derived from sulfur-rich kerogens can generate and expel oil at relatively low temperatures (< 90°C) and are immature relative to most oils: they are relatively high density, low

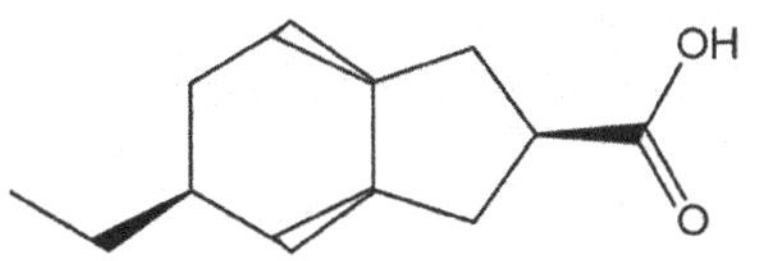

FIGURE 43.1 A simple naphthenic acid.

viscosity oils, with a high sulfur, nitrogen, and vanadium content, and a low volatile content (Tissot and Welte, 1984; Walters, 2017). Typical oils become increasingly enriched in hydrocarbons with increasing maturity, accompanied by decreasing bulk density, lower viscosity, and lower heteratom and metal content. With increasing maturation the expelled fluids become increasingly enriched in volatile saturated hydrocarbons that can separate into gas and liquid (condensate) phases when produced. Some kerogens generate waxy oils enriched in high molecular weight *n*-alkanes that may form gels at surface temperatures. If the paraffins solidify during production, they may plug the production wells and require periodic removal (Chi et al., 2019).

Secondary processes can greatly affect oil composition. For example, the late arrival of small alkanes can result in asphaltene precipitation in a reservoir, leading to the production of a lighter than expected oil. As discussed above, biodegradation occurs frequently in cooler reservoirs, depleting the hydrocarbons and generating naphthenic acids and thereby increasing density and viscosity. Thermochemical sulfate reduction can destroy hydrocarbons and generate CO_2 and H_2S. These reservoir alteration processes can be very limited or so extensive that they severely change the oil composition (Tissot and Welte, 1984; Walters, 2017).

Whatever the source and history, crude oils are complicated mixtures of many thousands of molecules. They are principally hydrocarbons, meaning molecules made exclusively of carbon and hydrogen, but all crudes and condensates contain some (1–30%) molecules with heteroatoms, principally sulfur, nitrogen, and/or oxygen. There are also trace levels of vanadium and nickel in most oils in the form of porphyrins (Furimsky, 2016), and other heteroatoms are present at low levels. These heteroatom-containing molecules contribute the color to crude oils; pure hydrocarbons are typically colorless or pale yellow.

43.5.1 Aromatic Hydrocarbons

The aromatic fraction of crude oil includes molecules with at least one aromatic ring that is most likely attached to other aromatic rings (the polyaromatic hydrocarbons) or to saturated rings (naphthenoaromatics) and/or alkyl substituents (Figure 43.2). Most crude oils contain significant amounts (up to several thousand ppm) of the simple aromatics: benzene with one ring, naphthalene and biphenyl with two, fluorene and phenanthrene with three, and chrysene with four. There are also traces of other simple aromatics such as anthracene (three rings), benz[a]anthracene and pyrene (four rings), and benzo[e]pyrene (five rings), but these are typically at the few ppm level. Far more abundant are the alkylated forms of these aromatics – the median phenanthrene ($C_{14}H_{10}$) has 16 carbon atoms appended, while the median pyrene ($C_{16}H_{10}$) has 18 (Cho et al., 2017). The most substituted forms detected by Cho et al. (2017) have 40 substituent carbons, and quite likely there are as yet undetected molecules with more. These additional carbons are present as methyl, linear, and cyclic alkyl forms, and the potential number of isomers is enormous. In total, aromatics typically make up about a third of the hydrocarbons of a crude oil (Tissot and Welte, 1984), although there is a substantial range.

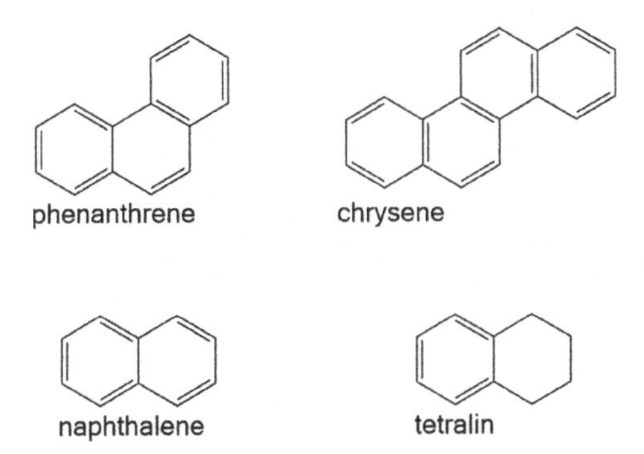

FIGURE 43.2 Representative aromatic hydrocarbon nuclei in crude oils.

Notes: Most crude oil molecules containing these motifs have substantial alkylation. Aromatic molecules are planar.

Several simple polycyclic aromatic hydrocarbons are on the US EPA list of priority pollutants (Keith and Telliard, 1979; Keith, 2015). The ones of most toxicological concern, such as benzo[*a*]pyrene and 5-methylchrysene (Hecht et al., 1975) are typically at the < 1 ppm level in crude oils (Grimmer et al., 1983). Such molecules are generated during partial combustion and hence are known as pyrogenics.

Some heteroatom-containing compounds are included in the aromatic fraction; these include the nitrogen-containing pyrroles, indoles and carbazoles, the sulfur-containing thiophenes (especially the benzo- and dibenzo-derivatives), and the oxygen-containing furans and their benzo- and dibenzo-derivatives. As with the aromatic hydrocarbons, alkylated forms are far more abundant than the parent compounds, and attached saturated rings are likely (Cho et al., 2017). Nevertheless, as we shall discuss below, most heteroatom-containing compounds are in the polar asphaltene and resin fractions.

43.5.2 Aliphatic Hydrocarbons

Aliphatic hydrocarbons exist in linear, branched, and cyclic forms (Figure 43.3). The linear uffins, the *n*-alkanes, are the most abundant single compounds in most non-degraded crude oils, ranging from methane to at least tetracontane ($C_{40}H_{82}$), with some waxy oils having alkanes with > 100 carbon atoms (Carbognani and Orea, 1999). Their total content in an oil can be > 20% in some waxy crudes. Individual compounds are rarely > 1% except in condensates.

The most abundant branched alkanes, *iso*-alkanes, are pristane and phytane (Figure 43.3), derived from the phytol of chlorophyll during maturation of kerogen under aerobic and anaerobic conditions respectively (Tissot and Welte, 1984). The branching lowers the boiling point of the alkane, so the pristane ($C_{19}H_{40}$) elutes just after the $C_{17}H_{36}$ *n*-heptadecane, and phytane ($C_{20}H_{42}$) just after the $C_{18}H_{38}$ *n*-octadecane in gas chromatography. As we shall see below, since microbes preferentially consume the linear alkane before the branched form, changes in the ratio of the linear

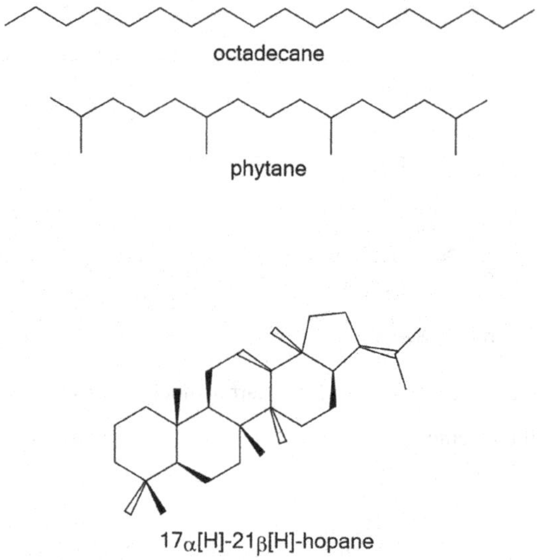

octadecane

phytane

17α[H]-21β[H]-hopane

FIGURE 43.3 Representative saturated hydrocarbons in crude oil.

Note: Cyclic saturated molecules are not planar, and many exhibit chirality.

to the branched compound provides an early indication that biodegradation has begun. Collectively, the total abundance of *iso*-alkanes is usually greater than that of the *n*-alkanes, but the vast number of isomers dilute the presence of individual compounds and obstruct their specific identification with most analytical methods.

About half the aliphatic hydrocarbons in most crude oils have one or more fully saturated rings, from cyclopentane and cyclohexane through decalin, perhydrochrysene, etc. The oil industry calls these compounds 'naphthenes'. As with the aromatics, most are decorated with alkylsubstituents, and again the vast number of potential isomers precludes their individual identification in most cases. An interesting group are the molecular fossils of cyclic microbial lipids known in the petroleum industry as biomarkers (Peters et al., 2007a, 2007b) (not to be confused with the medical use of this term). These, including the hopanes (Figure 43.3), steranes, and related molecules, were amongst the first to be clearly connected to molecules in living organisms, and are amongst the most resistant hydrocarbons to biodegradation. This, and their diversity in different oils, makes them useful internal markers for identifying both the oil itself, its origins, and any biodegradation and/or photooxidation that may have occurred since release to the environment (Prince et al., 1994).

Crude oils do not typically contain alkenes or alkynes, although these are generated during refinery processes, and can be found in refined products such as gasoline.

43.5.3 Resins

The resins in crude oils are defined by their insolubility in the solvents used to separate the aromatic and aliphatic compounds, while still being distinct from the asphaltenes. This family includes some of the larger (more than five rings) aromatics with substantial substituents, but is generally poorly characterized. Other components contain one or more sulfur, oxygen, or nitrogen atom and exhibit acidic or basic states. The majority of the acidic species contain one or more hydroxyl or carboxyl groups. Monoacids (O_2) associated with acyclic (fatty acids) and cyclic naphthenic (Figure 43.1) cores with two to five rings are the most common, but species with over eight oxygens have been detected. Acidic species can also include one or two additional heteroatoms (S or N) that are likely embedded in an aromatic ring. The majority of the basic compounds have nitrogen embedded in an aromatic ring, exposing a lone pair of electrons. Species related to substituted pyridine, quinoline, *iso*quinoline, and porphyrins are the most abundant resins (Walters and Higgins, 2018).

43.5.4 Asphaltenes

Asphaltenes are defined by their precipitation when small alkanes are added to an oil. While they are the least well characterized compounds within crude oil, they typically contain one or more hetreoatoms (usually sulfur, oxygen, and/or nitrogen and possibly other elements) in aromatic, naphthenoaromatic, and napthenic moieties with alkyl sidechains. Once thought to be of very high molecular weight, modern analytical techniques are revealing an enormous diversity with molecular weights mostly between ~200 and 1,200 (Chacon-Patino et al., 2017). Some of the large molecules are constructed around a single large polyaromatic core; others are smaller cores linked via alkyl chains in 'island and archipelago' motifs (Chacon-Patino et al., 2017).

There is limited evidence that asphaltenes are amenable to biodegradation (Jahromi et al., 2014), but once the hydrocarbons have been degraded the asphaltenes left behind lack the oiliness of whole oil and become difficult to distinguish from the fulvic and humic acids that are widespread in the environment (Schellekens et al., 2017).

43.6 SPILLED OIL IN THE ENVIRONMENT

Global estimates of the total amounts of oil released to the surface environment are likely imprecise. The National Research Council (2003) estimated that of a total of 1,300,000 tons which entered the sea every year in the 1990s 46% came from seeps, 37% from consumption and terrestrial run-off, and 12% from production and transportation. Catastrophic shipwrecks and well blowouts are fortunately rare, but they can be very large – for example, the 2010 Deepwater Horizon tragedy released 437,000 tons into the Gulf of Mexico (Barbier, 2015) – adding 73% to the annual estimates from seeps (National Research Council, 2003). As noted in the introduction, some terrestrial blowouts were enormous, for example the 1910 Lakeview Gusher in Kern County, California released an estimated 1.16 million tons (Sims and Frailing, 1950), and this was dwarfed by the outrageous war crime committed by Iraqi forces as they withdrew from Kuwait in 1991. The Iraqi army destroyed wellheads and released an enormous volume of oil onto the desert; some 52 million tons of oil were burnt as it emerged from destroyed wells (Khordagui and Al-Ajmi, 1993), and an unknown volume was released without combustion, resulting in lakes of oil (Kwarteng, 1999) extending over 49 km^2. It is no surprise that such enormous spills also influenced the local groundwater (Mukhopadhyay et al., 2008). A further 300,000–600,000 tons of oil were released to the sea (Khordagui and Al-Ajmi, 1993).

43.7 PROCESSES THAT ALTER AND REMOVE SPILLED OIL FROM THE ENVIRONMENT

Natural processes begin to alter the physical and chemical composition of spilled oil as soon as it is released into the environment. The oil begins to spread, primarily controlled by the viscosity at the spill temperature. It also soaks into porous soils. The smallest molecules evaporate and a small fraction dissolves if there is water available. Oil slicks at sea begin to incorporate water, and waves and currents break the slick into patches. If wave energy is sufficient, the oil will become entrained in the water as dispersed droplets, a process that can be enhanced by lowering the interfacial tension between oil and water with chemical dispersants. Photooxidation begins and can eventually polymerize the surface compounds. On land, oil penetrates porous soils, but also flows to lower ground, often spreading on streams and rivers. None of these processes removes oil from the environment – the only processes that do are combustion and biodegradation.

43.7.1 Evaporation

Evaporation is responsible for the removal of the most volatile lower molecular weight ($\leq$ 15 carbons) compounds and is controlled by surface area, wind velocity, and air and sea temperatures. The majority of evaporation usually occurs within the first 24–48 hours after a spill. Hydrocarbons with > 15 carbons still evaporate, but at slower rates; candles (median alkane hexacosane, $C_{26}H_{54}$) have no significant evaporation at room temperature until lit! Volatile components comprise ~20–50% of most crude oils, ~100% of gasoline, ~50% of diesel, but $\leq$ 10% of most lubricants and residual fuel oils. Evaporated hydrocarbons are either degraded in the atmosphere (Drozd et al., 2015; Smets et al., 2016), or are returned to earth in rain – in either case the major fate is likely to be biodegradation.

43.7.2 Dissolution

Very few hydrocarbons have significant solubility in water – the exceptions being the notorious BTEX (benzene toluene, ethylbenzene, and the xylenes), which dissolve from floating slicks and as water permeates oiled soils. These compounds can be > 10% of gasoline, but typically < 1% of a crude oil, and they are absent from diesel and heavier fuels. These soluble hydrocarbons are also very volatile, so dissolution is most likely only a significant fate in the subsurface, for example from leaking underground gasoline storage tanks. Some small polar molecules may also dissolve from crude oils. Such molecules are absent from refined fuels, but soluble oxygenates (e.g. ethanol, butanol, MTBE) are often added to gasolines.

43.7.3 Emulsification

Emulsions are mixtures of what at equilibrium would be immiscible fluids. Familiar examples of water-in-oil emulsions include butter and margarine, while mayonnaise is an example of an oil-in-water emulsion. In general, stable emulsions are made with mixing energy and require surfactant stabilizers.

Despite the adage that oil and water do not mix, this is not true for crude oil; with enough mixing energy, water readily absorbs into the oil phase as droplets stabilized by the resins and asphaltenes in the oil. This can eventually be so extensive that the mixture becomes a semi-solid – known in the oil-spill community as 'mousse' (Fingas, 2012), concomitant with perhaps a thousand-fold increase in viscosity (Fingas, 2015; Fingas and Fieldhouse, 2015). Most crude oil production is accompanied by water, and turbulence in the production system can generate emulsions during production (Wong et al., 2015). Such emulsions must be broken and the water removed before the oil can be shipped very far, and even with this preliminary treatment, the first operation at a refinery is to remove residual water (Abed et al., 2019). Fingas and Fieldhouse (2015) have characterized water in oil emulsions of many crude oils; all require energy, such as from production turbulence or wave action, to form. It is fair to say that all emulsions are problematic, increasing viscosity for pumping or collecting, and slowing biodegradation, photooxidation, and evaporation of spilled oil. Fortunately, chemical dispersants can break most emulsions, allowing the oil to disperse and become available for biodegradation (Lessard and DeMarco, 2000).

43.7.4 Dispersion

The reverse of a water-in-oil emulsion is an oil-in-water dispersion. In the environment, this can only occur to any significant extent when oil is spilled on water and there is wave energy for mixing. At equilibrium, oil will always float on water (until sufficient biodegradation has occurred, or dense particles have been absorbed, as we will discuss below). However, if the oil droplets are very small (< 100 μm) they become entrained in water by waves because their intrinsic buoyancy cannot overcome viscous drag. In confined media, such as laboratory flasks, the droplets coalesce and become large enough to rise to the surface over a few hours, but in the enormous volumes of rivers, lakes, and rivers, the droplets dilute to the sub-ppm level in hours (Bejarano et al., 2013; Lee et al., 2013), never find collision partners, and remain entrained. Lighter oils disperse more readily than heavier ones, but as they lose their volatile components, they become more viscous and less amenable to dispersion (Lessard and DeMarco, 2000). Dispersion can be enhanced by the addition of chemical dispersants, applied either at the wellhead or by aerial application to floating slicks at sea (Prince, 2015), even in the Arctic (Lewis and Prince, 2018). These complex mixtures of surfactants in a hydrocarbon solvent are designed to lower the interfacial tension between oil and water so that dispersion occurs with far less energy. Dispersants are not typically considered for rivers and lakes.

Dispersion is also facilitated by fine sediment particles such as clays, and perhaps by planktonic organism such as diatoms. Such aggregates may be a significant mechanism for the natural cleansing of oiled shorelines (Quigg et al., 2020).

43.7.5 Sedimentation

Sedimentation requires that the density of the oil droplet becomes greater than that of water. One process that might encompass this is the binding of mineral or biogenic particles (Quigg et al., 2020), although even here small aggregates may become entrained in water. Almost all hydrocarbons have densities < 1, but the asphaltenes, which are resistant to biodegradation, have higher densities (Powers et al., 2016), so as biodegradation removes the hydrocarbons, the density of the residuum increases.

43.7.6 Penetration into Soils and Sediments

Crude oils spilled from terrestrial wells and pipelines likely penetrate the surrounding soil, perhaps resting on any subterranean aquifer. Such spills are likely to be quite persistent (Baedecker et al., 2018).

43.7.7 Photooxidation

Photooxidation obviously requires light, and is most effective on unsaturated compounds such as aromatics (Garrett et al., 1998), but there is also evidence that secondary processes can oxidize saturated molecules as well (Hall et al., 2013). Since photooxidation is a surface phenomenon, it does not affect the bulk properties of thick layers of oil, but it can polymerize the surface, leading to long-lived pavements if the oil is on a shoreline, just as occurs with road materials (Owens et al., 2008).

43.7.8 Combustion

Terrestrial oil and gas seeps catch fire, and may have had a role in the origin of ancient religions (Etiope, 2015). Some oil tankers have caught fire, such as the tragic 1991 explosion on *MV Haven* as she was unloading offshore Genoa, Italy, which killed six crewmen (Kostianoy and Carpenter, 2018): about 100,000 tons of the total cargo of 140,000 tons burned. Nevertheless, it has proven difficult to deliberately ignite wrecks, as was found with the *Torrey Canyon* near the Isles of Scilly in 1969 (Burrows et al., 1974) and the *New Carissa* on the Oregon shore in 1999 (Gallagher et al., 2001). Prolonged combustion of oil floating on water necessitates that the oil be several millimeters thick so that the burning front can maintain a high enough temperature for volatility (Fingas, 2018). This is generally accomplished by corralling it with fire-resistant booms (Allen et al., 2011). *In situ* burning can also be a useful response option for cleaning oiled grasslands (Fingas, 2018).

43.7.9 Biodegradation

Biodegradation is by far the most significant route for spilled oil to leave the environment. Several hundred genera of bacteria (Prince et al., 2018), fungi (Prince, 2018), and even some archaea (Prince et al., 2018) are able to grow on hydrocarbons, and they are found in all environments amenable to life. Experiments routinely measure a > 80% loss of crude oil hydrocarbons, and the biodegradation pathways are fairly well understood (Prince and Walters, 2016). Microbes show a well-characterized preference for *n*-alkanes over *iso*-alkane and naphthenes, and parent aromatics before their alkylated congeners. A reasonable assumption is that initial metabolism converts 50% of the consumed oil to microbial biomass and 50% to CO_2 and H_2O, and that this conversion ratio is repeated as the degrading microbes are consumed in the food chain (Murray et al., 2020).

Three principal limitations to degradation can be very important. The first is that while rich in carbon and energy, crude oils do not provide the other elements essential for microbial growth – biologically available nitrogen, phosphorus, sulfur, iron, etc. Fortunately, seawater provides enough of these elements if the oil is dispersed as dilute small droplets (Bejarano et al., 2013). However, if oil is covering shoreline material it may vastly exceed natural levels of nutrients, and adding fertilizer nutrients may be required (Bragg et al., 1994). The second is that, since oil is so insoluble, biodegradation and dissolution must occur at the oil–water interface, and is therefore limited by the surface area of the oil available for microbial colonization. The biodegradation of oil in large droplets (> 1 mm), floating slicks, and oil on sediment is likely limited by the small surface available (Prince and Butler, 2014). Finally, although oil biodegradation in the absence of oxygen is quite well characterized (Prince and Walters, 2016), it is generally less extensive than under aerobic conditions. Thus if oil gets into anaerobic sediments, such as those in marshes, it is likely to be persistent.

But if oil is dispersed and allowed to dilute, as happens when an oil slick is treated with dispersants under typical sea states (or without dispersants in severe storms, such as following the 1993 *Braer* spill; Harris, 1995), then the rate of biodegradation is remarkably rapid – more than a dozen labs report 'half-lives' of 5–25 days (McFarlin and Prince, 2020; Prince et al., 2017) in environments ranging from the Arctic to the tropics. The key is dispersal – oil slicks are almost immune to biodegradation because so much of the oil is distant from the oil–water interface, and oil on shorelines is typically so concentrated that *in situ* nutrient concentrations severely limit biodegradation (Bragg et al., 1994).

43.8 OIL TOXICITY

Toxicity is a complicated topic because it has been known since at least the time of Paracelsus (1493–1541) that all things are toxic but it is the dose that defines the effect (Grandjean, 2016). This is further complicated by the fact that crude oil contains so many components, and each has its own individual toxicity (Di Toro et al., 2007). There is the additional concern that crude oil slicks are a physical insult, and that physical oiling can kill birds, mammals, and other animals by physically altering the insulating properties of their feathers or pelage so that they die of hypothermia (Haney et al., 2014). Oil droplets can adhere to plankton, again causing physical rather than toxic effects. Thus true aqueous toxicity has been assessed with 'water accommodated fractions' of oil. Acute toxicity is then seen to be by general narcosis, where hydrocarbons dissolve in, and disrupt, cellular membranes (Redman et al., 2017). The LC50 for crude oil on sensitive species is a few ppm in 48 or 96-hour tests (Hemmer et al., 2011). Fortunately oil dilutes to levels below a few ppm in much less than 48 hours (Bejarano et al., 2013; Wade et al., 2016), and as noted above biodegradation removes dispersed oil from the environment with a half-life of one to three weeks (Prince et al., 2017). No toxicity was detected during the biodegradation of 3 ppm oil (Prince et al., 2019).

Chronic toxicity is also a concern, although the speed of biodegradation likely limits the possibility for large scale effects from accidental spills if the oil is dispersed. Though oil toxicity is well studied, the relevancy of laboratory-generated data to accidental spills is questionable (Hodson et al., 2019). Marine oil seeps are known to host thriving ecosystems (Spies and Davis, 1979).

Crude oil spilled on soils can be toxic to plants and soil invertebrates, but some are more resistant than others are, and can be grown on contaminated soil as a remediation option known as phytoremediation (Fatima et al., 2018). Spills near production wells often include produced brine, and this poses additional issues for prompt bioremediation (Sublette et al., 2007).

REFERENCES

Abed, S. M., N. H. Abdurahman, R. M. Yunus, H. A. Abdulbari, and S. Akbari. 2019. Oil emulsions and the different recent demulsification techniques in the petroleum industry – A review. *IOP Conference Series: Materials Science and Engineering* 702: 012060.

Allen, A. A., D. Jaeger, N. J. Mabile, and D. Costanzo. 2011. The use of controlled burning during the Gulf of Mexico *Deepwater Horizon* MC-252 oil spill response. *International Oil Spill Conference Proceedings* 2011: 194.

Awan, A. R., R. Teigland, and J. Kleppe. 2008. A survey of North Sea enhanced-oil-recovery projects initiated during the years 1975 to 2005. *SPE Reservoir Evaluation & Engineering* 11: 497–512.

Baedecker, M. J., R. P. Eganhouse, H. Qi, I. M. Cozzarelli, J. J. Trost, and B. A. Bekins. 2018. Weathering of oil in a surficial aquifer. *Groundwater* 56: 797–809.

Barbier, C. J. 2015. MDL 2179. *Oil Spill by the oil rig "Deepwater Horizon" in the Gulf of Mexico*. US District Court Eastern District of Louisiana.

Bardi, U. 2019. Peak oil, 20 years later: Failed prediction or useful insight? *Energy Research & Social Science* 48: 257–261.

Bejarano, A. C., E. Levine, and A. J. Mearns. 2013. Effectiveness and potential ecological effects of offshore surface dispersant use during the *Deepwater Horizon* oil spill: A retrospective analysis of monitoring data. *Environmental Monitoring and Assessment* 185: 10281–10295.

BP. 2020a. *Approximate Conversion Factors: Statistical Review of World Energy*. London, British Petroleum.

BP. 2020b. *Crude Assays*. London, British Petroleum.

Bragg, J. R., R. C. Prince, E. J. Harner, and R. M. Atlas. 1994. Effectiveness of bioremediation for the *Exxon Valdez* oil spill. *Nature* 368: 413–418.

Burrows, P., C. Rowley, and D. Owen. 1974. *Torrey Canyon*: A case study in accidental pollution. *Scottish Journal of Political Economy* 21: 237–258.

Carbognani, L. and M. Orea. 1999. Studies on large crude oil alkanes. I. High temperature liquid chromatography. *Petroleum Science and Technology* 17: 165–187.

Chacon-Patino, M. L., S. M. Rowland, and R. P. Rodgers. 2017. Advances in asphaltene petroleomics. Part 1: Asphaltenes are composed of abundant island and archipelago structural motifs. *Energy & Fuels* 31: 13509–13518.

Chi, Y., J. Yang, C. Sarica, and N. Daraboina. 2019. A critical review of controlling paraffin deposition in production lines using chemicals. *Energy & Fuels* 33: 2797–2809.

Cho, Y., J. E. Birdwell, and M. Hur, et al. 2017. Extension of the analytical window for characterizing aromatic compounds in oils using a comprehensive suite of high-resolution mass spectrometry techniques and double bond equivalence versus carbon number plot. *Energy & Fuels* 31: 7874–7883.

Dehghani, K., D. J. Fischer, and M. Skalinski. 2008. Application of integrated reservoir studies and techniques to estimate oil volumes and recovery-Tengiz Field, Republic of Kazakhstan. *SPE Reservoir Evaluation & Engineering* 11: 362–378.

Di Toro, D. M., J. A. McGrath, and W. A. Stubblefield. 2007. Predicting the toxicity of neat and weathered crude oil: Toxic potential and the toxicity of saturated mixtures. *Environmental Toxicology and Chemistry* 26: 24–36.

Drozd, G. T., D. R. Worton, and C. Aeppli, et al. 2015. Modeling comprehensive chemical composition of weathered oil following a marine spill to predict ozone and potential secondary aerosol formation and constrain transport pathways. *JGR Oceans* 120: 7300–7315.

ECCC. 2020 *Physiochemical Properties of Petroleum Products*. Gatinau, QC: Environment and Climate Change Canada.

EIA. 2019. *Global Liquid Fuels*. Washington, DC: Energy Information Administration.

Etiope, G. 2015. Seeps in the ancient world: Myths, religions, and social development. In *Natural Gas Seepage*, ed. G. Etiope, 183–193. Cham: Springer.

Etiope, G., A. Drobniak, and A. Schimmelmann. 2013. Natural seepage of shale gas and the origin of "eternal flames" in the Northern Appalachian Basin, USA. *Marine and Petroleum Geology* 43: 178–186.

Exxon Mobil. 2020. *Crude Oil Assays*. Irving, TX: ExxonMobil.

Fatima, K., A. Imran, I. Amin, Q. M. Khan, and M. Afzal. 2018. Successful phytoremediation of crude-oil contaminated soil at an oil exploration and production company by plants-bacterial synergism. *International Journal of Phytoremediation* 20: 675–681.

Fingas, M. 2011. Introduction to oil chemical analysis. In *Oil Spill Science and Technology*, ed. M. F. Fingas, 87–109. Oxford: Gulf Professional Publishing.

Fingas, M. 2015. *Handbook of Oil Spill Science and Technology*. Chichester: Wiley

Fingas, M. 2018. *In-Situ Burning for Oil Spill Countermeasures*. Boca Raton, FL: CRC Press.

Fingas, M. and Fieldhouse, B. 2015. Water-in-oil emulsions: Formation and prediction. In *Handbook of Oil Spill Science and Technology*, ed. M. Fingas, 225–270. Chichester, Wiley.

Fingas, M. F. 2012. *The Basics of Oil Spill Cleanup*, 3rd edn. Boca Raton, FL: CRC Press.

Furimsky, E. 2016. On exclusivity of vanadium and nickel porphyrins in crude oil. *Energy & Fuels* 30: 9978–9980.

Gallagher, J. J., H. B. Hile, and J. A. Miller. 2001. The old *New Carissa*: A study in patience. *International Oil Spill Conference Proceedings*, 2001: 85–90

Garrett, R. M., I. J. Pickering, C. E. Haith, and R. C. Prince. 1998. Photooxidation of crude oils. *Environmental Science & Technology* 32: 3719–3723.

Grandjean, P. 2016. Paracelsus revisited: The dose concept in a complex world. *Basic & Clinical Pharmacology & Toxicology* 119: 126–132.

Gray, M. R. 2019. Fundamentals of partial upgrading of bitumen. *Energy & Fuels* 33: 6843–6856.

Grimmer, G., J. Jacob, and K. W. Naujack. 1983. Profile of the polycyclic aromatic compounds from crude oils. *Fresenius' Zeitschrift fur Analytische Chemie* 314: 29–36.

Hacquard, P., M. Simoen, and E. Hache. 2019. Is the oil industry able to support a world that consumes 105 million barrels of oil per day in 2025? *Oil & Gas Science and Technology–Revue D'IFP Energies Nouvelles* 74: 88.

Hall, B. R. and M. H. Link. 1990. Reservoir description of a miocene turbidite sandstone, Midway-Sunset Field, California. In *Sandstone Petroleum Reservoirs*, ed. J. H. Barwis, J. G. McPherson and J. R. J. Studlick, 509–533. New York, NY: Springer.

Hall, G. J., G. S. Frysinger, and C. Aeppli, et al. 2013. Oxygenated weathering products of *Deepwater Horizon* oil come from surprising precursors. *Marine Pollution Bulletin* 75: 140–149.

Haney, J. C., H. J. Geiger, and J. W. Short. 2014. Bird mortality from the *Deepwater Horizon* oil spill. II. Carcass sampling and exposure probability in the coastal Gulf of Mexico. *Marine Ecology Progress Series* 513: 239–252.

Harris, C. 1995. The *Braer* Incident: Shetland Islands, January 1993. *International Oil Spill Conference Proceedings* 1995: 813–819.

Headley, J. V., K. M. Peru, and M. P. Barrow 2016. Advances in mass spectrometric characterization of naphthenic acids fraction compounds in oil sands environmental samples and crude oil: A review. *Mass Spectrometry Reviews* 35: 311–328.

Hecht, S. S., M. Loy, R. R. Maronpot, and D. Hoffmann. 1975. A study of chemical carcinogenesis: Comparative carcinogenicity of 5-methylchrysene, benzo(a) pyrene, and modified chrysenes. *Cancer Letters* 1: 147–153.

Hemmer, M. J., M. G. Barron, and R. M. Greene. 2011. Comparative toxicity of eight oil dispersants, Louisiana sweet crude oil (LSC), and chemically dispersed LSC to two aquatic test species. *Environmental Toxicology and Chemistry* 30: 2244–2252.

Hodson, P. V., J. Adams, and R. S. Brown. 2019. Oil toxicity test methods must be improved. *Environmental Toxicology and Chemistry* 38: 302–311.

House, B. 1946. Spindletop. *Southwestern Historical Quarterly* 50: 36–43.

Jahromi, H., M. H. Fazaelipoor, S. Ayatollahi, and A. Niazi. 2014. Asphaltenes biodegradation under shaking and static conditions. *Fuel* 117: 230–235.

Johnson, R. J., B. D. Folwell, A. Wirekoh, M. Frenzel, and T. L. Skovhus. 2017. Reservoir Souring- Latest developments for application and mitigation. *Journal of Biotechnology* 256: 57–67.

Keith, L. H. 2015. The source of US EPA's sixteen PAH priority pollutants. *Polycyclic Aromatic Compounds* 35: 147–160.

Keith, L. H. and W. Telliard. 1979. ES & T special report: Priority pollutants: I-a perspective view. *Environmental Science & Technology* 13: 416–423.

Khordagui, H. and D. Al-Ajmi. 1993. Environmental impact of the Gulf War: An integrated preliminary assessment. *Environmental Management* 17: 557–562.

Kostianoy, A. G. and A. Carpenter. 2018. History, sources and volumes of oil pollution in the Mediterranean Sea. In *Oil Pollution in the Mediterranean Sea: Part I*, ed. A. Carpenter and A. G. Kostiano, 9–31. Cham: Springer.

Kwarteng, A. Y. 1999. Remote sensing assessment of oil lakes and oil-polluted surfaces at the Greater Burgan oil field, Kuwait. *International Journal of Applied Earth Observation and Geoinformation* 1: 36–47.

Lee, K., T. Nedwed, R. C. Prince, and D. Palandro. 2013. Lab tests on the biodegradation of chemically dispersed oil should consider the rapid dilution that occurs at sea. *Marine Pollution Bulletin* 73: 314–318.

Lessard, R. R. and G. DeMarco. 2000. The significance of oil spill dispersants. *Spill Science & Technology Bulletin* 6: 59–68.

Lewis, A. and R. C. Prince. 2018. Integrating dispersants in oil spill response in Arctic and other icy environments. *Environmental Science & Technology* 52: 6098–6112.

Liu, X. P., Z. J. Jin, and G. P. Bai, et al. 2017. Formation and distribution characteristics of Proterozoic–Lower Paleozoic marine giant oil and gas fields worldwide. *Petroleum Science* 14: 237–260.

Ma, Y. Z. and S. Holditch. 2015. *Unconventional Oil and Gas Resources Handbook: Evaluation and Development.* Oxford: Gulf Professional Publishing.

Manrique, E. J., C. P. Thomas, and R. Ravikiran et al. 2010. EOR: Current status and opportunities. In *SPE Improved Oil Recovery Symposium, 24–28 April, Tulsa, Oklahoma*, SPE130113. Richardson, TX: Society of Petroleum Engineers.

McFarlin, K. M. and R. C. Prince. 2020. Contradictory conclusions surrounding the effects of chemical dispersants on oil biodegradation. *International Oil Spill Conference Proceedings* 2020: 813–819.

McKinsey. 2020. *Crude Grades*. Washington, DC: McKinsey & Company.

Muggeridge, A., A. Cockin, and K. Webb. 2014. Recovery rates, enhanced oil recovery and technological limits. *Philosophical Transactions of the Royal Society A: Mathematical, Physical and Engineering Sciences* 372: 20120320.

Mukhopadhyay A, E. Al-Awadi, and M. Quinn, et al. 2008. Ground water contamination in Kuwait resulting from the 1991 Gulf War: A preliminary assessment. *Groundwater Monitoring & Remediation* 28: 81–93.

Murray, K. J., P. D. Boehm, and R. C. Prince. 2020. The importance of understanding transport and degradation of oil and gasses from deep-sea blowouts. In *Deep Oil Spills: Facts, Fate, and Effects*, ed. S. Murawski, C. Ainsworth, and S. Gilbert et al. 86–106. Cham: Springer.

National Research Council. 2003. *Oil in the Sea III: Inputs, Fates, and Effects*. Washington, DC: National Academies Press.

Owens, E. H., E. Taylor, and B. Humphrey. 2008. The persistence and character of stranded oil on coarse-sediment beaches. *Marine Pollution Bulletin* 56: 14–26.

Peters, K. E., C. C. Walters, and J. M. Moldowan. 2007a. *The Biomarker Guide. Volume 1. Biomarkers aned Isotopes in the Environment and Human History*, 2nd edn, Cambridge: Cambridge University Press.

Peters, K. E., Walters, C. C., and Moldowan, J. M. 2007b. *The Biomarker Guide Volume 2: Biomarkers and Isotopes in the Petroleum Exploration and Earth History*, 2nd edn. Cambridge: Cambridge University Press.

Powers, D. P., H. Sadeghi, H. W. Yarranton, and F. G. van Den Berg. 2016. Regular solution based approach to modeling asphaltene precipitation from native and reacted oils: Part 1, molecular weight, density, and solubility parameter distributions of asphaltenes. *Fuel* 178: 218–233.

Prince, R. C. 2015. Oil spill dispersants: Boon or bane? *Environmental Science and Technology* 49: 6376–6384.

Prince, R. C. 2018. Eukaryotic hydrocarbon degraders. In *Taxonomy, Genomics and Ecophysiology of Hydrocarbon-Degrading Microbes, Handbook of Hydrocarbon and Lipid Microbiology*, ed. T. J. McGenity, 1–20. Cham: Springer.

Prince, R. C., T. J. Amande, and T. J. McGenity. 2018. Prokaryotic hydrocarbon degraders. In *Taxonomy, Genomics and Ecophysiology of Hydrocarbon-Degrading Microbes*, ed. T. J. McGenity, 1–41. Cham: Springer.

Prince, R. C. and J. D. Butler. 2014. A protocol for assessing the effectiveness of oil spill dispersants in stimulating the biodegradation of oil. *Environmental Science and Pollution Research* 21: 9506–9510.

Prince, R. C., J. D. Butler, and A. D. Redman. 2017. The rate of crude oil biodegradation in the sea. *Environmental Science & Technology* 51: 1278–1284.

Prince, R. C., D. L. Elmendorf, and J. R. Lute et al. 1994. 17α(H),-21β(H)-hopane as a conserved internal marker for estimating the biodegradation of crude oil. *Environmental Science & Technology* 28: 142–145.

Prince, R. C., B. M. Hedgpeth, A. D. Redman, and J. D. Butler. 2019. The biodegradation of dispersed oil does not induce toxicity at environmentally-relevant concentrations. *Open Journal of Marine Science* 9: 113.

Prince, R. C. and C. C. Walters. 2016. Biodegradation of oil hydrocarbons and its implications for source identification. In *Standard Handbook Oil Spill Environmental Forensics, Fingerprinting and Source Identification*, 2 edn. S. Stout and Z. Wang, 869–916. New York, NY: Academic Press.

Quigg, A, U. Passow, and K. L. Daly et al., 2020. Marine Oil Snow Sedimentation and Flocculent Accumulation (MOSSFA) events: Learning from the past to predict the future. In *Deep Oil Spills: Facts, Fate, and Effects*, ed. S. Murawski, C. Ainsworth, and S. Gilbert et al., 196–220. Cham: Springer.

Redman, A. D., T. F. Parkerton, and M. Leon Paumen, et al. 2017. A re-evaluation of PETROTOX for predicting acute and chronic toxicity of petroleum substances. *Environmental Toxicology and Chemistry* 36: 2245–2252.

Schellekens, J., P. Buurman, and K. Kalbitz, et al. 2017. Molecular features of humic acids and fulvic acids from contrasting environments. *Environmental Science & Technology* 51: 1330–1339.

Simoneit, B. R., D. W. Deamer, and V. Kompanichenko. 2009. Characterization of hydrothermally generated oil from the Uzon caldera, Kamchatka. *Applied Geochemistry* 24: 303–309.

Sims, W. P. and W. G. Frailing. 1950. Lakeview Pool, Midway-Sunset field. *Journal of Petroleum Technology* 2: 7–18.

Smets, W., S. Moretti, S. Denys, and S. Lebeer. 2016. Airborne bacteria in the atmosphere: Presence, purpose, and potential. *Atmospheric Environment* 139: 214–221.

Spies, R. B. and P. H. Davis 1979. The infaunal benthos of a natural oil seep in the Santa Barbara Channel. *Marine Biology* 50: 227–237.

Sublette, K. L., J. B. Tapp, and J. B. Fisher et al. 2007. Lessons learned in remediation and restoration in the Oklahoma prairie: A review. *Applied Geochemistry* 22: 2225–2239.

Tissot, B. and D. Welte. 1984. *Petroleum Formation and Occurrence: A New Approach to Oil and Gas Exploration*. Cham: Springer.

Total Oil. 2020. *Crude Assays*. Paris: Total Oil.

UNCTAD. 2020. *Review of Maritime Transport*. Paris: United Nations Conference on Trade and Development.

Wade, T. L., J. L. Sericano, S. T. Sweet, A. H. Knap, and N. L. Guinasso. 2016. Spatial and temporal distribution of water column total polycyclic aromatic hydrocarbons (PAH) and total petroleum hydrocarbons (TPH) from the *Deepwater Horizon* (Macondo) incident. *Marine Pollution Bulletin* 103: 286–293.

Walters, C. C. 2017. Origin of petroleum. In *Handbook of Petroleum Technology*, ed. C. S. Hsu and P. R. Robinson, 359–379. Cham: Springer.

Walters, C. C. and M. B. Higgins. 2018. Petroleomics. In *Hydrocarbons, Oils and Lipids: Diversity, Origin, Chemistry and Fate*, ed. H. Wilkes, 1–28. Cham: Springer.

Whittemore, D. O. 2007. Fate and identification of oil-brine contamination in different hydrogeologic settings. *Applied Geochemistry* 22: 2099–2114.

Wilhelms, A., S. R. Larter, and I. Head, et al. 2001. Biodegradation of oil in uplifted basins prevented by deep-burial sterilization. *Nature* 411: 1034–1037.

Wong, S. F., J. S. Lim, and S. S. Dol. 2015. Crude oil emulsion: A review on formation, classification and stability of water-in-oil emulsions. *Journal of Petroleum Science and Engineering* 135: 498–504.

Zhou, S., H. Huang, and Y. Liu. 2008. Biodegradation and origin of oil sands in the Western Canada Sedimentary Basin. *Petroleum Science* 5: 87–94.

44 Oil Contaminated Soil
Understanding Biodegradation and Remediation Strategies

Kelly M. McFarlin
Roger C. Prince

CONTENTS

44.1 INTRODUCTION

Soil can be contaminated with hydrocarbons during the transportation, storage, and use of oil and petroleum-based products. Oil-contaminated soils have been reported on every continent, and contamination has sometimes been shown to persist for decades (Cho et al., 2019). Following a terrestrial oil spill, petroleum hydrocarbons can volatilize, penetrate into the soil, adsorb onto organic matter, and leach into the pore space and into the surrounding water. The ultimate fate of the hydrocarbons is biodegradation (or potentially combustion if the soil is exposed to intense forest fires), which will eventually transform into biomass, carbon dioxide, and water. Bioremediation technologies enhance natural processes by modifying those environmental factors that limit biodegradation, thereby minimizing the environmental impact of spills.

Soil consists of natural minerals, organic materials, mineral nutrients, trace metals, and inorganics that foster an abundance of microorganisms, including bacteria, fungi, archaea, and viruses (Pepper and Brusseau, 2019). Bacteria are quite abundant in soil, with approximately 10^8–10^{10} cells per gram of soil, approximately two orders of magnitude more than in surface seawater (Giovannoni and Stingl, 2005), and this number is even higher (2–20 times) in the presence of plants (Kuzyakov and Blagodatskaya, 2015). Soil bacteria and fungi are the foundation of most biogeochemical cycles by using organic or inorganic carbon for their cellular building blocks and elements such as oxygen, nitrogen, sulfur, and iron as electron acceptors for energy-generating oxidative reactions. Soil microorganisms then make nutrient sources available to plants, invertebrates, and protists in the soil environment through cell lysis and metabolic processes.

Bacterial populations are opportunists with a remarkable and diverse ability to take advantage of different substrates for growth and energy. Typically decaying organic carbon or carbon dioxide provide the substrates for bacterial growth in uncontaminated soils. In oil-contaminated soil, petrocarbon provides an additional carbon substrate, and bacteria that have acquired the ability to use this carbon source have an advantage over those that have not. Spills thus create a niche and an opportunity for oil-degraders to thrive beyond their microbial neighbors as they compete for carbon resources. The ability to consume oil is restricted to bacteria and some fungal and algal species, with over 300 genera of bacteria containing species known to degrade petroleum hydrocarbons (Prince et al., 2019), and a similar number of fungi (Prince, 2019).

Hydrocarbon biodegradation can take place under both aerobic and anaerobic conditions. In the presence of oxygen, oil-degrading microorganisms utilize various biodegradation genes to produce enzymes, such as monooxygenases and dioxygenases that perform the initial bond breakage of the hydrocarbon structure. Aerobic biodegradation typically proceeds faster than anaerobic biodegradation due to the metabolic advantage provided by oxygen when used as a terminal electron acceptor, though a number of anaerobic bacteria can break down hydrocarbons, including denitrifying bacteria and sulfate, iron, and carbon dioxide reducers (Kumari et al., 2020; Prince and Walters, 2016).

The response of microbial communities to the presence of hydrocarbons has been explored across a wide range of environments, including in the Arctic, Antarctic, deep-sea, mountaintops, and deserts. Although the microbial communities differ across these diverse environments, microorganisms with the ability to metabolize petroleum hydrocarbons have been found in every location that has been sampled and are therefore considered ubiquitous (Lawniczak et al., 2020). The majority of oil-degrading microorganisms are able to metabolize a range of carbon sources (including plant material) and are likely present in uncontaminated soil in low abundances. When soil becomes contaminated with oil, oil-degrading bacteria use the oil as a carbon and energy source, assimilating the carbon into cellular biomass and increasing in abundance. Individual species of microorganisms respond to the presence of oil by either increasing or decreasing in abundance, depending on whether they use the hydrocarbon (s) as a carbon source, if enough nutrients are available, or if an organism exhibits intrinsic sensitivity to the toxicity of small hydrocarbons (Lawniczak et al., 2020). The presence of oil usually results in decreased bacterial diversity as oil-degrading species are enriched in a stepwise response, influenced by the complexity of oil (Liang et al., 2016; Ribicic et al., 2018).

A detailed description of the chemical characteristics of crude oil and its formation is provided in Prince et al. (2021). Lower molecular weight hydrocarbons are generally biodegraded more readily than higher molecular weight hydrocarbons, with shorter straight-chain (*n*-) alkanes usually biodegrading prior to longer *n*-alkanes and branched alkanes, and parent polycyclic aromatic hydrocarbons (PAHs) biodegraded before their alkylated homologues (Prince and Walters, 2016). As biodegradation proceeds, the chemical structure of the parent hydrocarbon changes into a variety of metabolites; under aerobic conditions, the first intermediate is typically an alcohol, while in most cases anaerobic biodegradation initially yields a succinate-adduct (Prince and Walters, 2016). When spilled oil is exposed to environmental conditions and undergoes weathering (i.e. volatilization, degradation, oxidation, etc.), the bioavailability of individual hydrocarbons to oil-degraders changes over time. This makes it challenging to predict the microbial community response to oil pollution and the identity of specific oil-degraders at any spill location. Nevertheless, broad taxonomic groups such as the phyla Actinobacteria, Proteobacteria, and Firmicutes are generally enriched in the soil in the presence of crude oil in both culture-dependent and culture-independent assays (Auti et al., 2019; Fuentes et al., 2014), the specific response of the population depending on its composition prior to the spill and its genetic predisposition.

Functional genes for oil biodegradation can be used to infer a microbial community's ability to biodegrade oil. The presence of specific oil biodegradation genes can indicate the potential of the community to biodegrade petroleum hydrocarbons, while the expression of these genes indicates the act of primary biodegradation (i.e. the initial oxygenation of the hydrocarbon compound under aerobic conditions, or the addition of fumarate under anaerobic conditions). Mono- and dioxygenase biodegradation genes, such as *alkB* (Fuentes et al., 2014) and RHD (Paisse et al., 2012), are responsible for encoding the enzymes that facilitate aerobic biodegradation of some alkanes and aromatics, respectively. Numerous studies have reported the relative abundance of various alkane and aromatic genes in soil and their correlation to decreasing concentrations of their respective hydrocarbon class (Bengtsson et al., 2013; Singh et al., 2012). With the advent of molecular technologies over the past few decades, molecular techniques such as metagenomics, metatranscriptomics, and metaproteomics have identified microbial community shifts, expressed functional genes, and transcribed proteins, respectively, from environmental samples. These technologies have enabled microbial communities and their genes to be used as ecological indicators in monitoring polluted sites, allowing the identification of specific taxonomic groups and genes to act as biomarkers for petroleum hydrocarbon contamination (Pal et al., 2019). For example, the presence of phenanthrene increased the expression of PAH dioxygenases and genes associated with stress response and detoxification in soil, whereas genes associated with general metabolism were relatively unaffected (de Menezes et al., 2012).

The effectiveness of bioremediation has commonly been predicted under laboratory conditions. Such experiments were essential in providing a controlled environment to assess the impact of different environmental variables when bioremediation was first suggested (e.g. Atlas, 1975; National Research Council, 1993). Now that the technology has approached maturity, the majority of experiments are focusing on enhancement techniques and molecular analyses, recognizing that it is extremely challenging, if not impossible, to precisely mimic the natural environment in a laboratory setting. It is now clear that aerobic hydrocarbon biodegradation is significantly impacted by environmental factors such as soil type, moisture content, temperature, pH, salinity, and concentrations of nutrients and oxygen.

44.2 ENVIRONMENTAL FACTORS IMPACTING BIODEGRADATION

44.2.1 Soil Type

The type of soil (sand, clay, silt, course materials, etc.) and its environmental parameters (e.g. water content, pH, organic carbon content, temperature) influence the abundance and diversity of microorganisms (Pepper and Brusseau, 2019). Soil type determines the porosity and permeability of the soil (Neira et al., 2015), which in turn influence the water and oxygen content and the abundance of microorganisms. Porosity is not only influenced by the intrinsic texture of the soil, but also by the degree of compaction (Okoh et al., 2019). Clay soils generally have more pore spaces than sandy soil, though smaller pore sizes; and this affects the movement of oxygen, nutrients, and water (Pepper and Brusseau, 2019). Larger soil particles create larger pore spaces for the diffusion of gases, while smaller particles like sand will have smaller pore spaces. The morphology of the soil particles is also important as jagged edges and irregular shapes influence the size and frequency of the pore space. The importance of porosity was evident in an experiment by Kristensen et al. (2010), where the biodegradation potential of petroleum vapors was greatest in sandy loam, which has more pore space, followed by fine sand and limestone, which has less pore space.

Soil containing high concentrations of organic matter and clay may negatively affect biodegradation by influencing hydrocarbon bioavailability and limiting the available nutrients. Available nutrients are rapidly depleted by heterotrophic bacteria in the presence of carbon (Buerkert et al., 2012); petroleum hydrocarbons can sorb strongly onto clay particles limiting microbial access (Abdel-Shafy and Mansour, 2016). Nevertheless, while some studies have reported lower hydrocarbon biodegradation in the presence of clay compared to sand (Haghollahi et al., 2016), others have focused on the benefits of fine silt and clay compared to coarse soil fractions (Bengtsson et al., 2010; Chaerun et al., 2005). In highly contaminated soils, sorption of hydrocarbons to organic matter and clay may actually be beneficial to soil microorganisms, as sorption decreases local oil concentrations and potential toxicity, while simultaneously increasing the surface area for microbial adhesion. In addition, the high microbial abundance associated with clays (Taylor et al., 2002) may help to biodegrade adsorbed hydrocarbons.

44.2.2 Soil Moisture Content

Water is obviously essential for all life, and so it is no surprise that unirrigated contaminated desert soils show very little degradation except in the rainy season (Radwan et al., 1995). However, even in temperate soils, the amount of water can significantly influence the soil's microbial community's structure and abundance (Liu et al., 2020). Water assists in the movement of nutrients, oxygen, and hydrocarbons to oil-degrading microorganisms and low humidity may negatively affect the microbial community by causing cell dehydration. On the other hand, excess water will saturate the soil, filling all available pore spaces and significantly reducing gas (oxygen and carbon dioxide) permeability through the soil, and thereby limiting microbial respiration. The porosity of soil and its compaction determines the size and amount of available pore space for gaseous diffusion and water movement. Coarse materials within the soil can create large voids where oil may become trapped and be much less accessible to oil-degrading microbes.

Only a few hydrocarbons are soluble in water, principally the smallest alkanes ($< C_5$) and the notorious BTEX (benzene, toluene, ethylbenzene, and the xylenes), but these can dissolve out of contaminated soil (or a floating oil layer if spilled oil reaches an aquifer) and require remediation in the aqueous phase (Farhadian et al., 2008; Prince and Douglas, 2010). In contaminated aquifers, flowing water can be intercepted by bioactive barriers or biowalls (Obiri-Nyarko et al., 2014), where biodegradation is actively encouraged in an engineered zone so that contaminants are metabolized before they leave the barrier (Table 44.1).

44.2.3 Oxygen

Anoxic conditions can dominate the natural soil column and are mainly dictated by depth, water saturation, and porosity (Pepper and Brusseau, 2019). Well-aerated surface soils have similar oxygen contents (18% vol.) to the atmosphere (20% vol.), but slightly higher amounts of carbon dioxide (3% vs. 0.3%). Conversely, fine clay or compacted or saturated soils are often anaerobic and contain a higher composition of carbon dioxide than well-aerated surface soils (Pepper and Brusseau, 2019).

Anaerobic biodegradation can proceed in otherwise aerobic soils when the concentrations of hydrocarbons are very high and oxygen transfer from the surface is slower than its consumption by aerobes. Anaerobic microorganisms can thrive and mineralize petroleum hydrocarbons in wetlands, swamps, saturated soils, or deep underground soil and groundwater, by using nitrate, sulfate, or ferric ions (instead

TABLE 44.1
***In situ* and *Ex situ* Technologies for Stimulating Bioremediation of Oil-Contaminated Soil, as Well as Surrounding Air and Water Impacted by the Contaminated Soil**

Contaminated Air Emanating from Contaminated Soil		*In situ*	*Ex situ*
Biofiltration	Microbes inhabiting actively managed porous filters, such as peat, to degrade volatile hydrocarbons	X	
	Contaminated water emanating from contaminated soil		
Air stripping	Injection of air to stimulate volatilization – potentially combined with a biofilter to treat the effluent	X	X
Air sparging/aquifer sparging	Injection of air to stimulate aerobic biodegradation rather than volatilization	X	
Bioactive barrier, biowall	Physical diversion of groundwater flow so that all passes through an actively managed bioreactor – contaminant is fully degraded before water leaves the bioreactor	X	
Constructed wetland	Artificial marsh to stimulate biodegradation of contaminants	X	
Pump and treat	Pumping groundwater to the surface, treating water with a bioreactor, with potential reinjection		X
Phytoremediation	Useful for both stimulating biodegradation and also for intercep ting water flow so that the water is transpired by the plants	X	
Rhizofiltration	Floating plants to stimulate biodegradation, most likely by microbes on the plant roots that are themselves stimulated by phytochemicals and aerobic conditions	X	
	Non-aqueous phase liquid/oil layer floating on a subsurface aquifer		
Bioslurping	Vacuum extraction of the floating layer with excess airflow aerating the vadose zone	X	
	Contaminated soil		
Biofluffing	Augering soil to increase aeration	X	
Biopiles	Contaminated soil is excavated and placed on an impervious base in piles; active management of air, water, and nutrients maximize biodegradation		X
Bioventing/soil vapor extraction	Vacuum extraction of contaminant vapors from the vadose zone, thereby drawing in air to stimulate aerobic biodegradation of the residuum – potentially combined with a biofilter to treat the effluent	X	
Composting	Addition of biodegradable bulking agent to stimulate microbial activity		X
Landfarming	Active stimulation of aerobic biodegradation by tilling, with water and nutrient management		X
Phytoremediation/rhizoremediation	Plant growth to stimulate microbial degradation of contaminants and to stabilize soil to minimize wind-driven exposure	X	

of oxygen) as terminal electron acceptors (Kumari et al., 2020; Prince and Walters, 2016).

Since anaerobic biodegradation is generally slower than aerobic degradation, increasing the supply of oxygen to a contaminated soil will likely stimulate biodegradation. *In situ* technologies, such as bioventing (Mosco and Zytner, 2017), air sparging (Yao et al., 2017), and soil tilling (Nikolopoulou and Kalogerakis, 2016) have all been shown to be effective, usually with simultaneous control of moisture and inorganic nutrients (Table 44.1).

44.2.4 Temperature

Temperature has long been identified as having a significant impact on the fate of oil in contaminated environments (Atlas, 1975). In terrestrial environments, increased temperatures have been associated with increased microbial biomass and shifts in species composition and function (Dadrasnia et al., 2018). As temperature increases, so does the solubility of hydrocarbons (in water) and the bioavailability of petroleum compounds to oil-degrading bacteria. Optimum temperatures for oil biodegradation in soil environments have been reported to range from 30 to 40°C, while optimum temperature in some freshwater environments are a bit lower, ranging from 20 to 30°C, and even lower in marine environments, which ranged from 15 to 20°C (Das and Chadran, 2011; Jain et al., 2011). However, significant oil biodegradation does occur at temperatures < 5°C in both soil (McWatters et al., 2016; Rike et al., 2005) and seawater (McFarlin et al., 2014) when such temperatures are typical for the indigenous organisms.

Recently, there has been increasing evidence that temperature effects on oil biodegradation are more correlated to physical changes in oil behavior rather than bacterial metabolism (Brown et al., 2020). Specifically, lower temperatures increase oil viscosity and decrease the water solubility of hydrocarbons, which directly influences the bioavailability of hydrocarbon compounds to oil-degrading microbes. Therefore, temperature may affect oil-biodegradation rates by limiting bioavailability rather than limiting metabolic processes (Camenzuli and Freidman, 2015; Schulte, 2015). In fact, cold-loving bacteria (psychrophiles) have developed mechanisms to perform metabolic functions, such as oil-degradation, at low temperatures. For example, a common strategy for cold adaptation includes increased membrane flexibility by incorporating more unsaturated fatty acids in the cytoplasmic membrane (MacKelprang et al., 2017). These adaptations likely support the ability of indigenous bacterial populations to perform metabolic functions at similar rates in cold and warm environments.

44.2.5 Soil pH

Microorganisms are capable of thriving in a wide range of pHs, from alkaline deserts to areas influenced by acid mine drainage. The majority of oil-degrading microorganisms require a specific range of soil pH (between 6 and 8) for optimal crude oil biodegradation (EPA, 2017); and both acidic and basic pH extremes are expected to have a negative impact on oil biodegradation. Nevertheless, strongly alkaline soils can exhibit significant rates of hydrocarbon degradation (Sorokin et al., 2012); and moderately acidic soils (pH 5.4) have equivalent rates of biodegradation as standard neutral soils (Hamamura et al., 2006). Oil-degrading fungi are also sensitive to pH, but may be less sensitive than bacteria (Al-Hawash et al., 2020).

Soil pH does have a strong influence on the microbial community structure and the presence of functional genes associated with biodegradation (Al-Mailem et al., 2013). Soil microbial diversity generally decreases with increasing acidity, with increases in the genera *Acidiphilium*, *Acidobacteria*, and *Acidothermus* associated with acidic oil-contaminated soils (Liang et al., 2014). Uncertainty remains as to whether these microbial shifts are directly caused by pH or indirectly caused by factors associated with pH, such as nutrient availability, metal solubility, oxygen concentrations, and lower bioavailability of hydrocarbons. It is likely a combination of factors. During remediation, the soil pH is typically monitored and adjusted as necessary. Basic soils can be treated with ammonium sulfate or ammonium nitrate while acidic soils can be treated with calcium or magnesium carbonate (e.g. lime) (Alegbeleye, 2018), although as noted above, naturally acidic or alkaline soils seem to have oil-degrading microbes adapted to such conditions.

44.2.6 Salinity

Extreme fluctuations in salinity can have a negative effect on bacterial abundance and oil biodegradation. Salinity has been reported to have a greater impact on microbial community composition than other environmental parameters, such as temperature or pH (Lozupone and Knight, 2007). Hypersaline environments are known to reduce the water solubility of both hydrocarbons and oxygen, which potentially influences the ability of microorganisms to biodegrade oil and perform essential metabolic functions. Slight changes in salinity are not likely to impact the ability of microorganisms adapted to moderate salinity environments (2.5–10.0%) to biodegrade oil. For example, using saline soil (10%) from an oil field, Zhao et al. (2009) cultured a phenanthrene-degrading halophilic bacterial consortium and reported phenanthrene biodegradation at 5, 15, and 10%, but no biodegradation was reported at the two extremes of 0.1 and 20% salinity.

Even though the majority of bacterial populations from non-saline environments have shown inhibitory effects in response to salinization (e.g. decreased biomass; Gao et al., 2014), halophilic microorganisms have adapted mechanisms to thrive in high salinity environments and biodegrade petroleum hydrocarbons (Le Borgne et al., 2008; Sorokin et al., 2012). In dual stress environments with oil and high salinity, contaminated soil samples from the Yellow River Delta, China, were dominated by the Phyla Actinobacteria, Gamma-Proteobacteria, and Firmicutes, as well as *Deinococcus-Thermus* (Gao et al., 2014). Prince et al. (2019) reported that *Alcanivorax* and *Marinobacter* species are most commonly associated with oil contamination in saline environments. *Alcanivorax* strains (belonging to the order Oceanospirillales) are commonly associated with the biodegradation of alkanes in marine environments (Paniagua-Michel and Fathepure, 2018) and have been applied to saline soils to enhance alkane biodegradation (Dastgheib et al., 2011). In addition, *Marinobacter* strains have been reported to mineralize crude oil in hypersaline soil and water microcosms (Al-Mailem et al., 2013).

44.3 REMEDIATION TECHNOLOGIES

Petroleum hydrocarbons, including crude oil, can be removed from soil with various methods both *in situ* and *ex situ* (Table 44.1). Mechanical and chemical technologies that are commonly used for soil remediation include oxidation, evaporation, incineration, and stabilization/solidification (EPA, 2017). Biodegradation can be incorporated into these mechanical technologies as landfarming, biostimulation, bioventing, or bioaugmentation (Nagkirti et al., 2017). Bioremediation is advantageous due to its relatively low cost and low maintenance compared to more aggressive technologies (Azubuike et al., 2016; Dellagnezze et al., 2018). For bioremediation to be effective, petroleum hydrocarbons must be available to soil-dwelling oil-degrading microorganisms, along with adequate oxygen, water, and nutrients (Okoh et al., 2019). Therefore, the environmental conditions are typically adjusted to provide soil conditions that maximize contaminant loss and foster an abundant and active microbial community.

44.3.1 Nutrients and Biostimulation

Nutrients are crucial to microbial life as they support growth and all metabolic processes. Maintaining nutrient levels (nitrogen, phosphorus, and potassium (NPK)) in soil is exceedingly important during oil biodegradation because crude oil does not contain any of these nutrients and therefore introduces an imbalance in the carbon: nitrogen ratio. Without the required nutrients, hydrocarbons will likely persist in the contaminated soil as microorganisms and will not be able to increase in abundance and thereby increase the rate of biodegradation (Bell et al., 2013). Conversely, the growth of oil-degrading microorganisms and thus the loss of hydrocarbons can be significantly enhanced by the presence of nutrients (Okoh et al., 2019). Some nutrients are more important than others with respect to microbial growth and biodegradation. Relative to other nutrients, high concentrations of nitrogen and phosphorus stimulate oil biodegradation, but trace amounts of other nutrients, such as iron, are also important to microbial processes and the soil ecosystem (Prasad et al., 2017). Levels of nutrients are usually monitored because excessive NPK concentrations inhibit the biodegradation of hydrocarbons in soil (Chaineau et al., 2005; Ramadass et al., 2018). One way to determine if the soil has enough nitrogen is to compare the concentration of carbon to the concentration of nitrogen. The US Environmental Protection Agency (EPA, 2017) recommends an N: C ratio ranging from 1:10 to 1:100, depending upon the type of contaminant and initial soil conditions.

Biostimulation is usually defined as the addition of rate-limiting nutrients and electron acceptors, such as phosphorus, nitrogen, or oxygen, to enhance the activity of microorganisms and increase biodegradation in contaminated environments (Louati et al., 2013). Characterizing shifts in the structure of microbial communities and quantifying the change in abundance of specific taxa in response to amendments can assist in monitoring the success of biostimulation (Macchi et al., 2019). All Proteobacteria, except the 'Epsilon-Proteobacteria', have been shown to respond to biostimulation of hydrocarbon contaminated soil (Greer et al., 2010). Within Proteobacteria, Betaproteobacteria are strongly associated with the presence of nutrients in general (Labbe et al., 2007) and have been shown to be enriched in oil-contaminated high-organic matter soils (Bell et al., 2013). This class, which consists of approximately 75 genera and > 400 species, is likely capable of biodegrading a diverse mixture of hydrocarbons in soil. *Alcanovorax* species, members of the Gammaproteobacteria, are some of the most common bacteria associated with oil-contaminated seawater. Similar to members of Betaproteobacteria in soil environments, *Alcanovorax* has been shown to respond to nutrient supplementation in culture-dependent and independent methods. Hara et al. (2003) suggested that the success of *Alcanovorax* could be attributed to its ability to biodegrade branched alkanes, a trait that is likely shared among other taxonomic groups within Proteobacteria.

44.3.2 *Ex situ* Remediation Techniques

Ex situ bioremediation technologies, such as composting, landfarming, and biopiling, offers advantages such as containment and process control (Table 44.1). These technologies may enhance biodegradation rates by optimizing environmental conditions while implementing biostimulation and/or bioaugmentation techniques. Disadvantages associated with *ex situ* bioremediation technologies mainly include mechanical costs linked with unearthing, hauling, mixing, and containment.

44.3.2.1 Composting, Bulking Agents, and Biochar

In field applications, biostimulation of oil-contaminated soil is most commonly conducted with commercial fertilizers, although alternative nutrient sources have also been successful. For example, composting involves mixing contaminated soil under aerobic conditions with organic nutrient sources such as manure and agricultural wastes. Temperatures increase due to the exothermic reaction of biodegradation, which further enhances biodegradation. The addition of compost to oil-contaminated soil is a promising amendment as it can provide many benefits in addition to providing a nutrient source, including stabilizing the structure of the soil, improving oxygen diffusion, and water availability. The addition of compost can influence the indigenous microbial community structure within the soil by adding different strains of microorganisms, therefore potentially increasing microbial abundance and diversity.

Numerous composting studies have reported enhanced removal of hydrocarbons, with some studies showing > 90% removal of PAHs (Moretto et al., 2005). In one study, alkaline soil (pH ~12.8) highly contaminated with soot (200 mg/kg PAH) was treated in a large pilot reactor at temperatures up to 55°C with forced aeration. Addition of organic waste (e.g. sewage sludge and yard waste) increased the removal of total PAHs by 68% after 130 days (Moretto et al., 2005). The molecular response to compost in oil-contaminated soil has been observed at the laboratory scale after 36 weeks at 14°C, where yard waste and organic household biowaste significantly stimulated the abundance and diversity of *alkB* harboring bacteria, and the relative abundance of various classes of Proteobacteria, including the genera *Shewanella*, *Hydrocarboniphaga* and *Agrobacterium* (Wallish et al., 2014). Similar to other amendments, compost influences the microbial community structure and abundance, and has been shown to increase the loss of hydrocarbons (mainly PAHs), although it remains unclear if these increases in hydrocarbon loss are due to the added compost microbes, the nutrients from the compost, or a combination of both.

Other amendments, such as bulking agents, have been shown to enhance oil biodegradation in soil by increasing the porosity. Straube et al. (2003) reported ~86% removal of total PAH in soil from a wood treatment facility, which included moderate losses of pyrene and 'benzo(a)pyrene' in microcosms containing indigenous microorganisms, ground rice hulls as a bulking agent, pelletized dried blood as a nitrogen source, and a biosurfactant-producing/non-oil-degrading *Pseudomonas*. In a laboratory study, peanut hull powder (a bulking agent and an immobilization agent for indigenous bacteria) enhanced oil removal from crude-oil-contaminated soil by increasing the bioavailability of water, oxygen, nutrients, and hydrocarbons (Xu and Lu, 2010).

The addition of biochar to oil-contaminated soil increased remediation efficiency, especially for the PAH fraction (Kong et al., 2018). Biochar is a carbon-rich material produced by the pyrolysis of biomass (e.g. woody material, livestock manure, crop straw) under oxygen-limited conditions. The majority of studies have focused on biochar's ability to sequester carbon, improve the quality of agricultural soils, and remediate heavy metal-contaminated soil due to biochar's large surface area and enhanced pore structure, abundant oxygen-containing functional groups, and high cation exchange capacity (Dai et al., 2020; Zhu et al., 2017). The pore structure within biochar can provide a habitat for microbial colonization while improving soil conditions that benefit the growth and function of microorganisms (Zhu et al., 2017). The high adsorption capacity of biochar can increase water-holding capacity and thus soil fertility as leaching of soluble nutrients is reduced (Abbaspour et al., 2020). Biochar increases soil microbial abundance and activity (Cao et al., 2016; Hardy et al., 2019), and enhances PAH biodegradation in mesocosm and pot experiments (Kong et al., 2018; Zhang et al., 2018a). Others have reported negative effects of biochar on PAH biodegradation due to the adsorption of PAHs and the subsequent reduced bioavailability to PAH degraders (Han et al., 2016; Zhang et al., 2013), but this seems to be a function of the pyrolysis process and its effect on the biochar structure (Beesley et al., 2010; Kong et al., 2018; Zhang et al., 2018b), and the presence of soil amendments, such as nutrients and PAH-degrading bacteria or fungi (El-Naggar et al., 2019).

Bao et al. (2020) reported that biochar, mushroom residue, and corn straw significantly enhanced the biodegradation of PAHs in contaminated soil and attributed it to increased bioavailability of PAHs and increases in PAH-degrading microorganisms in the soil. Furthermore, the combination of rice straw biochar inoculated with fungi (*Mycobacterium gilvum*) and nutrients was shown to significantly enhance the loss of phenanthrene, fluoranthene, and pyrene in coke plant soil (Xiong et al., 2017).

44.3.2.2 Landfarming

Landfarming is one of the oldest technologies for petroleum hydrocarbon bioremediation (Lukic et al., 2017), although its use seems to be declining worldwide. Briefly, the process consists of mixing nutrients with contaminated soil and spreading the soil onto an appropriate area of minimally contaminated soil, and tilling to mix the contaminant into the upper meter of soil. Contaminated soils can also be excavated and spread onto lined beds (Kuppusamy et al., 2016). Frequent tilling and appropriate fertilizer addition maintains aeration and enhances the biodegradation and volatilization of petroleum hydrocarbons. Both volatilization and biodegradation are responsible for the removal of low molecular weight hydrocarbons, while higher molecular weight hydrocarbons are predominately removed by biodegradation. Large amounts of land are required for landfarming, which can unfortunately generate large amounts of dust as well as volatiles that may require treating in the surrounding air. The addition of moisture (40–80%) can minimize dust while maintaining an active microbial community (Kuppusamy et al., 2016). Marin et al. (2005) reported that landfarming removed 80% of total hydrocarbons over 11 months from soil contaminated with refinery sludge. Landfarming could take as little as six months or over two years depending on the amount of high molecular weight hydrocarbons (Kuppusamy et al., 2016).

44.3.2.3 Biopiles

Biopiles combine techniques utilized in both landfarming and composting. Unlike landfarming which spreads out the contaminated soil, biopiles are treated as large mounds of contaminated soil (commonly 1–3 m in height) that can be enclosed with impervious covers to promote soil heating and prevent run-off, evaporation, and volatilization. Similar to other *ex situ* remediation techniques, the efficiency of a biopile to remediate oil-contaminated soil depends upon soil and oil characteristics, and enhancing microbial processes through aeration, water content, and nutrients. Aeration is typically carried out by artificially forced air or mechanical devices such as skid steer loaders that periodically mix the piles. Biopiles are more efficient with sandy soils compared to clay soils, and low porosity soils are commonly amended with bulking agents to enhance the bioavailability of hydrocarbons, water, nutrients, and oxygen to oil-degrading microorganisms (Kuppusamy et al., 2016). Compared to landfarming, the environmental variables (moisture, pH, temperature, aeration, and nutrients) within a biopile can be simpler to engineer and maintain. Therefore, it is not surprising that biopiles and similar technologies have been shown to be successful in remediating petroleum hydrocarbons in Arctic and Antarctic contaminated sites (Camenzuli and Freidman, 2015; Greer and Juck, 2017). Alvarez et al. (2020) reported over 75% hydrocarbon removal in Antarctic biopiles (0.5 ton; 94% sand) contaminated with diesel oil (~2,000 mg/kg) and biostimulated with nutrients (C:N:P ratio of 100:17.6:1.73) over 40 days, compared to a ~50% loss in control unamended plots. Even though mean temperatures in the biopiles were approximately 6.5°C and the piles were covered with a geomembrane, significant loss of hydrocarbons occurred in the unamended plots due to biodegradation (i.e. natural attenuation) and volatilization. In another study, 77% of high molecular weight hydrocarbons were removed when landfarming was applied as a pretreatment for hydrocarbon-contaminated soil in Mexico, followed by the implementation of biopiles (Iturbe and Lopez, 2015). As biopiles continue to be implemented worldwide, field studies are focusing on increasing biodegradation rates by coupling *in situ* treatments with biopiling (Kuppusamy et al., 2016).

44.3.2.4 Bioaugmentation

Bioaugmentation is a controversial technique that involves introducing degradative microorganisms to a contaminated site. Many microbial cultures are available commercially with claims they will enhance biodegradation; however, these mixtures are not regulated and have shown varying results. The scientific literature does not support bioaugmentation in marine or freshwater environments, where research has long shown that indigenous populations consistently outperform introduced oil-degraders (Venosa et al., 1992). Although several laboratory studies have claimed the success of bioaugmentation in initially enhancing the biodegradation of petroleum hydrocarbons in soil (Abena et al., 2019; Nwankwegu and Onwosi, 2017; Pacwa-Płociniczak et al., 2019; Poorsoleiman et al., 2020), others have reported no enhancement (Anza et al., 2019; Radwan et al., 2019).

There is increased interest in applying indigenous adapted consortia to contaminated field sites to ensure that the added inoculum is compatible with the environmental conditions. Laboratory mesocosms indicate that biostimulation with nitrogen and phosphorus outperforms bioaugmentation with native oil-degrading bacteria (Wu et al., 2019), but limited field studies are reported in the literature. In a recent field experiment, the addition of a bacterial consortium resulted in significant loss of weathered TPH (total petroleum hydrocarbons)-contaminated soil after 63 days (a 92–96% loss), though the control also lost 63% due to aeration (Poi et al., 2017) and microbial analyses weren't reported, making it impossible to determine the fate and thus the effect of the added inoculum. In a different study, sequence analysis indicated that the bioaugmentation inoculum did not survive in the contaminated soil and did not enhance the rate of PAH biodegradation compared to the unamended control at 22 days (Piubeli et al., 2018).

The likely advantage of adding a designed consortium of oil-degrading microorganisms to oil-contaminated soil is to promote the initial rate of biodegradation. Small-scale contaminated sites that have low microbial biomass are good candidates for bioaugmentation and have shown promising results in the field (Mishra et al., 2001). When implementing bioaugmentation, it is important to incorporate strains that can withstand the environmental conditions, predation, and high concentrations of petroleum hydrocarbons. Furthermore, it would be advantageous to select strains that are fast growing and a consortium that can biodegrade a range of contaminants (i.e. large and small linear, branched, cyclic, and aromatic hydrocarbons).

Microorganisms selected for bioaugmentation may be isolated from the contaminated site or selected from other sites with similar soil conditions and contaminants. It is well known that a consortium of oil-degrading bacteria is more effective at removing hydrocarbons than a single strain (Pelz et al., 1999; Zaida and Piakong, 2018), with no microorganism universally applicable to bioaugmentation. An indigenous microbial community can utilize different degradation pathways and facilitate the biodegradation of hydrocarbon metabolites when intermediates produced by one strain are further degraded by another strain (Heinaru et al., 2005). The majority of bioaugmentation experiments have been carried out using: Gram-negative bacteria belonging to the genera *Pseudomonas* and *Flavobacterium*; Gram-positive bacteria belonging to the genera *Rhodococcus*, *Mycobacterium*, and *Bacillus*; and fungi such as *Aspergillus*, *Penicillium*, and *Verticillium* (Pacwa-Płociniczak et al., 2019; Poi et al., 2017; Zaida and Piakong, 2018).

Bioaugmentation can also influence the success of other remediation technologies when used in conjunction. For example, the addition of plant growth-promoting bacteria, such as *Bacillus paramycoides* ST9, improved phytoremediation by improving plant survival and phytostabilization (Saran et al., 2020). In challenging environmental conditions, such as saline soils and cold environments, combinations of bioaugmentation with phytoremediation or biostimulation have shown significant enhancement of petroleum hydrocarbon loss (Chaudhary and Kim, 2019; Sima et al., 2019).

Soil fungi are also active in the bioremediation of petroleum hydrocarbons. Many types of fungi have developed the ability to utilize aliphatic and aromatic hydrocarbons as the sole carbon source (Dacco et al., 2020; Horel and Schiewer, 2020; Prince, 2019), with cyclic aliphatics being the most resistant to fungal attack (Varjani and Upasani, 2017). Fungal degradation usually occurs in the presence of oxygen, but anaerobic biodegradation by yeasts is common, and recently a filamentous fungi was reported to anaerobically biodegrade PAHs (Aydin et al., 2017).

44.3.3 Biosurfactants

Bioavailability of oil to oil-degrading microorganisms is one of the main rate-limiting factors involved in biodegradation (Ron and Rosenberg, 2002); oil biodegradation in soil may be enhanced by indigenous bacterial strains that produce biosurfactant molecules. Different bacteria synthesize a wide diversity of biosurfactant chemical structures that may be classified as glycolipids, phospholipids, lipopeptides, or polymeric surfactants (Vater et al., 2002). Soil bacteria secrete biosurfactants into the pore space of the soil matrix to enhance solubility of hydrophobic hydrocarbons and increase the contact area between the oil droplet and bacteria (Zenati et al., 2018). Soil bacteria can also produce biosurfactants that remain near the cell surface to facilitate the uptake of oil into the cell through direct cellular contact (Noudeh et al., 2010). The presence of biosurfactant producing bacteria in soil may increase the bioavailability of oil, as surfactants can aid in the uptake of hydrophobic compounds and/or promote cellular protection from toxicological effects. Surfactants can also increase the bioavailability of oil to degradative bacteria by increasing the water-retention capacity of oil-polluted soil and facilitating the transport of hydrophobic oil through the pore space and ultimately through the cell wall.

Several studies have reported increased removal (60–80 %) of PAHs by rhamnolipids (Reis et al., 2013). Surfactin, a cyclic lipopeptide produced by different strains of *Bacillus subtilis*, has been widely studied and shown to accelerate the biodegradation of *n*-alkanes, phenanthrene, and recently alkylated aromatic compounds (Wang et al., 2020). Bezza and Chirwa (2015) reported that a lipopeptide biosurfactant isolated from *Bacillus subtilis* doubled PAH biodegradation. In another laboratory study, a rhamnolipid biosurfactant (at 1.5 g/L) produced by *Pseudomonas aeruginosa* SR17 enhanced the biodegradation of TPH in soil significantly more than the synthetic surfactant sodium dodecyl sulfate (SDS) (Patowary et al., 2018). Further, 'gas chromatography mass spectrometry' (GC/MS) analysis indicated that the rhamnolipid eliminated fluoranthene, benzo(b) fluorene, and benzo(d)anthracene completely within six months and the remaining PAHs were depleted up to 60–80%. Enhancement was attributed to increased abundance of heterotrophic bacteria (Patowary et al., 2018). The effects of two common commercial surfactants on the biodegradation of crude oil by a bacterial consortium in a soil-aqueous system was evaluated by Xu et al. (2018) who reported that 'Tween 80' was generally superior to SDS. Other have reported that biosurfactants were more efficient at TPH removal in contaminated soil than Tween 80 and SDS (Hu et al., 2020). However, logistics and the availability of surfactant amendments must also be considered when planning large-scale remediation projects.

44.3.4 Phytoremediation

Knowledge pertaining to the effects of phytotechnologies has been increasing over the past 70 years (Zhang et al., 2020), and these relatively inexpensive solutions have proven to be effective at removing petroleum hydrocarbons from soil (Alves et al., 2018; Tripathi et al., 2020; Yavari et al., 2015). Due to the chemical nature of petroleum hydrocarbons, such as low water solubility and high soil sorption (i.e. log K_{ow} > 4), relatively few petroleum hydrocarbons are likely to become sequestered in plant tissue (i.e. phytoextraction) or transpired through leaves (i.e. phytovolatilization). Phytoextraction of low molecular weight hydrocarbons was reported by Boonsaner et al. (2011), who identified the bioaccumulation of BTEX in the shots of canna lily.

The most effective phytoremediation technology for petroleum-hydrocarbon-contaminated soil seems to be rhizodegradation, which involves complex interactions between plant roots, root exudates, the surrounding soil, and the microbial community. The presence of plants can increase the abundance of oil-degrading microorganisms in contaminated soil orders of magnitude higher than unplanted controls (Lim et al., 2016). Trees and grasses with extensive root systems are the most effective plants for phytoremediation of petroleum-hydrocarbon-contaminated soil (Cook and Hesterberg, 2013). Large root structures can influence the bulk soil by releasing root exudates that shift the bacterial community structure and enhance the biodegradation of PAHs (Musilova et al., 2016; Wang et al., 2020). Plants secrete secondary metabolites (e.g. phenolic compounds) into the rhizosphere as the root cells lyse during root turnover (Fulekar and Fulekar, 2020). Secondary metabolites, such as salicylate, have been shown to induce oil-degrading microorganisms to turn on specific biodegradation pathways and increase PAH degradation (Page et al., 2015). At high oil concentrations (7,300 mg/kg), in southern Finland, poplars were able to remove over 50% of total measurable hydrocarbons after one year and 78% after four years, which was 22% more than the unplanted control (Lopez-Echartea et al., 2020). High concentrations of hydrocarbons can influence the presence and growth of vegetation due to their phytotoxic properties (Alves et al., 2018). The ideal plant for phytoremediation should be tolerant to the environmental conditions, soil type and contaminant concentration, capable of producing large amounts of above and below-ground biomass, and resistant to bioaccumulation and trophic transfer.

Phytoremediation of petroleum-contaminated soil can be enhanced through biostimulation or bioaugmentation. The addition of nitrogen and phosphorus to the bulk soil can reduce competition between plants and microbes and increase plant and microbial biomass (Yavari et al., 2015). Many laboratory and field studies have shown that certain bacteria favor fertilizer amendments (Leewis et al., 2016; Wang et al., 2020). Among these bacteria are *Pseudomonas* and *Pseudoxanthomonas* which are commonly found in oil-polluted soil and known to produce biosurfactants (rhamnolipids) that promote the biodegradation of high molecular weight PAHs (Singleton et al., 2016). Commercial or biobased surfactants (e.g. SDS or rhamnolipids) can be added to contaminated soil to enhance the solubility of hydrocarbons and the success of phytoremediation. A recent greenhouse study reported that the addition of the surfactant Tween 80 to crude-oil-contaminated soil increased nutrient uptake and decreased the toxic effects of crude oil, allowing the 'vetiver grass' to tolerate higher concentrations of crude oil (Keshavarz et al., 2020). Inoculating the bulk soil and rhizosphere with arbuscular mycorrhizal fungi, hydrocarbon-degrading microorganisms, or plant growth-promoting bacteria can promote rhizoremediation by increasing nutrient availability, plant biomass, and the abundance of degradative enzymes (Yavari et al., 2015). Over a seven-year field study in south Finland, rhizoremediation of contaminated soil from an oil tanker truck accident using poplar trees decreased TPH concentration by 78% after three years, without any nutrient additions (Lopez-Echartea et al., 2020). Leewis et al. (2013) showed that even without active site management, plants could remediate diesel and crude-oil-contaminated soil in sub-arctic Alaska.

44.4 CONCLUSION

When soil becomes contaminated with oil, many options for remediation are available (Table 44.1) and no single remediation technology is suitable for all petroleum hydrocarbons and environments. Response options include *ex situ* mechanical methods involving haul and treat and *in situ* biological enhancement methods that involve bioremediation. Chemical characterization of the spilled oil combined with environmental characterization of the soil will help to guide the most efficient remediation strategy. *Ex situ* bioremediation technologies, such as composting, landfarming, and biopiles, may offer advantages such as containment and process control. Other amendments, such as bulking agents and biochar, have been shown to enhance oil biodegradation in soil. Biostimulation and bioaugmentation techniques strive to optimize environmental conditions and microbial populations to enhance the bioavailability of hydrocarbons to oil-degrading microorganisms and increase biodegradation. A potential drawback to bioremediation is the amount of time required to achieve compliance compared to mechanical techniques that either combust the contaminant or move it to a hazardous waste landfill.

The bioavailability of petroleum hydrocarbons is the main limiting factor to bioremediation. Contaminated soils with favorable water content and high porosity that allows sufficient exchange of oxygen and nutrients will increase hydrocarbon bioavailability to oil-degrading microorganisms and enhance biodegradation in soils. Soils at near neutral pH, with low salinity, optimal water content, and low concentrations of organic matter and clay, are most ideal for biodegradation.

Many studies have indicated that consortia of indigenous oil-degrading bacteria are more effective at removing hydrocarbons than a subset of known oil-degraders. Bioaugmentation with select oil-degraders has been shown to enhance the initial removal of petroleum hydrocarbons from soil, but the majority of studies indicate that bioremediation using indigenous soil bacteria is more reliable and easier to implement and maintain. Recent literature supports that bioremediation using indigenous bacteria can be further enhanced with biostimulation strategies such as nutrients, biochar, and surfactants. When ideal soil conditions are maintained, the abundance of indigenous oil-degraders will increase naturally in the presence of hydrocarbons to allow significant loss of petroleum hydrocarbons.

Bioremediation strategies are commonly considered low-cost green solutions for soil remediation compared to traditional mechanical options. Nonetheless, environmental variables cannot be completely controlled in the field, and the success of biodegradation is never guaranteed. Often the socioeconomic benefits of bioremediation are outweighed by requirements to rapidly attain rigorous clean-up standards, which then force the implementation of mechanical techniques over bioremediation strategies. Bioremediation remains an important tool for remediating oil-contaminated soil and its efficacy will continue to increase over time as researchers implement strategies that increase the rate and extent of biodegradation across a wide range of environmental conditions.

REFERENCES

Abbaspour, A., F. Zohrabi, V. Dorostkar, A. Faz, and J. A. Acosta. 2020. Remediation of an oil-contaminated soil by two native plants treated with biochar and mycorrhizae. *Journal of Environmental Management* 254: 109755.

Abdel-Shafy, H. I. and M. S. M. Mansour. 2016. A review on polycyclic aromatic hydrocarbons: Source, environmental impact, the effect on human health and remediation. *Egyptian Journal of Petroleum* 25: 107–123.

Abena, M. T. B., T. Li, M. N. Shah, and W. Zhong. 2019. Biodegradation of total petroleum hydrocarbons (TPH) in highly contaminated soils by natural attenuation and bioaugmentation. *Chemosphere* 234: 864–874.

Alegbeleye, O. O. 2018. Petroleum microbiology under extreme conditions. In *Microbial Action on Hydrocarbons*, ed. V. Kumar, M. Kumar, and R. Prasad, 441–484. Singapore: Springer.

Al-Hawash, A. B., W. S. Al-Qurnawi, and H. A. Abbood, et al. 2020. Pyrene-degrading fungus *Ceriporia lacerata* RF-7 from contaminated soil in Iraq. *Polycyclic Aromatic Compounds* 2020: 1713183.

Al-Mailem, D. M., M. Eliyas, and S. S. Radwan. 2013. Oil-bioremediation potential of two hydrocarbonoclastic, diazotrophic *Marinobacter* strains from hypersaline areas along the Arabian Gulf coasts. *Extremophiles* 17: 463–470.

Alvarez, L. M., L. A. M. Ruberto, J. M. Gurevich, and W. P. Mac Cormack. 2020. Environmental factors affecting reproducibility of bioremediation field assays in Antarctica. *Cold Regions Science and Technology* 169: 102915.

Alves, W. S., E. A. Manoel, and N. S. Santos, et al. 2018. Phytoremediation of polycyclic aromatic hydrocarbons (PAH) by cv. *Crioula: A Brazilian alfalfa cultivar. International Journal of Phytoremediation* 20: 747–755.

Anza, M., O. Salazar, and L. Epelde, et al. 2019. Remediation of organically contaminated soil through the combination of assisted phytoremediation and bioaugmentation. *Applied Sciences* 9: 4757.

Atlas, R. M. 1975. Effects of temperature and crude oil composition on petroleum biodegradation. *Applied Microbiology* 30: 396–403.

Auti, A. M., N. P. Narwade, N. M. Deshpande, and D. P. Dhotre. 2019. Microbiome and imputed metagenome study of crude and refined petroleum-oil-contaminated soils: Potential for hydrocarbon degradation and plant-growth promotion. *Journal of Biosciences* 44: 114.

Aydin, S., H. A. Karacay, and A. Shahi, et al. 2017. Aerobic and anaerobic fungal metabolism and omics insights for increasing polycyclic aromatic hydrocarbons biodegradation. *Fungal Biology Reviews* 31: 61–72.

Azubuike, C. C., C. B. Chikere, and G. C. Okpokwasili. 2016. Bioremediation techniques–classification based on site of application: Principles, advantages, limitations and prospects. *World Journal of Microbiology and Biotechnology* 32: 180.

Bao, H., J. Wang, and H. Zhang, et al. 2020. Effects of biochar and organic substrates on biodegradation of polycyclic aromatic hydrocarbons and microbial community structure in PAHs-contaminated soils. *Journal of Hazardous Materials* 385: 121595.

Beesley, L., E. Moreno-Jimenez, and J. L. Gomez-Eyles. 2010. Effects of biochar and greenwaste compost amendments on mobility, bioavailability and toxicity of inorganic and organic contaminants in a multi-element polluted soil. *Environmental Pollution* 158: 2282–2287.

Bell, T. H., F. Yergeau, and C. Maynard, et al. 2013. Predictable bacterial composition and hydrocarbon degradation in Arctic soils following diesel and nutrient disturbance. *ISME Journal* 7: 1200–1210.

Bengtsson, G., N. Torneman, J. G. de Lipthay, and S. L. Sorensen. 2013. Microbial diversity and PAH catabolic genes tracking spatial heterogeneity of PAH concentrations. *Microbial Ecology* 65: 91–100.

Bengtsson, G., N. Torneman, and X. Yang. 2010. Spatial uncoupling of biodegradation, soil respiration, and PAH concentration in PAH contaminated soil. *Environmental Pollution* 158: 2865–2871.

Bezza, F. A. and E. M. N. Chirwa. 2015. Production and applications of lipopeptide biosurfactant for bioremediation and oil recovery by *Bacillus subtilis* CN2. *Biochemical Engineering Journal* 101:168–178.

Boonsaner, M., S. Borrirukwisitsak, and A. Boonsaner. 2011. Phytoremediation of BTEX contaminated soil by *Canna×generalis*. *Ecotoxicology and Environmental Safety* 74: 1700–1707.

Brown, D. M., L. Camenzuli, and A. Redman, et al. 2020. Is the Arrhenius-correction of biodegradation rates, as recommended through REACH guidance, fit for environmentally relevant conditions? An example from petroleum biodegradation in environmental systems. *Science of the Total Environment* 732:139293.

Buerkert, A., R. G. Joergensen, B. Ludwig, and E. Schlecht. 2012. Nutrient and carbon fluxes in terrestrial agro-ecosystems. In *Marschner's Mineral Nutrition of Higher Plants*, 3rd edn, ed. P. Marschner, 473–482. Cambridge: Academic Press.

Camenzuli, D. and B. L. Freidman. 2015. On-site and in situ remediation technologies applicable to petroleum hydrocarbon contaminated sites in the Antarctic and Arctic. *Polar Research* 34: 24492.

Cao, Y., B. Yang, and Z. Song, et al. 2016. Wheat straw biochar amendments on the removal of polycyclic aromatic hydrocarbons (PAHs) in contaminated soil. *Ecotoxicology and Environmental Safety* 130: 248–255.

Chaerun, S. K., K. Tazaki, R. Asada, and K. Kogure. 2005. Interaction between clay minerals and hydrocarbon-utilizing indigenous microorganisms in high concentrations of heavy oil: Implications for bioremediation. *Clay Minerals* 40:105–114.

Chaineau, C. H., G. Rougeux, C. Yepremian, and J. Oudot. 2005. Effects of nutrient concentration on the biodegradation of crude oil and associated microbial populations in the soil. *Soil Biology and Biochemistry* 37: 1490–1497.

Chaudhary, D. K. and J. Kim. 2019. New insights into bioremediation strategies for oil-contaminated soil in cold environments. *International Biodeterioration & Biodegradation* 142: 58–72.

Cho, E., M. Park, and M. Hur, et al. 2019. Molecular-level investigation of soils contaminated by oil spilled during the Gulf War. *Journal of Hazardous Materials* 373: 271–277.

Cook, R. L. and D. Hesterberg. 2013. Comparison of trees and grasses for rhizoremediation of petroleum hydrocarbons. *International Journal of Phytoremediation* 15: 844–860.

Dacco, C., C. Girometta, and M. D. Asemoloye, et al. 2020. Key fungal degradation patterns, enzymes and their applications for the removal of aliphatic hydrocarbons in polluted soils: A review. *International Biodeterioration & Biodegradation* 147: 104866.

Dadrasnia, A., M. M. Usman, T. Alinejad, B. Motesharezadeh, and S. M. Mousavi. 2018. Hydrocarbon degradation assessment: Biotechnical approaches involved. In *Microbial Action on Hydrocarbons*, ed. V. Kumar, M. Kumar, and R. Prasad, 63–95. Singapore: Springer.

Dai, Y., H. Zheng, Z. Jiang, and B. Xing. 2020. Combined effects of biochar properties and soil conditions on plant growth: A meta-analysis. *Science of the Total Environment* 713: 136635.

Das, N. and P. Chadran. 2011. Microbial degradation of petroleum hydrocarbon contaminants: An overview. *Biotechnology Research International* 2011: 941810.

Dastgheib, S. M. M., M. A. Amoozegar, K. Khajeh, and A. Ventosa. 2011. A halotolerant *Alcanivorax* sp. strain with potential application in saline soil remediation. *Applied Microbiology and Biotechnology* 90: 305–312.

de Menezes, A., N. Clipson, and E. Doyle. 2012. Comparative metatranscriptomics reveals widespread community responses during phenanthrene degradation in soil. *Environmental Microbiology* 14: 2577–2588.

Dellagnezze, B. M., M. B. Gomes, and V. M. de Oliveira. 2018. Microbes and petroleum bioremediation. In *Microbial Action on Hydrocarbons*, ed. V. Kumar, M. Kumar, and R. Prasad, 97–123. Amsterdam: Springer.

El-Naggar, A., S. S. Lee, and J. Rinklebe, et al. 2019. Biochar application to low fertility soils: A review of current status, and future prospects. *Geoderma* 337: 536–554.

EPA. 2017. *How to Evaluate Alternative Cleanup Technologies for Underground Storage Tank Sites: A Guide for Corrective Action Plan Reviewers*. Washington, DC: Environmental Protection Agency.

Farhadian, M., C. Vachelard, D. Duchez, and C. Larroche. 2008. *In situ* bioremediation of monoaromatic pollutants in groundwater: A review. *Bioresource Technology* 99: 5296–5308.

Fuentes, S., V. Mendez, P. Aguila, and M. Seeger. 2014. Bioremediation of petroleum hydrocarbons: Catabolic genes, microbial communities, and applications. *Applied Microbiology and Biotechnology* 98: 4781–4794.

Fulekar, M. H. and I. Fulekar. 2020. Rhizosphere bioremediation: Green technology to clean up the environment. In *Bioremediation Technology: Hazardous Waste Management*, ed. M. H. Fulekar, and B. Pathak, 227–241. Baco Raton, FL: CRC Press.

Gao, Y. C., S. H. Guo, and J. N. Wang, et al. 2014. Effects of different remediation treatments on crude oil contaminated saline soil. *Chemosphere* 117: 486–493.

Giovannoni, S. J. and U. Stingl. 2005. Molecular diversity and ecology of microbial plankton. *Nature* 437: 343–348.

Greer, C. W. and D. F. Juck. 2017. Bioremediation of petroleum hydrocarbon spills in cold terrestrial environments. In *Psychrophiles: From Biodiversity to Biotechnology*, ed. R. Margesin, 645–660. Cham: Springer.

Greer, C. W., L. G. Whyte, and T. D. Niederberger. 2010. Microbial communities in hydrocarbon contaminated temperate, tropical, alpine and polar soils. In *Handbook of Hydrocarbon and lipid microbiology*, ed. K. N. Timmis. 2013–2328. Berlin: Springer.

Haghollahi, A., M. H. Fazaelipoor, and M. Schaffie. 2016. The effect of soil type on the bioremediation of petroleum contaminated soils. *Journal of Environmental Management* 180:197–201.

Hamamura, N., S. H. Olson, D. M. Ward, and W. P. Inskeep. 2006. Microbial population dynamics associated with crude-oil biodegradation in diverse soils. *Applied and Environmental Microbiology* 72: 6316–6324.

Han, T., Z. Zhao, M. Bartlam, and Y. Wang. 2016. Combination of biochar amendment and phytoremediation for hydrocarbon removal in petroleum-contaminated soil. *Environmental Science and Pollution Research* 23: 21219–21228.

Hara, A., K. Syutsubo, and S. Harayama. 2003. *Alcanivorax* which prevails in oil-contaminated seawater exhibits broad substrate specificity for alkane degradation. *Environmental Microbiology* 5: 746–753.

Hardy, B., S. Sleutel, J. E. Dufey, and J.-T. Cornelis. 2019. The long-term effect of biochar on soil microbial abundance, activity and community structure is overwritten by land management. *Frontiers in Environmental Science* 7: 110.

Heinaru, E., M. Merimaa, and S. Viggor, et al. 2005. Biodegradation efficiency of functionally important populations selected for bioaugmentation in phenol-and oil-polluted area. *FEMS Microbiology Ecology* 51: 363–373.

Horel, A. and S. Schiewer. 2020. Microbial degradation of different hydrocarbon fuels with mycoremediation of volatiles. *Microorganisms* 8: 163.

Hu, X., Y. Qiao, and L.-Q. Chen, et al. 2020. Enhancement of solubilization and biodegradation of petroleum by biosurfactant from *Rhodococcus erythropolis* HX-2. *Geomicrobiology Journal* 37: 159–169.

Iturbe, R. and J. Lopez. 2015. Bioremediation for a soil contaminated with hydrocarbons. *J. Petroleum & Environmental Biotechnology* 6: 208.

Jain, P. K., V. K. Gupta, and R. K. Gaur, et al. 2011. Bioremediation of petroleum oil contaminated soil and water. *Research Journal of Environmental Toxicology* 5: 1–26.

Keshavarz, S., R. Ghasemi-Fasaei, A. Ronaghi, and M. Zarei. 2020. Nutritional composition, growth and enzymatic response of vetiver grass grown under crude oil contamination as influenced by gibberellic acid and surfactant. *Journal of Plant Nutrition* 43: 418–428.

Kong, L., Y. Gao, Q. Zhou, X. Zhao, and Z. Sun. 2018. Biochar accelerates PAHs biodegradation in petroleum-polluted soil by biostimulation strategy. *Journal of Hazardous Materials* 343: 276–284.

Kristensen, A. H., K. Henriksen, L. Mortensen, K. M. Scow, and P. Moldrup. 2010. Soil physical constraints on intrinsic biodegradation of petroleum vapors in a layered subsurface. *Vadose Zone Journal* 9: 137–147.

Kumari, B., G. Singh, G. Sinam, and D. P. Singh. 2020. Microbial remediation of crude oil-contaminated sites. In *Environmental Concerns and Sustainable Development*, ed. V. Shukla, and N. Kumar, 333–351. Singapore: Springer.

Kuppusamy, S., T. Palanisami, M. Megharaj, K. Venkateswarlu, and R. Naidu. 2016. *In-situ* remediation approaches for the management of contaminated sites: A comprehensive overview. *Reviews of Environmental Contamination and Toxicology* 236:1–115.

Kuzyakov, Y. and E. Blagodatskaya. 2015. Microbial hotspots and hot moments in soil: Concept & review. *Soil Biology and Biochemistry* 83: 184–199.

Labbe, D., R. Margesin, F. Schinner, L. G. Whyte, and C. W. Greer. 2007. Comparative phylogenetic analysis of microbial communities in pristine and hydrocarbon-contaminated Alpine soils. *FEMS Microbiology Ecology* 59: 466–475.

Lawniczak, L., M. Wozniak-Karczewska, A. P. Loibner, H. J. Heipieper, and L. Chrzanowski, 2020. Microbial degradation of hydrocarbons-Basic principles for bioremediation: A review. *Molecules* 25: 856.

Le Borgne, S., D. Paniagua, and R. Vazquez-Duhalt. 2008. Biodegradation of organic pollutants by halophilic bacteria and archaea. *Journal of Molecular Microbiology and Biotechnology* 15: 74–92.

Leewis, M. C., C. M. Reynolds, and M. B. Leigh. 2013. Long-term effects of nutrient addition and phytoremediation on diesel and crude oil contaminated soils in subarctic Alaska. *Cold Regions Science and Technology* 96:129–138.

Leewis, M. C., O. Uhlik, and S. Fraraccio, et al. 2016. Differential impacts of willow and mineral fertilizer on bacterial communities in diesel oil-polluted soil. *Frontiers in Microbiology* 7: 837.

Liang, Y., H. Zhao, and Y. Deng, et al. 2016. Long-term oil contamination alters the molecular ecological networks of soil microbial functional genes. *Frontiers in Microbiology* 7: 60.
Liang, Y., H. Zhao, X. Zhang, J. Zhou, and G. Li. 2014. Contrasting microbial functional genes in two distinct saline-alkali and slightly acidic oil-contaminated sites. *Science of the Total Environment* 487: 272–278.
Lim, M. W., E. von Lau, and P. E. Poh. 2016. A comprehensive guide of remediation technologies for oil contaminated soil-present works and future directions. *Marine Pollution Bulletin* 109: 14–45.
Liu, H., H. Gao, and M. Wu, et al. 2020. Distribution characteristics of bacterial communities and hydrocarbon degradation dynamics during the remediation of petroleum-contaminated soil by enhancing moisture content. *Microbial Ecology* 2020:1476–1477.
Lopez-Echartea, E., M. Strejcek, S. Mukherjee, O. Uhlik, and K. Yrjala. 2020. Bacterial succession in oil-contaminated soil under phytoremediation with poplars. *Chemosphere* 243: 125242.
Louati, H., O. B. Said, and P. Got, et al. 2013. Microbial community responses to bioremediation treatments for the mitigation of low-dose anthracene in marine coastal sediments of Bizerte lagoon (Tunisia). *Environmental Science and Pollution Research* 20: 300–310.
Lozupone, C. A. and R. Knight. 2007. Global patterns in bacterial diversity. *Proceedings of the National Academy of Sciences of the United States of America* 104: 11436–11440.
Lukic, B., A. Panico, and D. Huguenot, et al. 2017. A review on the efficiency of landfarming integrated with composting as a soil remediation treatment. *Environmental Technology Reviews* 6: 94–116.
Macchi, M., S. Festa, N. E. Vega-Vela, I. S. Morelli, and B. M. Coppotelli. 2019. Assessing interactions, predicting function, and increasing degradation potential of a PAH-degrading bacterial consortium by effect of an inoculant strain. *Environmental Science and Pollution Research* 26: 25932–25944.
MacKelprang, R., A. Burkert, and M. P. Haw, et al. 2017. Microbial survival strategies in ancient permafrost: Insights from metagenomics. *ISME Journal* 11: 2305–2318.
Marin, J. A., T. Hernandez, and C. Garcia. 2005. Bioremediation of oil refinery sludge by landfarming in semiarid conditions: Influence on soil microbial activity. *Environmental Research* 98: 185–195.
McFarlin, K. M., R. C. Prince, R. Perkins, and M. B. Leigh. 2014. Biodegradation of dispersed oil in Arctic seawater at −1°C. *Plos One* 9: 84297.
McWatters, R. S., D. Wilkins, and T. Spedding, et al. 2016. On site remediation of a fuel spill and soil reuse in Antarctica. *Science of the Total Environment* 571: 963–973.
Mishra, S., J. Jyot, R. C. Kuhad, and B. Lal. 2001. *In situ* bioremediation potential of an oily sludge-degrading bacterial consortium. *Current Microbiology* 43: 328–335.
Moretto, L. M., S. Silvestri, and P. Ugo, et al. 2005. Polycyclic aromatic hydrocarbons degradation by composting in a soot-contaminated alkaline soil. *Journal of Hazardous Materials* 126:141–148.
Mosco, M. J. and R. G. Zytner. 2017. Large-scale bioventing degradation rates of petroleum hydrocarbons and determination of scale-up factors. *Bioremediation Journal* 21:149–162.
Musilova, L., J. Ridl, M. Polivkova, T. Macek, and O. Uhlik. 2016. Effects of secondary plant metabolites on microbial populations: Changes in community structure and metabolic activity in contaminated environments. *International Journal of Molecular Sciences* 17: 1205.
Nagkirti, P., A. Shaikh, G. Vasudevan, V. Paliwal, and P. Dhakephalkar. 2017. Bioremediation of terrestrial oil spills: Feasibility assessment. In *Optimization and Applicability of Bioprocesses*, ed. H. J. Purohit, V. C. Kalia, A. N. Vaidya, and A. A. Khardenavis, 141–173. Singapore: Springer.
National Research Council. 1993. *In Situ Bioremediation: When Does it Work?* Washington, DC: National Academies Press.

Neira, J., M. Ortiz, L. Morales, and E. Acevedo. 2015. Oxygen diffusion in soils: Understanding the factors and processes needed for modeling. *Chilean Journal of Agricultural Research* 75: 35–44.

Nikolopoulou, M. and N. Kalogerakis. 2016. *Ex situ* bioremediation treatment (Landfarming). In *Hydrocarbon and Lipid Microbiology Protocols*, ed. T. McGenity, K. Timmis, and B. Nogales, 195–220. Heidelberg: Springer.

Noudeh, G. D., A. D. Noodeh, and M. H. Moshafi, et al. 2010. Investigation of cellular hydrophobicity and surface activity effects of biosynthesed biosurfactant from broth media of PTCC 1561. *African Journal of Microbiology Research* 4: 1814–1822.

Nwankwegu, A. S. and C. O. Onwosi. 2017. Bioremediation of gasoline contaminated agricultural soil by bioaugmentation. *Environmental Technology & Innovation* 7: 1–11.

Obiri-Nyarko, F., S. J. Grajales-Mesa, and G. Malina. 2014. An overview of permeable reactive barriers for *in situ* sustainable groundwater remediation. *Chemosphere* 111: 243–259.

Okoh, E., Z. R. Yelebe, B. Oruabena, E. S. Nelson, and O. P. Indiamaowei. 2019. Clean-up of crude oil-contaminated soils: Bioremediation option. *International Journal of Environmental Science and Technology* 17: 1185–1198.

Pacwa-Płociniczak, M., J. Czapla, T. Płociniczak, and Z. Piotrowska-Seget. 2019. The effect of bioaugmentation of petroleum-contaminated soil with *Rhodococcus erythropolis* strains on removal of petroleum from soil. *Ecotoxicology and Environmental Safety* 169: 615–622.

Page, A. P., E. Yergeau, and C. W. Greer. 2015. *Salix purpurea* stimulates the expression of specific bacterial xenobiotic degradation genes in a soil contaminated with hydrocarbons. *Plos One* 10: 0132062.

Paisse, S., M. Goni-Urriza, T. Stadler, H. Budzinski, and R. Duran. 2012. Ring-hydroxylating dioxygenase (RHD) expression in a microbial community during the early response to oil pollution. *FEMS Microbiology Ecology* 80: 77–86.

Pal, S., A. Roy, and S. K. Kazy. 2019. Exploring microbial diversity and function in petroleum hydrocarbon associated environments through omics approaches. In *Microbial Diversity in the Genomic Era*, ed. S. Dash, and H. R. Dash, 171–194. Cambridge: Academic Press.

Paniagua-Michel, J. and B. Z. Fathepure. 2018. Microbial consortia and biodegradation of petroleum hydrocarbons in marine environments. In *Microbial Action on Hydrocarbons*, ed. V. Kumar, M. Kumar, and R. Prasad, 1–20. Singapore: Springer.

Patowary, R., K. Patowary, M. C. Kalita, and S. Deka. 2018. Application of biosurfactant for enhancement of bioremediation process of crude oil contaminated soil. *International Biodeterioration & Biodegradation* 129: 50–60.

Pelz, O., M. Tesar, and R.-M. Wittich, et al. 1999. Towards elucidation of microbial community metabolic pathways: Unravelling the network of carbon sharing in a pollutant-degrading bacterial consortium by immunocapture and isotopic ratio mass spectrometry. *Environmental Microbiology* 1:167–174.

Pepper, I. L. and M. L. Brusseau. 2019. Physical-chemical characteristics of soils and the subsurface. In *Environmental and Pollution Science*, ed. M. L. Brousseau, I. L. Pepper, and C. P. Gerda, 9–22. Cambridge: Academic Press.

Piubeli, F. A., L. G. dos Santos, E. N. Fernandez, et al. 2018. The emergence of different functionally equivalent PAH degrading microbial communities from a single soil in liquid PAH enrichment cultures and soil microcosms receiving PAHs with and without bioaugmentation. *Polish Journal of Microbiology* 67: 365–375.

Poi, G., A. Aburto-Medina, P. C. Mok, A. S. Ball, and E. Shahsavari. 2017. Large scale bioaugmentation of soil contaminated with petroleum hydrocarbons using a mixed microbial consortium. *Ecological Engineering* 102: 64–71.

Poorsoleiman, M. S., S. A. Hosseini, A. Etminan, H. Abtahi, and A. Koolivand. 2020. Effect of two-step bioaugmentation of an indigenous bacterial strain isolated from oily waste sludge on petroleum hydrocarbons biodegradation: Scaling-up from a liquid mineral medium to a two-stage composting process. *Environmental Technology & Innovation* 17: 100558.

Prasad, M., M. Chaudhary, M. Choudhary, T. K. Kumar, L. K. Jat. 2017. Rhizosphere microorganisms towards soil sustainability and nutrient acquisition. In *Agriculturally Important Microbes for Sustainable Agriculture*, ed. V. P. Meena, J. Mishra, A. Bisht, and A. Pattanayak, 31–49. Singapore: Springer.

Prince, R. C. 2019. Eukaryotic hydrocarbon degraders. In *Taxonomy, Genomics and Ecophysiology of Hydrocarbon-Degrading Microbes*, ed. T. J. Mac Genity, 53–72. Cham: Springer.

Prince, R. C., T. J. Amande, and T. J. McGenity. 2019. Prokaryotic hydrocarbon degraders. In *Taxonomy, Genomics and Ecophysiology of Hydrocarbon-Degrading Microbes*, ed. T. J. McGenity, 1–39. Cham: Springer.

Prince, R. C. and G. S. Douglas. 2010. Remediation of petrol and diesel in subsurface from petrol station leaks. In *Handbook of Hydrocarbon and Lipid Microbiology*, ed. K. N. Timmis, 2598–2608. Berlin: Springer.

Prince, R. C., B. M. Hedgpeth, and K. M. McFarlin. 2021. Crude oils and their fate in the environment. In *Handbook of Biodiesel and Petrodiesel Fuels: Science, Technology, Health, and Environment. Volume 3. Petrodiesel Fuels: Science, Technology, Health, and Environment*, ed. O. Konur. Boca Raton, FL: CRC Press.

Prince, R. C. and C. C. Walters. 2016. Biodegradation of oil hydrocarbons and its implications for source identification. In *Standard Handbook Oil Spill Environmental Forensics*, ed. S. A. Stout, and Z. Wang, 869–916. Amsterdam: Elsevier.

Radwan, S. S., D. M. Al-Mailem, and M. K. Kansour. 2019. Bioaugmentation failed to enhance oil bioremediation in three soil samples from three different continents. *Scientific Reports* 9: 19508.

Radwan, S. S., N. A. Sorkhoh, F. Fardoun, and R. H. Al-Hasan. 1995. Soil management enhancing hydrocarbon biodegradation in the polluted Kuwaiti desert. *Applied Microbiology and Biotechnology* 44: 265–270.

Ramadass, K., M. Megharaj, K. Venkateswarlu, and R. Naidu. 2018. Bioavailability of weathered hydrocarbons in engine oil-contaminated soil: Impact of bioaugmentation mediated by *Pseudomonas* spp. on bioremediation. *Science of the Total Environment* 636: 968–974.

Reis, R. S., G. J. Pacheco, A. G. Pereira, and D. M. G. Freire. 2013. Biosurfactants: Production and applications. In *Biodegradation: Life of Science*, ed. R. Chamy, and F. Rosenkranz, 31–36. London: Intech Open.

Ribicic, D., K. M. McFarlin, and R. Netzer, et al. 2018. Oil type and temperature dependent biodegradation dynamics: Combining chemical and microbial community data through multivariate analysis. *BMC Microbiology* 18: 1221–1229.

Rike, A. G., K. B. Haugen, and B. Engene. 2005. *In situ* biodegradation of hydrocarbons in arctic soil at sub-zero temperatures-field monitoring and theoretical simulation of the microbial activation temperature at a Spitsbergen contaminated site. *Cold Regions Science and Technology* 41:189–209.

Ron, E. Z. and E. Rosenberg. 2002. Biosurfactants and oil bioremediation. *Current Opinion in Biotechnology* 13: 249–252.

Saran, A., V. Imperato, and L. Fernandez, et al. 2020. Phytostabilization of polluted military soil supported by bioaugmentation with PGP-trace element tolerant bacteria isolated from *Helianthus petiolaris*. *Agronomy* 10: 204.

Schulte, P. M. 2015. The effects of temperature on aerobic metabolism: Towards a mechanistic understanding of the responses of ectotherms to a changing environment. *Journal of Experimental Biology* 218: 1856–1866.

Sima, N. A. K., A. Ebadi, N. Reiahisamani, and B. Rasekh. 2019. Bio-based remediation of petroleum-contaminated saline soils: Challenges, the current state-of-the-art and future prospects. *Journal of Environmental Management* 250:109476.

Singh, S. N., B. Kumari, and S. Mishra. 2012. Microbial degradation of alkanes. In *Microbial Degradation of Xenobiotics*, ed. S. N. Singh. Heidelberg: Springer.

Singleton, D. R., A. C. Adrion, and M. D. Aitken. 2016. Surfactant-induced bacterial community changes correlated with increased polycyclic aromatic hydrocarbon degradation in contaminated soil. *Applied Microbiology and Biotechnology* 100:10165–10177.

Sorokin, D. Y., A. J. Janssen, and G. Muyzer, G. 2012. Biodegradation potential of halo (alkali) philic prokaryotes. *Critical Reviews in Environmental Science and Technology* 42: 811–856.

Straube, W. L., C. C. Nestler, and L. D. Hansen, et al. 2003. Remediation of polyaromatic hydrocarbons (PAHs) through landfarming with biostimulation and bioaugmentation. *Acta Biotechnologica* 23: 179–196.

Taylor, J. P., B. Wilson, M. S. Mills, and R. G. Burns. 2002. Comparison of microbial numbers and enzymatic activities in surface soils and subsoils using various techniques. *Soil Biology and Biochemistry* 34: 387–401.

Tripathi, S., V. K. Singh, and P. Srivastava, et al. 2020. Phytoremediation of organic pollutants: Current status and future directions. In *Abatement of Environmental Pollutants*, ed. P. Singh, A. Kumar, and A. Borthakur, 81–105. Amsterdam: Elsevier.

Varjani, S. J. and V. N. Upasani. 2017. A new look on factors affecting microbial degradation of petroleum hydrocarbon pollutants. *International Biodeterioration & Biodegradation* 120: 71–83.

Vater, J., B. Kablitz, and C. Wilde, et al. 2002. Matrix-assisted laser desorption ionization-time of flight mass spectrometry of lipopeptide biosurfactants in whole cells and culture filtrates of *Bacillus subtilis* C-1 isolated from petroleum sludge. *Applied and Environmental Microbiology* 68: 6210–6219.

Venosa, A. D., J. R. Haines, and D. M. Allen. 1992. Efficacy of commercial inocula in enhancing biodegradation of weathered crude oil contaminating a Prince William Sound beach. *Journal of Industrial Microbiology* 10:1–11.

Wallish, S., T. Gril, and X. Dong, et al. 2014. Effects of different compost amendments on the abundance and composition of *alk B* harboring bacterial communities in a soil under industrial use contaminated with hydrocarbons. *Frontiers in Microbiology* 5: 96.

Wang, X., T. Cai, and W. Wen, et al. 2020. Surfactin for enhanced removal of aromatic hydrocarbons during biodegradation of crude oil. *Fuel* 267: 117272.

Wu, M., J. Wu, X. Zhang, and X. Ye. 2019. Effect of bioaugmentation and biostimulation on hydrocarbon degradation and microbial community composition in petroleum-contaminated loessal soil. *Chemosphere* 237: 124456.

Xiong, B., Y. Zhang, and Y. Hou, et al. 2017. Enhanced biodegradation of PAHs in historically contaminated soil by *M. gilvum* inoculated biochar. *Chemosphere* 182: 316–324.

Xu, R., Z. Zhang, L. Wang, N. Yin, and X. Zhan. 2018. Surfactant-enhanced biodegradation of crude oil by mixed bacterial consortium in contaminated soil. *Environmental Science and Pollution Research* 25: 14437–14446.

Xu, Y. and M. Lu. 2010. Bioremediation of crude oil-contaminated soil: Comparison of different biostimulation and bioaugmentation treatments. *Journal of Hazardous Materials* 183: 395–401.

Yao, M., X. Kang, and Y. Zhao, et al. 2017. A mechanism study of airflow rate distribution within the zone of influence during air sparging remediation. *Science of the Total Environment* 609: 377–384.

Yavari, S., A. Malakahmad, and N. B. Sapari. 2015. A review on phytoremediation of crude oil spills. *Water, Air, & Soil Pollution* 226: 279.

Zaida, Z. N., and M. T. Piakong. 2018. Bioaugmentation of petroleum hydrocarbon in contaminated soil: A review. In *Microbial Action on Hydrocarbons*, ed. V. Kumar, M. Kumar, and R. Prasad, 415–439. Singapore: Springer.

Zenati, B., A. Chebbi, and A. Badis, et al. 2018. A non-toxic microbial surfactant from *Marinobacter hydrocarbonoclasticus* SdK644 for crude oil solubilization enhancement. *Ecotoxicology and Environmental Safety* 154: 100–107.

Zhang, G. X., X. F. Guo, and Y. E. Zhu, et al. 2018a. The effects of different biochars on microbial quantity, microbial community shift, enzyme activity, and biodegradation of polycyclic aromatic hydrocarbons in soil. *Geoderma* 328: 100–108.

Zhang, L., Y. Jing., Y. Xiang, R. Zhang, and H. Lu. 2018b. Responses of soil microbial community structure changes and activities to biochar addition: A meta-analysis. *Science of the Total Environment* 643: 926–935.

Zhang, X., H. Wang, and L. He, et al. 2013. Using biochar for remediation of soils contaminated with heavy metals and organic pollutants. *Environmental Science and Pollution Research* 20: 8472–8483.

Zhang, Y., C. Li, and X. Ji, et al. 2020. The knowledge domain and emerging trends in phytoremediation: A scientometric analysis with CiteSpace. *Environmental Science and Pollution Research* 2020: 7642–7646.

Zhao, B., H. Wang, X. Mao, and R. Li. 2009. Biodegradation of phenanthrene by a halophilic bacterial consortium under aerobic conditions. *Current Microbiology* 58: 205–210.

Zhu, X., B. Chen, L. Zhu, and B. Xing. 2017. Effects and mechanisms of biochar-microbe interactions in soil improvement and pollution remediation: A review. *Environmental Pollution* 227: 98–115.

45 Biotechnology Applications in Oil Recovery

Moon Sik Jeong
Kun Sang Lee

CONTENTS

45.1 INTRODUCTION

In the petroleum industry, the development of eco-friendly and economical technology is an important issue. Existing chemical processes show high efficiency in oil production, while causing environmental problems. In this respect, 'microbial enhanced oil recovery' (MEOR) can be an excellent alternative. This technology combines biotechnology with the oil recovery process and utilizes the various microbes and their microbial products (Belyaev et al., 1983; Lazar et al., 2007). This technology changes the physical properties of the reservoir and its fluids, as well as the chemical properties (Guo et al., 2015). Typically, porosity and the permeability interfacial tension between water and oil, and oil viscosity, are the main properties changed by the bioproducts. This technique is not a wholly new concept. In 1926, Beckman introduced the use of microorganisms to extract oil in a porous media (Lazar et al., 2007). Zobell (1947) used 'sulfate-reducing bacteria' (SRB) to improve oil recovery.

In the MEOR process, various microorganisms and their products are used to contribute to 'enhanced oil recovery' (EOR) (Mukherjee and Das, 2010). These microbial products can be classified into seven groups as biomass, biopolymers, biosurfactants, biogases, bioacids, biosolvents, and emulsifiers (Patel et al., 2015; Safdel et al., 2017). The biomass is adsorbed in the high permeability layers, which helps to increase the oil production in the unswept zone by changing the path of the water flow. The biopolymers, like biomass, adsorb on rock surfaces to reduce permeability, or dissolve in water to increase water viscosity to improve the mobility ratio. The biosurfactants reduce the surface and interfacial tension between reservoir fluids to increase the displacement efficiency and affect wettability alteration. The biogases, which are generated by certain microorganisms, can repressurize the reservoir or dissolve in oil to reduce the viscosity. Bioacids and biosolvents can increase the porosity and permeability by dissolving some part of the reservoir rock, which can help produce some of the trapped oil. Oil emulsification occurs when the microbial emulsifiers form a stable emulsion with hydrocarbons, which are produced in oil-in-water (Patel et al., 2015). Table 45.1 summarizes the microbial product groups and their effects on the oil recovery process (Safdel et al., 2017).

Two methods for applying the MEOR process, *in situ* and *ex situ*, have also been studied. In the *ex situ* method, the externally generated target bioproducts are injected into the reservoir to increase oil recovery. Such a method allows direct control in that the composition and combination of the injected bioproducts can be selected. The microorganisms used in the *ex situ* process can be either grown or engineered in the laboratories to increase sweep and/or displacement efficiencies. The microbial products such as biopolymers and biosurfactants are mixed into the injected water, or sometimes combined with synthetic chemicals (Patel et al., 2015).

There have been many concerns, though the *ex situ* method seems viable. First, it is very expensive to manufacture the *ex situ* bioproducts. Though the use of crude bioproducts may greatly reduce this concern, it does not completely solve this problem, so the *ex situ* method can be actively used in the petroleum industry (Pornsunthorntawee et al., 2008b; Zheng et al., 2012). Furthermore, the injected bacteria are not competitive compared to the indigenous bacteria adapted to the reservoir environment (Patel et al., 2015).

The *in situ* method stimulates the indigenous bacteria to generate the desired bioproducts. Proper nutrient injection stimulates the indigenous microbes to produce biopolymers, biosurfactants, bioacids, and biosolvents. While the *ex situ* method can somewhat predict the results, the *in situ* method has a high uncertainty in the results depending on the field applications. Nevertheless, the *in situ* method is more attractive in the petroleum industry because of its high potential (Bao et al., 2009; Gudina et al., 2012; Mukherjee and Das, 2010; Patel et al., 2015; Youssef et al., 2013).

This chapter deals with the various mechanisms used for oil production in the MEOR process. The main strategies for the application of MEOR are determined by the main substances generated from microbial activity. In particular, selective plugging by biomass/biopolymers and wettability alteration by biosurfactants are described as the main mechanisms for oil recovery. The effects of biosolvents, bioacids, and biogases are also briefly presented. Additionally, the degradation of hydrocarbons by microorganisms and the effect of improving oil recovery are discussed. This review aims to increase the understanding of how microbial technology can be applied to oil recovery.

TABLE 45.1
Major Microbial Products, their Effects on Oil Recovery, and Applicable Type of Reservoir

Microbial Products					
Product Class	**Sample Products**	**Microorganisms**	**Recovery Problems**	**Effect**	**Type of Formation/ Reservoir**
Biomass	Microbial cells and EPS (mainly exopolysaccharides)	*Bacillus, Licheniformis, Leuconostoc mesenteroides Xanthomonas Campestris*	Poor microscopic displacement efficiency	Permeability reduction in thief zone; Selective and nonselective plugging; Emulsification; Alteration of mineral surfaces to water-wet; Oil viscosity reduction; Oil pour point desulfurization reduction; Oil degradation	Stratified reservoirs with different permeability
Biopolymers	Xanthan gum, Pullulan, Levan, Curdlan, Dextran, Scleroglucan	*Xanthomonas* sp., *Aureobasidium* sp., *Bacillus* sp., *Alcaligeness* sp., *Leuconostoc* sp., *Sclerotium* sp., *Brevibacterium*	Poor volumetric sweep efficiency	Permeability reduction in water-swept regions, Modification of injectivity profile and viscosity, Control of mobility ratio, Selective or nonselective plugging	Stratified reservoirs with different permeability
Biosurfactants	Emulsan, Alasan, Surfactin, Rhamnolipid Lichenysin, Glycolipids, Viscosin, Trehaloselipids	*Bacillus* sp., *Pseudomonads, Rhodococcus* sp., *Acinetobacter, Arthrobacter*	Poor microscopic displacement efficiency	IFT reduction, emulsification, Improvement of pore scale displacement, Alteration of mineral surfaces to water-wet	Sandstone or carbonate reservoirs with moderate temperatures (<50°C) and relatively light oil (API>25)
Gases	CO_2, CH_4, H_2, N_2	Fermentative bacteria, Methanogens, *Clostridium, Enterobacter*	Heavy oil	Reservoir repressurization, Oil swelling, Permeability increase, IFT and viscosity reduction, Flow characteristics improvement	Heavy oil-bearing formations (API<15)
Acids	Propionic acid, Butyric acid	Fermentative bacteria, *Clostridium, Enterobacter,* Mixed acidogens	Low porosity, Poor drainage, Formation damage	Dissolution of carbonaceous minerals or deposits, Permeability and porosity increase, Emulsification, CO_2 production from reaction between acids and carbonate minerals, Oil viscosity reduction	Carbonate reservoirs

(Continued)

TABLE 45.1
(Continued)

Microbial Products					
Product Class	**Sample Products**	**Microorganisms**	**Recovery Problems**	**Effect**	**Type of Formation/ Reservoir**
Solvents	Alcohols and ketones that are typical cosurfactants, Acetone, Butanol, Propan-2-diol	Fermentative bacteria, *Clostridium, Zymomonas, Klebsiella*	Heavy oil, Poor microscopic displacement efficiency	Oil viscosity reduction, Alteration of mineral surfaces to water-wet, Permeability increase, Heavy, long chain hydrocarbons removal from pore throats, Emulsification, IFT reduction	Heavy oil-bearing formations (API<15), strongly oil-wet, waterflooded reservoirs
Emulsifiers	Some kinds	*Acinetobacter* sp., *Candida, Pseudomonas* sp., *Bacillus* sp.	Paraffin and oil sludge deposition, Poor microscopic displacement efficiency	Oil emulsification	Waxy oil (>C_{22} alkanes); paraffinic oil, and asphaltene-bearing formations
Hydrocarbon metabolism	Some kinds	Aerobic hydrocarbon degraders	• Paraffin deposition, Poor microscopic displacement efficiency	Paraffin deposits removal, Oil mobility improvement	Paraffin deposited wells, mature waterflooded reservoirs

45.2 SELECTIVE PLUGGING

One of the major causes of adverse effects on oil recovery is the presence of high permeability zones in the reservoir, known as the "thief zone". This thief zone is the major cause of the bypassed flow of injected water, making it difficult to recover oil in low permeability zones. The thief zone causes a channeling effect associated with early water breakthrough. Selective plugging during MEOR occurs in this high permeability zone, directing the flow path of the injected water into the low permeability zone and allowing additional oil to be recovered. This phenomenon is mainly caused by biomass and biopolymer.

As microorganisms grow in the reservoir, the volume they occupy in the pores increases and surface molecules tend to attach to the injected nutrients. In this way, they form a biofilm in the porous media, which in turn lowers the flow of fluids (Karimi et al., 2012). The formation of these colonies and biomass groups has advantages in terms of their survival and evolution (Xavier and Foster, 2007). This selective plugging is usually performed by injecting selected bacteria or stimulating indigenous bacteria. As the injected water increases the area of contact with the oil, the sweep efficiency also increases, which positively affects oil recovery. The formation of such biomass can help oil recovery in another way as well as permeability reduction. In addition to diverting the injected water flow (Satyanarayana et al., 2012), biomass also changes the wettability of the rock surface, creating a favorable condition for oil recovery (Karimi et al., 2012).

Jenneman (1989) proposed the following four criteria for the successful implementation of microbial plugging: the microorganisms must be sized to pass through the porous media; suitable nutrients should be injected for microbial growth; the microorganisms must generate the bioproducts necessary for selective plugging; and the growth rate of microorganisms should not be so fast as to clog the wells.

An experiment injecting *Bacillus licheniformis* BNP29 into the low permeability core confirmed that the bacteria met all of the above criteria (Yakimov et al., 1997). These species show a higher selective plugging efficiency than SRB, which releases dangerous substances during the MEOR process. Therefore, rather than stimulating all the microorganisms present in the reservoir, it is very important to selectively stimulate a specific species which shows high efficiency for microbial plugging (Patel et al., 2015).

Biopolymers generated by microorganisms are another substance that causes selective plugging. The microorganisms produce surface molecules that form biopolymers during their growth. Many biopolymers are exopolysaccharides, which increase cell adhesion and protect cells from the outside environment (Poli et al., 2011).

Some biopolymers such as xanthan gum act as an agent to increase the water viscosity in the MEOR process (Sandvik and Maerker, 1977; Yakimov et al., 1997; Mukherjee and Das, 2010; Patel et al., 2015). To apply these biopolymers to the MEOR process, they are generated *ex situ* and injected into the reservoir. Alternatively, biopolymers can be produced by stimulating certain injected microorganisms or indigenous microorganisms.

Microbial species which produce biopolymers that can be used to enhance oil recovery include *Xanthomonous*, *Aureobasidium*, and *Bacillus* (Mukherjee and Das, 2010). Among the produced biopolymers, xanthan gum and curdlan are very useful in various industries. Xanthan gum especially is most widely used in food, cosmetics, chemicals, and many other industries (Palaniraj and Jayaraman, 2011). It is very resistant to high temperatures and salinity, and can therefore be utilized in drilling operations. As a result, researchers have tried to find mutant strains that can produce large amounts of the polymer, or strains that can grow with low cost nutrients (Kurbanoglu and Kurbanoglu, 2007; Rottava et al., 2009; Palaniraj and Jayaraman, 2011). Xanthan gum is also used to increase water viscosity, helping to achieve higher oil recovery than conventional waterflooding. Generally, the polymer is used by dissolving directly into injected water. However, the polymer produced by stimulating the indigenous bacteria is also used.

Curdlan is applicable to the MEOR process by reducing the permeability of the porous media. In an experiment, several biopolymers were injected with acid-producing bacteria into Berea sandstone cores. The mixtures decreased the permeability from 850 to 3 md and from 904 to 5 md, respectively. In addition, the residual resistance factors were reduced from 334 to 186 (Fink, 2011).

The common application of biopolymers in the MEOR process is to mix them directly into the injected water rather than to stimulate the injected or indigenous bacteria to generate them (Fox et al., 2003). The biopolymers and other microbial products can also cause selective plugging. Indeed, microbial products, such as emulsions, have a plugging effect due to their high viscosity (Zheng et al., 2012).

45.3 WETTABILITY ALTERATION

The wettability of the reservoir is an important factor in determining oil recovery. Large amounts of oil are deposited in carbonate reservoirs, which are usually mixed or oil-wet conditions. Under these environments, the displacement efficiency is low because it is difficult to extract oil from the reservoir rock (Mukherjee and Das, 2010). In order to increase oil productivity, it is necessary to increase the water wetness in the rock (Armstron and Wildenschild, 2010; Lazar et al., 2007; Mukherjee and Das, 2010). For this purpose, a method of injecting a surfactant or adsorbing microorganism on the rock surface may be used. This technique improves oil production by increasing the displacement efficiency. An experiment using the *Enterobacter cloacae* strain showed the wettability alteration of the glass surface. In this study, several microbial products, including a biosurfactant, influenced the wettability change, and biofilm formation had the greatest effect (Karimi et al., 2012).

In general, however, biosurfactants have the greatest effect on the change in wettability among the bioproducts. According to one laboratory study (Salehi et al., 2008), when the ion charge and polarity compounds were added to the biosurfactant, its wettability-altering ability was improved. In addition, it was confirmed that surfactin, a biosurfactant, was more efficient than a synthetic surfactant. In other words, the surface altering materials generated by the microorganisms can change the reservoir rock to be hydrophilic, which is associated with an increase in oil recovery.

45.4 SURFACE TENSION ALTERATION

Biosurfactants have received the greatest attention among the products used in the MEOR process. Because these amphipathic molecules are biodegradable, pH-hardy, and temperature tolerant, they can replace conventional surfactants used in oil recovery. Like conventional surfactants, biosurfactants also reduce surface and interfacial tensions to improve recovery (Gudina et al., 2012; Mukherjee and Das, 2010; Pornsunthorntawee et al., 2008a). If crude biosurfactants are used instead of expensive biosurfactants made for medical purposes, the problem of biosurfactant cost can be solved (Mukherjee and Das, 2010; Pornsunthorntawee et al., 2008b; Zheng et al., 2012). Biosurfactants produced by microorganisms not only change the surface tension, but also adsorb on the immiscible interfaces to emulsify the oil and increase the microbial mobility (Lazar et al., 2007; Mukherjee and Das, 2010; Yan et al., 2012). Indeed, biosurfactants are known to aid the migration of microbial colonies to other environmental areas (Deziel et al., 2003). In addition, biosurfactants are known to exhibit similar levels of effect at lower concentrations than conventional chemical surfactants (Gudina et al., 2012; Mukherjee and Das, 2010).

To apply the biosurfactants to MEOR, they must first be screened, developed, and economically scaled up. The optimal biosurfactant must meet the following conditions: strong surfactant capacity, low 'critical micelle concentration' (CMC), high resistance to temperature and pH, high solubility, and high emulsification ability (Walter et al., 2010). This microbial material can be produced by an *ex situ* process, or by stimulating the indigenous or injected microorganisms. In some cases, the addition of metal ions to the biosurfactant may result in an increase in its ability by way of the polar interactions between them (Thimon et al., 1992). The MEOR process using the biosurfactants showed more than 95% oil recovery in a sandpack experiment (Banat, 1993).

In addition to interfacial tension reduction in oil production, other studies on biosurfactant effects have also been conducted (Zheng et al., 2012). Such effects change wettability during the oil recovery process, and they purify contaminated soils because they help degrade long alkyl chins (Gudina et al., 2012; Mukherjee and Das, 2010; Pornsunthorntawee et al., 2008a; Yan et al., 2012). This suggests that several aspects of microbial surfactants should be considered simultaneously in developing a MEOR application strategy. Because these different factors interact, the cascade effects, which occur in the field application of biosurfactants, must be considered.

Through experimental studies, the effectiveness and hardness of biosurfactants are well known. Gudina et al. (2012) performed an experiment using biosurfactants produced from *Bacillus* strains. When the interfacial tension of the water was reduced from 72 to 30 mN/m, the temperature condition was 40°C when the biosurfactant was used with bacteria and 121°C when the biosurfactant was used alone. These results indicated that, considering the resistance to temperature conditions, the use of a biosurfactant alone mixed in the injected water could increase the potential for success. The biosurfactants may also perform better than chemically manufacturing surfactants (Pornsunthornatawee et al., 2008a).

Depending on the bacteria, the time to generate the microbial surfactants and the performance of produced surfactant are different. Therefore, understanding the individual characteristics and selecting the appropriate type of bacteria and surfactants are critical to MEOR success. There are various types of biosurfactants, which can be classified as glycolipids, fatty acid biosurfactants, lipopeptides, emulsifying proteins, and particulate biosurfactants (Mukherjee and Das, 2010). Among them, rhamnolipids belonging to the glycolipid type and lipopeptides are most frequently used in the MEOR process. They can lower the interfacial tension between water and oil phases to less than 0.1 mN/m (Youssef et al., 2013).

45.5 BIOACIDS, BIOSOLVENTS, AND BIOGASES

Microbes can generate a variety of acids, solvents, and gases that can be used in the MEOR process. The extracellular acids generated by the microorganisms can dissolve some part of the rock in a carbonate reservoir and improve oil recovery (Lazar et al., 2007; Mukherjee and Das, 2010). Furthermore, microbial acids can dissolve the rocks and help recover the trapped oil. Acetic and propionic acids are most frequently used for this purpose (Mukherjee and Das, 2010). *Clostridium* species are known to produce acetate and butyrate, and biosurfactant-generating *Bacillus* species can produce bioacids (Harner et al., 2011).

Biosolvents such as ethanol and acetone also have the ability to dissolve carbonate rocks (Bryant, 1987). Like bioacids, biosolvents also change the porosity and permeability of the reservoir, enabling trapped oil recovery. The biosolvent-producing microorganisms are *Zymomonas mobilis*, *Clostridium acetobutylicum*, and *Clostridium pasteurianum*. These biosolvents are mainly generated by stimulating the injected or indigenous bacteria (Bryant, 1990). The extensive production of biosolvents such as ethanol utilizes a new technique (Pinilla et al., 2011), but the amount is not enough to be injected directly into the reservoir.

Applying additional pressure to the reservoir may be one of the ways to produce additional oil with a mechanism similar to primary recovery. To induce this phenomenon, the gas generating microorganisms in the reservoir can be stimulated. Microorganisms ferment carbohydrates and generate gases such as methane, carbon dioxide, and hydrogen (Mukherjee and Das, 2010). Bioacids and biosolvents can also be produced by the same fermentation process. The microbial gases not only repressurize, but also dissolve in the oil to reduce its viscosity. This contributes to produce additional oil (Lazar et al., 2007; Mukherjee and Das, 2010).

Field applications of microbial acids, solvents, and gases are rarely found in the literature. This is because the MEOR efficiency of these three materials is lower than selective plugging or biosurfactant processes. Therefore, it is more effective to use them as an auxiliary method to increase the efficiency of other MEOR processes rather than using them alone. Other processes still remain more effective, controllable, and efficient (Patel et al., 2015).

45.6 BIODEGRADATION AND CLEAN-UP

Many indigenous bacteria in reservoirs consume hydrocarbons as nutrients. In this process, they degrade the long alkyl chains of heavy oils, resulting in an increase of light oil potions in the reservoir (Gudina et al., 2012; Mukherjee and Das, 2010; Patel et al., 2015). These light oils have an advantage in production because of their low viscosity. The ability of microbes to decompose oil is one of the most attractive and practical methods in the MEOR process. In order to carry out this process, the indigenous bacteria are stimulated, or degrading bacteria are injected. These microorganisms tend to migrate to the rock surface and alter the wettability (Kowalewski et al., 2006). In an experimental study, certain strains of *Bacillus subtilis* were found to degrade the long alkyl chains at 40°C, resulting in reduced oil viscosity and increased productivity (Gudina et al., 2012). The use of microorganisms with the ability to decompose oil and produce biosurfactants guarantees great success for MEOR (Gudina et al., 2012).

These microorganisms are attractive in that they are used not only in oil production but also in other areas. Oily sludge, which occurs after the drilling operation, has the risk of contaminating soil, air, and groundwater (Yan et al., 2012). These bacteria break down the oily substances in contaminated areas and alleviate environmental problems. Among various ways to remove these toxic substances, the use of suitable microorganisms is the most environmentally friendly method (Yan et al., 2012). Using the rhamnolipid is one successful bioremediation technique.

In another study, oil-contaminated soils were cleaned up using biosurfactants with the characteristics of low 'carboxymethyl cellulose' (CMC), high miscibility, and low molecular weights (Urum and Pekdemir, 2004). These microorganisms can also be utilized to maintain the availability of industrial equipment. These bacteria can be used to keep tanks and storages clean and can be used to remove plugging in oil wells (Yakimov et al., 1997; Yan et al., 2012; Zheng et al., 2012). One study has successfully used microbial consortiums designed to degrade precipitated paraffins (Lazar et al., 1999). Due to its environmentally friendly characteristics, microorganisms can be an alternative to the use of harsh chemicals in the equipment maintenance area.

45.7 CONCLUSIONS

In response to unstable oil prices and tightening environmental regulations, the need for new technologies in the petroleum industry continues to increase. MEOR is a suitable alternative with low operating cost and environmental benefits. The microorganisms used in MEOR can grow with low cost nutrients; the microbial products are biodegradable and nontoxic. Microbial products such as biopolymers, biosurfactants, bioacids, biosolvents, and biogases can help with additional oil recovery. In particular, studies on biopolymers and biosurfactants are quite advanced, and even pilot scale applications have been carried out. The hydrocarbon degrading bacteria can also be used for environmental engineering. However, the understanding of microbial activities in reservoir conditions is still insufficient for practical applications. In addition, the uncertainty of the MEOR process still remains a major problem in field application. More extensive research is needed to reduce the uncertainty of MEOR and improve its performance.

REFERENCES

Armstrong, R. T. and D. Wildenschild. 2010. *Designer-wet micromodels for studying potential changes in wettability during microbial enhanced oil recovery*. In *Proceedings of the American Geophysical Union Fall Meeting*, San Francisco, California, USA, 13–17 December, 2010.

Banat, I. M. 1993. The isolation of a thermophilic biosurfactant producing *Bacillus* sp. *Biotechnology Letters* 15: 591–594.

Bao, M., X. Kong, G. Jiang, X. Wang, and X. Li. 2009. Laboratory study on activating indigenous microorganisms to enhance oil recovery in Shengli oilfield. *Journal of Petroleum Science and Engineering* 66: 42–46.

Belyaev, S. S., R. Wolkin, and W. R. Kenealy, et al. 1983. Methanogenic bacteria from the Bondyuzhskoe oil field: General characterization and analysis of stable-carbon isotopic fractionation. *Applied and Environmental Microbiology* 45: 691–697.

Bryant, R. S.. 1987. Potential uses of microorganisms in petroleum recovery technology. *Proceedings of the Oklahoma Academy of Science* 67: 97–104.

Bryant, R. S., 1990. *Microbial Enhanced Oil Recovery and Compositions therefor*. Patent no: US 4905761. Chicago, IL: Illinois Institute of Technology.

Deziel, E., F. Lepine, S. Milot, and R. Villemur. 2003. *rhl A* is required for the production of a novel biosurfactant promoting swarming motility in *Pseudomonas aeruginosa*: 3-(3-hydroxyalkanoyloxy) alkanoic acids (HAAs), the precursors of rhamnolipids. *Microbiology* 149: 2005–2013.

Fink, J. 2011. *Petroleum Engineer's Guide to Oil Field Chemicals and Fluids*. Houston, TX: Gulf Professional Publishing.

Fox, S. L., X. Xie, K. D. Schaller, E. P. Robertson, and G. A. Bala. 2003. *Permeability Modification using a Reactive Alkaline Soluble Biopolymer*. Idaho Falls, ID: Idaho National Laboratory.

Gudina, E. J., J. F. B. Pereira, L. R. Rodrigues, J. A. P. Coutinho, and J. A. Teixeira. 2012. Isolation and study of microorganisms from oil samples for application in microbial enhanced oil recovery. *International Biodeterioration & Biodegradation* 68: 56–64.

Guo, H., Y. Li, and Z. Yiran, et al. 2015. *Progress of microbial enhanced oil recovery in China*. In: *Proceedings of the SPE Asia Pacific Enhanced Oil Recovery Conference*, Kuala Lumpur, Malaysia, 11–13 August, 2015.

Harner, N. K., T. L. Richardson, and K. A. Thompson, et al. 2011. Microbial processes in the Athabasca Oil Sands and their potential applications in microbial enhanced oil recovery. *Journal of Industrial Microbiology & Biotechnology* 38: 1761.

Jenneman, G. E.. 1989. The potential for *in-situ* microbial applications. *Developments in Petroleum Science* 22: 37–74.

Karimi, M., M. Mahmoodi, A. Niazi, Y. Al-Wahaibi, and S. Ayatollahi. 2012. Investigating wettability alteration during MEOR process, a micro/macro scale analysis. *Colloids and Surfaces B: Biointerfaces* 95: 129–136.

Kowalewski, E., I. Rueslatten, K. H. Steen, G. Bodtker, and O. Torsaeter. 2006. Microbial improved oil recovery-bacterial induced wettability and interfacial tension effects on oil production. *Journal of Petroleum Science and Engineering* 52: 275–286.

Kurbanoglu, E. B. and N. I. Kurbanoglu. 2007. Ram horn hydrolysate as enhancer of xanthan production in batch culture of *Xanthomonas campestris* EBK-4 isolate. *Process Biochemistry* 42: 1146–1149.

Lazar, I., I. G. Petrisor, and T. F. Yen. 2007. Microbial enhanced oil recovery (MEOR). *Petroleum Science and Technology* 25: 1353–1366.

Lazar, I., A. Voicu, and C. Nicolescu, et al. 1999. The use of naturally occurring selectively isolated bacteria for inhibiting paraffin deposition. *Journal of Petroleum Science and Engineering* 22: 161–169.

Mukherjee, A. K. and K. Das. 2010. Microbial surfactants and their potential applications: An overview. In *Biosurfactants*, ed. R. Sen, 56–64. Berlin: Springer.

Palaniraj, A. and V. Jayaraman. 2011. Production, recovery and application of xanthan gum by *Xanthomonas campestris*. *Journal of Food Engineering* 106: 1–12.

Patel, J., S. Borgohain, and M. Kumar, et al. 2015. Recent developments in microbial enhanced oil recovery. *Renewable and Sustainable Energy Reviews* 52: 1539–1558.

Pinilla, L., R. Torres, and C. Ortiz. 2011. Bioethanol production in batch mode by a native strain of *Zymomonas mobilis*. *World Journal of Microbiology and Biotechnology* 27: 2521–2528.

Poli, A., P. Di Donato, G. R. Abbamondi, and B. Nicolaus. 2011. Synthesis, production, and biotechnological applications of exopolysaccharides and polyhydroxyalkanoates by archaea. *Archaea* 2011: 693253.

Pornsunthorntawee, O., N. Arttaweeporn, and S. Paisanjit, et al. 2008a. Isolation and comparison of biosurfactants produced by *Bacillus subtilis* PT2 and *Pseudomonas aeruginosa* SP4 for microbial surfactant-enhanced oil recovery. *Biochemical Engineering Journal* 42: 172–179.

Pornsunthorntawee, O., P. Wongpanit, S. Chavadej, M. Abe, and R. Rujiravanit. 2008b. Structural and physicochemical characterization of crude biosurfactant produced by *Pseudomonas aeruginosa* SP4 isolated from petroleum-contaminated soil. *Bioresource Technology* 99: 1589–1595.

Rottava, I., G. Batesini, and M. F. Silva, et al. 2009. Xanthan gum production and rheological behavior using different strains of *Xanthomonas* sp. *Carbohydrate Polymers* 77: 65–71.

Safdel, M., M. A. Anbaz, A. Daryasafar, and M. Jamialahmadi. 2017. Microbial enhanced oil recovery, a critical review on worldwide implemented field trials in different countries. *Renewable and Sustainable Energy Reviews* 74: 159–172.

Salehi, M., S. J. Johnson, and J. T. Liang. 2008. Mechanistic study of wettability alteration using surfactants with applications in naturally fractured reservoirs. *Langmuir* 24: 14099–14107.

Sandvik, E. I. and J. M. Maerker. 1977. Application of xanthan gum for enhanced oil recovery. *ACS Symposium Series* 45: 242–264.

Satyanarayana, T., B. N. Johri, and A. Prakash. 2012. *Microorganisms in Sustainable Agriculture and Biotechnology*. Berlin: Springer.

Thimon, L., F. Peypoux, and G. Michel. 1992. Interactions of surfactin, a biosurfactant from *Bacillus subtilis*, with inorganic cations. *Biotechnology Letters* 14: 713–718.

Urum, K. and T. Pekdemir. 2004. Evaluation of biosurfactants for crude oil contaminated soil washing. *Chemosphere* 57: 1139–1150.

Walter, V., C. Syldatk, and R. Hausmann. 2010. Screening concepts for the isolation of biosurfactant producing microorganisms. In *Biosurfactants*, ed. R. Sen, 1–13. Berlin: Springer.

Xavier, J. B. and K. R. Foster. 2007. Cooperation and conflict in microbial biofilms. *Proceedings of the National Academy of Sciences of the United States* 104: 876–881.

Yakimov, M. M., M. M. Amro, and M. Bock, et al. 1997. The potential of *Bacillus licheniformis* strains for in situ enhanced oil recovery. *Journal of Petroleum Science and Engineering* 18: 147–160.

Yan, P., M. Lu, and Q. Yang, et al. 2012. Oil recovery from refinery oily sludge using a rhamnolipid biosurfactant-producing *Pseudomonas*. *Bioresource Technology* 116: 24–28.

Youssef, N., D. R. Simpson, M. J. McInerney, and K. E. Duncan. 2013. *In-situ* lipopeptide biosurfactant production by *Bacillus* strains correlates with improved oil recovery in two oil wells approaching their economic limit of production. *International Biodeterioration & Biodegradation* 81: 127–132.

Zheng, C., L. Yu, L. Huang, J. Xiu, and Z. Huang. 2012. Investigation of a hydrocarbon-degrading strain, *Rhodococcus ruber* Z25, for the potential of microbial enhanced oil recovery. *Journal of Petroleum Science and Engineering* 81: 49–56.

Zobell, C. E.. 1947. Microbial transformation of molecular hydrogen in marine sediments, with particular reference to petroleum. *American Association of Petroleum Geologists Bulletin* 31: 1709–1751.

Part X

Petrodiesel Fuels in General

46 Petrodiesel Fuels: A Scientometric Review of the Research

Ozcan Konur

CONTENTS

46.1 INTRODUCTION

Crude oils have been primary sources of energy and fuels, such as petrodiesel. However, significant public concerns about the sustainability, price fluctuations, and adverse environmental impact of crude oils have emerged since the 1970s (Ahmadun et al., 2009; Atlas, 1981; Babich and Moulijn, 2003; Kilian, 2009; Perron, 1989). Thus, biooils (Bridgwater and Peacocke, 2000; Czernik and Bridgwater, 2004; Gallezot, 2012; Mohan et al., 2006) and biooil-based biodiesel fuels (Chisti, 2007; Hill et al., 2006; Lapuerta et al., 2008; Mata et al., 2010; Zhang et al., 2003) have emerged as alternatives to crude oils and crude oil-based petrodiesel fuels in recent decades. Despite these developments, petrodiesel fuels (Birch and Cary, 1996; Khalili et al., 1995; Koebel et al., 2000; Rogge et al., 1993; Salvi et al., 1999; Schauer et al., 1999; Song, 2003; Song and Ma, 2003; Stanislaus et al., 2010; Zhu et al., 2002) have still been a major source of energy and fuels, besides biodiesel fuels (Konur, 2021a–ag).

However, for the efficient progression of the research in this field, it is necessary to develop efficient incentive structures for the primary stakeholders and to inform these stakeholders about the research (Konur, 2000, 2002a–c, 2006a–b, 2007a–b; North, 1991a–b).

Scientometric analysis offers ways to evaluate the research in a respective field (Garfield, 1955, 1972; Konur, 2011, 2012a–n, 2015, 2016a–f, 2017a–f, 2018a–b, 2019a–b). However, there has been no current scientometric study of the research on petrodiesel fuels.

This chapter presents a study on the scientometric evaluation of the research on petrodiesel fuels using two datasets. The first dataset includes the 100-most-cited papers ($n = 100$ sample papers) whilst the second set includes population papers ($n =$ over 16,000 population papers) published between 1980 and 2019.

The data on the indices, document types, authors, institutions, funding bodies, source titles, 'Web of Science' subject categories, keywords, research fronts, and citation impact are presented and discussed.

46.2 MATERIALS AND METHODOLOGY

The search for the literature was carried out in the 'Web of Science' (WOS) database in January 2020. It contains the 'Science Citation Index-Expanded' (SCI-E), the 'Social Sciences Citation Index' (SSCI), the 'Book Citation Index-Science' (BCI-S), the 'Conference Proceedings Citation Index-Science' (CPCI-S), the 'Emerging Sources Citation Index' (ESCI), the 'Book Citation Index-Social Sciences and Humanities' (BCI-SSH), the 'Conference Proceedings Citation Index-Social Sciences and Humanities' (CPCI-SSH), and the 'Arts and Humanities Citation Index' (A&HCI).

The keywords for the search of the literature are collated from the screening of abstract pages for the first 1,000 highly cited papers. This keyword set is provided in the Appendix.

Two datasets are used for this study. The highly cited 100 papers comprise the first dataset (sample dataset, $n = 100$ papers) whilst all the papers form the second dataset (population data set, over 16,000 papers).

The data on the indices, document types, publication years, institutions, funding bodies, source titles, countries, 'Web of Science' subject categories, citation impact, keywords, and research fronts are collated from these datasets. The key findings are provided in the relevant tables and figure, supplemented with explanatory notes in the text. The findings are discussed and a number of conclusions are drawn and a number of recommendations for further study are made.

46.3 RESULTS

46.3.1 Indices and Documents

There are over 23,000 papers in this field in the 'Web of Science' as of January 2020. This original population dataset is refined for the document type (article, review, book chapter, book, editorial material, note, and letter) and language (English), resulting in over 16,000 papers comprising over 70.2% of the original population dataset.

The primary index is the SCI-E for both the sample and population papers; 94.6% of the population papers are indexed by the SCI-E database. Additionally 6.1, 3.8, and 0.9% of these papers are indexed by the CPCI-S, ESCI, and BCI-S databases, respectively. The papers on the social and humanitarian aspects of this field are relatively negligible with 1.8 and 0.1% of the population papers indexed by the SSCI and A&HCI, respectively.

Brief information on the document types for both datasets is provided in Table 46.1. The key finding is that article types of documents are the primary documents for both sample and population papers, whilst reviews form 16% of the sample papers.

46.3.2 Authors

Brief information on the most-prolific 21 authors with at least three sample papers each is provided in Table 46.2. Around 340 and 30,400 authors contribute to the sample and population papers, respectively.

The most-prolific authors are 'Anders Blomberg', 'David Diaz-Sanchez', and 'Thomas Sandstrom' with five sample papers each, working primarily on 'respiratory illnesses' and 'vascular illnesses' caused by diesel exhaust particles. The other prolific authors are 'David B. Kittelson', 'Ning Li', 'Andrew E. Nel', and 'Andrew Saxon' with five sample papers each, mostly working on 'respiratory illnesses' caused by diesel exhaust particles.

TABLE 46.1
Document Types

	Document Type	Sample Dataset (%)	Population Dataset (%)	Difference (%)
1	Article	82	95.0	−13.0
2	Review	16	1.7	14.3
3	Book chapter	0	0.9	−0.9
4	Proceeding paper	7	6.1	0.9
5	Editorial material	0	1.8	−1.8
6	Letter	0	1.0	−1.0
7	Book	0	0.1	−0.1
8	Note	0	0.5	−0.5

TABLE 46.2
Authors

	Author	Sample Papers (%)	Population Papers (%)	Surplus (%)	Institution	Country	Research Front
1	Blomberg, Anders	6	0.2	5.8	Umea Univ.	Sweden	Vascular, respiratory illnesses
2	Diaz-Sanchez, David	6	0.2	5.8	Univ. Calif. L. A.	USA	Respiratory illnesses
3	Sandstrom, Thomas	6	0.3	5.7	Umea Univ.	Sweden	Vascular, respiratory illnesses
4	Kittelson, David B.	5	0.2	4.8	Univ. Minnesota	USA	Exhaust particles
5	Li, Ning	5	0.1	4.9	Univ. Calif. L. A.	USA	Respiratory illnesses
6	Nel, Andrew E.*	5	0.1	4.9	Univ. Calif. L. A.	USA	Respiratory illnesses
7	Saxon, Andrew	5	0.1	4.9	Univ. Calif. L. A.	USA	Respiratory illnesses
8	Ma, Xiaoliang	4	0.1	3.9	Pennsylvania State Univ.	USA	Desulfurization
9	Makkee, Michiel	4	0.2	3.8	Delft Univ. Technol.	Netherlands	Exhaust particles
10	McMurry Peter H.	4	0.1	3.9	Univ. Minnesota	USA	Exhaust particles
11	Moulijn, Jacob A.	4	0.2	3.8	Delft Univ. Technol.	Netherlands	Exhaust particles
12	Bernal-Agustin, Jose L.	3	0.1	2.9	Univ. Zarazoga	Spain	Power generation
13	Donaldson, Ken	3	0.1	2.9	Univ. Edinburgh	UK	Vascular, respiratory illnesses
14	Dufo-Lopez, Rodolfo	3	0.1	2.9	Univ. Zarazoga	Spain	Power generation
15	Mills, Nicholas L.	3	0.1	2.9	Univ. Edinburgh	UK	Vascular, respiratory illnesses
16	Newby, David E.*	3	0.1	2.9	Univ. Edinburgh	UK	Vascular, respiratory illnesses
17	Park, Kihong	3	0.1	2.9	Univ. Minnesota	USA	Exhaust particles
18	Robinson, Simon D.	3	0.1	2.9	Victoria Heart Inst.	Australia	Vascular, respiratory illnesses
19	Song, Chunsuan	3	0.1	2.9	Pennsylvania State Univ.	USA	Desulfurization
20	Tornqvist, Hakan	3	0.1	2.9	Umea Univ.	Sweden	Vascular, respiratory illnesses
21	Wang, Meiying F.	3	0.1	2.9	Univ. Calif. L. A.	USA	respiratory illnesses

*'Highly Cited Researchers' in 2019 (Clarivate Analytics, 2019).

On the other hand, a number of authors have a significant presence in the population papers: 'Constantine D. Rakopoulos', 'Rolf D. Reitz', 'Raul Payri', 'Masaru Sagai', 'Jingping Liu', 'Hirohisa Takano', 'Zuohua Huang', 'Zhen Zhao', 'F. Javier Salvador', 'Debora Fino', 'Chang Sik Lee', 'Suhan Park', 'Jian Zhang', 'Jesus Benajes', 'Evangelos G. Giakoumis', 'Antonio Garcia', 'Dimitrios T. Hountalas', 'Chun Shun Cheng', 'Chunde Yao', 'Haixia Zhao', 'Ye Wang', 'Jose M. Desantes', 'Zuohua Huang', 'Myoungho Sunwoo', 'Mingfa Yao', 'Takamichi Ichinose', 'Kyoungdoug Min', 'Aijun Duan', 'Samad Jafarmadar', 'Magin Lapuerta', 'Fushui Liu', 'Jaime Gimeno', 'Vito Specchia', and 'Octavio Armas' have at least 0.22% of the population papers each.

The most-prolific institution for these top authors is the 'University of California Los Angeles' of the USA with five sample papers. The other prolific institutions with three authors are 'Umea University' of Sweden, the 'University of Edinburgh' of the UK, and the 'University of Minnesota' of the USA. Thus, in total, eight institutions house these top authors.

It is notable that two of these top researchers are listed in the 'Highly Cited Researchers' (HCR) in 2019 (Clarivate Analytics, 2019; Docampo and Cram, 2019).

The most-prolific country for these top authors is the USA with ten. The other prolific countries with three authors are the UK and Sweden and those with two authors are the Netherlands and Spain. Thus, in total, six countries contribute to these top papers.

There are three key topical research fronts for these top researchers: 'respiratory illnesses' and 'vascular illnesses' caused by diesel exhaust particles, and 'exhaust particles' with 12, 7, and 5 sample papers, respectively. There are also two papers related to 'power generation' by diesel fuels and 'desulfurization' of diesel fuels.

It is further notable that there is a significant gender deficit among these top authors as only one of them is female (Xie and Shauman, 1998; Lariviere et al., 2013).

The authors with the most impact are 'David Diaz-Sanchez', 'Anders Blomberg', and 'Thomas Sandstrom' with 5.8, 5.8, and 5.7% publication surpluses, respectively. The other authors with the most impact are 'Ning Li', 'Andrew E. Nel', and 'Andrew Saxon' with a 4.9% publication surplus each. On the other hand, there are ten authors with the least impact with a 2.9% publication surplus each.

46.3.3 Publication Years

Information about the publication years for both datasets is provided in Figure 46.1. This figure shows that 6, 26, 56, and 12% of the sample papers and 6.6, 10.9, 25.9, and 56.5 % of the population papers were published in the 1980s, 1990s, 2000s, and 2010s, respectively.

Similarly, the most-prolific publication years for the sample dataset are 2004, 2005, 2003, and 2001 with 11, 8, 8, and 7 papers, respectively. On the other hand, the most-prolific publication years for the population dataset are 2019, 2018, 2017, 2016, and 2015 with 8.2, 7.0, 6.9, 6.3, and 6.2% of the population papers, respectively. It is notable that there is a sharply rising trend for the population papers starting in the 1990s.

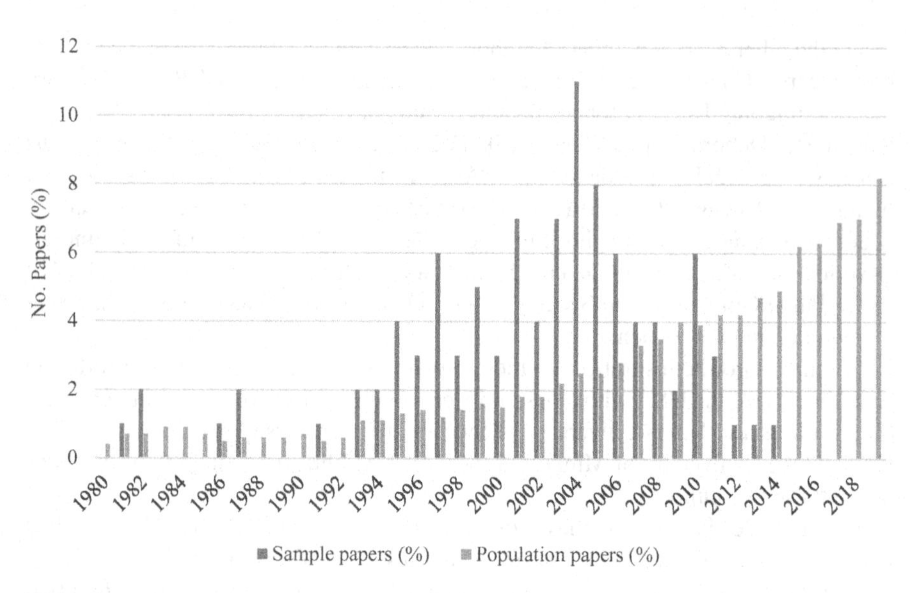

FIGURE 46.1 The research output between 1980 and 2019.

46.3.4 Institutions

Brief information on the top 16 institutions with at least 3% of the sample papers each is provided in Table 46.3. In total, around 120 and 6,300 institutions contribute to the sample and population papers, respectively. Additionally, 6.1% of the population papers have no institutional information on their abstract pages.

These top institutions publish 70.0 and 10.9% of the sample and population papers, respectively. The top institution is the 'University of California Los Angeles' of the USA with 12.0 and 0.4% of the sample and population papers, respectively. This top institution is followed by 'Umea University' of Sweden with six sample papers. The US 'Department of Energy', 'Pennsylvania State University', and the 'University of Minnesota' are the other prolific institutions with five sample papers each.

The most-prolific country for these top institutions is the USA with nine sample papers. The other prolific country is Spain with two sample papers.

The institutions with the most impact are the 'University of California Los Angeles' and 'Umea University' with 11.6 and 5.7% publication surpluses, respectively. On the other hand, the institutions with the least impact are the 'University Polytechnics Valencia' of Spain and the 'National Institute for Environmental Studies' of Japan with a 1.9% publication surplus each.

It is notable that some institutions have a heavy presence in the population papers: 'Tianjin University', the 'Indian Institute of Technology', 'Shanghai Jiao Tong University', the 'National Technical University of Athens', the 'National Scientific Research Center', 'Hanyang University', 'Tsinghua University', the 'University of Wisconsin Madison', the US 'Environmental Protection Agency', the 'Beijing Institute of Technology', the 'Polytechnic University of Turin', the 'National

TABLE 46.3
Institutions

	Institution	Country	No. of Sample Papers (%)	No. of Population Papers (%)	Difference (%)
1	Univ. Calif. L. A.	USA	12	0.4	11.6
2	Umea Univ.	Sweden	6	0.3	5.7
3	Dept. Energ.	USA	5	2.1	2.9
4	Penn. State Univ.	USA	5	0.4	4.6
5	Univ. Minnesota	USA	5	0.4	4.6
6	Univ. Polytech. Valencia	Spain	4	2.1	1.9
7	Ford Motor Co.	USA	4	0.9	3.1
8	Univ. Calif. Riverside	USA	4	0.4	3.6
9	Delft Univ. Technol.	Netherlands	4	0.3	3.7
10	Natl. Inst. Env. Stud.	Japan	3	1.1	1.9
11	Chinese Acad. Sci.	China	3	0.8	2.2
12	Sandia Natl. Labs.	USA	3	0.6	2.4
13	Lovelace Resp. Res. Inst.	USA	3	0.5	2.5
14	University of Edinburgh	UK	3	0.2	2.8
15	Univ. So. California	USA	3	0.2	2.8
16	Univ. Zaragoza	Spain	3	0.2	2.8

Research Council' of Italy, the 'China University of Petroleum', the 'General Motors Company', the US 'Centers for Disease Control Prevention', the US 'National Institute for Occupational Safety Health', 'Jiangsu University', the 'Helmholtz Association', 'West Virginia University', 'Castilla La Mancha University', and the 'University of North Carolina' with at least a 0.7% presence in the population papers each.

46.3.5 Funding Bodies

Brief information about the top six funding bodies with at least 2% of the sample papers each is provided in Table 46.4. It is significant that only 23.0 and 39.7% of the sample and population papers declare any funding, respectively. Around 70 and 5,400 funding bodies fund the research for the sample and population papers, respectively.

The top funding body is the US 'National Institute of Allergy Infectious Diseases' with 11.0 and 0.2% of the sample and population papers, respectively. The other prolific bodies are the 'Medical Research Council' of the UK and the 'National Institute of Environmental Health Sciences' of the USA with three sample papers each.

The most-prolific countries for these funding bodies are the USA and the UK with three and two funding bodies, respectively. In total, these funding bodies are from two countries and Europe only.

It is notable that some top funding agencies have a heavy presence in the population studies. Some of them are the 'National Natural Science Foundation', the 'National Basic Research Program', and the 'Fundamental Research Funds for the

TABLE 46.4
Funding Bodies

	Institution	Country	No. of Sample Papers (%)	No. of Population Papers (%)	Difference (%)
1	National Institute of Allergy Infectious Diseases	USA	11	0.2	10.8
2	Medical Research Council	UK	3	0.2	2.8
3	National Institute of Environmental Health Sciences	USA	3	0.8	2.2
4	British Heart Foundation	UK	2	0.2	1.8
5	European Union	EU	2	1.1	0.9
6	Department of Energy	USA	2	1.4	0.6

Central Universities' of China, the 'National Science Foundation' of the USA, the 'National Council for Scientific and Technological Development' of Brazil, the 'Engineering Physical Sciences Research Council' of the UK, the 'Natural Sciences and Engineering Research Council' of Canada, the US 'Environmental Protection Agency', the 'Ministry of Education Culture Sports Science and Technology' of Japan, and 'CAPES' of Brazil with at least 0.5% of the population papers each. These funding bodies are located in China, Brazil, Canada, the UK, and the USA.

The funding body with the most impact is the 'National Institute of Allergy Infectious Diseases' of the USA with a 10.8% publication surplus. On the other hand, the funding bodies with the least impact are the 'European Union' and the US 'Department of Energy' with 0.9 and 0.6% publication surpluses, respectively.

46.3.6 Source Titles

Brief information about the top 12 source titles with at least three sample papers each is provided in Table 46.5. In total, around 20 and 5,400 source titles publish the sample and population papers, respectively. On the other hand, these top 12 journals publish 57 and 14% of the sample and population papers, respectively.

The top journals are 'Fuel', 'Environmental Science Technology', 'Catalysis Today', and the 'Journal of Allergy and Clinical Immunology', publishing six sample papers each. These top journals are followed by 'Atmospheric Environment' and the 'Journal of Immunology' with five sample papers each.

Although these journals are indexed by 15 subject categories, the top category is 'Engineering Chemical' with six journals. The other most-prolific subject categories are 'Energy Fuels', 'Environmental Sciences', and 'Chemistry Physical' with three journals each.

The journals with the most impact are the 'Journal of Allergy and Clinical Immunology', 'Catalysis Today', and the 'Journal of Immunology' with 5.8, 5.0, and 4.8% publication surpluses, respectively. On the other hand, the journals with the least impact are 'Fuel' and 'Energy Conversion and Management' with 0.7 and 1.9% publication surpluses, respectively.

TABLE 46.5
Source Titles

	Source Title	Wos Subject Category	No. of Sample Papers (%)	No. of Population Papers (%)	Difference (%)
1	Fuel	Ener. Fuels, Eng. Chem.	6	5.3	0.7
2	Environmental Science Technology	Eng. Env., Env. Sci.	6	1.8	4.2
3	Catalysis Today	Chem. Appl., Chem. Phys., Eng. Chem.	6	1.0	5.0
4	Journal of Allergy and Clinical Immunology	Allerg., Immun.	6	0.2	5.8
5	Atmospheric Environment	Env. Sci., Meteor. Atmosph. Sci.	5	1.2	3.8
6	Journal of Immunology	Immun.	5	0.2	4.8
7	Energy Conversion and Management	Therm., Ener. Fuels, Mechs.	4	2.1	1.9
8	Applied Catalysis B Environmental	Chem. Phys., Eng. Env., Eng. Chem.	4	1.1	2.9
9	Journal of Aerosol Science	Eng. Chem., Eng. Mech., Env. Sci., Meteor. Atmosph. Sci.	4	0.6	3.4
10	American Journal of Respiratory and Critical Care Medicine	Crit. Care Med., Respir. Syst.	4	0.2	3.8
11	Progress in Energy and Combustion Science	Therm., Ener. Fuels, Eng. Chem., Eng. Mech.	4	0.1	3.9
12	Journal of Catalysis	Chem. Phys., Eng. Chem.	3	0.2	2.8

It is notable that some journals have a heavy presence in the population papers. Some of them are 'Energy Fuels', 'Proceedings of the Institution of Mechanical Engineers Part D Journal of Automobile Engineering', the 'Journal of Engineering for Gas Turbines and Power Transactions of the ASME', 'Energy', 'Applied Thermal Engineering', the 'International Journal of Engine Research', 'Applied Energy', 'Chemistry and Technology of Fuels and Oils', 'Diesel Progress North American Edition', and the 'International Journal of Hydrogen Energy' with at least a 1.0% presence in the population papers each.

46.3.7 Countries

Brief information about the top 12 countries with at least two sample papers each is provided in Table 46.6. In total, around 20 and 120 countries contribute to the sample and population papers, respectively.

TABLE 46.6
Countries

	Country	No. of Sample Papers (%)	No. of Population Papers (%)	Difference (%)
1	USA	47	20.2	26.8
2	Japan	9	6.6	2.4
3	UK	9	6.4	2.6
4	Spain	9	4.3	4.7
5	Sweden	7	2.1	4.9
6	China	6	15.3	–9.3
7	Netherlands	5	1.4	3.6
8	France	3	3.2	–0.2
9	Switzerland	3	1.4	1.6
10	S. Korea	2	4.5	–2.5
11	Germany	2	4.2	–2.2
12	Greece	2	1.8	0.2
	Europe-8	40	24.8	15.2
	Asia-3	17	26.4	–9.4

The top country is the USA, publishing 47.0 and 20.2% of the sample and population papers, respectively. Japan, the UK, and Spain follow the USA with nine sample papers each. The other prolific countries are Sweden, China, and the Netherlands.

On the other hand, the European and Asian countries represented in this table publish altogether 40 and 17% of the sample papers and 24.8 and 26.4% of the population papers, respectively.

It is notable that the publication surplus for the USA and these European and Asian countries is 26.8, 15.2, and –9.8%, respectively. On the other hand, the countries with the most impact are the USA, Sweden, and Spain with 26.8, 4.9, and 4.7% publication surpluses, respectively. Furthermore, the countries with the least impact are China, South Korea, and Germany with –9.3, –2.5, and –2.3% publication deficits, respectively.

It is also notable that some countries have a heavy presence in the population papers. The major producers of the population papers are India, Italy, Canada, Australia, Turkey, Brazil, Iran, Russia, Poland, Malaysia, Finland, and Taiwan with at least 1.0% of the population papers each.

46.3.8 'Web of Science' Subject Categories

Brief information about the top 17 'Web of Science' subject categories with at least four sample papers each is provided in Table 46.7. The sample and population papers are indexed by around 30 and 180 subject categories, respectively.

For the sample papers, the top subject is 'Engineering Chemical' with 35.0 and 24.6% of the sample and population papers, respectively. This top subject category is followed by 'Energy Fuels', 'Environmental Sciences', and 'Chemistry Physical'

TABLE 46.7
Web of Science Subject Categories

	Subject	No. of Sample Papers (%)	No. of Population Papers (%)	Difference (%)
1	Engineering Chemical	35	24.6	10.4
2	Energy Fuels	26	30.9	–4.9
3	Environmental Sciences	22	13.3	8.7
4	Chemistry Physical	17	7.6	9.4
5	Meteorology Atmospheric Sciences	13	3.0	10.0
6	Engineering Mechanical	12	21.9	–9.9
7	Engineering Environmental	12	5.6	6.4
8	Thermodynamics	11	12.7	–1.7
9	Immunology	11	0.7	10.3
10	Chemistry Applied	7	3.1	3.9
11	Allergy	6	0.4	5.6
12	Toxicology	5	4.9	0.1
13	Respiratory System	5	0.7	4.3
14	Mechanics	4	5.5	–1.5
15	Chemistry Multidisciplinary	4	2.3	1.7
16	Green Sustainable Science Technology	4	2.2	1.8
17	Critical Care Medicine	4	0.2	3.8

with 26, 22, and 17% of the sample papers, respectively. The other prolific subject categories are 'Meteorology Atmospheric Sciences', 'Engineering Mechanical', 'Engineering Environmental', 'Thermodynamics', and 'Immunology' with at least 17 sample papers each.

It is notable that the publication surplus is most significant for 'Engineering Chemical', 'Immunology', and 'Meteorology Atmospheric Sciences' with 10.4, 10.3, and 10.0% publication surpluses, respectively. On the other hand, the subjects with least impact are 'Engineering Mechanical', 'Energy Fuels', 'Thermodynamics', and 'Mechanics' with –9.9, –4.9, –1.7, and –1.5% publication deficits, respectively. This latter group of subject categories are under-represented in the sample papers.

Additionally, some subject categories also have a heavy presence in the population papers: 'Transportation Science Technology', 'Engineering Multidisciplinary', 'Engineering Petroleum', 'Materials Science Multidisciplinary', 'Engineering Electrical Electronic', 'Public Environmental Occupational Health', 'Chemistry Analytical', 'Engineering Civil', 'Biotechnology Applied Microbiology', 'Automation Control Systems', 'Electrochemistry', 'Engineering Marine', 'Instruments Instrumentation', 'Multidisciplinary Sciences', and 'Pharmacology Pharmacy' have at least a 1.0% presence in the population papers each.

46.3.9 Citation Impact

These sample papers received about 35,000 citations as of January 2020. Thus, the average number of citations per paper is about 350.

46.3.10 Keywords

Although a number of keywords are listed in the Appendix for the datasets related to this field, some of them are more significant for the sample papers.

The most-prolific keyword for the keyword set related to diesel fuels is '*diesel*' with 105 occurrences. The other prolific keywords are '*sulfur*' (16), 'emission*' (20), 'exhaust*' (38), '*particle*' (20), 'particulate*' (14), 'engine*' (27), 'combustion' (15), '*fuel*' (35), '*cataly*' (11), 'gasoline' (8), 'aromatic*' (7), 'performance*' (5), 'ige' (6), 'macrophage' (5), 'inflam*' (5), 'human*' (7), 'inhal*' (6), 'toxic*' (6), 'exposure*' (7), 'blend*' (6), 'injection*' (5), and 'soot*' (9).

46.3.11 Research Fronts

Brief information about the key research fronts is provided in Table 46.8. There are eight major topical research fronts for these sample papers. 'Diesel fuel exhaust emissions' is the most prolific research front with 37 papers. In this front, 'diesel fuel exhaust particle emissions' is the most-prolific research subfront with 28 papers (Birch and Cary, 1996; Schauer et al., 1999; Zhu et al., 2002). The other prolific research subfronts are 'NO_x emissions' (Kim et al., 2010; Koebel et al., 2000), 'exhaust in general' (Khalili et al., 1995; Miguel et al., 1998), and 'aerosol emissions' (Miguel et al., 1998; Rogge et al., 1993).

This top research front is followed by 'health impact of diesel fuel exhaust emissions' with 28 papers. In this research front the most-prolific research subfront is 'respiratory illnesses' caused by diesel fuel exhaust emissions with 11 papers

TABLE 46.8
Research Fronts

	Research Front	No. of Sample Papers (%)
1	**Desulfurization**	**14**
2	**Emissions**	**37**
2.1	Exhaust particles	28
2.2	NO_x emissions	4
2.3	Exhaust in general	3
2.4	Aerosol emissions	2
3	**Fuels**	**14**
4	**Power generation**	**4**
5	**Health impact**	**28**
5.1	Respiratory illnesses	11
5.2	Vascular illnesses	2
5.3	Toxicity	4
5.4	Cancer	6
5.5	Allergy	3
5.6	Health in general	2
6	**Engine**	**6**
7	**Soil bioremediation**	**2**
8	**Environment**	**7**
9	**Production**	2

(McCreanor et al., 2007; Salvi et al., 1999). The other prolific research subfronts are 'allergy' (Diaz-Sanchez et al.,1994; Nel et al., 1998), 'cancer' (Schuetzle et al., 1981, Xiao et al., 2003), and 'vascular illnesses' (Mills et al., 2005, 2007) caused by diesel fuel exhaust emissions, toxicity (Aoki et al., 2001), and the general health impact of these emissions (Sagai et al., 1993; Sydbom et al., 2001).

The other research fronts are 'desulfurization' of diesel fuels (14) (Song, 2003; Song and Ma, 2003, Stanislaus et al., 2010), 'combustion and properties of diesel fuels' (14) (Hansen et al., 2005; Zheng et al., 2004), 'power generation' by diesel fuels (4) (Dufo-Lopez and Bernal-Agustin, 2005; Wang et al., 2010), 'diesel engine' (6) (Lloyd and Cackette, 2001; Sahoo et al., 2009), the impact of diesel fuels on the 'environment' (7) (Lloyd and Cackette, 2001; Namkoong et al., 2002), 'bioremediation of the soils' contaminated with diesel fuels (2) (Bento et al., 2005; Namkoong et al., 2002), and 'production of diesel fuels' (2) (Stanislaus et al., 2010).

46.4 DISCUSSION

The size of the research in this field has increased to over 16,000 papers as of January 2020. It is expected that the number of the population papers in this field will exceed 30,000 by the end of the 2020s.

The research has developed more in the technological aspects of this field, rather than the social and humanitarian pathways, as evidenced by the negligible number of population papers in the indices of the 'Web of Science', SSCI, and A&HCI.

The article types of documents are the primary documents for both datasets and reviews are over-represented by 14.3% in the sample papers, whilst articles are under-represented by 13.0% (Table 46.1). Thus, the contribution of reviews by 16% of the sample papers in this field is highly exceptional (cf. Konur, 2011,2012a–n, 2015, 2016a–f, 2017a–f, 2018a–b, 2019a–b).

Twenty-one authors from eight institutions have at least three sample papers each (Table 46.2). Ten, three, three, two, and two of these authors are from the USA, the UK, Sweden, the Netherlands, and Spain, respectively.

There are three key topical research fronts for these top researchers: 'respiratory illnesses' and 'vascular illnesses' caused by diesel exhaust particles, and 'exhaust particles' with 12, 7, and 5 sample papers, respectively. There are also two papers related to 'power generation' by diesel fuels and 'desulfurization of diesel fuels'.

It is significant that there is ample 'gender deficit' among these top authors as only one of them is female (Xie and Shauman, 1998; Lariviere et al., 2013).

The population papers have built on the sample papers, primarily published in the 2000s and to a lesser extent in the 1990s (Figure 46.1). Following the rising trend, particularly in the 2000s and 2010s, it is expected that the number of population papers will reach 30,000 by the end of the 2020s, nearly doubling the current size.

The engagement of the institutions in this field at the global scale is significant as around 120 and 6,300 institutions contribute to the sample and population papers, respectively.

Sixteen top institutions publish 70.0 and 10.9% of the sample and population papers, respectively (Table 46.3). The top institutions are the 'University of California Los Angeles' of the USA and 'Umea University' of Sweden.

The most-prolific countries for these top institutions are the USA and Spain. It is notable that some institutions with a heavy presence in the population papers are under-represented in the sample papers.

It is significant that only 23.0 and about 39.7% of the sample and population papers declare any funding, respectively. It is notable that the most-prolific countries for these funding bodies are the USA and the UK (Table 46.4). It is further notable that some top funding agencies for the population studies do not enter this top funding body list.

However, the lack of Chinese funding bodies in this top funding body table is notable. This finding is in contrast with the studies showing the heavy research funding in China, where the NSFC is the primary funding agency (Wang et al., 2012).

The sample and population papers are published by around 20 and 5,400 journals, respectively. It is significant that the top 12 journals publish 57 and 14% of the sample and population papers, respectively (Table 46.5).

The top journals are 'Fuel', 'Environmental Science Technology', 'Catalysis Today', and the 'Journal of Allergy and Clinical Immunology'. These top journals are followed by 'Atmospheric Environment' and the 'Journal of Immunology'.

The top categories for these journals are 'Engineering Chemical', 'Energy Fuels', 'Environmental Sciences', and 'Chemistry Physical'. It is notable that some journals with a heavy presence in the population papers are relatively under-represented in the sample papers.

In total, around 20 and 120 countries contribute to the sample and population papers, respectively. The top country is the USA (Table 46.6). This finding is in line with the studies arguing that the USA is not losing ground in science and technology (Leydesdorff and Wagner, 2009).

The other prolific countries are Japan, the UK, and Spain and to a lesser extent Sweden, China, and the Netherlands. These findings are in line with the studies showing heavy research activity in these countries in recent decades (Bordons et al., 2015; Hammarfelt et al., 2016; Hu et al., 2018; Negishi et al., 2004; Rinia, 2000; Zhou and Leydesdorff, 2006).

On the other hand, the European and Asian countries represented in this table publish altogether 40 and 17% of the sample papers and 24.8 and 26.4% of the population papers, respectively. These findings are in line with the studies showing that both European and Asian countries have superior publication performance in science and technology (Bordons et al., 2015; Glanzel and Schlemmer, 2007; Okubo et al., 1998; Youtie et al., 2008).

It is notable that the publication surplus for the USA and these European and Asian countries is 26.8, 15.2, and –9.8%, respectively. On the other hand, the countries with the most impact are the USA, Sweden, and Spain. Furthermore, the countries with the least impact are China, South Korea, and Germany.

China's presence in this top table is notable. This finding is in line with China's efforts to be a leading nation in science and technology (Guan and Ma, 2007; Youtie et al., 2008; Zhou and Leydesdorff, 2006).

It is also notable that some countries have a heavy presence in the population papers. The major producers of the population papers are India, Italy, Canada,

Australia, Turkey, Brazil, Iran, Russia, Poland, Malaysia, Finland, and Taiwan with at least 1.0% of the population papers each (Glanzel et al., 2006; Hassan et al., 2012; Huang et al., 2006; Kumar and Jan, 2014; Prathap, 2017).

The sample and population papers are indexed by around 30 and 180 subject categories, respectively. For the sample papers, the top subject is 'Engineering Chemical' with 35.0 and 24.6% of the sample and population papers, respectively (Table 46.7). This top subject category is followed by 'Energy Fuels', 'Environmental Sciences', and 'Chemistry Physical'. The other prolific subject categories are 'Meteorology Atmospheric Sciences', 'Engineering Mechanical', 'Engineering Environmental', 'Thermodynamics', and 'Immunology'.

It is notable that the publication surplus is most significant for 'Engineering Chemical', 'Immunology', and 'Meteorology Atmospheric Sciences'. On the other hand, the subjects with least impact are 'Engineering Mechanical', 'Energy Fuels', 'Thermodynamics', and 'Mechanics'. This latter group of subject categories are under-represented in the sample papers.

These sample papers receive about 35,000 citations as of January 2020. Thus, the average number of citations per paper is about 350. Hence, the citation impact of these top 100 papers in this field is significant.

Although a number of keywords are listed in the Appendix for the datasets related to this field, some of them are more significant for the sample papers.

The most-prolific keyword for the keyword set related to diesel fuels is '*diesel*'. The other most-prolific keywords are '*sulfur*', 'emission*', 'exhaust*', '*particle*', 'particulate*', 'engine*', 'combustion', '*fuel*', '*cataly*', 'gasoline', 'aromatic*', 'performance*', 'ige', 'macrophage', 'inflam*', 'human*', 'inhal*', 'toxic*', 'exposure*', 'blend*', 'injection*', and 'soot*'. As expected, these keywords provide valuable information about the pathways of the research in this field.

There are eight major topical research fronts for these sample papers. 'Diesel fuel exhaust emissions' is the most prolific research front. In this front, 'diesel fuel exhaust particle emissions' is the most-prolific research subfront. The other prolific research subfronts are 'NO_x emissions', 'exhaust in general', and 'aerosol emissions'.

This top research front is followed by 'health impact of diesel fuel exhaust emissions'. In this research front the most-prolific research subfront is 'respiratory illnesses' caused by diesel fuel exhaust emissions. The other prolific research subfronts are 'allergy', 'cancer', and 'vascular illnesses' caused by diesel fuel exhaust emissions, toxicity, and the general health impact of these emissions.

The other research fronts are 'desulfurization of diesel fuels', 'combustion and properties of diesel fuels', 'power generation by diesel fuels', 'diesel engine', 'environmental impact of diesel fuels', 'bioremediation of the soils' contaminated with diesel fuels, and 'production of diesel fuels'.

The key emphasis in these research fronts is the exploration of the structure–processing–property relationships of diesel fuels obtained from crude oils (Cheng and Ma, 2011; Konur and Matthews, 1989; Rogers and Hopfinger, 1994; Scherf and List, 2002).

46.5 CONCLUSION

This chapter has mapped the research on petrodiesel fuels using a scientometric method.

The size of over 16,000 population papers shows the public importance of this interdisciplinary research field. However, it is significant that the research has developed more in the technological aspects in this field, rather than the social and humanitarian pathways.

Articles dominate both the sample and population papers. The population papers, primarily published in the 2010s, build on these sample papers, primarily published in the 2000s.

The data presented in the tables and figure show that a small number of authors, institutions, funding bodies, journals, keywords, research fronts, subject categories, and countries have shaped the research in this field.

It is notable that the authors, institutions, and funding bodies from the USA, Sweden, Spain, the UK, the Netherlands, and Japan dominate the research in this field. Furthermore, it is also notable that some countries have a heavy presence in the population papers. The major producers of the population papers are India, Italy, Canada, Australia, Turkey, Brazil, Iran, Russia, Poland, Malaysia, Finland, and Taiwan. Additionally, China, South Korea, and Germany are under-represented significantly in the sample papers.

These findings show the importance of the progression of efficient incentive structures for the development of the research in this field as in other fields. It seems that some countries (such as the USA, Sweden, Spain, the UK, the Netherlands, and Japan) have efficient incentive structures for the development of the research in this field, contrary to India, India, Italy, Canada, Australia, Turkey, Brazil, Iran, Russia, Poland, Malaysia, Finland, Taiwan, China South Korea, and Germany.

It further seems that although the research funding is a significant element of these incentive structures, it might not be a sole solution for increasing the incentives for the research in this field as in the case of India, India, Italy, Canada, Australia, Turkey, Brazil, Iran, Russia, Poland, Malaysia, Finland, Taiwan, China South Korea, and Germany.

On the other hand, it seems there is more to do to reduce the significant gender deficit in this field as in other fields of science and technology (Lariviere et al., 2013; Xie and Shauman, 1998).

The data on the research fronts, keywords, source titles, and subject categories provide valuable evidence for the interdisciplinary (Lariviere and Gingras, 2010; Morillo et al., 2001) nature of the research in this field.

There is ample justification for the broad search strategy employed in this study due to the interdisciplinary nature of this research field as evidenced by the top subject categories. The search strategy employed in this study is in line with the search strategies employed for related and other research fields (Konur, 2011, 2012a–n, 2015, 2016a–f, 2017a–f, 2018a–b, 2019a–b). It is particularly noted that only 26.0 and 30.9% of the sample and population papers are indexed by the 'Energy Fuels' subject category, respectively.

There are eight major topical research fronts for these sample papers. 'Diesel fuel exhaust emissions' is the most prolific research front. In this front, 'diesel fuel exhaust particle emissions' is the most-prolific research subfront. The other prolific research subfronts are 'NO_x emissions', 'exhaust emissions in general', and 'aerosol emissions'.

This top research front is followed by 'health impact of diesel fuel exhaust emissions'. In this research front the most-prolific research subfront is 'respiratory illnesses' caused by diesel fuel exhaust emissions. The other prolific research subfronts are 'allergy', 'cancer', and 'vascular illnesses' caused by diesel fuel exhaust emissions, toxicity, and the general health impact of these emissions.

The other research fronts are 'desulfurization of diesel fuels', 'combustion and properties of diesel fuels', 'power generation by diesel fuels', 'diesel engine', 'environmental impact of diesel fuels', 'bioremediation of the soils' contaminated with diesel fuels, and 'production of diesel fuels'.

It is recommended that further scientometric studies are carried out for each of these research fronts building on the pioneering studies in these fields.

ACKNOWLEDGMENTS

The contribution of the highly cited researchers in the fields of petrodiesel fuels is greatly acknowledged.

46.A APPENDIX

46.A.1 The Keyword Set for Petrodiesel Fuels

TI = (diesel) NOT TI = (fat or fats or vegetable or biodiesel or "*bio-diesel" or triglyceride* or renewable or jatropha or "bio-oil*" or waste* or olive* or *seed* or "methyl-ester*" or karanj* or *alga* or polanga or palm or rape* or canola or soy* or mahua or cooking or honge or sesame or *edible or frying or cotton* or sunflower* or soapstock* or tallow or pongamia or camelina or *cellulos* or eucalyptus or "fatty-acid*" or *ester* or *pyroly* or poon or myco* or fung* or green or *nut* or linseed* or jojoba or croton or castor or tall or "fish oil*" or moringa or calophyllum or ceiba or hydrodeoxygenation or tung or "pine oil*" or biomass or corn or olive)

REFERENCES

Ahmadun, F. R., A. Pendashteh, and L. C. Abdullah, et al. 2009. Review of technologies for oil and gas produced water treatment. *Journal of Hazardous Materials* 170:530–551.

Clarivate Analytics. 2019. *Highly cited researchers: 2019 Recipients*. Philadelphia, PA: Clarivate Analytics. https://recognition.webofsciencegroup.com/awards/highly-cited/2019/ (accessed January 3, 2020).

Aoki, Y., H. Sato, and N. Nishimura, et al. 2001. Accelerated DNA adduct formation in the lung of the Nrf2 knockout mouse exposed to diesel exhaust. *Toxicology and Applied Pharmacology* 173:154–160.

Atlas, R. M. 1981. Microbial degradation of petroleum hydrocarbons: An environmental perspective. *Microbiological Reviews* 45:180–209.

Babich, I. V. and J. A. Moulijn. 2003. Science and technology of novel processes for deep desulfurization of oil refinery streams: A review. *Fuel* 82:607–631.

Bento, F. M., F. A. O. Camargo, B. C. Okeke, and W. T. Frankenberger. 2005. Comparative bioremediation of soils contaminated with diesel oil by natural attenuation, biostimulation and bioaugmentation. *Bioresource Technology* 96:1049–1055.

Birch, M. E. and R. A. Cary. 1996. Elemental carbon-based method for monitoring occupational exposures to particulate diesel exhaust. *Aerosol Science and Technology* 25:221–241.

Bordons, M., B. Gonzalez-Albo, J. Aparicio, and L. Moreno. 2015. The influence of R&D intensity of countries on the impact of international collaborative research: Evidence from Spain. *Scientometrics* 102:1385–1400.

Bridgwater, A. V. and G. V. C. Peacocke. 2000. Fast pyrolysis processes for biomass. *Renewable & Sustainable Energy Reviews* 4:1–73.

Cheng, Y. Q. and E. Ma. 2011. Atomic-level structure and structure–property relationship in metallic glasses. *Progress in Materials Science* 56:379–473.

Chisti, Y. 2007. Biodiesel from microalgae. *Biotechnology Advances* 25:294–306.

Czernik, S. and A. V. Bridgwater. 2004. Overview of applications of biomass fast pyrolysis oil. *Energy & Fuels* 18:590–598.

Diaz-Sanchez, D., A. R. Dotson, H. Takenaka, and A. Saxon. 1994. Diesel exhaust particles induce local IgE production *in vivo* and alter the pattern of IgE messenger-RNA isoforms. *Journal of Clinical Investigation* 94:1417–1425.

Docampo, D. and L. Cram. 2019. Highly cited researchers: A moving target. *Scientometrics* 118:1011–1025.

Dufo-Lopez, R. and J. L. Bernal-Agustin. 2005. Design and control strategies of PV-Diesel systems using genetic algorithms. *Solar Energy* 79:33–46.

Gallezot, P. 2012. Conversion of biomass to selected chemical products. *Chemical Society Reviews* 41:1538–1558.

Garfield, E. 1955. Citation indexes for science. *Science* 122:108–111.

Garfield, E. 1972. Citation analysis as a tool in journal evaluation. *Science* 178:471–479.

Glanzel, W., J. Leta, and B. Thijs. 2006. Science in Brazil. Part 1: A macro-level comparative study. *Scientometrics* 67:67–86.

Glanzel, W. and B. Schlemmer. 2007. National research profiles in a changing Europe (1983–2003): An exploratory study of sectoral characteristics in the Triple Helix. *Scientometrics* 70:267–275.

Guan, J. C. and N. Ma. 2007. China's emerging presence in nanoscience and nanotechnology: A comparative bibliometric study of several nanoscience 'giants'. *Research Policy* 36:880–886.

Hammarfelt, B., G. Nelhans, P. Eklund, and F. Astrom. 2016. The heterogeneous landscape of bibliometric indicators: Evaluating models for allocating resources at Swedish universities. *Research Evaluation* 25:292–305.

Hansen, A. C., Q. Zhang, and P. W. L. Lyne. 2005. Ethanol-diesel fuel blends: A review. *Bioresource Technology* 96:277–285.

Hassan, S. U., P. Haddawy, P. Kuinkel, A. Degelsegger, and C. Blasy. 2012. A bibliometric study of research activity in ASEAN related to the EU in FP7 priority areas. *Scientometrics* 91:1035–1051.

Hill, J., E. Nelson, D. Tilman, S. Polasky, and D. Tiffany. 2006. Environmental, economic, and energetic costs and benefits of biodiesel and ethanol biofuels. *Proceedings of the National Academy of Sciences of the United States of America* 103:11206–11210.

Hu, Z., G. Lin, T. Sun, T., and X. Wang. 2018. An EU without the UK: Mapping the UK's changing roles in the EU scientific research. *Scientometrics* 115:1185–1198.

Huang, M. H., H. W. Chang, and D. Z. Chen. 2006. Research evaluation of research-oriented universities in Taiwan from 1993 to 2003. *Scientometrics* 67:419–435.
Khalili, N. R., P. A. Scheff, and T. M. Holsen. 1995. PAH source fingerprints for coke ovens, diesel and gasoline-engines, highway tunnels, and wood combustion emissions. *Atmospheric Environment* 29:533–542.
Kilian, L. 2009. Not all oil price shocks are alike: Disentangling demand and supply shocks in the crude oil market. *American Economic Review* 99:1053–1069.
Kim, C. H., G. S. Qi, K. Dahlberg, and W. Li. 2010. Strontium-doped perovskites rival platinum catalysts for treating NO_x in simulated diesel exhaust. *Science* 327:1624–1627.
Koebel, M., M. Elsener, and M. Kleemann. 2000. Urea-SCR: A promising technique to reduce NO_x emissions from automotive diesel engines. *Catalysis Today* 59:335–345.
Konur, O. 2000. Creating enforceable civil rights for disabled students in higher education: An institutional theory perspective. *Disability & Society* 15:1041–1063.
Konur, O. 2002a. Access to Nursing Education by disabled students: Rights and duties of nursing programs. *Nurse Education Today* 22:364–374.
Konur, O. 2002b. Assessment of disabled students in higher education: Current public policy issues. *Assessment and Evaluation in Higher Education* 27:131–152.
Konur, O. 2002c. Access to employment by disabled people in the UK: Is the Disability Discrimination Act working? *International Journal of Discrimination and the Law* 5:247–279.
Konur, O. 2006a. Participation of children with dyslexia in compulsory education: Current public policy issues. *Dyslexia* 12:51–67.
Konur, O. 2006b. Teaching disabled students in Higher Education. *Teaching in Higher Education* 11:351–363.
Konur, O. 2007a. A judicial outcome analysis of the Disability Discrimination Act: A windfall for the employers? *Disability & Society* 22:187–204.
Konur, O. 2007b. Computer-assisted teaching and assessment of disabled students in higher education: The interface between academic standards and disability rights. *Journal of Computer Assisted Learning* 23:207–219.
Konur, O. 2011. The scientometric evaluation of the research on the algae and bio-energy. *Applied Energy* 88:3532–3540.
Konur, O. 2012a. Evaluation of the research on the social sciences in Turkey: A scientometric approach. *Energy Education Science and Technology Part B: Social and Educational Studies* 4:1893–1908.
Konur, O. 2012b. Prof. Dr. Ayhan Demirbas' scientometric biography. *Energy Education Science and Technology Part A: Energy Science and Research* 28:727–738.
Konur, O. 2012c. The evaluation of the biogas research: A scientometric approach. *Energy Education Science and Technology Part A: Energy Science and Research* 29:1277–1292.
Konur, O. 2012d. The evaluation of the educational research: A scientometric approach. *Energy Education Science and Technology Part B: Social and Educational Studies* 4:1935–1948.
Konur, O. 2012e. The evaluation of the global energy and fuels research: A scientometric approach. *Energy Education Science and Technology Part A: Energy Science and Research* 30:613–628.
Konur, O. 2012f. The evaluation of the research on the Arts and Humanities in Turkey: A scientometric approach. *Energy Education Science and Technology Part B: Social and Educational Studies* 4:1603–1618.
Konur, O. 2012g. The evaluation of the research on the biodiesel: A scientometric approach. *Energy Education Science and Technology Part A: Energy Science and Research* 28:1003–1014.

Konur, O. 2012h. The evaluation of the research on the bioethanol: A scientometric approach. *Energy Education Science and Technology Part A: Energy Science and Research* 28:1051–1064.

Konur, O. 2012i. The evaluation of the research on the biofuels: A scientometric approach. *Energy Education Science and Technology Part A: Energy Science and Research* 28:903–916.

Konur, O. 2012j. The evaluation of the research on the biohydrogen: A scientometric approach. *Energy Education Science and Technology Part A: Energy Science and Research* 29:323–338.

Konur, O. 2012k. The evaluation of the research on the microbial fuel cells: A scientometric approach. *Energy Education Science and Technology Part A: Energy Science and Research* 29:309–322.

Konur, O. 2012l. The scientometric evaluation of the research on the production of bioenergy from biomass. *Biomass and Bioenergy* 47:504–515.

Konur, O. 2012m. The scientometric evaluation of the research on the deaf students in higher education. *Energy Education Science and Technology Part B: Social and Educational Studies* 4:1573–1588.

Konur, O. 2012n. The scientometric evaluation of the research on the students with ADHD in higher education. *Energy Education Science and Technology Part B: Social and Educational Studies* 4:1547–1562.

Konur, O. 2015. Current state of research on algal biodiesel. In *Marine Bioenergy: Trends and Developments*, ed. S. K. Kim, and C. G. Lee, 487–512. Boca Raton, FL: CRC Press.

Konur, O. 2016a. Scientometric overview in nanobiodrugs. In *Nanoarchitectonics for Smart Delivery and Drug Targeting*, ed. A. M. Holban and A. M. Grumezescu, 405–428. Amsterdam: Elsevier.

Konur, O. 2016b. Scientometric overview regarding nanoemulsions used in the food industry. In *Emulsions: Nanotechnology in the Agri-Food Industry*, ed. A. M. Grumezescu, 689–711. Amsterdam: Elsevier.

Konur, O. 2016c. Scientometric overview regarding the nanobiomaterials in antimicrobial therapy. In *Nanobiomaterials in Antimicrobial Therapy*, ed. A. M. Grumezescu, 511–535. Amsterdam: Elsevier.

Konur, O. 2016d. Scientometric overview regarding the nanobiomaterials in dentistry. In *Nanobiomaterials in Dentistry*, ed. A. M. Grumezescu, 425–453. Amsterdam: Elsevier.

Konur, O. 2016e. Scientometric overview regarding the surface chemistry of nanobiomaterials. In *Surface Chemistry of Nanobiomaterials*, ed. A. M. Grumezescu, 463–486. Amsterdam: Elsevier.

Konur, O. 2016f. The scientometric overview in cancer targeting. In *Nanoarchitectonics for Smart Delivery and Drug Targeting*, ed. A. M. Holban and A. Grumezescu, 871–895. Amsterdam; Elsevier.

Konur, O. 2017a. Recent citation classics in antimicrobial nanobiomaterials. In *Nanostructures for Antimicrobial Therapy*, ed. A. Ficai and A. M. Grumezescu, 669–685. Amsterdam: Elsevier.

Konur, O. 2017b. Scientometric overview in nanopesticides. In *New Pesticides and Soil Sensors*, ed. A. M. Grumezescu. 719–744. Amsterdam: Elsevier.

Konur, O. 2017c. Scientometric overview regarding oral cancer nanomedicine. In *Nanostructures for Oral Medicine*, ed. E. Andronescu, A. M. Grumezescu, 939–962. Amsterdam: Elsevier.

Konur, O. 2017d. Scientometric overview regarding water nanopurification. In *Water Purification*, ed. A. M. Grumezescu, 693–716. Amsterdam: Elsevier.

Konur, O. 2017e. Scientometric overview in food nanopreservation. In *Food Preservation*, ed. A. M. Grumezescu, 703–729. Amsterdam: Elsevier.

Konur, O. 2017f. The top citation classics in alginates for biomedicine. In *Seaweed Polysaccharides: Isolation, Biological and Biomedical Applications*, ed. J. Venkatesan, S. Anil, S. K. Kim, 223–249. Amsterdam: Elsevier.

Konur, O. 2018a. Scientometric evaluation of the global research in spine: An update on the pioneering study by Wei et al. *European Spine Journal* 27:525–529.

Konur, O. 2018b. Bioenergy and biofuels science and technology: Scientometric overview and citation classics. In *Bioenergy and Biofuels*, ed. O. Konur, 3–63. Boca Raton: CRC Press.

Konur, O. 2019a. Cyanobacterial bioenergy and biofuels science and technology: A scientometric overview. In *Cyanobacteria: From Basic Science to Applications*, ed. A. K. Mishra, D. N. Tiwari, and A. N. Rai, 419–442. Amsterdam: Elsevier.

Konur, O. 2019b. Nanotechnology applications in food: A scientometric overview. In *Nanoscience for Sustainable Agriculture*, ed. R. N. Pudake, N. Chauhan, and C. Kole, 683–711. Cham: Springer.

Konur, O., ed. 2021a. *Handbook of Biodiesel and Petrodiesel Fuels: Science, Technology, Health, and Environment.* Boca Raton, FL: CRC Press.

Konur, O., ed. 2021b. *Handbook of Biodiesel and Petrodiesel Fuels: Science, Technology, Health, and Environment. Volume 1. Biodiesel Fuels: Science, Technology, Health, and Environment.* Boca Raton, FL: CRC Press.

Konur, O., ed. 2021c. *Handbook of Biodiesel and Petrodiesel Fuels: Science, Technology, Health, and Environment. Volume 2. Biodiesel Fuels based on the Edible and Nonedible Feedstocks, Wastes, and Algae: Science, Technology, Health, and Environment.* Boca Raton, FL: CRC Press.

Konur, O., ed. 2021d. *Handbook of Biodiesel and Petrodiesel Fuels: Science, Technology, Health, and Environment. Volume 3. Petrodiesel Fuels: Science, Technology, Health, and Environment.* Boca Raton, FL: CRC Press.

Konur, O. 2021e. Biodiesel and petrodiesel fuels: Science, technology, health, and environment. In *Handbook of Biodiesel and Petrodiesel Fuels: Science, Technology, Health, and Environment. Volume 1. Biodiesel Fuels: Science, Technology, Health, and Environment*, ed. O. Konur. Boca Raton, FL: CRC Press.

Konur, O. 2021f. Biodiesel and petrodiesel fuels: A scientometric review of the research. In *Handbook of Biodiesel and Petrodiesel Fuels: Science, Technology, Health, and Environment. Volume 1. Biodiesel Fuels: Science, Technology, Health, and Environment*, ed. O. Konur. Boca Raton, FL: CRC Press.

Konur, O. 2021g. Biodiesel and petrodiesel fuels: A review of the research. In *Handbook of Biodiesel and Petrodiesel Fuels: Science, Technology, Health, and Environment. Volume 1. Biodiesel Fuels: Science, Technology, Health, and Environment*, ed. O. Konur. Boca Raton, FL: CRC Press.

Konur, O. 2021h Nanotechnology applications in the diesel fuels and the related research fields: A review of the research. In *Handbook of Biodiesel and Petrodiesel Fuels: Science, Technology, Health, and Environment. Volume 1. Biodiesel Fuels: Science, Technology, Health, and Environment*, ed. O. Konur. Boca Raton, FL: CRC Press.

Konur, O. 2021i. Biooils: A scientometric review of the research. In *Handbook of Biodiesel and Petrodiesel Fuels: Science, Technology, Health, and Environment. Volume 1. Biodiesel Fuels: Science, Technology, Health, and Environment*, ed. O. Konur. Boca Raton, FL: CRC Press.

Konur, O. 2021j. Characterization and properties of biooils: A review of the research. In *Handbook of Biodiesel and Petrodiesel Fuels: Science, Technology, Health, and Environment. Volume 1. Biodiesel Fuels: Science, Technology, Health, and Environment*, ed. O. Konur. Boca Raton, FL: CRC Press.

Konur, O. 2021k. Biomass pyrolysis and pyrolysis oils: A review of the research. In *Handbook of Biodiesel and Petrodiesel Fuels: Science, Technology, Health, and Environment.*

Volume 1. Biodiesel Fuels: Science, Technology, Health, and Environment, ed. O. Konur. Boca Raton, FL: CRC Press.

Konur, O. 2021l. Biodiesel fuels: A scientometric review of the research. In *Handbook of Biodiesel and Petrodiesel Fuels: Science, Technology, Health, and Environment. Volume 1. Biodiesel Fuels: Science, Technology, Health, and Environment*, ed. O. Konur. Boca Raton, FL: CRC Press.

Konur, O. 2021m. Glycerol: A scientometric review of the research. In *Handbook of Biodiesel and Petrodiesel Fuels: Science, Technology, Health, and Environment. Volume 1. Biodiesel Fuels: Science, Technology, Health, and Environment*, ed. O. Konur. Boca Raton, FL: CRC Press.

Konur, O. 2021n. Propanediol production from glycerol: A review of the research. In *Handbook of Biodiesel and Petrodiesel Fuels: Science, Technology, Health, and Environment. Volume 1. Biodiesel Fuels: Science, Technology, Health, and Environment*, ed. O. Konur Boca Raton, FL: CRC Press.

Konur, O. 2021o. Edible oil-based biodiesel fuels: A scientometric review of the research. In *Handbook of Biodiesel and Petrodiesel Fuels: Science, Technology, Health, and Environment. Volume 2. Biodiesel Fuels based on the Edible and Nonedible Feedstocks, Wastes, and Algae: Science, Technology, Health, and Environment*, ed. O. Konur. Boca Raton, FL: CRC Press.

Konur, O. 2021p. Palm oil-based biodiesel fuels: A review of the research. In *Handbook of Biodiesel and Petrodiesel Fuels: Science, Technology, Health, and Environment. Volume 2. Biodiesel Fuels based on the Edible and Nonedible Feedstocks, Wastes, and Algae*, ed. O. Konur. Boca Raton, FL: CRC Press.

Konur, O. 2021q. Rapeseed oil-based biodiesel fuels: A review of the research. In *Handbook of Biodiesel and Petrodiesel Fuels: Science, Technology, Health, and Environment. Volume 2. Biodiesel Fuels based on the Edible and Nonedible Feedstocks, Wastes, and Algae*, ed. O. Konur. Boca Raton, FL: CRC Press.

Konur, O. 2021r. Nonedible oil-based biodiesel fuels: A scientometric review of the research. In *Handbook of Biodiesel and Petrodiesel Fuels: Science, Technology, Health, and Environment. Volume 2. Biodiesel Fuels based on the Edible and Nonedible Feedstocks, Wastes, and Algae: Science, Technology, Health, and Environment*, ed. O. Konur. Boca Raton, FL: CRC Press.

Konur, O. 2021s. Waste oil-based biodiesel fuels: A scientometric review of the research. In *Handbook of Biodiesel and Petrodiesel Fuels: Science, Technology, Health, and Environment. Volume 2. Biodiesel Fuels based on the Edible and Nonedible Feedstocks, Wastes, and Algae: Science, Technology, Health, and Environment*, ed. O. Konur. Boca Raton, FL: CRC Press.

Konur, O. 2021t. Algal biodiesel fuels: A scientometric review of the research. In *Handbook of Biodiesel and Petrodiesel Fuels: Science, Technology, Health, and Environment. Volume 2. Biodiesel Fuels based on the Edible and Nonedible Feedstocks, Wastes, and Algae: Science, Technology, Health, and Environment*, ed. O. Konur. Boca Raton, FL: CRC Press.

Konur, O. 2021u. Algal biomass production for biodiesel production: A review of the research. In *Handbook of Biodiesel and Petrodiesel Fuels: Science, Technology, Health, and Environment. Volume 2. Biodiesel Fuels based on the Edible and Nonedible Feedstocks, Wastes, and Algae*, ed. O. Konur Boca Raton, FL: CRC Press.

Konur, O. 2021v. Algal biomass production in wastewaters for biodiesel production: A review of the research. In *Handbook of Biodiesel and Petrodiesel Fuels: Science, Technology, Health, and Environment. Volume 2. Biodiesel Fuels based on the Edible and Nonedible Feedstocks, Wastes, and Algae*, ed. O. Konur. Boca Raton, FL: CRC Press.

Konur, O. 2021x. Algal lipid production for biodiesel production: A review of the research. In *Handbook of Biodiesel and Petrodiesel Fuels: Science, Technology, Health, and Environment. Volume 2. Biodiesel Fuels based on the Edible and Nonedible Feedstocks, Wastes, and Algae*, ed. O. Konur Boca Raton, FL: CRC Press.

Konur, O. 2021y. Crude oils: A scientometric review of the research. In *Handbook of Biodiesel and Petrodiesel Fuels: Science, Technology, Health, and Environment. Volume 3. Petrodiesel Fuels: Science, Technology, Health, and Environment*, ed. O. Konur. Boca Raton, FL: CRC Press.

Konur, O. 2021z. Petrodiesel fuels: A scientometric review of the research. In *Handbook of Biodiesel and Petrodiesel Fuels: Science, Technology, Health, and Environment. Volume 3. Petrodiesel Fuels: Science, Technology, Health, and Environment*, ed. O. Konur. Boca Raton, FL: CRC Press.

Konur, O. 2021aa. Bioremediation of petroleum hydrocarbons in the contaminated soils: A review of the research. In *Handbook of Biodiesel and Petrodiesel Fuels: Science, Technology, Health, and Environment. Volume 3. Petrodiesel Fuels: Science, Technology, Health, and Environment*, ed. O. Konur. Boca Raton, FL: CRC Press.

Konur, O. 2021ab. Desulfurization of diesel fuels: A review of the research. In *Handbook of Biodiesel and Petrodiesel Fuels: Science, Technology, Health, and Environment. Volume 3. Petrodiesel Fuels: Science, Technology, Health, and Environment*, ed. O. Konur. Boca Raton, FL: CRC Press.

Konur, O. 2021ac. Diesel fuel exhaust emissions: A scientometric review of the research. In *Handbook of Biodiesel and Petrodiesel Fuels: Science, Technology, Health, and Environment. Volume 3. Petrodiesel Fuels: Science, Technology, Health, and Environment*, ed. O. Konur. Boca Raton, FL: CRC Press.

Konur, O. 2021ad. The adverse health and safety impact of diesel fuels: A scientometric review of the research. In *Handbook of Biodiesel and Petrodiesel Fuels: Science, Technology, Health, and Environment. Volume 3. Petrodiesel Fuels: Science, Technology, Health, and Environment*, ed. O. Konur. Boca Raton, FL: CRC Press.

Konur, O. 2021ae. Respiratory illnesses caused by the diesel fuel exhaust emissions: A review of the research. In *Handbook of Biodiesel and Petrodiesel Fuels: Science, Technology, Health, and Environment. Volume 3. Petrodiesel Fuels: Science, Technology, Health, and Environment*, ed. O. Konur. Boca Raton, FL: CRC Press.

Konur, O. 2021af. Cancer caused by the diesel fuel exhaust emissions: A review of the research. In *Handbook of Biodiesel and Petrodiesel Fuels: Science, Technology, Health, and Environment. Volume 3. Petrodiesel Fuels: Science, Technology, Health, and Environment*, ed. O. Konur. Boca Raton, FL: CRC Press.

Konur, O. 2021ag. Cardiovascular and other illnesses caused by the diesel fuel exhaust emissions: A review of the research. In *Handbook of Biodiesel and Petrodiesel Fuels: Science, Technology, Health, and Environment. Volume 3. Petrodiesel Fuels: Science, Technology, Health, and Environment*, ed. O. Konur. Boca Raton, FL: CRC Press.

Konur, O. and F. L. Matthews. 1989. Effect of the properties of the constituents on the fatigue performance of composites: A review. *Composites* 20:317–328.

Kumar, S. and J. M. Jan. 2014. Research collaboration networks of two OIC nations: Comparative study between Turkey and Malaysia in the field of 'Energy Fuels', 2009–2011. *Scientometrics* 98:387–414.

Lapuerta, M., O. Armas, and J. Rodriguez-Fernandez. 2008. Effect of biodiesel fuels on diesel engine emissions. *Progress in Energy and Combustion Science* 34:198–223.

Lariviere, V. and Y. Gingras. 2010. On the relationship between interdisciplinarity and scientific impact. *Journal of the American Society for Information Science and Technology* 61:126–131.

Lariviere, V., C. Ni, Y. Gingras, B. Cronin, B., and C. R. Sugimoto. 2013. Bibliometrics: Global gender disparities in science. *Nature News* 504:211–213.

Leydesdorff, L. and C. Wagner. 2009. Is the United States losing ground in science? A global perspective on the world science system. *Scientometrics* 78:23–36.

Lloyd, A. C. and T. A. Cackette. 2001. Diesel engines: Environmental impact and control. *Journal of the Air & Waste Management Association* 51:809–847.

Mata, T. M., A. A. Martins, and N. S. Caetano. 2010. Microalgae for biodiesel production and other applications: A review. *Renewable & Sustainable Energy Reviews* 14:217–232.

McCreanor, J., P. Cullinan, and M. J. Nieuwenhuijsen, et al. 2007. Respiratory effects of exposure to diesel traffic in persons with asthma. *New England Journal of Medicine* 357:2348–2358.

Miguel, A. H., T. W. Kirchstetter, R. A. Harley, and S. V. Hering. 1998. On-road emissions of particulate polycyclic aromatic hydrocarbons and black carbon from gasoline and diesel vehicles. *Environmental Science & Technology* 32:450–455.

Mills, N. L., H. Tornqvist, and M. C. Gonzalez. 2007. Ischemic and thrombotic effects of dilute diesel-exhaust inhalation in men with coronary heart disease. *New England Journal of Medicine* 357:1075–1082.

Mills, N. L., H. Tornqvist, and S. D. Robinson, et al. 2005. Diesel exhaust inhalation causes vascular dysfunction and impaired endogenous fibrinolysis. *Circulation* 112:3930–3936.

Mohan, D., C. U. Pittman, and P. H. Steele. 2006. Pyrolysis of wood/biomass for bio-oil: A critical review. *Energy & Fuels* 20:848–889.

Morillo, F., M. Bordons, and I. Gomez. 2001. An approach to interdisciplinarity through bibliometric indicators. *Scientometrics* 51:203–222.

Namkoong, W., E. Y. Hwang, J. S. Park, and J. Y. Choi. 2002. Bioremediation of diesel-contaminated soil with composting. *Environmental Pollution* 119:PII S02697491(01): 00328–1.

Negishi, M., Y. Sun, and K. Shigi. 2004. Citation database for Japanese papers: A new bibliometric tool for Japanese academic society. *Scientometrics* 60:333–351.

Nel, A. E., D. Diaz-Sanchez, D. Ng, T. Hiura, and A. Saxon. 1998. Enhancement of allergic inflammation by the interaction between diesel exhaust particles and the immune system. *Journal of Allergy and Clinical Immunology* 102:539–554.

North, D. C. 1991a. *Institutions, Institutional Change and Economic Performance*. Cambridge, Mass.: Cambridge University Press.

North, D. C. 1991b. Institutions. *Journal of Economic Perspectives* 5:97–112.

Okubo, Y., J. C. Dore, T. Ojasoo, and J. F. Miquel. 1998. A multivariate analysis of publication trends in the 1980s with special reference to South-East Asia. *Scientometrics* 41:273.

Perron, P. 1989. The great crash, the oil price shock, and the unit root hypothesis. *Econometrica: Journal of the Econometric Society* 57:1361–1401.

Prathap, G. 2017. A three-dimensional bibliometric evaluation of recent research in India. *Scientometrics* 110:1085–1097.

Rinia, E. J. 2000. Scientometric studies and their role in research policy of two research councils in the Netherlands. *Scientometrics* 47:363–378.

Rogers, D. and A. J. Hopfinger. 1994. Application of genetic function approximation to quantitative structure-activity relationships and quantitative structure-property relationships. *Journal of Chemical Information and Computer Sciences* 34:854–866.

Rogge, W. F., L. M. Hildemann, M. A. Mazurek, G. R. Cass, and B. R. T. Simoneit. 1993. Sources of fine organic aerosol. 2. Noncatalyst and catalyst-equipped automobiles and heavy-duty diesel trucks. *Environmental Science & Technology* 27:636–651.

Sagai, M., H. Saito, T. Ichinose, M. Kodama, and Y. Mori. 1993. Biological effects of diesel exhaust particles. 1. *In vitro* production of superoxide and *in vivo* toxicity in mouse. *Free Radical Biology and Medicine* 14:37–47.

Sahoo, B. B., N. Sahoo, and U. K. Saha. 2009. Effect of engine parameters and type of gaseous fuel on the performance of dual-fuel gas diesel engines-A critical review. *Renewable & Sustainable Energy Reviews* 13:1151–1184.

Salvi, S., A. Blomberg, and B. Rudell, et al. 1999. Acute inflammatory responses in the airways and peripheral blood after short-term exposure to diesel exhaust in healthy human

volunteers. *American Journal of Respiratory and Critical Care Medicine* 159:702–709.
Schauer, J. J., M. J. Kleeman, G. R. Cass, and B. R. T. Simoneit. 1999. Measurement of emissions from air pollution sources. 2. C_1 through C_{30} organic compounds from medium duty diesel trucks. *Environmental Science & Technology* 33:1578–1587.
Scherf, U. and E. J. List. 2002. Semiconducting polyfluorenes-towards reliable structure–property relationships. *Advanced Materials* 14:477–487.
Schuetzle, D., F. S. C. Lee, T. J. Prater, and S. B. Tejada. 1981. The identification of polynuclear aromatic hydrocarbon (PAH) derivatives in mutagenic fractions of diesel particulate extracts. *International Journal of Environmental Analytical Chemistry* 9:93–144.
Song, C. and X. L. Ma. 2003. New design approaches to ultra-clean diesel fuels by deep desulfurization and deep dearomatization. *Applied Catalysis B-Environmental* 41:207–238.
Song, C. S. 2003. An overview of new approaches to deep desulfurization for ultra-clean gasoline, diesel fuel and jet fuel. *Catalysis Today* 86:211–263.
Stanislaus, A., A. Marafi, and M. S. Rana. 2010. Recent advances in the science and technology of ultra-low sulfur diesel (ULSD) production. *Catalysis Today* 153:1–68.
Sydbom, A., A. Blomberg, and S. Parnia, et al. 2001. Health effects of diesel exhaust emissions. *European Respiratory Journal* 17:733–746.
Wang, D. H., C. V. Nayar, and C. Wang. 2010. *Modeling of stand-alone variable speed diesel generator using doubly-fed induction generator. IEEE PEDG 2010: The 2nd International Symposium on Power Electronics for Distributed Generation Systems* 2010:1–6.
Wang, X., D. Liu, K. Ding, K., and X. Wang. 2012. Science funding and research output: A study on 10 countries. *Scientometrics* 91:591–599.
Xiao, G. G., M. Y. Wang, N. Li, J. A. Loo, and A. E. Nel. 2003. Use of proteomics to demonstrate a hierarchical oxidative stress response to diesel exhaust particle chemicals in a macrophage cell line. *Journal of Biological Chemistry* 278:50781–50790.
Xie, Y. and K. A. Shauman. 1998. Sex differences in research productivity: New evidence about an old puzzle. *American Sociological Review* 63:847–870.
Youtie, J, P. Shapira, and A. L. Porter. 2008. Nanotechnology publications and citations by leading countries and blocs. *Journal of Nanoparticle Research* 10:981–986.
Zhang, Y., M. A. Dube, D. D. McLean, and M. Kates. 2003. Biodiesel production from waste cooking oil: 1. Process design and technological assessment. *Bioresource Technology* 89:1–16.
Zheng, M., G. T. Reader, and J. G. Hawley. 2004. Diesel engine exhaust gas recirculation: A review on advanced and novel concepts. *Energy Conversion and Management* 45:883–900.
Zhou, P. and L. Leydesdorff. 2006. The emergence of China as a leading nation in science. *Research Policy* 35:83–104.
Zhu, Y. F., W. C. Hinds, S. Kim, S. Shen, and C. Sioutas. 2002. Study of ultrafine particles near a major highway with heavy-duty diesel traffic. *Atmospheric Environment* 36:4323–4335.

47 Combustion and Formation of Emissions in Compression Ignition Engines and Emission Reduction Techniques

Shyam Pandey
Amit Kumar Sharma

CONTENTS

47.1 INTRODUCTION

The purpose of 'internal combustion (IC) engines' is to transform the chemical energy of fuel into thermal energy by combustion or oxidation in an engine. The working fluid is an air–fuel mixture before combustion, and combustible gases after combustion. Since the combustible gases only undergo an expansion of work generation, the term 'IC engine' is used. Nevertheless, a combustion product and fluid which undergoes expansion during energy transformation is different in the case of the 'external combustion (EC) engine', i.e. a closed-cycle gas turbine and steam sterling engine. J. J. E. Lenoir developed the first engine in 1860, which used a coal-gas air mixture without any compression mechanism. The induction of the charge and air takes place first and then the charge is ignited by the spark whilst the subsequent expansion produces the work output. This was the first commercial engine, and nearly 5,000 were sold from 1860 to 1865.

Nicolaus A. Otto and Eugen Langen invented an atmospheric engine which operated up to 11% efficiency in 1867, and which is considered a more successful attempt than the previous one. Later on, to overcome the low efficiency and excessive weight shortcomings, four-stroke prototype engines were built in 1876 by Otto, which is considered as one of the great breakthroughs in engine invention, and almost 50,000 units were sold by the end of 1890.

Rudolf Diesel, a German engineer, had come out with a totally different concept of injecting liquid fuel into the hot air environment, resulting in a start of combustion where no spark-plug is required. The advantage was demonstrated in terms of efficiency, which was more feasible than other engines. Later on, it was named a 'diesel engine' and became inevitably an engine for all spheres of application.

47.2 COMBUSTION IN DIESEL ENGINES

The diesel fuel is injected at very high velocities into the combustion chamber through one or more orifices or nozzles in the injector tip. The mechanism of diesel fuel injection transforms the large pressure energy of the fuel into higher kinetic energy, and the fuel is atomized into small drops. The next stage of combustion begins with the combining of fuel particles with air as the vaporization of the atomized fine fuel particles begins. Latent heat vaporization (h_{fg}) is given off by compressed air in the cylinder, which has already attained a high pressure and temperature due to compression. However, the autoignition point of the fuel is lower than the cylinder pressure (P_{cy}) and temperature. The three stages of combustion (atomization and vaporization, fuel–air mixing, and burning) are repeated till the last portion of fuel is injected into the cylinder. High speed photography studies were carried out under normal operating conditions and are shown in Figure 47.1 (Heywood, 1988).

It is clear from the above discussion that autoignition does not start until a fraction of the fuel vaporizes and is mixed with air. The lean 'air–fuel ratio' (A/F) in an engine always leads to a variation of the 'equivalence ratio' (φ) in a combustion chamber

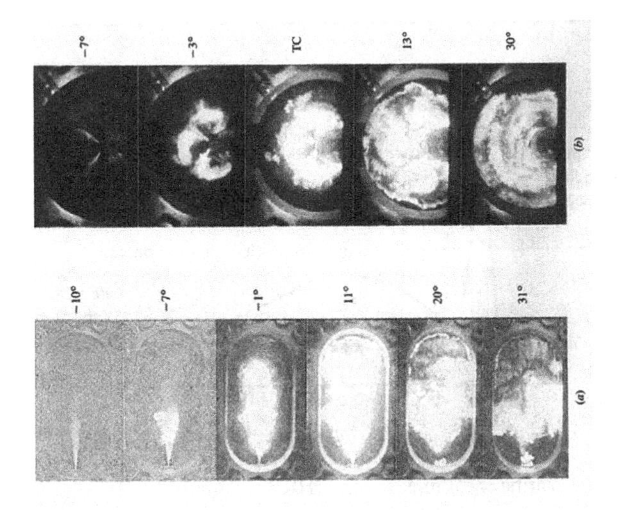

FIGURE 47.1 Combustion of four sprays in 'direct injection' (DI) diesel engine with swirling.

Source: Heywood (1988).

(Stone, 1999). The heterogeneous characteristics of diesel combustion are mainly due to the heterogeneous mixture of the air–fuel in a cylinder, and this is the main cause of prominent pollutants like 'particulate matter' (PM) and 'nitrogen oxides' (NO_x).

47.2.1 Stages of Diesel Engine Combustion

Figure 47.2 shows a typical heat-release diagram which depicts four different stages of 'direct injection' (DI) diesel combustion.

47.2.2 Ignition Delay (ab)

Diesel fuel is injected into the cylinder at the end of compression and done at high pressure that leads to the formation of a fuel jet, which breaks the liquid fuel into small droplets which start vaporizing in the hot-air environment. The time period between the start of fuel injection and combustion is known as 'ignition delay' (ID), although combustion does not start until a flame is visible and a pressure rise is detected inside the cylinder (Taylor, 1985). ID depends on several parameters, such as I_{pr}, I_t, and the CN of fuel. Further, fuels with poor autoignition properties exhibit longer delay periods (DPs).

47.2.3 Rapid Combustion (Premixed Combustion (bc))

The mixture after completion of the DP undergoes fuel-vapor and air mixing, and subsequent reactions within the charge lead to ignition spontaneously. This results in a rapid rise in P_{cy} and temperature due to combustion. Therefore, the rate of burning

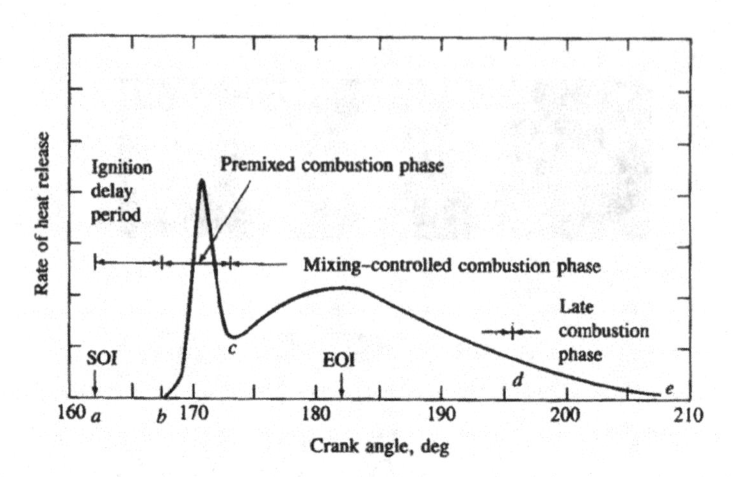

FIGURE 47.2 DI-engine 'heat release rate' (HRR) diagram.

Source: Heywood (1988).

and duration of the premixed phase are mainly related to the extent of DP. The cloud of charge, which ignites spontaneously, causes sharp increase in the P_{cy} and may lead to a knocking sound. This depends on the other parameters in the diesel engine. Thus, the premixed phase is characterized by a sharp increase in pressure and HRR. This phase occurs in a premixed flame, which is characterized by high temperature and a fuel-rich region. The soot predecessors are originated before nucleation of soot and grow in a rich air–fuel region of the subsequently diffusing flame.

47.2.4 Mixing Controlled Combustion (CD)

After completion of the premixed combustion phase, the remaining *un*burned portions of the fuel get compressed and the preparation time (ID) gets shortened. Therefore, in the mixing controlled phase, the combustion reactions are controlled by the rate at which the F/A mixture is formed, or the diffusion of fuel into air, or vice versa, takes place. This phase is also known as the diffusion combustion phase. The diffusion flames are responsible for soot particle formation.

47.2.5 Late Combustion Phase (DE)

As described by Heywood (1988), due to drop in the temperature of the in-cylinder gases, the reaction rates are slowed down during this phase. As the last stage of expansion there is a possibility that a small fraction of energy is yet to be released from the fuel energy associated with the soot particles and fuel-rich gases. Therefore, some of the fuel is burnt, which leads to the oxidation of soot and fuel-rich combustion. This is also known as the late combustion phase.

47.3 DIESEL ENGINE EXHAUST EMISSIONS

In spite of the high η_{tH} and durability of the 'compression ignition (CI) engine' there is a drawback in the form of emissions, which results in environmental concerns. As mentioned previously, NO_x emissions are still a major problem for CI engines. NO_x emissions in DI engines mainly consist of 'nitrogen monoxide' (NO) and 'nitrogen dioxide' (NO_2). The 'spark-ignition (SI) engine' as compared to the DI engine produces more NO_x and PM. However, a DI engine emits less 'carbon monoxide' (CO) and 'unburned hydrocarbon' (UHC).

47.3.1 NO Formation in CI Engines

In diesel engines, combustion starts after completion of the ID. Fuel and air that are mixed together form flammable mixtures around stoichiometric compositions. In the 'premixed combustion phase' the net HRR becomes positive and the first peak is observed in the 'mixing-controlled combustion phase; combustion also occurs in the region where the local φ is close to stoichiometric. NO formation rates depend on the local φ and temperature.

NO forms during combustion. Its formation is explained below (Mellor et al., 1998) and involves thermal, prompt, and nitrous oxide mechanisms.

47.3.1.1 Thermal Mechanism

NO_x formation occurs in the hot combustion gases at sufficiently high temperature (> 1800 K). In the presence of free radicals, when atomic oxygen reacts with free radicals (O, N, H, and OH), NO formation takes place. The Zeldovich mechanism (1946) clearly explains this by the following reactions (Heywood, 1988; Stone, 1999):

$$O + N_2 \leftrightharpoons NO + N \tag{47.1}$$

$$N + O_2 \leftrightharpoons NO + O \tag{47.2}$$

$$N + OH \leftrightharpoons NO + H \tag{47.3}$$

Lavoie et al. (1970) introduced the third reaction (Equation 47.3) whereas the first two were suggested by Zeldovich.

The formation of NO mainly occurs in the high temperature zone, in excess oxygen regions through reaction (Equation 47.1), leaving free atoms of N_2. These combine later with O_2 (Equation 47.2) or with the OH radical (Equation 47.3), initially present in the combustion process, to form NO.

The NO forms can directly be converted to NO_2 by reaction (Equation 47.4). Afterwards, it can be also be converted back to NO by reaction (Equation 47.5; Heywood, 1988).

$$NO + HO_2 \leftrightharpoons NO_2 + OH \tag{47.4}$$

$$NO_2 + O \leftrightharpoons NO + O_2 \tag{47.5}$$

NO_x emissions are harmful for both the environment and health, as they destroy the ozone layer, leading to more UV radiation on earth. Nitric acid vapor formed under favorable conditions affects human health in the form of respiratory and heart disease (Nightingale et al., 2000).

47.3.1.2 Prompt Mechanism

NO is formed in the flame as CH, CH_2, etc. The CH and CH_2 radicals react with N_2 to give the intermediate species, HCN and CN, by the reactions (Mellor et al., 1998):

$$CH + N_2 \leftrightharpoons HCN + N \tag{47.6}$$

$$CH_2 + N_2 \leftrightharpoons HCN + NH \tag{47.7}$$

Thermal NO formation in the burned gas behind the flame front is much higher compared to any NO in the flame front. Therefore, prompt NO formation is small in the case of CI engines. However, prompt NO may be significant (Sher, 1998) in diesel engines with 'exhaust gas recirculation' (EGR).

47.3.1.3 Nitrous Oxide (N_2O) Mechanism

In diesel engines, high-pressure combustion involves decomposition and formation reactions. Therefore, the 'Zeldovich mechanism' alone does not predict exact NO formation. N_2O formation and decomposition may also lead to significant amounts of NO generation. This could be generated in both the premixed and diffusion-controlled combustion (Mellor et al.,1998) phases:

$$N_2O + O \rightleftharpoons 2NO \tag{47.8}$$

$$N_2 + O + M \rightleftharpoons N_2O + M \tag{47.9}$$

By adding the reaction:

$$O_2 + M \rightleftharpoons 2O + M \tag{47.10}$$

The overall balanced chemical reaction is:

$$2O_2 + 2N_2 \rightleftharpoons 4NO \tag{47.11}$$

47.3.1.4 NO_2 Formation

NO_2 emissions from SI engines are almost negligible, since NO is of the order of several thousand ppm while NO_2 is approximately 60 to 70 ppm. However, in the case of CI engines, the fraction of NO_2 emissions typically varies from 10 to 30% of the total NO_x emissions. Hilliard and Wheeler (1979) and Merryman and Levy (1975) presented the following mechanism for the formation of NO_2 in DI engines, in which NO is converted to NO_2 by the reaction of NO with HOO^- radicals. The reactions are:

$$NO + HO_2 \rightleftharpoons NO_2 + OH \tag{47.12}$$

and, in the post flame region, NO_2 reacts with atomic oxygen and is converted into NO and O_2 (produces NO again):

$$NO_2 + O \rightleftharpoons NO + O_2 \tag{47.13}$$

If high turbulence prevails in the engine, mixing of the burnt gases with colder air or an air–fuel mixture may quench the decomposition reactions, which leads to suppression of the conversion of NO_2 into NO. In diesel engines at low loads, the ratio of NO_2/NO becomes high due to the freezing of the decomposition reactions (Hilliard and Wheeler, 1979). The combustion model proposed by Dec (1997) suggested premixed combustion occurs in the rich mixture ($\phi = 2$ to 4). Thus, neither thermal NO nor prompt NO is likely to be formed in significant concentrations during premixed combustion. Most of the NO is formed during the mixing controlled combustion phase. The NO formation region is located around the periphery of the spray.

47.3.2 Particulate Matter (PM)

PMs are the suspended solid and liquid particles which get collected on dilution of the exhaust. In addition, the temperature of the filtering is not more than 52°C. PM besides NO_x emissions are a major pollutant from combustion in CI engines. Incomplete combustion of hydrocarbon fuels results in PM, some of which is contributed by the lubricating oil (Heywood, 1988). Particulates in CI engines are due to deposition of organic compounds on carbonaceous material. PM is a mixture of C, HC, S compound, and ash, and they may vary in size, composition, solubility, and toxicity (Burtscher, 2005). The soot formation model was proposed by Dec (1997) and is illustrated in Figure 47.3.

Particles from CI engines can be divided into 'soot particles' (SOF) and 'volatile fractions' (VOF) . Solid fractions include carbonaceous agglomerates (soot) and inorganic ash, while volatile fractions include organic compounds (HC) and inorganic compounds (H_2SO_4 and sulfates).

Sulfur reacts with water vapor to form H_2SO_4 and along with SOF/VOF starts condensing on solid particles at lower temperatures. Figure 47.4 shows a composition of PM as presented by Kittelson (1998).

The mass of total particles is mainly due to the solid fraction and its constituents (soot and ash) due to the combustion process. The most characteristic pollutants emitted by CI engines are soot whose formation starts with the fuel and air reaction in the premixed flame. High temperature and the absence of O_2 cause large hydrocarbon molecules to undergo thermal decomposition called cracking. The process of cracking is similar to pyrolysis, which produces precursors to soot formation. During combustion, the formation of PM starts with a few molecules (12 to 22 C atoms and an H/C ratio of about 2) and ends up with a particle diameter of about 100 nm which consist of 20–30 nm spherules containing 10^5 C (carbon) atoms and an H/C ratio of about 0.1 (Heywood, 1988). Unsaturated HCs (hydrocarbons) such as C_2H_4, C_2H_2, and polycyclic aromatic HCs are generally present in unsaturated HCs. These conditions are generally characterized by a 1,000–2,800 K temperature and a 50–100 atm pressure (Heywood, 1988). The kinetics of reaction and the diffusion process cause soot particles to oxidize on the surface of the spray. Overall, a 90% oxidation of soot

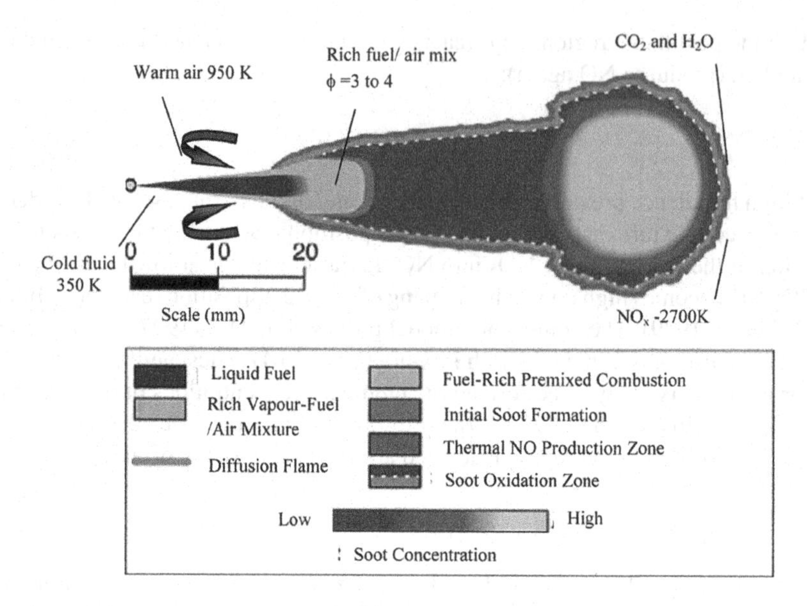

FIGURE 47.3 Dec combustion model for DI diesel engine.

Source: Dec (1997).

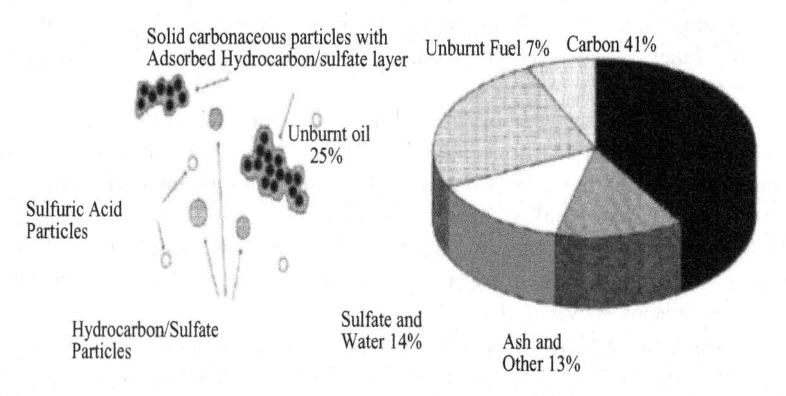

FIGURE 47.4 Soot particle composition.

Source: Kittelson (1998).

has been suggested by the different studies. A higher rate of oxidation decreases tail-pipe soot emissions.

The measurement of PM can be done by a smoke meter for smoke opacity measuring the filtration method, smoke number (SN) -Bosch Scale or transmission electron microscopy (TEM)-particle-size, aggregate size and morphology of soot-particles.

However, various means are adopted to cut down the hazardous effect of PM emissions. A 'common rail and unit injection system' is a high pressure, electrically controlled, fuel injection system capable of precisely controlling the fuel and air mixing process in the cylinder which, in turn, reduces PM (smoke in particular) from the tail pipe. Oxidation catalysts and particulate traps are effective only at high temperatures (500–600°C). Fuel oxygenates and fuel-borne oxygen reduce the formation of soot which is the main component of PM. Oxygenates increase the fraction of free-radicals such as O, OH, and HCO. This causes carbon to oxidize into CO and CO_2. This reduces the availability of C to form a soot precursor. Free radical concentration can also oxidize the soot precursor in the diffusion flame, limiting PAH growth and inhibiting soot particles. Benvenutti et al. (2005) reported that methyl radicals also lead to the formation of soot oxidizing radicals.

47.3.3 'Unburned Hydrocarbons' (UHC) and CO

The UHC emissions from CI engines are comparatively much smaller than from SI engines. As the boiling points of hydrocarbon compounds associated with diesel fuel are higher, they have higher molecular weights than gasoline. This results in complex UHC products. Diesel engine fuel spray shows wide variation in the 'equivalence ratio' which occurs from the spray cone to the outer boundary. The fuel close to the spray boundary has a low equivalence ratio ($\phi = 0.3$) which is leaner than the limit of combustion. Therefore, it will not autoignite the fast reaction front. The magnitude of UHC depends on the quantity of fuel injected during ID, the mixing rate with air, and whether cylinder conditions are favorable to autoignition. Therefore, over-mixing (over-leaning) is one of the important causes of UHC formation during normal operation, which becomes more predominant in a light-load operation when the ID is longer (Greeves and Khan, 1977; Yu and Shahed, 1981).

However, under-mixing which results in over-rich mixtures is also considered as one of the major causes of UHC emission. The fuel that leaves the injector at low velocity, relatively late in the combustion process, is another important source of UHC (Heywood, 1988). This mechanism occurs frequently, especially when injection of fuel is being done after the ID period and during over-fueling conditions, such as while accelerating at high loads. Advancing the fuel-injection timing promotes oxidation of HCs due to the high combustion temperature and high pressure. CO is a product of incomplete combustion, usually when the engine is operating at very rich air–fuel mixture conditions. Accordingly, the CO emissions from IC engines are controlled by the fuel–air φ ratio, since SI engines operate close to the stoichiometric fuel–air equivalence ratio at part load, but are fuel rich at full load. CO concentrations in the exhaust increase with the increasing fuel–air φ ratio, as the excess fuel increases (Harrington and Shishu, 1973). Thus, it needs to be controlled at high engine loads. To control CO emission in the case of SI engines, exhaust gas treatment is required, such as oxidation catalysts. Combustion in a lean region causes a very low level of CO emissions. In the case of well-designed DI engines, CO emissions are low enough to be considered irrelevant.

47.4 NO_X-REDUCING TECHNIQUES

NO_x emissions can be reduced by employing certain common techniques in DI engines.

47.4.1 Injection Timing

This technique is used to control NO_x emissions. In DI engines, NO_x emissions can be reduced by retarding the 'injection timing' (IT). Thus, the combustion process is shifted toward the expansion stroke, causing a reduction in the peak pressure and temperature, which slows down the reaction to form NO_x. However, this method of reducing NO_x also increases fuel consumption along with the formation of PM. The duration of the diffusion combustion phase is extended by delaying the injection timing, which results in a lower temperature during expansion. The change in fuel injection timing provides a classic example of the NO_x–PM trade-off which is a characteristic of CI engines.

47.4.2 EGR

In this technique, NO_x emissions can be reduced by recirculating a part of the exhaust gas. However, at high loads, PM emissions can be observed; thus there is a trade-off between NO_x and smoke emission. Hence, this technique is mainly used in low-load and low-speed conditions. While recirculating a fraction of the exhaust gases, the O_2 concentration in the cylinder decreases and acts as a diluent agent in the combustion chamber. The pollutants can be minimized by changing the thermodynamic properties and O_2 concentration without affecting too much the power and thermal efficiency; and this is the principal reason for applying EGR in CI engines. At stoichiometric combustion, a very small fraction of oxygen is left as a combustion product, while at idle-load conditions about 20% oxygen is available at the tail pipe of modern diesel engines. Thus, with the increase in the engine load, the concentration of CO_2 and H_2O and the specific heat of the exhaust gases increase. The flame temperature and the maximum temperature of the working fluid will be lowered with the increase in CO_2 and H_2O. Therefore, at low engine load operation, higher rates of EGR can be tolerated by the engine, whereas a small rate of EGR is sufficient at high engine load operations.

The decrease in O_2 concentration and combustion temperatures reduces the rate of oxidation of species, including soot, UHC, and CO (Jacobs et al., 2005). Studies conducted on diesel engines have shown that the EGR can reduce NO_x substantially (Lapuerta et al., 2000). Ladommatos et al. (1996a–b, 1997) worked extensively on the effect of EGR and have published three papers which suggested the following mechanisms are responsible for NO_x control:

1. Dilution mechanism: Increased level of inert gas concentration in the charge increases ID or requires a greater time for premixing and long burn duration, which is known as EGR's dilution effect. This results in lowered flame temperatures. Moreover, the partial pressure of oxygen is reduced at low temperatures and the reduced availability of O_2 is expected to suppress NO_x formation.
2. Thermal mechanism: Higher specific heat capacity and a thermal mass of recirculated CO_2 and H_2O compared to O_2 and N_2 present in the fresh air. Thus, it decreases the temperature rise in the combustion chamber after the same heat

release. This also helps to reduce heat transfer losses, which occur during and after the combustion event, at that time when the surface area and volume ratio of the engine is at a maximum. Thus, the heat loss to the coolant reduces due to a low peak combustion temperature.

3. Chemical mechanism: As the combustion process progresses, water vapor and carbon dioxide molecules undergo dissociation (endothermic dissociation of water vapor); the rate of heat release is modified. This leads to NO_x formation, and the heat absorbed during dissociation results in a decrease in the flame temperature.

47.5 DIESEL EXHAUST AFTER-TREATMENT TECHNOLOGIES

The available technology and measures used to control combustion inside the cylinder and hence to cut down emissions cannot provide a large reduction in harmful pollutants. So, to meet the stringent emissions norms, the after-treatment of exhaust gases has been used for the efficient conversion of harmful pollutants into harmless substances. Therefore, in CI engines, to reduce emissions of harmful pollutants like CO, UHC, NO and soot or particulates, different techniques are used. Thus, 'diesel oxidation catalysts' (DOCs), 'diesel particulate filters' (DPFs), as well as the highly efficient NO_x after-treatment technologies will become mandatory in the future.

47.5.1 DPFs

This method is based on filtration of PM by using ceramic materials such as cordierite ($2MgO_2$. Al_2O_3. $5SiO_2$), aluminum titanate, and SiC. Figure 47.5 shows a honeycomb ceramic monolith that traps the PM as the gas flows through the porous

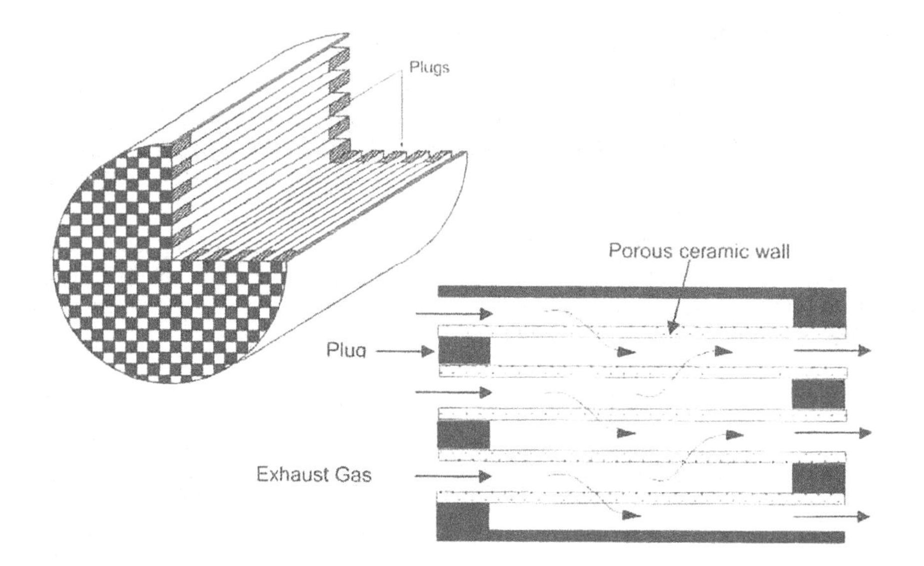

FIGURE 47.5 Ceramic wall flow filter for diesel particulates.

walls of its cells; thus it is also known as a 'ceramic wall flow filter'. The flow channels are arranged in such a manner that alternate ends of these cells are plugged at one end and kept open at the opposite end. Therefore, exhaust gas enters from the upstream end, and while flowing along the length of the channel flow the direction is changed and it enters an adjacent channel to leave the filter. It is evident that the long-term application of DPFs leads to produce a back pressure due to the blockage of the passage by PM which in turn manifests in terms of a fuel penalty. So, there is a need to clean accumulated PM by burning or oxidizing, which is also known as regeneration. Hot burning gas raises the temperature of the DPF which converts soot into CO_2.

47.5.2 DOCs

Since early 1990s, DOCs have been widely used in light-duty diesel commercial vehicles in European countries. DOCs consists of a ceramic honeycomb monolith which has a large number of parallel and straight open channels for a flow of exhaust gases. The flow through these channels is laminar. The inorganic material (Al_2O_3, SiO_2, TiO_2), ceramic, or metal with dispersed Pt, Pd, or Rh is used on its surface (van Basshuysen et al., 2004). The oxidation of UHC and CO are as follows:

$$C_yH_n + \left(1 + \frac{n}{4}\right)_y \rightarrow O_2yCO_2 + \frac{n}{2}H_2O \tag{47.14}$$

$$CO + \frac{1}{2}O_2 \rightarrow CO_2 \tag{47.15}$$

$$CO + H_2O \rightarrow CO_2 + H_2 \tag{47.16}$$

47.5.3 NO_x-Reducing After-Treatment Technologies

Two 'after-treatment technologies' could be used to reduce NO_x emissions from CI engines. As diesel engines operate with excess air, the exhaust gas is oxygen rich. Therefore, in the oxygen rich environment, conversion of NO_x needs additional reducing agents, termed 'reductants'. So, 'selective catalytic reduction' (SCR) and a 'lean NO_x trap' are discussed below.

The SCR of NO_x employs anhydrous ammonia or urea as a reducing agent. Urea is injected into the exhaust, which undergoes thermal decomposition and hydrolysis so as to become converted into ammonia.

Urea hydrolysis:

$$(NH_2)_2CO + H_2O \rightarrow CO_2 + 2NH_3 \tag{47.17}$$

NO_x conversion:

$$4NO + 4NH_3 + O_2 \rightarrow 4N_2 + 6H_2O \tag{47.18}$$

$$6NO_2 + 8NH_3 \rightarrow 7N_2 + 12H_2O \tag{47.19}$$

A lean NO_x trap that is used to reduce NO_x emissions is also known as an 'NSC' or a 'NO_x absorber'. This converter catalyzes the reduction of NO to N_2. Hence, platinum is used as a catalyst where the NSC converter contains barium compounds as storage components, since barium has a very high affinity for NO_x.

47.6 EMISSIONS AND THEIR HEALTH EFFECTS

The exhaust from diesel engines is a mixture of a number of components in gas or particle form. Gaseous components of diesel engine exhaust include CO_2, O_2, N_2, water vapor, CO, nitrogen compounds, sulfur compounds, and numerous low-molecular-weight HCs. PM consists of fine and ultra-fine particles, which have a large surface area and a good medium for absorbing organic compounds. The ultra-fine particles are respirable and can reach deep into the lungs. This indicates a human health hazard when people are exposed to engine exhaust. The hazards include non-cancer respiratory effects and lung cancer. NO_x, when it reacts with moisture, forms small particles, which penetrate inside the lungs and worsen the respiratory system causing emphysema and bronchitis. Nightingale et al. (2000) found that sulfur oxides (SO_x) nitrogen oxides (NO_x) when mixed with water vapor (H_2O) present in the air form sulfuric acid and nitric acid, resulting in acid rain. This causes air pollution and adversely affects the respiratory system of those prone to diseases like chronic obstructive pulmonary disease (pulmonary ventilation provides the air flow function between the atmosphere and lungs), asthma, pneumonia, tuberculosis, and lung cancer. Kahn et al. (1988) found severe effects from NO_x and sulfur compounds on the nervous system and which increased chronic bronchitis and asthma. In another study, breathing problems, nausea, and an increase in the sensation of burning of the eyes were observed (Gamble et al., 1987).

Edling and Axelson (1984) revealed that PMs with the metal composition of iron, zinc, vanadium, and nickel have a direct effect on the pulmonary and cardiovascular system, cause variations in the heart rate, and cause arrhythmias that may lead ultimately to death due to blockage of *atrioventricular*. Ultrafine particles (< 100 nm) present in PM cause severe effects of inflammation, damage the epithelium, and cause phagocytosis inhibition.

47.7 ADDITIVES FOR DIESEL ENGINES

Compounds containing oxygen can be used as fuel oxygenates to reduce soot emission or to supply oxygen to form CO_2 and reduce carbon rich particles. Various oxygenates generally used along with diesel are alcohol (Bilgin et al., 2002; Huang and Lu, 2004a–b), methyl or ethyl esters or biodiesel (Jimenez-Espadafor et al., 2012; Pandey et al., 2012; Sharp et al., 2000), ethers (Liotta and Montalvo, 1993; Youn et al., 2011), and ethylene glycol monoacetate (Lin and Huang, 2003).

The capacity of oxygenates to reduce smoke capacity and PM, CO, and NO_x emissions depends on the chemical structure and amount of oxygen in the fuel.

Straight-chained vegetable oils have long been considered as alternative fuel to replace petrodiesel fuels (Shahid and Jamal, 2008). A start was made by Rudolf

Diesel when he demonstrated the use of 100% peanut oil to run his engine at an exhibition in Paris. Since vegetable oils are usually triglycerides (having branched hydrocarbon chains of fatty acids), they possess some unfavorable physicochemical properties which impede their path to be directly used as fuel for CI engines. The viscosity of vegetable oils (35 cSt to 60 cSt at 40°C) is much higher compared to that of diesel (4 cSt at 40°C), and this is due to the large molecular weight and complex molecular structure of vegetable oils. To overcome these difficulties for use successfully in diesel engines, vegetable oils or triglycerides undergo dilution, pyrolysis, microemulsions, and transesterification processes (Ali and Hanna, 1994). Transesterification is the process of splitting vegetable oils into light molecules. The merits of using biodiesel instead of petrodiesel are its comparable energy density, cetane number, heat of vaporization, higher combustion efficiency, low sulfur and aromatic content, renewability, and stoichiometric air/fuel ratio (Agarwal, 2007). Biodiesel is nontoxic and its biodegradation is faster than petrodiesel. Greenhouse gas effects are least observed in the case of biodiesel (Balat and Balat, 2008; Murugesan et al., 2009).

Oxygenates are generally used as a blend with petrodiesel fuel. Depending on the level of engine modifications, the required alcohol can be used through fumigation, dual fuel injection, and even as neat alcohol. However, the last option requires extensive engine modification, including the injection system, fueling, and injector. It may be noted that fumigation is an attractive method as it allows the utilization of alcohol in engines with only minor modification and without issues of miscibility of alcohol (methanol and ethanol in particular) with petrodiesel fuel.

47.8 OXYGENATES AND DIESEL COMBUSTION

Extensive investigation has been done to understand the effect of oxygenates on the performance and emissions of DI engines. Kozak et al. (2008, 2009) have evaluated the potential of oxygenates to reduce PM emissions. Investigators have found that the molecular structure and oxygen content of oxygenates primarily control PM emissions. Ironically, there is still a lack of clarity regarding the parameter (molecular structure or oxygen content) that will predominantly reduce PM emissions. The mechanisms by which oxygenated compounds affect pollutant formation in CI engines are still widely debated by researchers.

Nabi et al. (2000) investigated the effect of O_2 content on smoke emissions. Six different oxygenates were studied to understand their effect on emissions of single cylinder diesel engines. They found that smoke emissions decreased linearly with the oxygen content of oxygenates which reached almost zero at oxygen contents of 38% (wt) or more. Miyamoto et al. (1998) and Cheng et al. (2002) reported that the reduction in the PM level was largely influenced by the oxygen content of the blends rather than the chemical structure of the oxygenates.

However, several investigators believe that in addition to oxygen content, the chemical structure of an oxygenate plays a significant role in determining its effectiveness in reducing PM emissions. Yeh et al. (2001) examined the effect of molecular structure on PM emissions by using 14 different oxygenates where the blends were prepared keeping the same oxygen content. Oxygenates found to be more effective were alcohol; more volatile compounds showed a more superior performance than heavier ones. They found that ethers, esters, and carbonates outperformed in PM reduction. However, the effect of fuel bound oxygen on NO_x emissions is not clear. A few studies have reported a small increase in NO_x whereas others have concluded that the emissions of oxides of nitrogen remain essentially unchanged. Although the addition of oxygen modifies the fuel as a whole, these changes in NO_x formation cannot be characterized independently of oxygen content. In addition to that, testing conditions and engine characteristics may have a significant influence on NO_x emissions.

REFERENCES

Agarwal, A. K. 2007. Biofuels (alcohols and biodiesel) applications as fuels for internal combustion engines. *Progress in Energy and Combustion Science* 33: 233–271.

Ali, Y. and M. Hanna. 1994. Alternative diesel fuels from vegetable oils. *Bioresource Technology* 50: 153–163.

Balat, M. and H. Balat. 2008. Critical review of bio-diesel as a vehicular fuel. *Energy Conversion and Management* 49: 2727–2741.

Benvenutti, L. H., C. S. T. Marques, and C. A. Bertran. 2005. Chemiluminescent emission data for kinetic modeling of ethanol combustion. *Combustion Science and Technology* 177: 1–26.

Bilgin, A., O. Durgun, and Z. Sahin. 2002. The effects of diesel-ethanol blends on diesel engine performance. *Energy Sources* 24: 431–440.

Burtscher, H. 2005. Physical characterization of particulate emissions from diesel engines: A review. *Journal of Aerosol Science* 36: 896–932.

Cheng, A. S., R. W. Dibble, and B. A. Buchholz. 2002. The effect of oxygenates on diesel engine particulate matter. *SAE Technical Paper* 2002-01-1705.

Dec, J. 1997. A conceptual model of DI Diesel combustion based on laser-sheet imaging. *SAE Technical Paper* 970873.

Edling, C. and O. Axelson. 1984. Risk factors of coronary heart disease among personnel in a bus company. *International Archives of Occupational and Environmental Health* 54: 181–183

Gamble, J., W. Jones, and S. Minshall. 1987. Epidemiological-environmental study of diesel workers: Chronic effects of diesel exhaust on the respiratory system. *Environmental Research* 44: 6–17

Greeves, G. and I. Khan, C. H. T. Wang, and I. Fenne. 1977. Origins of hydrocarbon emissions from diesel engines. *SAE Technical Paper* 770259.

Harrington, J. A. and R. C. Shishu. 1973. A single-cylinder engine study of the effects of fuel type, fuel stoichiometry, and hydrogen-to-carbon ratio and CO, NO, and HC exhaust emissions. *SAE Technical Paper* 730476.

Heywood, J. B. 1988. *Internal Combustion Engines Fundamentals*. New York, NY: McGraw-Hill Education.

Hilliard, J. C. and R. W. Wheeler. 1979. Nitrogen dioxide in engine exhaust. *SAE Technical Paper* 790691.

Huang, Z., H. Lu, and D. Jiang, et al. 2004a. Combustion behaviors of a compression-ignition engine fuelled with diesel/methanol blends under various fuel delivery advance angles. *Bioresource Technology* 95: 331–341.

Huang, Z. H., H. B. Lu, and D. M. Jiang, et al. 2004b. Engine performance and emissions of a compression-ignition engine operating on the diesel-methanol blends. *Proceedings of the Institution of Mechanical Engineers, Part D: Journal of Automobile Engineering* 218: 435–447.

Jacobs, T. J. and S. V. Bohac, D. N. Assanis, and P. G. Szymkowicz. 2005. Lean and rich premixed compression ignition combustion in a light-duty diesel engine. *SAE Technical Paper* 2005-01-0166.

Jimenez-Espadafor, F. J., M. Torres, and J. A. Velez, E. Carvajal, and J. A. Becerra. 2012. Experimental analysis of low temperature combustion mode with diesel and biodiesel fuels: A method for reducing NO_x and soot emissions. *Fuel Processing Technology* 103: 57–63.

Kahn, G., P. Orris, and J. Weeks 1988. Acute overexposure to diesel exhaust: Report of 13 cases. *American Journal of Industrial Medicine* 13: 405–406.

Kittelson, D. B. 1998. Engines and nanoparticles: A review. *Journal of Aerosol Science.* 29: 575–588.

Kozak, M., J. Merkisz, P. Bielaczyc and A. Szczotka. 2008. The Influence of synthetic oxygenates on Euro IV diesel passenger car exhaust emissions. Part 3. *SAE Technical Paper* 2008-01-2387.

Kozak, M., J. Merkisz, P. Bielaczyc and A. Szczotka. 2009. The influence of oxygenated diesel fuels on a diesel vehicle PM/NO_x emission trade-off. *SAE Technical Paper* 2009-01-2696.

Ladommatos, N., S. Abdelhalim, H. Zhao, and Z. Hu. 1996a. The dilution, chemical, and thermal effects of exhaust gas recirculation on diesel engine emissions. Part 1: Effect of reducing inlet charge oxygen. *SAE Technical Paper* 961165.

Ladommatos, N., S. Abdelhalim, H. Zhao, and Z. Hu. 1996b. The dilution, chemical, and thermal effects of exhaust gas recirculation on diesel engine emissions. Part 2: Effects of carbon dioxide. *SAE Technical Paper* 961167.

Ladommatos, N., S. Abdelhalim, H. Zhao, and Z. Hu. 1997. The Dilution, chemical, and thermal effects of exhaust gas recirculation on diesel engine emissions. Part 3: Effects of water vapor. *SAE Technical Paper* 971659.

Lapuerta, M. and J. J. Hernandez, and F. Gimenez. 2000. Evaluation of exhaust gas re-circulation as a technique for reducing diesel engine NOx emissions. *Proceedings of the Institution of Mechanical Engineers, Part D: Journal of Automobile Engineering* 214: 85–93.

Lavoie, G. A., J. B. Heywood and J. C. Keck. 1970. Experimental and theoretical investigation of nitric oxide formation in internal combustion engines. *Combustion Science and Technology* 1: 313–326.

Lin, C. Y. and J. C. Huang. 2003. An oxygenating additive for improving the performance and emission characteristics of marine diesel engines. *Ocean Engineering* 30: 1699–1715.

Liotta, F. J. and D. Montalvo. 1993. The effect of oxygenated fuels on emissions from a modern heavy-duty diesel engine. *SAE Technical Paper* 932734.

Mellor, A. M., J. P. Mello, K. P. Duffy, W. L. Easley, and J. C. Faulkner. 1998. Skeletal mechanism for NO_x chemistry in diesel engines. *SAE Technical Paper* 981450.

Merryman, E. L. and A. Levy. 1975. Nitrogen oxide formation in flames: The roles of NO_2 and fuel nitrogen. *Symposium (International) on Combustion* 15: 1073–1083.

Miyamoto, N., H. Ogawa, and N. M. Nurun, et al. 1998. Smokeless, low NO_x, high thermal efficiency, and low noise diesel combustion with oxygenated agents as main fuel. *SAE Technical Paper* 980506.

Murugesan A, C. Umarani, R. Subramanian, and N. Nedunchezhian. 2009. Bio-diesel as an alternate fuel for diesel engines: A review. *Renewable and Sustainable Energy Reviews* 3: 653–662.

Nabi, M., M. Minami, H. Ogawa, and N. Miyamoto. 2000. Ultra low emission and high performance diesel combustion with highly oxygenated fuel. *SAE Technical Paper* 2000-01-0231.

Nightingale, J. A., R. Maggs, and P. D. Cullinan, et al. 2000. Airway inflammation after controlled exposure to diesel exhaust particulates. *American Journal of Respiratory and Critical Care Medicine* 162: 161–166.

Pandey, S, A. Sharma, and P. K. Sahoo. 2012. Experimental investigation on the performance and emission characteristics of a diesel engine fuelled with ethanol, diesel and Jatropha based biodiesel blends. *International Journal of Advances in Engineering & Technology* 4: 341–353.

Shahid, E. M. and Y. Jamal. 2008. A review of biodiesel as vehicular fuel. *Renewable and Sustainable Energy Reviews* 12: 2484–2494.

Sharp, C., S. Howell, and J. Jobe. 2000. The effect of biodiesel fuels on transient emissions from modern diesel engines. Part 1: Regulated emissions and performance. *SAE Technical Paper* 2000-01-1967.

Sher, E. 1998. *Handbook of Air Pollution from Internal Combustion Engines: Pollutant Formation and Control*. London: Academic Press.

Stone, R. 1999. *Introduction to Internal Combustion Engines*. London: Palgrave.

Taylor, C. F. 1985. *The Internal-Combustion Engine in Theory and Practice: Combustion, Fuels, Materials, Design*, 2nd edn. Cambridge, MA: MIT Press.

van Basshuysen, R. and F. Schaefer. 2004. *Internal Combustion Engine Handbook*. Warrendale, PA: Society of Automotive Engineers.

Yeh, L., D. Rickeard, and J. Duff, et al. 2001. Oxygenates: An evaluation of their effects on diesel emissions. *SAE Technical Paper* 2001-01-2019.

Youn, I. M., S. H. Park, H. G. Roh, and C. S. Lee. 2011. Investigation on the fuel spray and emission reduction characteristics for dimethyl ether (DME) fueled multi-cylinder diesel engine with common-rail injection system. *Fuel Processing Technology* 92: 1280–1287.

Yu, R. and S. Shahed 1981. Effects of injection timing and exhaust gas recirculation on emissions from a D.I. diesel engine. *SAE Technical Paper* 811234.

48 Bioremediation of Petroleum Hydrocarbons in Contaminated Soils

A Review of the Research

Ozcan Konur

CONTENTS

48.1 INTRODUCTION

Crude oils have been primary sources of energy and fuels, such as petrodiesel. However, significant public concerns about the sustainability, price fluctuations, and adverse environmental impact of crude oils have emerged since the 1970s (Ahmadun et al., 2009; Atlas, 1981; Babich and Moulijn, 2003; Kilian, 2009; Perron, 1989). Thus, biooils (Bridgwater et al., 1999; Bridgwater and Peacocke, 2000; Czernik and Bridgwater, 2004) and biooil-based biodiesel fuels (Chisti, 2007; Hill et al., 2006; Hu et al., 2008) have emerged as alternatives to crude oils and crude oil-based petrodiesel fuels, respectively, in recent decades. Nowadays, although biodiesel fuels are being used increasingly, petrodiesel fuels are still used extensively in the transportation and power sectors (Konur, 2021a–ag). Therefore, there has been great public interest in petrodiesel fuels (Birch and Cary, 1996; Khalili et al., 1995; Rogge et al., 1993; Song, 2003; Song and Ma, 2003).

However, petroleum hydrocarbons contaminate soil (Maliszewska-Kordybach, 1996; Nadal et al., 2004; Wilcke, 2000) with significant ecotoxicity to terrestrial and aquatic organisms (Agarwal et al., 2009; Eom et al., 2007; Sverdrup et al., 2002), resulting in much public concern regarding the potential loss of biodiversity (Brussaard, 1997; Brussaard et al., 2007; Wall and Moore, 1999; Wall et al., 2015). Therefore, it has been necessary to reduce the amount of toxic petroleum hydrocarbons in contaminated soil (Gan et al., 2009; Johnsen et al., 2005; Wilson and Jones, 1993).

Furthermore, for the efficient progression of the research in this field, it is necessary to develop efficient incentive structures for the primary stakeholders and to inform these stakeholders about the research (Konur, 2000, 2002a–c, 2006a–b, 2007a–b; North, 1991a–b).

Although there have been a number of reviews and book chapters in this field (Gan et al., 2009; Johnsen et al., 2005; Wilson and Jones, 1993), there has been no review of the 20-most-cited articles. Thus, this chapter reviews these articles by highlighting their key findings on the bioremediation of petroleum hydrocarbons in contaminated soil. Then, it discusses these key findings.

48.2 MATERIALS AND METHODOLOGY

The search for the literature was carried out in the 'Web of Science' (WOS) database in February 2020. It contains the 'Science Citation Index Expanded' (SCI-E), the 'Social Sciences Citation Index' (SSCI), the 'Book Citation Index–Science' (BCI-S), the 'Conference Proceedings Citation Index-Science' (CPCI-S), the 'Emerging Sources Citation Index' (ESCI), the 'Book Citation Index-Social Sciences and Humanities' (BCI-SSH), the 'Conference Proceedings Citation Index-Social Sciences and Humanities' (CPCI-SSH), and the 'Arts and Humanities Citation Index' (A&HCI).

The keywords for the search of the literature are collated from the screening of abstract pages for the first 1,000 highly cited papers on petrodiesel fuels. These keywords sets are provided in the Appendix of the related chapter (Konur, 2021p).

The 20-most-cited articles are selected for this review and the key findings of the review are presented and discussed briefly.

48.3 RESULTS

Bento et al. (2005) evaluate the comparative bioremediation of soils contaminated with 'total petroleum hydrocarbons' (TPH) and diesel fuels by 'natural attenuation', 'biostimulation', and 'bioaugmentation' in a paper with 358 citations. Additionally, they monitored the number of diesel-degrading microorganisms and microbial activity as indexed by the dehydrogenase assay. They collected soils contaminated with diesel oil in the field from Long Beach, California, USA and Hong Kong, China. After 12 weeks of incubation, they observed that all three treatments showed differing effects on the degradation of light (C_{12}–C_{23}) and heavy (C_{23}–C_{40}) fractions of TPH in the soil samples. Bioaugmentation of the Long Beach soil showed the greatest degradation in the light (72.7%) and heavy (75.2%) fractions of TPH. Natural attenuation was more effective than biostimulation (addition of nutrients), most notably in the Hong Kong soil. They observed the greatest microbial activity (dehydrogenase activity) with bioaugmentation of the Long Beach soil (3.3-fold) and with natural attenuation of the Hong Kong sample (4.0-fold). The number of diesel fuel-degrading microorganisms and heterotrophic population was not influenced by the bioremediation treatments. They conclude that soil properties and the indigenous soil microbial population affected the degree of biodegradation with the implication that detailed site-specific characterization studies were needed prior to deciding on the proper bioremediation method.

Weissenfels et al. (1992) study the degradation of petroleum hydrocarbons, 'polycyclic petroleum hydrocarbons' (PAHs), in contaminated soils, from two different industrial sites under simulated land treatment conditions in a paper with 338 citations. They observed that soil samples from a former impregnation plant (soil A) showed high degradation rates of PAHs by autochthonous microorganisms, whereas PAHs in material of a closed-down coking plant (soil B) were not degraded even after inoculation with bacteria known to effectively degrade PAHs. As they observed rapid PAH biodegradation in soil B after PAHs were extracted and restored into the extracted soil material, the kind of PAH binding in soil B completely prevented biodegradation. Sorption of PAHs onto extracted material of soil B followed a two-phase process (fast and slow); the latter was discussed in terms of migration of PAHs into soil organic matter, representing less accessible sites within the soil matrix. Such adsorbed PAHs were non-bioavailable and thus non-biodegradable. By eluting soil B with water, they detected no biotoxicity, assayed as inhibition of bioluminescence, in the aqueous phase. When treating soil A analogously, they observed a distinct toxicity, which was reduced relative to the amount of activated carbon added to the soil material. They conclude that sorption of organic pollutants onto soil organic matter significantly affected biodegradability as well as biotoxicity.

Das and Mukherjee (2007) compare the efficiency of *Bacillus subtilis* DM-04 and *Pseudomonas aeruginosa* M and NM strains isolated from a petroleum contaminated soil sample from North-East India for the biodegradation of petroleum hydrocarbons in soil and a shake flask study in a paper with 317 citations. These bacterial strains could utilize these hydrocarbons as a sole source of carbon and energy. They observed that bioaugmentation of a TPH contaminated microcosm with *P. aeruginosa* M and NM consortia and a *B. subtilis* strain showed a significant reduction of TPH levels in treated soil as compared to control soil at the end of the experiment (120 d). *P. aeruginosa* strains were more efficient than *B. subtilis* strains in reducing the TPH content from the medium. The plate count technique indicated expressive growth and biosurfactant production by exogenously seeded bacteria in petroleum rich soil. They conclude that *B. subtilis* DM-04 and *P. aeruginosa* M and NM strains could be effective for *in situ* bioremediation of petroleum hydrocarbons.

Vinas et al. (2005) study the bacterial community dynamics and biodegradation processes in a highly creosote-contaminated soil undergoing a range of laboratory-based bioremediation treatments in a paper with 261 citations. They monitored the dynamics of the eubacterial community, the number of heterotrophs and PAH degraders, and TPH and PAH concentrations during the bioremediation process. They observed that TPH and PAHs were significantly degraded in all treatments (72–79% and 83–87%, respectively), and the biodegradation values were higher when nutrients were not added, especially for benzo(a)anthracene and chrysene. The moisture content and aeration were the key factors associated with PAH bioremediation. Neither biosurfactant addition, bioaugmentation, nor ferric octate addition led to differences in PAH or TPH biodegradation compared to biodegradation with nutrient treatment. All treatments resulted in a high first-order degradation rate during the first 45 days, which was markedly reduced after 90 days. They observed a sharp increase in the size of the heterotrophic and PAH-degrading microbial populations, which coincided with the highest rates of TPH and PAH biodegradation. At the end of the incubation period, PAH degraders were more prevalent in samples to which nutrients had not been added. There was a remarkable shift in the composition of the bacterial community due to both the biodegradation process and the addition of nutrients. At early stages of biodegradation, the α-Proteobacteria group (*Sphingomonas* and *Azospirillum*) was the dominant group in all treatments. At later stages, the γ-Proteobacteria group (*Xanthomonas*), the α-Proteobacteria group (*Sphingomonas*), and the Cytophaga-Flexibacter-Bacteroides group (Bacteroidetes) were the dominant groups in the nonnutrient treatment, while the γ-Proteobacteria group (*Xanthomonas*), the β-Proteobacteria group (*Alcaligenes* and *Achromobacter*), and the α-Proteobacteria group (*Sphingomonas*) were the dominant groups in the nutrient treatment. They conclude that specific bacterial phylotypes were associated both with different phases of PAH degradation and with nutrient addition in a preadapted PAH-contaminated soil. There were complex interactions between bacterial species and medium conditions that influenced the biodegradation capacity of the microbial communities involved in bioremediation processes.

Namkoong et al. (2002) study the appropriate mix ratio of organic amendments for enhancing diesel fuel degradation during contaminated soil composting in a paper with 232 citations. They added sewage sludge or compost as an amendment for supplementing organic matter for composting contaminated soil. The ratios of contaminated soil to organic amendments were 1:0.1, 1:0.3, 1:0.5, and 1:1 as wet weight bases. They spiked diesel fuel at 10,000 mg/kg sample on a dry weight basis. They observed that the degradation of diesel fuels was significantly enhanced by the addition of these organic amendments relative to straight soil. Degradation rates of TPH and n-alkanes were the greatest at a ratio of 1:0.5 of contaminated soil to organic amendments on a wet weight basis. They observed preferential degradation of n-alkanes over TPH regardless of the kind and the amount of organic amendments. The first order degradation constant of n-alkanes was about twice the TPH degradation constant. Normal alkanes could be divided in two groups (C_{10}–C_{15} versus C_{16}–C_{20}) based on the first order kinetic constant. Volatilization loss of TPH was only about 2% of initial TPH. Normal alkanes lost by volatilization were mainly the compounds C_{10}–C_{16}. They found high correlations among the TPH degradation rate, amount of CO_2 evolved, and dehydrogenase activity. They conclude that the degradation of diesel fuels was significantly enhanced by the addition of these organic amendments relative to straight soil.

Ghazali et al. (2004) study the bioremediation of petroleum hydrocarbons in contaminated soils by mixed cultures of hydrocarbon-degrading bacteria in a paper with 218 citations. The mixtures of bacteria, Consortium I and Consortium 2, consisted of three and six bacterial strains, respectively. They used bacterial strains from the collection of strains at a local university, and isolated them from hydrocarbon-contaminated soil samples by enrichment on either crude oil or individual hydrocarbons as the sole carbon source. They selected the strains based on the criteria that they would be able to display good growth in crude oil, individual hydrocarbon compounds, or both. They evaluated their ability to degrade hydrocarbons in the environment using soil samples that were contaminated with diesel, crude oil, or engine oil. They observed that Consortium 2, which consisted of six bacterial strains, was more efficient at removing the medium and long-chain alkanes in the diesel-contaminated soil compared to Consortium 1. Further, Consortium 2 could effectively remove the medium and long-chain alkanes in the engine oil such that the alkanes were undetectable after a 30-day incubation period. Consortium 2 consisted predominantly of *Bacillus* and *Pseudomonas* spp. They conclude that Consortium 2 including *Bacillus* and *Pseudomonas* spp. was more efficient at removing the medium and long-chain alkanes in the diesel-contaminated soil compared to Consortium 1.

Mishra et al. (2001) perform a full-scale study evaluating an inoculum addition to stimulate *in situ* bioremediation of oily-sludge-contaminated soil at an oil refinery where the indigenous population of hydrocarbon-degrading bacteria in the soil was very low (10^3–10^4 CFU/g of soil) in a paper with 212 citations. In a feasibility study, application of a bacterial consortium and nutrients resulted in maximum biodegradation of TPH in 120 days. Therefore, they selected this treatment for the full-scale study. In the full-scale study, they treated plots A and B with a bacterial consortium and nutrients, which resulted in a 92.0 and 89.7% removal of TPH, respectively, in one year, compared to a 14.0% removal of TPH in the control plot C. In plot A, they observed that the alkane fraction of TPH was reduced by 94.2%, the aromatic fraction of TPH was reduced by 91.9%, and NSO (N-, S-, and O containing compound) and asphaltene fractions of TPH were reduced by 85.2% in one year. Similarly, in plot B the degradation of alkane, aromatic, and NSO plus asphaltene fractions of TPH was 95.1, 94.8, and 63.5%, respectively, in 345 days. However, in plot C, removal of alkane (17.3%), aromatic (12.9%), and NSO plus asphaltene (5.8%) fractions was much less. The population of introduced *Acinetobacter baumannii* strains in plots A and B was stable even after one year. Physical and chemical properties of the soil at the bioremediation site improved significantly in one year. They conclude that the application of a bacterial consortium and nutrients stimulated *in situ* bioremediation of oily-sludge-contaminated soils.

Margesin and Schinner (2001) study the feasibility of bioremediation as a treatment option for a chronically diesel-fuel-polluted soil in an alpine glacier area at an altitude of 2,875 m above sea level in a paper with 205 citations. To examine the efficiencies of 'natural attenuation' and 'biostimulation', they used field-incubated lysimeters (mesocosms) with unfertilized and fertilized (N-P-K) soil. For three summer seasons (July 1997 to September 1999), they monitored changes in hydrocarbon concentrations in soil and soil leachate and the accompanying changes in soil microbial counts and activity. A significant reduction in the diesel oil level was achieved. At the end of the third summer season (after 780 days), they observed that the initial level of contamination (2,612 μg of hydrocarbons g [dry weight] of soil^{-1}) was reduced by 50 and 70% in the unfertilized and fertilized soil, respectively. Nonetheless, the residual levels of contamination (1,296 and 774 μg of hydrocarbons g [drg weight] of soil^{-1} in the unfertilized and fertilized soil, respectively) were still high.

Most of the hydrocarbon loss occurred during the first summer season (42% loss) in the fertilized soil and during the second summer season (41% loss) in the unfertilized soil. In the fertilized soil, all biological parameters (microbial numbers, soil respiration, catalase and lipase activities) were significantly enhanced and correlated significantly with each other, as well as with the residual hydrocarbon concentration, pointing to the importance of biodegradation. The effect of biostimulation of the indigenous soil microorganisms declined with time. The microbial activities in the unfertilized soil fluctuated around background levels during the whole study. They conclude that bioremediation as a treatment option for a chronically diesel-fuel-polluted soil was feasible in this alpine glacier area.

Urum and Pekdemir (2004) evaluate the ability of aqueous biosurfactant solutions (aescin, lecithin, rhamnolipid, saponin, and tannin) for possible applications in washing crude-oil-contaminated soil in a paper with 204 citations. They measured the biosurfactants' behavior in soil-water, water-oil, and oil-soil systems (such as foaming, solubilization, sorption to soil, emulsification, surface and interfacial tension) and compared with a well-known chemical surfactant ('sodium dodecyl sulfate', SDS) at varying concentrations. They observed that the biosurfactants were able to remove a significant amount of crude oil from the contaminated soil at different solution concentrations, for instance rhamnolipid and SDS removed up to 80% oil and lecithin by about 42%. The performance of water alone in crude oil removal was equally as good as those of the other biosurfactants. Oil removal was due to mobilization, caused by the reduction of surface and interfacial tensions. Solubilization and emulsification effects in oil removal were negligible due to the low crude oil solubilization of 0.11%. They conclude that knowledge of biosurfactants' behavior across different systems is paramount before their use in the practical application of oil removal.

Huang et al. (2004) develop a multi-process phytoremediation system composed of physical (volatilization), photochemical (photooxidation) and microbial remediation, and phytoremediation (plant-assisted remediation) processes using creosote as a test contaminant in a paper with 204 citations. The techniques were land-farming (aeration and light exposure), introduction of contaminant degrading bacteria, 'plant growth promoting rhizobacteria' (PGPR), and plant growth of contaminant-tolerant tall fescue (*Festuca arundinacea*). Over a four-month period, they observed that the average efficiency of removal of 16 priority PAHs by the multi-process remediation system was twice that of land-farming, 50% more than bioremediation alone, and 45% more than phytoremediation by itself. Importantly, the multi-process system was capable of removing most of the highly hydrophobic, soil-bound PAHs from soil. The key elements for successful phytoremediation were the use of plant species that had the ability to proliferate in the presence of high levels of contaminants and strains of PGPR that increase plant tolerance to contaminants and accelerate plant growth in heavily contaminated soils. They conclude that the synergistic use of these approaches resulted in rapid and massive biomass accumulation of plant tissue in contaminated soil, putatively providing more active metabolic processes, leading to more rapid and more complete removal of PAHs.

Gunther et al. (1996) study the effects of growing ryegrass (*Lolium perenne* L.) on the biodegradation of petroleum hydrocarbons in laboratory scale soil columns in a paper with 194 citations. They determined the degradation of hydrocarbons as well as bacterial numbers, soil respiration rates, and soil dehydrogenase activities. In the rhizosphere soil system, they observed that aliphatic hydrocarbons disappeared faster than in unvegetated columns. Abiotic loss by evaporation was of minor significance. Elimination of pollutants was accompanied by an increase in microbial numbers and activities. The microbial plate counts and soil respiration rates were substantially higher in the rhizosphere than in the bulk soil. They conclude that biodegradation of hydrocarbons in the rhizosphere was stimulated by plant roots.

Sarkar et al. (2005) compare two methods of biostimulation in a laboratory incubation study with 'monitored natural attenuation' (MNA) for TPH degradation in diesel-contaminated Tarpley clay soil with low carbon content in a paper with 193 citations. One method utilized rapid-release inorganic fertilizers rich in N and P, and the other used sterilized, slow-release biosolids, which added C in addition to N and P. After eight weeks of incubation, they observed that both biostimulation methods degraded approximately 96% of TPH compared to MNA, which degraded by 93.8%. However, in the first week of incubation, biosolid-amended soils showed a linear two orders of magnitude increase in microbial population compared to MNA, whereas, in the fertilizer-amended soils, only a one order of magnitude increase was noted. In the following weeks, the microbial population in the fertilizer-amended soils dropped appreciably, suggesting a toxic effect owing to fertilizer-induced acidity and/or NH_3 overdosing. They conclude that biosolid addition was a more effective soil amendment method for biostimulation than the commonly practiced inorganic fertilizer application, because of the abilities of biosolids to supplement carbon. As there was no statistically significant difference between the biostimulation methods and MNA, MNA could be a viable remediation strategy in certain soils with high native microbial population.

Salanitro et al. (1997) determine the limits and extent of hydrocarbon biodegradation, earthworm and plant toxicity, and waste leachability of crude-oil-containing soils in a paper with 190 citations. They mix three oils (heavy, medium, and light of API gravity 14, 30, and 55, respectively) into silty loamy soils containing low (0.3%) or high (4.7%) organic carbon at 4,000–27,000 mg/kg TPH. They observed that hydrocarbon bioremediation in these artificially weathered oily soils usually followed first-order removal rates in which 50–75% and 10–90% of the TPH were degraded in three to four months for the low and high organic soils, respectively. After bioremediation, hydrocarbons in oily soils decreased from 70 to 90%, from 40 to 60%, and from 35 to 60% for those carbon number species in the range of C_{11}–C_{22}, C_{23}–C_{32}, and C_{35}–C_{44}, respectively. Most oily soils were initially toxic to earthworms in which few animals survived 14-day bioassays. In a solid phase Microtox test, most oily soils had EC50 values that were $\leq$ 50%. Seed germination and plant growth (21-day test, wheat and oat but not corn) were also significantly reduced (0–25% of controls) in untreated soils containing the medium and light crude oils but not the heavy oil. Bioremediated soils were neither toxic to earthworms, inhibitory in the Microtox assay, nor did they inhibit seed germination after 5 (high organic soil) or 10–12 (low organic soil) months of treatment. Water-soluble hydrocarbons (e.g. O&G and BTEX) could leach from pretreated soils (medium and light crude oily soils) in column or batch extraction experiments. However, after bioremediation, most of the aromatic compounds were no longer leachable from the soils. Treated oily soils lost their toxicity and potential to leach significant amounts of BTEX. These nontoxic soils contained 1,000–8,600 mg/kg residual hydrocarbons as TPH. Furthermore, the remaining petroleum compounds might be bound or unavailable. They provide a basis for a framework in which petroleum hydrocarbon-containing soils can be evaluated by ecological assessment methods such as biodegradability, ecotoxicity, and the leaching potential of regulated substances.

Ghosh et al. (1995) study the role of surfactants in the desorption of soil-bound PAHs and 'polychlorinated biphenyls' (PCBs) in a paper with 183 citations. They observed that the solubilization of individual PAHs in an extract of a weathered, coal-tar-contaminated soil containing a mixture of PAHs and other petroleum derivatives was significantly less than that for pure compounds. Batch soil washing with Triton X-100 (a commercial, nonionic 'alkyl phenol ethoxylate') increased the effective diffusion rate of PAHs from the contaminated soil by four orders of magnitude compared to that obtained by gas purging, when the results were analyzed using a radial diffusion model. At concentrations of up to 24 times its 'critical micelle concentration' (CMC), Triton X-100 did not enhance hydrocarbon degradation in the coal-tar-contaminated soil. However, the biosurfactant rhamnolipid R1, at a concentration of 50x CMC, increased the rate of mineralization of 4,4'-chlorinated biphenyl, mobilized from a laboratory-contaminated soil by more than 60 times. They conclude that surfactants have a critical role in the desorption of soil-bound PAHs and PCBs.

Lai et al. (2009) develop a screening method to evaluate the oil removal capability of biosurfactants for oil-contaminated soils collected from a heavy oil-polluted site in a paper with 179 citations. They identified the ability of removing TPH from soil by two biosurfactants and compared that with synthetic surfactants. They observed that biosurfactants exhibited a much higher TPH removal efficiency than the synthetic ones examined. By using a 0.2 mass% of rhamnolipids, surfactin, Tween 80, and Triton X-100, the TPH removal for the soil contaminated with ca. 3,000 mg TPH/kg dry soil was 23, 14, 6, and 4%, respectively, while the removal efficiency increased to 63, 62, 40, and 35%, respectively, for the soil contaminated with ca. 9,000 mg TPH/kg dry soil. The TPH removal efficiency also increased with an increase in biosurfactant concentration (from 0 to 0.2 mass %) but it did not vary significantly for the contact time of one and seven days. They conclude that biosurfactants exhibit much higher TPH removal efficiency than the synthetic ones examined for oil-contaminated soils.

Jorgensen et al. (2000) perform composting of lubricating-oil-contaminated soil in a field scale (5 × 40 m^3) using bark chips as the bulking agent, and test two commercially available mixed microbial inocula as well as the effect of the level of added nutrients (N, P, K) in a paper with 178 citations. They also performed composting of diesel-oil-contaminated soil performed at one level of nutrient addition and with no inoculum. They observed that the mineral oil degradation rate was most rapid during the first few months, and followed a typical first-order degradation curve. Over five months, composting of the mineral oil decreased in all piles with lubrication oil from approximately 2,400 to 700 mg (kg dry w)$^{-1}$, which was about 70% of the mineral oil content. Correspondingly, the mineral oil content in the pile with diesel-oil-contaminated soil decreased with 71% from 700 to 200 mg (kg dry w)$^{-1}$. In this type of treatment with the addition of a large amount of organic matter,, the general microbial activity as measured by soil respiration was enhanced and no particular effect of added inocula was observed. They conclude that bioremediation of petroleum-hydrocarbon-contaminated soil by composting in biopiles was efficient.

Whang et al. (2008) study the potential application of two biosurfactants, surfactin (SF) and rhamnolipid (RL), for enhanced biodegradation of diesel-contaminated water and soil with a series of bench-scale experiments in a paper with 172 citations. They produced RL, a commonly isolated glycolipid biosurfactant, by *Pseudomonas aeruginosa* J4, while they produced the surfactin, a lipoprotein type biosurfactant, by *Bacillus subtilis* ATCC 21332. They observed that both biosurfactants reduced surface tension to less than 30 dynes/cm from 72 dynes/cm with CMC values of 45 and 50 mg/L for SF and RL, respectively. In addition, the results of diesel dissolution experiments also demonstrated their ability in increasing diesel solubility with increased biosurfactant addition. In diesel/water batch experiments, an addition of 40 mg/L of SF significantly enhanced biomass growth (2,500 mg VSS/L) as well as increased diesel biodegradation percentage (94%), compared to batch experiments with no SF addition (1,000 mg VSS/L and 40% biodegradation). The addition of SF of more than 40 mg/L, however, decreased both biomass growth and diesel biodegradation efficiency, with a worse diesel biodegradation percentage (0%) at 400 mg/L of SF addition. They observed similar trends for both specific rate constants of biomass growth and diesel degradation, as SF addition increased from 0 to 400 mg/L. The addition of RL to diesel/water systems from 0 to 80 mg/L substantially increased biomass growth and diesel biodegradation percentage from 1,000 to 2,500 mg VSS/L and 40 to 100%, respectively. RL addition at a concentration of 160 mg/L provided similar results to those of an 80 mg/L addition. Finally, they also studied the potential application of SF and RL in stimulating indigenous microorganisms for enhanced bioremediation of diesel-contaminated soil. They confirmed the enhancing capability of both SF and RL on both the efficiency and rate of diesel biodegradation in diesel/soil systems.

Kirk et al. (2005) measure changes in microbial communities caused by the addition of two species of plants in a soil contaminated with 31,000 ppm of total petroleum hydrocarbons in a paper with 170 citations. They observed that perennial ryegrass and/or alfalfa increased the number of rhizosphere bacteria in the hydrocarbon-contaminated soil. These plants also increased the number of bacteria capable of petroleum degradation as estimated by the most probable number (MPN) method. Eco-Biolog plates did not detect changes in metabolic diversity between bulk and rhizosphere samples but 'denaturing gradient gel electrophoresis' (DGGE) analysis of PCR-amplified partial 16S rDNA sequences indicated a shift in the bacterial community in the rhizosphere samples. 'Dice coefficient' matrices derived from DGGE profiles showed similarities between the rhizospheres of alfalfa and perennial ryegrass/alfalfa mixture in the contaminated soil at week seven. Perennial ryegrass and perennial ryegrass/alfalfa mixture caused the greatest change in the rhizosphere bacterial community as determined by DGGE analysis. They conclude that plants altered the microbial population and that these changes were plant-specific and could contribute to the degradation of petroleum hydrocarbons in contaminated soil.

Gallego et al. (2001) find that microbiological and chemical analyses and a suitable bioreactor design were very useful for suggesting the best ways to improve biodegradation extents in a diesel-enriched soil in a paper with 170 citations. They observed that biostimulation with inorganic N and P produced the best results in a simple bioreactor, with biodegradation extents higher than 90% after 45 days. In addition, the addition of activated sludge from a domestic wastewater plant increased the degradation rate largely. In both cases, microbiological studies showed the presence of *Acinetobacter* sp. degrading most of the hydrocarbons. Simultaneously, they studied a diesel fuel release (approximately 400,000 1). Samples taken in polluted soil and water revealed that bacteria from *Acinetobacter* were predominant. In plate studies, *Acinetobacter* colonies produced a whitish substance with the characteristics of a biosurfactant. Remarkably, the presence of this product was evident at the field site, both in the riverbanks and in the physical recovery plant. The study of the similarities between laboratory results and the diesel spill site strongly suggested that natural conditions at the field site allowed the implementation of *in situ* bioremediation after physical removal of light nonaqueous-phase liquids. They conclude that microbiological and chemical analyses and a suitable bioreactor design were very useful for suggesting the best ways to improve biodegradation extents in a diesel-enriched soil.

Oberbremer et al. (1990) study the effect of the addition of microbial surfactants on hydrocarbon degradation in a soil population in a stirred reactor in a paper with 166 citations. They observed that the hydrocarbon degradation rate could be doubled by the addition of 'sophorose lipids' as biosurfactants in a model system containing 10% soil and a 1.35% hydrocarbon mixture of tetradecane, pentadecane, hexadecene, '1,2,4-trimethylcyclohexane', pristane ('2,6,10,14-tetramethylpentadecane') phenyldecane, and naphthalene suspended in a mineral salts medium. The adaptation phases for two degradation phases were shortened, and the extent of degradation and final biomass were increased. The added biosurfactants were degraded after they had facilitated the degradation of all hydrocarbon components. They conclude that the addition of microbial surfactants on hydrocarbon degradation in a soil population in a stirred reactor was beneficial.

48.4 DISCUSSION

48.4.1 Bioremediation of Petroleum Hydrocarbons in Contaminated Soils

Bento et al. (2005) evaluate the comparative bioremediation of soils contaminated with TPH, diesel fuels, by natural attenuation, biostimulation, and bioaugmentation in a paper with 358 citations. They conclude that soil properties and the indigenous soil microbial population affected the degree of biodegradation with the implications that detailed site-specific characterization studies were needed prior to deciding on the proper bioremediation method.

Weissenfels et al. (1992) study the degradation of petroleum hydrocarbons, PAHs, in contaminated soils, from two different industrial sites under simulated land treatment conditions in a paper with 338 citations. They conclude that sorption of organic pollutants onto soil organic matter significantly affected biodegradability as well as biotoxicity.

Das and Mukherjee (2007) compare the efficiency of *Bacillus subtilis* DM-04 and *Pseudomonas aeruginosa* M and NM strains isolated from a petroleum contaminated soil sample from North-East India for the biodegradation of petroleum hydrocarbons in soil and shake flask study in a paper with 317 citations. They conclude that *B. subtilis* DM-04 and *P. aeruginosa* M and NM strains could be effective for *in situ* bioremediation of petroleum hydrocarbons.

Vinas et al. (2005) study the bacterial community dynamics and biodegradation processes in a highly creosote-contaminated soil undergoing a range of laboratory-based bioremediation treatments in a paper with 261 citations. They conclude that specific bacterial phylotypes were associated both with different phases of PAH degradation and with nutrient addition in a preadapted PAH-contaminated soil. There were complex interactions between bacterial species and medium conditions that influenced the biodegradation capacity of the microbial communities involved in bioremediation processes.

Namkoong et al. (2002) study the appropriate mix ratio of organic amendments for enhancing diesel fuel degradation during contaminated soil composting in a paper with 232 citations. They conclude that the degradation of diesel fuels was significantly enhanced by the addition of these organic amendments relative to straight soil.

Ghazali et al. (2004) study the bioremediation of petroleum hydrocarbons in contaminated soils by mixed cultures of hydrocarbon-degrading bacteria in a paper with 218 citations. They conclude that Consortium 2 including *Bacillus* and *Pseudomonas* spp. was more efficient at removing the medium and long-chain alkanes in the diesel-contaminated soil compared to Consortium 1.

Mishra et al. (2001) perform a full-scale study evaluating an inoculum addition to stimulate *in situ* bioremediation of oily-sludge-contaminated soil at an oil refinery where the indigenous population of hydrocarbon-degrading bacteria in the soil was very low (10^3–10^4 CFU/g of soil) in a paper with 212 citations. They conclude that the application of a bacterial consortium and nutrients stimulated *in situ* bioremediation of oily-sludge-contaminated soils.

Margesin and Schinner (2001) study the feasibility of bioremediation as a treatment option for a chronically diesel-fuel-polluted soil in an alpine glacier area in a paper with 205 citations. They conclude that the bioremediation as a treatment option for a chronically diesel-fuel-polluted soil was feasible in this alpine glacier area.

Urum and Pekdemir (2004) evaluate of the ability of aqueous biosurfactant solutions (aescin, lecithin, rhamnolipid, saponin, and tannin) for possible applications in washing crude-oil-contaminated soil in a paper with 204 citations. They conclude that knowledge of biosurfactants' behavior across different systems was paramount before their use in the practical application of oil removal.

Huang et al. (2004) develop a multi-process phytoremediation system composed of physical (volatilization), photochemical (photooxidation) and microbial remediation, and phytoremediation processes using creosote as a test contaminant in a paper with 204 citations. They conclude that the synergistic use of these approaches resulted in rapid and massive biomass accumulation of plant tissue in contaminated soil, putatively providing more active metabolic processes, leading to more rapid and more complete removal of PAHs.

Gunther et al. (1996) study the effects of growing ryegrass (*Lolium perenne* L.) on the biodegradation of petroleum hydrocarbons in laboratory-scale soil columns in a paper with 194 citations. They conclude that biodegradation of hydrocarbons in the rhizosphere was stimulated by plant roots.

Sarkar et al. (2005) compare two methods of biostimulation in a laboratory incubation study with MNA for TPH degradation in diesel-contaminated Tarpley clay soil with low carbon content in a paper with 193 citations. They conclude that biosolids addition was a more effective soil amendment method for biostimulation than the commonly practiced inorganic fertilizer application, because of the abilities of biosolids to supplement carbon. As there was no statistically significant difference between the biostimulation methods and MNA, MNA could be a viable remediation strategy in certain soils with high native microbial population.

Salanitro et al. (1997) determine the limits and extent of hydrocarbon biodegradation, earthworm and plant toxicity, and waste leachability of crude-oil-containing soils in a paper with 190 citations. They provide a basis for a framework in which petroleum hydrocarbon-containing soils can be evaluated by ecological assessment methods such as biodegradability, ecotoxicity, and the leaching potential of regulated substances.

Ghosh et al. (1995) study the role of surfactants in the desorption of soil-bound PAHs and polychlorinated biphenyls (PCBs) in a paper with 183 citations. They conclude that surfactants have a critical role in the desorption of soil-bound PAHs and PCBs.

Lai et al. (2009) develop a screening method to evaluate the oil removal capability of biosurfactants for oil-contaminated soils collected from a heavy oil-polluted site in a paper with 179 citations. They conclude that biosurfactants exhibited much higher TPH removal efficiency than the synthetic ones examined for crude-oil-contaminated soils.

Jorgensen et al. (2000) perform composting of lubricating-oil-contaminated soil in a field scale (5×40 m^3) using bark chips as the bulking agent, and test two commercially available mixed microbial inocula as well as the effect of the level of added nutrients (N, P, K) in a paper with 178 citations. They conclude that bioremediation of petroleum-hydrocarbon-contaminated soil by composting in biopiles was efficient.

Whang et al. (2008) study the potential application of two biosurfactants, surfactin and rhamnolipid, for enhanced biodegradation of diesel-contaminated water and soil with a series of bench-scale experiments in a paper with 172 citations. They conclude that plants altered the microbial population; these changes were plant-specific and could contribute to degradation of petroleum hydrocarbons in contaminated soil.

Kirk et al. (2005) measure changes in microbial communities caused by the addition of two species of plants in a soil contaminated with 31,000 ppm of total petroleum hydrocarbons in a paper with 170 citations. They conclude that plants altered the microbial population; these changes were plant-specific and could contribute to the degradation of petroleum hydrocarbons in contaminated soil.

Gallego et al. (2001) find that microbiological and chemical analyses and a suitable bioreactor design were very useful for suggesting the best ways to improve biodegradation extents in a diesel-enriched soil in a paper with 170 citations. They conclude that microbiological and chemical analyses and a suitable bioreactor design were very useful for suggesting the best ways to improve biodegradation extents in a diesel-enriched soil.

Oberbremer et al. (1990) study the effect of the addition of microbial surfactants on hydrocarbon degradation in a soil population in a stirred reactor in a paper with 166 citations. They conclude that the effect of the addition of microbial surfactants on hydrocarbon degradation in a soil population in a stirred reactor was beneficial.

48.4.2 Societal Implications

The contamination of soils with petroleum hydrocarbons has been well established (Maliszewska-Kordybach, 1996; Nadal et al., 2004; Wilcke, 2000). This includes contamination by petrodiesel fuels and crude oils. The ecotoxicity of these soils contaminated with petroleum hydrocarbons to both terrestrial and aquatic organisms has also been well established (Agarwal et al., 2009; Eom et al., 2007; Sverdrup et al., 2002).

Thus, the bioremediation of these toxic petroleum hydrocarbons in oils has been necessary to reduce the ecotoxicity of these compounds on terrestrial and aquatic organisms, meeting the great public concern regarding this matter, such as the potential loss of biodiversity in soil (Brussaard, 1997; Brussaard et al., 2007; Wall and Moore, 1999; Wall et al., 2015).

The prolific studies presented in this chapter provide the key findings on the bioremediation of these toxic petroleum compounds in soil. These studies show that bioremediation of these toxic compounds has been efficient by using a number of microbes and plants.

48.5 CONCLUSION

This chapter has presented the key findings of the 20-most-cited article papers in this field.

The contamination of soils with petroleum hydrocarbons has been well established (Maliszewska-Kordybach, 1996; Nadal et al., 2004; Wilcke, 2000). This includes contamination by petrodiesel fuels and crude oils. The ecotoxicity of these soils contaminated with petroleum hydrocarbons to both terrestrial and aquatic organisms has also been well established (Agarwal et al., 2009; Eom et al., 2007; Sverdrup et al., 2002).

Thus, the bioremediation of these toxic petroleum hydrocarbons in oil has been necessary to reduce the ecotoxicity of these compounds on terrestrial and aquatic organisms, meeting the great public concern regarding this matter, such as the potential loss of biodiversity in soil (Brussaard, 1997; Brussaard et al., 2007; Wall and Moore, 1999; Wall et al., 2015).

The prolific studies presented in this chapter provide the key findings on the bioremediation of the toxic petroleum compounds in soil. These studies show that bioremediation of these toxic compounds has been efficient using a number of microbes and plants.

ACKNOWLEDGMENTS

The contribution of the highly cited researchers in this field is greatly acknowledged.

REFERENCES

Agarwal, T., P. S. Khillare, V. Shridhar, and S. Ray. 2009. Pattern, sources and toxic potential of PAHs in the agricultural soils of Delhi, India. *Journal of Hazardous Materials* 163:1033–1039.

Ahmadun, F. R., A. Pendashteh, and L. C. Abdullah, et al. 2009. Review of technologies for oil and gas produced water treatment. *Journal of Hazardous Materials* 170:530–551.

Atlas, R. M.. 1981. Microbial degradation of petroleum hydrocarbons: An environmental perspective. *Microbiological Reviews* 45:180–209.

Babich, I. V. and J. A. Moulijn. 2003. Science and technology of novel processes for deep desulfurization of oil refinery streams: A review. *Fuel* 82:607–631.

Bento, F. M., F. A. O. Camargo, B. C. Okeke, and W. T. Frankenberger. 2005. Comparative bioremediation of soils contaminated with diesel oil by natural attenuation, biostimulation and bioaugmentation. *Bioresource Technology* 96:1049–1055.

Birch, M. E. and R. A. Cary. 1996. Elemental carbon-based method for monitoring occupational exposures to particulate diesel exhaust. *Aerosol Science and Technology* 25:221–241.

Bridgwater, A. V., D. Meier, and D. Radlein. 1999. An overview of fast pyrolysis of biomass. *Organic Geochemistry* 30:1479–1493.

Bridgwater, A. V. and G. V. C. Peacocke. 2000. Fast pyrolysis processes for biomass. *Renewable & Sustainable Energy Reviews* 4:1–73.

Brussaard, L. 1997. Biodiversity and ecosystem functioning in soil. *Ambio* 26:563–570.

Brussaard, L., P. C. De Ruiter, and G. G. Brown. 2007. Soil biodiversity for agricultural sustainability. *Agriculture, Ecosystems & Environment* 121:233–244.

Chisti, Y. 2007. Biodiesel from microalgae. *Biotechnology Advances* 25:294–306.

Czernik, S. and A. V. Bridgwater. 2004. Overview of applications of biomass fast pyrolysis oil. *Energy & Fuels* 18:590–598.

Das, K. and A. K. Mukherjee. 2007. Crude petroleum-oil biodegradation efficiency of *Bacillus subtilis* and *Pseudomonas aeruginosa* strains isolated from a petroleum-oil contaminated soil from North-East India. *Bioresource Technology* 98:1339–1345.

Eom, I. C., C. Rast, A. M. Veber, and P. Vasseur. 2007. Ecotoxicity of a polycyclic aromatic hydrocarbon (PAH)-contaminated soil. *Ecotoxicology and Environmental Safety* 67:190–205.

Gallego, J. L. R., J. Loredo, J. F. Llamas, F. Vazquez, and J. Sanchez. 2001. Bioremediation of diesel-contaminated soils: Evaluation of potential *in situ* techniques by study of bacterial degradation. *Biodegradation* 12:325–335.

Gan, S., E. V. Lau, and H. K. Ng. 2009. Remediation of soils contaminated with polycyclic aromatic hydrocarbons (PAHs). *Journal of Hazardous Materials* 172:532–549.

Ghazali, F. M., R. N. Z. A. Rahman, A. B. Salleh, and M. Basri. 2004. Biodegradation of hydrocarbons in soil by microbial consortium. *International Biodeterioration & Biodegradation* 54:61–67.

Ghosh, M. M., I. T. Yeom, Z. Shi, C. D. Cox, and K. G. Robinson. 1995. Surfactant-enhanced bioremediation of PAH- and PCB-contaminated soils. *Bioremediation Series* 3:15–23.

Gunther, T., U. Dornberger, and W. Fritsche. 1996. Effects of ryegrass on biodegradation of hydrocarbons in soil. *Chemosphere* 33:203–215.

Hill, J., E. Nelson, D. Tilman, S. Polasky, and D. Tiffany. 2006. Environmental, economic, and energetic costs and benefits of biodiesel and ethanol biofuels. *Proceedings of the National Academy of Sciences of the United States of America* 103:11206–11210.

Hu, Q., M. Sommerfeld, and E. Jarvis, et al. 2008. Microalgal triacylglycerols as feedstocks for biofuel production: Perspectives and advances. *Plant Journal* 54:621–639.

Huang, X. D., Y. El-Alawi, D. M. Penrose, B. R. Glick, and B. M. Greenberg. 2004. A multi-process phytoremediation system for removal of polycyclic aromatic hydrocarbons from contaminated soils. *Environmental Pollution* 130:465–476.

Johnsen, A. R., L. Y. Wick, and H. Harms. 2005. Principles of microbial PAH-degradation in soil. *Environmental Pollution* 133:71–84.

Jorgensen, K. S., J. Puustinen, and A. M. Suortti. 2000. Bioremediation of petroleum hydrocarbon-contaminated soil by composting in biopiles. *Environmental Pollution* 107:245–254.

Khalili, N. R., P. A. Scheff, and T. M. Holsen. 1995. PAH source fingerprints for coke ovens, diesel and gasoline-engines, highway tunnels, and wood combustion emissions. *Atmospheric Environment* 29:533–542.

Kilian, L. 2009. Not all oil price shocks are alike: Disentangling demand and supply shocks in the crude oil market. *American Economic Review* 99:1053–1069.

Kirk, J. L., J. N. Klironomos, H. Lee, and J. T. Trevors. 2005. The effects of perennial ryegrass and alfalfa on microbial abundance and diversity in petroleum contaminated soil. *Environmental Pollution* 133:455–465.

Konur, O. 2000. Creating enforceable civil rights for disabled students in higher education: An institutional theory perspective. *Disability & Society* 15:1041–1063.

Konur, O. 2002a. Access to Nursing Education by disabled students: Rights and duties of nursing programs. *Nurse Education Today* 22:364–374.

Konur, O. 2002b. Assessment of disabled students in higher education: Current public policy issues. *Assessment and Evaluation in Higher Education* 27:131–152.

Konur, O. 2002c. Access to employment by disabled people in the UK: Is the Disability Discrimination Act working? *International Journal of Discrimination and the Law* 5:247–279.

Konur, O. 2006a. Participation of children with dyslexia in compulsory education: Current public policy issues. *Dyslexia* 12:51–67.

Konur, O. 2006b. Teaching disabled students in Higher Education. *Teaching in Higher Education* 11:351–363.

Konur, O. 2007a. A judicial outcome analysis of the Disability Discrimination Act: A windfall for the employers? *Disability & Society* 22:187–204.

Konur, O. 2007b. Computer-assisted teaching and assessment of disabled students in higher education: The interface between academic standards and disability rights. *Journal of Computer Assisted Learning* 23:207–219.

Konur, O., ed. 2021a. *Handbook of Biodiesel and Petrodiesel Fuels: Science, Technology, Health, and Environment*. Boca Raton, FL: CRC Press.

Konur, O., ed. 2021b. *Handbook of Biodiesel and Petrodiesel Fuels: Science, Technology, Health, and Environment. Volume 1. Biodiesel Fuels: Science, Technology, Health, and Environment*. Boca Raton, FL: CRC Press.

Konur, O., ed. 2021c. *Handbook of Biodiesel and Petrodiesel Fuels: Science, Technology, Health, and Environment. Volume 2. Biodiesel Fuels based on the Edible and Nonedible Feedstocks, Wastes, and Algae: Science, Technology, Health, and Environment*. Boca Raton, FL: CRC Press.

Konur, O., ed. 2021d. *Handbook of Biodiesel and Petrodiesel Fuels: Science, Technology, Health, and Environment. Volume 3. Petrodiesel Fuels: Science, Technology, Health, and Environment*. Boca Raton, FL: CRC Press.

Konur, O. 2021e. Biodiesel and petrodiesel fuels: Science, technology, health, and environment. In *Handbook of Biodiesel and Petrodiesel Fuels: Science, Technology, Health, and Environment. Volume 1. Biodiesel Fuels: Science, Technology, Health, and Environment*, ed. O. Konur. Boca Raton, FL: CRC Press.

Konur, O. 2021f. Biodiesel and petrodiesel fuels: A scientometric review of the research. In *Handbook of Biodiesel and Petrodiesel Fuels: Science, Technology, Health, and Environment. Volume 1. Biodiesel Fuels: Science, Technology, Health, and Environment*, ed. O. Konur. Boca Raton, FL: CRC Press.

Konur, O. 2021g. Biodiesel and petrodiesel fuels: A review of the research. In *Handbook of Biodiesel and Petrodiesel Fuels: Science, Technology, Health, and Environment. Volume 1. Biodiesel Fuels: Science, Technology, Health, and Environment*, ed. O. Konur. Boca Raton, FL: CRC Press.

Konur, O. 2021h Nanotechnology applications in the diesel fuels and the related research fields: A review of the research. In *Handbook of Biodiesel and Petrodiesel Fuels: Science, Technology, Health, and Environment. Volume 1. Biodiesel Fuels: Science, Technology, Health, and Environment*, ed. O. Konur. Boca Raton, FL: CRC Press.

Konur, O. 2021i. Biooils: A scientometric review of the research. In *Handbook of Biodiesel and Petrodiesel Fuels: Science, Technology, Health, and Environment. Volume 1. Biodiesel Fuels: Science, Technology, Health, and Environment*, ed. O. Konur. Boca Raton, FL: CRC Press.

Konur, O. 2021j. Characterization and properties of biooils: A review of the research. In *Handbook of Biodiesel and Petrodiesel Fuels: Science, Technology, Health, and Environment. Volume 1. Biodiesel Fuels: Science, Technology, Health, and Environment*, ed. O. Konur. Boca Raton, FL: CRC Press.

Konur, O. 2021k. Biomass pyrolysis and pyrolysis oils: A review of the research. In *Handbook of Biodiesel and Petrodiesel Fuels: Science, Technology, Health, and Environment. Volume 1. Biodiesel Fuels: Science, Technology, Health, and Environment*, ed. O. Konur. Boca Raton, FL: CRC Press.

Konur, O. 2021l. Biodiesel fuels: A scientometric review of the research. In *Handbook of Biodiesel and Petrodiesel Fuels: Science, Technology, Health, and Environment. Volume 1. Biodiesel Fuels: Science, Technology, Health, and Environment*, ed. O. Konur. Boca Raton, FL: CRC Press.

Konur, O. 2021m. Glycerol: A scientometric review of the research. In *Handbook of Biodiesel and Petrodiesel Fuels: Science, Technology, Health, and Environment. Volume 1. Biodiesel Fuels: Science, Technology, Health, and Environment*, ed. O. Konur. Boca Raton, FL: CRC Press.

Konur, O. 2021n. Propanediol production from glycerol: A review of the research. In *Handbook of Biodiesel and Petrodiesel Fuels: Science, Technology, Health, and Environment. Volume 1. Biodiesel Fuels: Science, Technology, Health, and Environment*, ed. O. Konur. Boca Raton, FL: CRC Press.

Konur, O. 2021o. Edible oil-based biodiesel fuels: A scientometric review of the research. *In Handbook of Biodiesel and Petrodiesel Fuels: Science, Technology, Health, and Environment. Volume 2. Biodiesel Fuels based on the Edible and Nonedible Feedstocks, Wastes, and Algae: Science, Technology, Health, and Environment*, ed. O. Konur. Boca Raton, FL: CRC Press.

Konur, O. 2021p. Palm oil-based biodiesel fuels: A review of the research. In *Handbook of Biodiesel and Petrodiesel Fuels: Science, Technology, Health, and Environment. Volume 2. Biodiesel Fuels based on the Edible and Nonedible Feedstocks, Wastes, and Algae*, ed. O. Konur. Boca Raton, FL: CRC Press.

Konur, O. 2021q. Rapeseed oil-based biodiesel fuels: A review of the research. In *Handbook of Biodiesel and Petrodiesel Fuels: Science, Technology, Health, and Environment.*

Volume 2. Biodiesel Fuels based on the Edible and Nonedible Feedstocks, Wastes, and Algae, ed. O. Konur. Boca Raton, FL: CRC Press.

Konur, O. 2021r. Nonedible oil-based biodiesel fuels: A scientometric review of the research. In *Handbook of Biodiesel and Petrodiesel Fuels: Science, Technology, Health, and Environment. Volume 2. Biodiesel Fuels based on the Edible and Nonedible Feedstocks, Wastes, and Algae: Science, Technology, Health, and Environment*, ed. O. Konur. Boca Raton, FL: CRC Press.

Konur, O. 2021s. Waste oil-based biodiesel fuels: A scientometric review of the research. In *Handbook of Biodiesel and Petrodiesel Fuels: Science, Technology, Health, and Environment. Volume 2. Biodiesel Fuels based on the Edible and Nonedible Feedstocks, Wastes, and Algae: Science, Technology, Health, and Environment*, ed. O. Konur. Boca Raton, FL: CRC Press.

Konur, O. 2021t. Algal biodiesel fuels: A scientometric review of the research. In *Handbook of Biodiesel and Petrodiesel Fuels: Science, Technology, Health, and Environment. Volume 2. Biodiesel Fuels based on the Edible and Nonedible Feedstocks, Wastes, and Algae: Science, Technology, Health, and Environment*, ed. O. Konur. Boca Raton, FL: CRC Press.

Konur, O. 2021u. Algal biomass production for biodiesel production: A review of the research. In *Handbook of Biodiesel and Petrodiesel Fuels: Science, Technology, Health, and Environment. Volume 2. Biodiesel Fuels based on the Edible and Nonedible Feedstocks, Wastes, and Algae*, ed. O. Konur Boca Raton, FL: CRC Press.

Konur, O. 2021v. Algal biomass production in wastewaters for biodiesel production: A review of the research. In *Handbook of Biodiesel and Petrodiesel Fuels: Science, Technology, Health, and Environment. Volume 2. Biodiesel Fuels based on the Edible and Nonedible Feedstocks, Wastes, and Algae*, ed. O. Konur. Boca Raton, FL: CRC Press.

Konur, O. 2021x. Algal lipid production for biodiesel production: A review of the research. In *Handbook of Biodiesel and Petrodiesel Fuels: Science, Technology, Health, and Environment. Volume 2. Biodiesel Fuels based on the Edible and Nonedible Feedstocks, Wastes, and Algae*, ed. O. Konur. Boca Raton, FL: CRC Press.

Konur, O. 2021y. Crude oils: A scientometric review of the research. In *Handbook of Biodiesel and Petrodiesel Fuels: Science, Technology, Health, and Environment. Volume 3. Petrodiesel Fuels: Science, Technology, Health, and Environment*, ed. O. Konur. Boca Raton, FL: CRC Press.

Konur, O. 2021z. Petrodiesel fuels: A scientometric review of the research. In *Handbook of Biodiesel and Petrodiesel Fuels: Science, Technology, Health, and Environment. Volume 3. Petrodiesel Fuels: Science, Technology, Health, and Environment*, ed. O. Konur. Boca Raton, FL: CRC Press.

Konur, O. 2021aa. Bioremediation of petroleum hydrocarbons in the contaminated soils: A review of the research. In *Handbook of Biodiesel and Petrodiesel Fuels: Science, Technology, Health, and Environment. Volume 3. Petrodiesel Fuels: Science, Technology, Health, and Environment*, ed. O. Konur. Boca Raton, FL: CRC Press.

Konur, O. 2021ab. Desulfurization of diesel fuels: A review of the research. In *Handbook of Biodiesel and Petrodiesel Fuels: Science, Technology, Health, and Environment. Volume 3. Petrodiesel Fuels: Science, Technology, Health, and Environment*, ed. O. Konur. Boca Raton, FL: CRC Press.

Konur, O. 2021ac. Diesel fuel exhaust emissions: A scientometric review of the research. In *Handbook of Biodiesel and Petrodiesel Fuels: Science, Technology, Health, and Environment. Volume 3. Petrodiesel Fuels: Science, Technology, Health, and Environment*, ed. O. Konur. Boca Raton, FL: CRC Press.

Konur, O. 2021ad. The adverse health and safety impact of diesel fuels: A scientometric review of the research. In *Handbook of Biodiesel and Petrodiesel Fuels: Science,*

Technology, Health, and Environment. Volume 3. Petrodiesel Fuels: Science, Technology, Health, and Environment, ed. O. Konur. Boca Raton, FL: CRC Press.

Konur, O. 2021ae. Respiratory illnesses caused by the diesel fuel exhaust emissions: A review of the research. In *Handbook of Biodiesel and Petrodiesel Fuels: Science, Technology, Health, and Environment. Volume 3. Petrodiesel Fuels: Science, Technology, Health, and Environment*, ed. O. Konur. Boca Raton, FL: CRC Press.

Konur, O. 2021af. Cancer caused by the diesel fuel exhaust emissions: A review of the research. In *Handbook of Biodiesel and Petrodiesel Fuels: Science, Technology, Health, and Environment. Volume 3. Petrodiesel Fuels: Science, Technology, Health, and Environment*, ed. O. Konur. Boca Raton, FL: CRC Press.

Konur, O. 2021ag. Cardiovascular and other illnesses caused by the diesel fuel exhaust emissions: A review of the research. In *Handbook of Biodiesel and Petrodiesel Fuels: Science, Technology, Health, and Environment. Volume 3. Petrodiesel Fuels: Science, Technology, Health, and Environment*, ed. O. Konur. Boca Raton, FL: CRC Press.

Lai, C. C., Y. C. Huang, Y. H. Wei, and J. S. Chang. 2009. Biosurfactant-enhanced removal of total petroleum hydrocarbons from contaminated soil. *Journal of Hazardous Materials* 167:609–614.

Maliszewska-Kordybach, B. 1996. Polycyclic aromatic hydrocarbons in agricultural soils in Poland: Preliminary proposals for criteria to evaluate the level of soil contamination. *Applied Geochemistry* 11:121–127.

Margesin, R. and F. Schinner. 2001. Bioremediation (natural attenuation and biostimulation) of diesel-oil-contaminated soil in an alpine glacier skiing area. *Applied and Environmental Microbiology* 67:3127–3133.

Mishra, S., J. Jyot, R. C. Kuhad, and B. Lal. 2001. Evaluation of inoculum addition to stimulate *in situ* bioremediation of oily-sludge-contaminated soil. *Applied and Environmental Microbiology* 67:1675–1681.

Nadal, M., M. Schuhmacher, and J. L. Domingo. 2004. Levels of PAHs in soil and vegetation samples from Tarragona County, Spain. *Environmental Pollution* 132:1–11.

Namkoong, W., E. Y. Hwang, J. S. Park, and J. Y. Choi. 2002. Bioremediation of diesel-contaminated soil with composting. *Environmental Pollution* 119:23–31.

North, D. C. 1991a. *Institutions, Institutional Change and Economic Performance*. Cambridge, Mass: Cambridge University Press.

North, D. C. 1991b. Institutions. *Journal of Economic Perspectives* 5:97–112.

Oberbremer, A., R. Mullerhurtig, and F. Wagner. 1990. Effect of the addition of microbial surfactants on hydrocarbon degradation in a soil population in a stirred reactor. *Applied Microbiology and Biotechnology* 32:485–489.

Perron, P. 1989. The great crash, the oil price shock, and the unit root hypothesis. *Econometrica: Journal of the Econometric Society* 57:1361–1401.

Rogge, W. F., L. M. Hildemann, M. A. Mazurek, G. R. Cass, and B. R. T. Simoneit. 1993. Sources of fine organic aerosol. 2. Noncatalyst and catalyst-equipped automobiles and heavy-duty diesel trucks. *Environmental Science & Technology* 27:636–651.

Salanitro, J. P., P. B. Dorn, and M. H. Huesemann, et al. 1997. Crude oil hydrocarbon bioremediation and soil ecotoxicity assessment. *Environmental Science & Technology* 31:1769–1776.

Sarkar, D., M. Ferguson, R. Datta, and S. Birnbaum. 2005. Bioremediation of petroleum hydrocarbons in contaminated soils: Comparison of biosolids addition, carbon supplementation, and monitored natural attenuation. *Environmental Pollution* 136:187–195.

Song, C. and X. L. Ma. 2003. New design approaches to ultra-clean diesel fuels by deep desulfurization and deep dearomatization. *Applied Catalysis B-Environmental* 41:207–238.

Song, C. S. 2003. An overview of new approaches to deep desulfurization for ultra-clean gasoline, diesel fuel and jet fuel. *Catalysis Today* 86:211–263.

Sverdrup, L. E., T. Nielsen, and P. H. Krogh. 2002. Soil ecotoxicity of polycyclic aromatic hydrocarbons in relation to soil sorption, lipophilicity, and water solubility. *Environmental Science & Technology* 36:2429–2435.

Urum, K. and T. Pekdemir. 2004. Evaluation of biosurfactants for crude oil contaminated soil washing. *Chemosphere* 57:1139–1150.

Vinas, M., J. Sabate, M. J. Espuny, and A. M. Solanas. 2005. Bacterial community dynamics and polycyclic aromatic hydrocarbon degradation during bioremediation of heavily creosote-contaminated soil. *Applied and Environmental Microbiology* 71:7008–7018.

Wall, D. H. and J. C. Moore. 1999. Interactions underground: Soil biodiversity, mutualism, and ecosystem processes. *BioScience* 49:109–117.

Wall, D. H., U. N. Nielsen, and J. Six. 2015. Soil biodiversity and human health. *Nature* 528:69–76.

Weissenfels, W. D., H. J. Klewer, and J. Langhoff. 1992. Adsorption of polycyclic aromatic-hydrocarbons (PAHs) by soil particles: Influence on biodegradability and biotoxicity. *Applied Microbiology and Biotechnology* 36:689–696.

Whang, L. M., P. W. G. Liu, C. C. Ma, and S. S. Cheng. 2008. Application of biosurfactants, rhamnolipid, and surfactin, for enhanced biodegradation of diesel-contaminated water and soil. *Journal of Hazardous Materials* 151:155–163.

Wilcke, W. 2000. Polycyclic aromatic hydrocarbons (PAHs) in soil: A review. *Journal of Plant Nutrition and Soil Science* 163:229–248.

Wilson, S. C. and K. C. Jones. 1993. Bioremediation of soil contaminated with polynuclear aromatic-hydrocarbons (PAHs): A review. *Environmental Pollution* 81:229–249.

49 Petrodiesel and Biodiesel Fuels for Marine Applications

Mohamad Issa
Adrian Ilinca

CONTENTS

49.1 INTRODUCTION

The total supply of liquid fuel in the world currently stands at about 4,000 megatons (MT) per annum. At 300–400 MT per annum, marine liquid fuels make up a significant proportion. A large proportion of marine fuel consumption (around 77%) is low-quality, low-price, residual fuel also known as 'heavy fuel oil' (HFO), which appears to be high in sulfur and is consumed almost exclusively by large, cargo-carrying ships. The most important pollutants produced by diesel engine driven vessels are 'carbon dioxides' (CO_2), 'carbon monoxide' (CO), 'particulate matter' (PM), 'nitrogen oxides' (NO_x), and 'sulfur oxides' (SO_x). About 14–31, 4–9, and 3–6% of global NO_X, SO_x, and CO_2 emissions are from marine vessels, respectively (Issa et al., 2019b). Marine transport faces tougher demands to comply with the 'Paris Agreement' and lower 'greenhouse gas' (GHG) emissions regarding fuel quality and exhaust emissions as stricter regulations are imposed in different regions around the world. The implementation by the 'International Maritime Organization' (IMO) of 'Sulfur Emission Control Areas' (SECAs) with a maximum permissible sulfur of 0.1% in marine fuel since 2015 would boost demand for low sulfur fuels. The control of emissions of NO_X has also been implemented in a systematic manner since 2016, raising the pressure to make a low NO_X energy conversion. Additionally, shipping raises questions about climate change to reduce its GHG emissions. Both the need for low sulfur fuels and the need to minimize emissions of NO_X and GHG have created a challenge for ships to operate more efficiently and in an environmentally friendly way. Reducing emissions of sulfur dioxide, nitrogen oxides, and GHGs in accordance with current legislation and reducing the effects on climate change would entail a substantial shipping propulsion change. Thus, changing fuels has become an interesting option.

Biofuels are seen as the most viable option for reducing emissions from shipping. Bio-derived fuels are environmentally friendly, sustainable, and clean compared with marine fuel oil and marine diesel currently used. In addition, the fuel properties and the combustion properties are identical to fossil fuels: HFO, 'marine diesel oil' (MDO) and 'liquefied natural gas' (LNG). 'Straight vegetable oils' (SVOs), biodiesel, and 'bio-liquid natural gas' (bio-LNG) are the promising green alternatives to HFO, distillate fuels, and LNG in the short to medium term, respectively. The other promising alternatives are pyrolysis crude, 'Fisher–Tropsch diesel', and biomethanol, but these are for the longer term. All these oils are completely sulfur-free and may be used to comply with sulfur content and other restrictions on emissions.

Changing fuels, however, can require changes in engine technology, e.g. to gas or dual-fuel engines, but can also be achieved with modern fuels that can be used with small modifications and improvements in old engines. In addition, they can either be used in conjunction with traditional, oil-based marine fuels, thus only covering part of the energy demand of a vessel, or to replace conventional fuels completely.

A number of studies have assessed the performance of the currently used fossil marine fuels. Winebrake et al. (2007) also included biofuels, but only biodiesel based on soybean. Numerous studies have also investigated alternative (bio-based) terrestrial transport fuels. However, there are some things that vary between shipping and road-based transport. Firstly, the basis for comparison varies, since the fuels actually used in shipping (mainly HFOs) vary from those used in road vehicles (gasoline and diesel). Facilities and storage requirements often vary, just as the engines do. Thus, it is likely that fuels not well suited for road transport might be beneficial as marine fuels and vice versa. Contrary to the positive production of biofuels such as biodiesel used in land-based transportation, there are still many barriers to the promotion of biofuels as an alternative fuel for marine vessels, such as technological development, technical incorporation, relatively higher marine diesel or bunker fuel oil prices, availability of feedstock, and operational problems.

A widespread change from fossil to biofuel in marine applications would increase marine fuel quality and also improve air quality, protect the natural environment, and improve the health of individuals.

49.2 CURRENT MARINE FUELS

In refineries, marine fuels are refined from crude petroleum oil. It is usually stored at bunker stations located in port areas prior to use. HFO is mostly used in today's shipping industry to power the main engine, while marine diesel and 'marine gas oil' (MGO) are usually used for auxiliary engines and harbor service. ISO 8217:2017 (International Standards Organization) sets out the specifications for fuel in marine diesel engines and boilers prior to standard on-board treatment (settling, centrifuging, and filtration). The IMO reports that 'residual fuel oil' (RFO) represents 77% of all marine use. Use of high-sulfur RFOs concentrates on the largest long-haul vessels. Figure 49.1 illustrates a standard mix of fuel sold in Singapore, one of the busiest seaports in the world.

Considering the Singapore Port offers practically no MGO or MDO, the sale is true only for heavy fuels (i.e. 77.2% of the market). Therefore, distillate fuels would

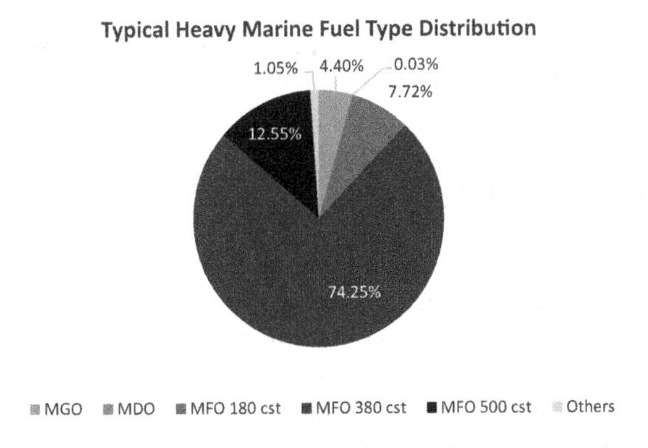

FIGURE 49.1 Typical heavy fuel type distribution.

Source: MPAS (2019).

TABLE 49.1
Global Marine Fuel use Estimated from Imo and Singapore Port Bunkering Statistics 2019

Fuel Type	Other Names	Market (%)	Megatons per year
Heavy fuel 500 cst	HSFO 500 cst, RFO, RMG 500, IFO 500, MFO 500	10	33
Heavy fuel 300 cst	HSFO 380 cst, RFO, RMG 380, IFO 380, MFO 380	60	200
Heavy fuel 180 cst	HSFO 180 cst, RFO, RMG 180, IFO 180, MFO 380	6	20
Distillate fuels	Diesel, marine diesel, MGO, MDO, LFO	23	77
Others		1	3
Total		100	333

Notes: RMG: residual marine gas; IFO: 'intermediate fuel oil'.
Source: MPAS (2019).

add up to 22.8%. The corresponding fuel statistics are shown in Table 49.1, including lighter types of fuels not sold in Singapore.

49.2.1 Marine Fuel Standards and Classification

There are five types of marine fuels that are categorized according to their blends and viscosity:

- MGO: The diesel fuel used in land vehicles is similar to automotive diesel.
- MDO: This contains a mixture of heavy fuel oil and MGO, has low viscosity, and needs no preheating prior to use.
- IFO: Almost similar to MDO – a mixture of residual oil or HFO with MGO.
- 'Marine fuel oil' (MFO): This is a mixture of HFO and MGO which contains less gas oil than IFO.
- HFO: The lowest marine fuel category. It is a residual oil, of high viscosity, and needs preheating before use.

It is also possible to group and classify marine fuel into three basic types, namely distillate, intermediate, and residual, as shown in Table 49.2. In addition to viscosity, density is also an important attribute for grading marine fuel, since it is filtered before use to remove water and soil. Most of the marine fuel products are produced in compliance with ISO 8217 specifications. This norm currently does not require biodiesel composition but plans to do so in the future.

On the other hand, the ASTM D975 specifications require 5% biodiesel blends known as B5 fuels that are currently available on most countries' markets. The basic properties of biodiesel, vehicle, marine, and heavy oil fuels are illustrated in Table 49.3.

TABLE 49.2
Form of and Grades of Marine Fuel

Fuel Type	Fuel Grades	Common Industry Name
Distillate	*DMX, DMA, DMB, DMZ	Gas oil or marine gas oil, marine diesel oil
Intermediate	IFO 180, 380	Intermediate fuel oil
Residual	**RMA-RMK	Fuel oil or residual fuel oil

*DMX is a special light distillate, which is primarily intended for use in emergency engines. DMA (also known as 'marine gas oil', MGO) is a marine distillate of general use that must be free of traces of residual fuel. DMB ('marine diesel oil', MDO) has traces of residual fuel that can be heavy in sulfur although DMZ, the fourth class of distillate, must not contain residual fuel constituents and has a higher aromatic content and slightly higher viscosity at 40°C compared to other distillate fuels.

**RMA-RMK: Based on their viscosity (kinematic viscosity) – RMA, RMB, RMD, RME, RMG, and RMK – residue fuels are classified into six types of fuel according to ISO 8217.

Source: Kołwzan and Narewski (2012).

TABLE 49.3
Form of and Grades of Marine Fuel

Properties	Marine Diesel ISO 8217	Heavy Fuel Oil ISO 8217	Automotive Diesel EN590	Biodiesel EN4214
Density/15°C, kg/m^3	<900	975–1010	820–845	860–900
Viscosity/40°C, cSt	<11	<700/50°C	2.0–4.5	3.5–5.0
Flashpoint	>60	>60	>55	>120
Cetane no.	>35	>20	>51	>51
Ash content (%)	<0.01	<0.2	<0.01	<0.01
Water, ppm	<300	<5,000	<200	<500
Sulfur, ppm	<200,000	<50,000	<350	<10
Calorific value, Mj/kg	42	40	43	37.5

Source: Noor et al. (2018).

49.2.2 Combustion of Marine Fuels

Marine fuels contain a far larger range of chemical compounds relative to biofuels, including much more sulfur. The combustion of gasoline and diesel fuels releases host contaminants and heavy metals that influence local and regional air quality and these are well connected to global warming problems. The air pollution associated with transport contributes to reduced visibility, damage to plants and buildings, and increased incidence of human disease and premature death (Witherby Seamanship, 2013).

Table 49.4 summarizes the key environmental and health impacts of primary petroleum combustion products, including CO_2, CO, 'unburned hydrocarbons' (UHC), NO_X, SO_X, and PM.

TABLE 49.4
The Environmental and Safety Impacts of Petroleum Combustion Emissions

Combustion Product	Impacts
CO	Evidence of incomplete combustion or incineration. Within the atmosphere, CO reacts with oxygen to form ozone, a molecule that is highly reactive and damages plant leaves and human and animal lungs.
CO_2	Contributes to climate change and global warming.
NO and NO_X	A precursor to ozone, they also react with atmospheric water and cause acid rain.
SO_2 and SO_3	Precursor of acid rain.
Benzene	The least acidic hydrocarbon and the most toxic carcinogen.
Lead	Has been phased out of petrol in most countries but is still being used as an octane enhancer.
Particulate matter	Particulates, which are made from SOx, Nox, and hydrocarbons, contribute to the production of ozone and impact visibility and global warming.

Source: Issa et al. (2019a).

49.2.3 The Marine Fuel Market: A View of World Fuel Consumption by Class of Vessel

Of the about 53,000 merchant ships that trade globally, around 11,000 are bulk carriers. As of January 1, 2019, general cargo ships comprised the majority of ships in the world merchant fleet, which consists of approximately 52% cargo ships, 28% tankers, and 20% container ships, according to 'Statista Transportation & Logistics' (Wagner, 2020); see Figure 49.2. Of the total MFO sold in 2019, 58% is referred to as RFO. However, with the entry into force of 'MARPOL Annex VI' in January 2020, the US Energy Information Administration (EIA) projects that the US ocean-going bunker fuel market's share of high-sulfur residual fuel oil decreased from 58% in 2019 to 3% in 2020, before recovering to 24% in 2022 (Figure 49.3). The number of vessels fitted with scrubbers that continue using high-sulfur RFO remains small, despite a recent rise in scrubber installation and orders. Consequently, the 'Annual Energy Outlook' (AEO) 2019 expected a significant but brief rise in the share of distillate fuel oil and low-sulfur residual fuel oil in 2019 and soon after in 2020 (EIA, 2019).

49.2.4 Marine Fuel Regulations

Global regulations restricting sulfur in ocean-going vessel oils, which are scheduled to take effect in January 2020, have consequences for vessel owners, refiners, and global oil markets. Stakeholders will respond to these regulations in different ways, increasing confusion on both short and long-term prices for crude oil and petroleum products.

When sulfur is burned, it produces sulfur dioxide in marine fuel, a precursor to acid rain. Thanks to increasingly strict regulations imposed by individual countries or groups of countries, the sulfur content of transport fuels has been declining for

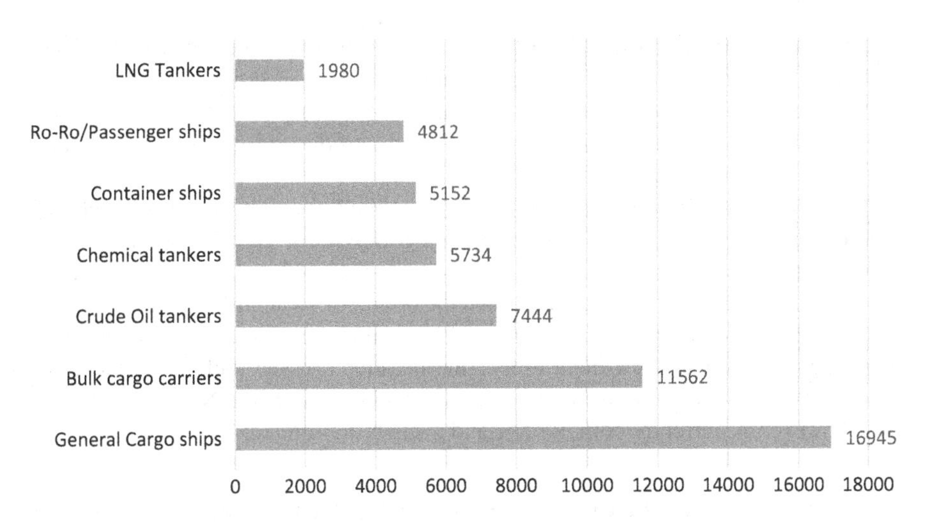

FIGURE 49.2 Number of ships in the world merchant fleet as of January 1, 2019 by type.

Source: Wagner (2020).

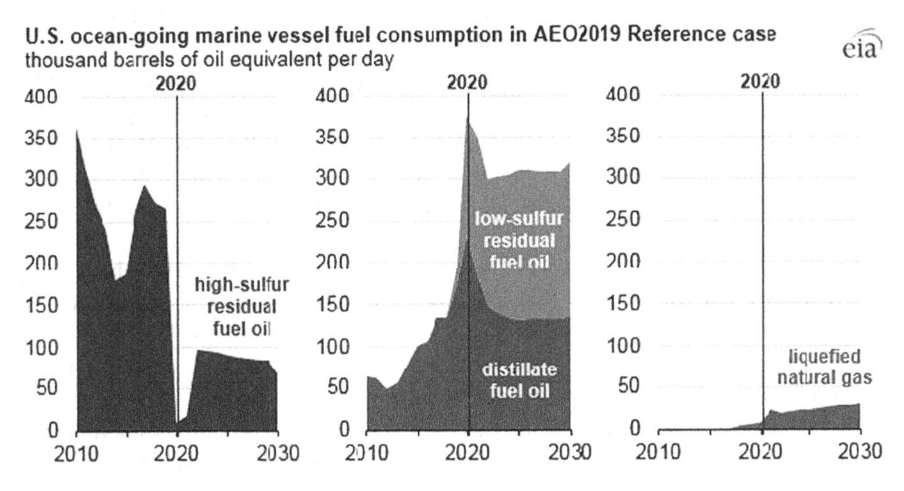

FIGURE 49.3 International marine shipping consumption by ocean-going vessel bunkering at US ports (thousand barrels of oil equivalent per day).

Source: EIA (2019).

several years. In the USA, federal and state laws restrict the sulfur content of motor gasoline, diesel fuel, and heating fuel. According to the International Energy Agency, the upcoming 2020 guidelines extend across jurisdictions of various countries to fuels used in the open ocean, representing the largest portion of the global marine fuel market of about 3.9 million barrels per day.

The 'International Maritime Organization' (IMO), the United Nations organization of 171 member states that sets shipping requirements, is set to reduce the overall sulfur content (by percentage weight) in marine fuels used in the open seas from 3.5 to 0.5% by 2020. The aim of these regulations is to eliminate sulfur dioxide, nitrogen oxides, and other emissions from global vessel exhausts. The 2020 sulfur level reduction follows a number of similar cuts in marine fuel sulfur levels, such as those that decreased sulfur content in IMO-designated 'Emission Control Areas' from 1.0 to 0.1% in 2015. Specific sulfur limits have been introduced for other areas around ports in Europe and parts of China.

Vessel operators have several choices to meet the latest IMO sulfur restrictions. One choice is to move to an IMO-compliant lower-sulfur petrol. The cost, widespread availability, and specifications of a new fuel to be used in marine engines, however, are still unclear. Even ships have the option of switching to non-petroleum based fuels. Some newer ships and some currently under construction have dual-fuel engines which would allow them to use non-petroleum-based fuels like LNG after limited modifications. The infrastructure to support the use of LNG as a shipping fuel, however, is currently both in scale and availability limited.

49.3 MARINE BIOFUELS AND CONVERSION TECHNOLOGIES

When combined with fossil oils, biofuels provide possible synergistic advantages by reducing the sulfur content and providing significantly lower ash and emission profiles. Biofuels can be low in sulfur and nitrogen, but also low in carbon intensity, depending on the biomass feedstock and processing conditions.

Biomass services range from maize kernels to maize stalks, soybean, and canola oils to animal fats, and from prairie grasses to hardwoods, including algae. In the long run, we will need multiple technologies to make use of these different sources of energy. Some technologies are already being developed; there are others waiting to be so. Today biochemical, electrical, and thermochemical processes are the most common technologies.

Candidates for biofuels include (1) 'oxygenated biofuels' such as SVO, biodiesel, 'fast pyrolysis' (FP) biooil, and 'hydrothermal liquefaction' (HTL) biocrude; and (2) hydrocarbon biofuels like 'renewable diesel', Fischer-Tropsch (F-T) diesel, 'fully upgraded (deoxygenated) biooil', and biocrude. Tables 49.5 and 49.6 provide a list of the forms, characteristics, and properties of these biofuels.

Each candidate has different properties that determine the form of marine fuel that can be combined, displaced, or partially displaced. Both of them however have extremely low sulfur levels. Oxygenated biofuels have lower energy densities than 'liquid hydrocarbon fuels'; however, bound oxygen atoms serve as oxidizers when combusted to minimize PM formation. Of the biofuels of nonoxygenated hydrocarbons, biooils/biocrudes derived from hydrotreated FP and hydrothermal HTL are more difficult to boil than traditional diesel and may require fractionation depending on application. However, the consistency criteria and the need for further evaluation of blending biofuels, such as the possible need to eliminate water (as oxygenated fuels can be hydrophilic in nature) or any residual solids, remain important unknowns.

Of the biofuels mentioned in Table 49.5, only the 'marine distillate fuels' are miscible with biodiesel, F-T diesel, renewable diesel, and upgraded biooil. Only biodiesel (at concentrations up to 7% vol.) is currently approved for use with MGO as a marine fuel, and studies by ExxonMobil, MARAD, and others have shown substantial reductions in PM when biodiesel is combined with MGO (Kass et al., 2018).

TABLE 49.5
Main Features of Marine Biofuel Blendstocks

Fuel Type	Availability	Feedstock	Advantages	Disadvantages
SVO	Commercial, but research required for marine use	Vegetable oil	Inexpensive	Oxidation stability and shelf life
Renewable diesel	Commercial	Vegetable oil, animal fats	Miscible with MGO; mature technology near zero O_2	High production costs
F-T diesel	Commercial, not presently from biomass	Woody and other biomass	Miscible with MGO; excellent combustion properties	Complex processing and expensive
F-P biooil	Commercial	Woody and other biomass	Possible low PM formation; miscible with butanol and butanol blends	Incompatibilities with infrastructure and not miscible with neat MGO
Biodiesel or fatty acid methyl esters	Commercial	Vegetable oil, animal fats	Miscible with MGO; mature technology; approved for use with MGO	Oxidation stability and shelf life
HTL biocrude	Research stage	Woody and other biomass	Improved heating value Vs FP biooil	Demonstration scale only
HFO baseline	Commercial	Petroleum crude	Mature technology; low cost; Existing infrastructure	High SO_X and PM formation and nonrenewable; requires onboard processing
Upgraded FP biooil	Research stage	—	- Miscible with MGO; - good heating value	Bench scale only

TABLE 49.6
Main Fuel Properties of the Biofuels Selected for Marine Use

Property	Biodiesel	Renewable Diesel	FT-diesel	Fp Biooil (Woody Feeds)	Upgraded Biooil	Htl Biocrude (Woody Feeds)
Specific gravity	0.88	0.78	0.765	1.1–1.3	0.84	1.1
Kinematic viscosity (40°C) cSt	4–6	2–4	2	40–100	—	—
Cetane number	47–65	>70	>70	—	—	—
Lubricity, μm		650	371	—	—	—
Lower heating value, Mj/kg	37.2	44.1	43	16	—	32
Cloud point, °C	−3 to 15	−5 to −34	−18			
Water content, mass%	Nil	Nil	Nil	20–35	0.1	8
Oxygen content, %	11	0	0	34–45	0.5	10–13
Sulfur content, %	<0.0015	<0.0005	<0.1	0–0.05	<0.005	0

PM reduction is an immediate environmental advantage of oxygenated fuels and can often be accomplished at relatively low levels of blend (< 10%). Since biooils cannot be mixed directly with distillates, any attempt to combine biooil with MGO would involve the use of surfactants to shape an emulsified fuel mixture. Emulsified fuels are vulnerable to separation over time, even though they exist as microemulsions, and thus usually have a poor shelf life. The effect of biooil-drained water on the combustion cycle is unknown. Nevertheless, the introduction of water to combustion is a proven approach to reducing emissions of PM and, in some situations, even emissions of NOx. The straight replacement of MGO with a biofuel demands that the amount of biofuel production be sufficiently high to fulfil consumption needs. It is uncertain whether the current production of biofuels will effectively displace MGO. Nevertheless, as blends, biofuels provide incentives for reducing both PM and CO_2; and as demand for MGOs rises, biofuels may also provide an economic opportunity.

49.3.1 Biomass-Derived Diesel Fuels

SVOs, also known as 'pure vegetable oils' (PVOs), are derived from plants for use as fuel only. Such oils do not undergo any intermediate processing steps, but will be used directly from extraction in diesel engines. Studies have shown that they can be used in low speed engines (all sizes of carriers and cargo ships) to replace IFO or heavy oil (Florentinus et al., 2012), although functional fuels for large scale or long-term use are not commonly considered. Because of their higher viscosity and high boiling point, SVOs reduce engine lifespan due to carbon deposit accumulation inside the engine and engine lubricant damage. Because of the possibility of engine damage and lubricating oil gelling, it is therefore not recommended that vegetable oils be used as raw untreated oil.

A process called transesterification creates refined biodiesel where specific oils (triglycerides) are converted into methyl esters. Glycerol and water are processed as side products which are later discarded as unwanted products. Biodiesel is often widely referred to as 'methyl ester fatty acid' (FAME), derived from vegetable oil or animal fat that was transesterified with methanol or ethanol. Sodium methylate appears to be used as a catalyst. FAME is a more suitable fuel for diesel engines than SVO, with a lower boiling point and viscosity than SVO, which results in better engine efficiency. Biodiesel has a higher flash point (149°C) and level of cetane than traditional diesel, which rapidly degrades in water. Yet FAME has a high cloud point that can cause filter clogging and poor fuel flow at temperatures below 32°C.

Biodiesel can be used to replace MDO and MGO in diesel engines of low to medium speed (tug boats, small carriers, and container ships), but it is more widely used as a fuel additive and can be directly (drop-in) poured into fuel tanks. FAME as a fuel has strong properties including ignition and lubricity. Theoretically, diesel vehicles can be run with 100% confidence, but they require engine modifications as well as engine manufacturers' approval. FAME blends with petrodiesel of up to 20% are also commonly marketed on the diesel fuel market because it can be used in diesel equipment with little to no engine modifications.

One of the key benefits of biodiesel is that it restores engine lubricity and decreases the smoke, soot, and burnt diesel odor from engine exhausts while protecting against wear in fuel and injector pumps. Biodiesel's main technological drawback over petrodiesel is its lower thermal energy content, as it has a higher oxygen content compared to traditional hydrocarbon fuels.

While biofuels are not yet widely used in the maritime field, it is possible that marine biofuels could be developed and produced to be technologically compatible with marine engines based on existing biofuel technologies. Thus, they could be used as drop-in fuels in shipping vessels. Additionally, marine diesel engines' very high fuel versatility opens them up for the production of new biofuel processes mixing various grades and biofuel forms. Figure 49.4 shows an overview of the different feedstock conversion routes to marine biofuels (IEA, 2017).

49.3.2 Drop-in Fuels

Engine manufacturers find it very costly to configure a new marine engine for a new fuel, so ship owners do not turn to a different fuel unless their ship's engine's fuel supply is assured for life. Therefore, if new fuels could be made functionally identical to those already in operations, they would be completely compliant with existing fuel infrastructure without the need for significant investment in modifications to it.

'Drop-in biofuels' are classified as liquid biohydrocarbons that functionally equate to 'petroleum-derived fuels' and are fully compatible with the current

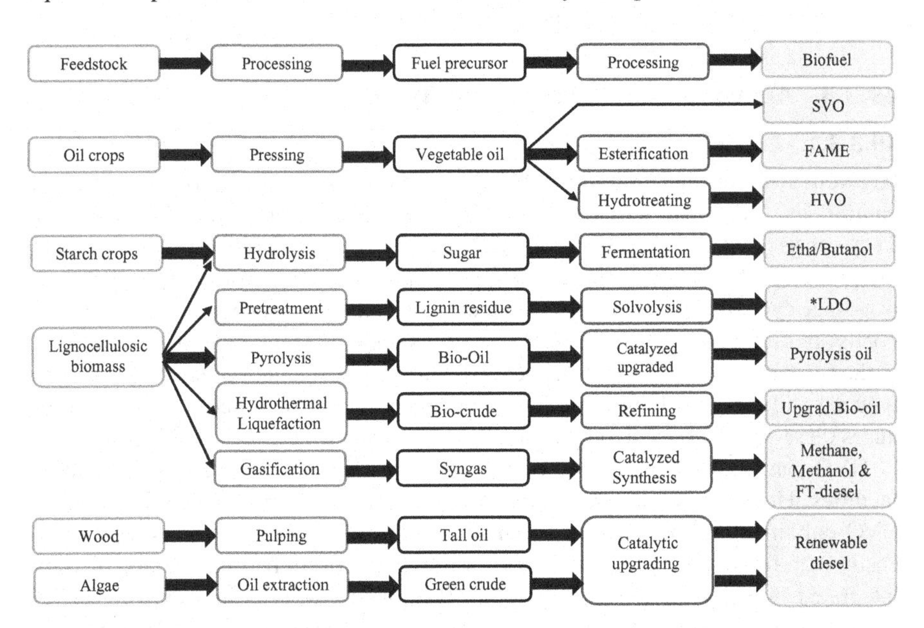

FIGURE 49.4 Overview of the different routes for conversion of feedstock to marine biofuels for both conventional and advanced biofuels.

Note: LDO: 'Lignin diesel oil'.

petroleum structure. The following bulk property criteria must be met by definition: miscibility with petroleum fuels, compliance with performance specifications, reasonable storability, transportability with existing logistics systems, and usability within existing engines. Furthermore, they must also be very compliant with already existing fuel injection systems.

Chemically, drop-in biofuels consist of a mixture of several different hydrocarbons and share similar combustion properties as conventional fuels to be compatible with the oil infrastructure. This implies that the fuel will consist mainly of carbon and hydrogen by mass, with small traces of nitrogen, oxygen, sulfur, and metals. The difficulty of generating drop-in fuels from biomass is that the high biomass O content is not appropriate for direct conversion, and thus requires additional processing measures such as deoxygenation to achieve a sufficient H: C ratio. Another solution would be to intensively hydrotreat, i.e. add H_2, which is expensive and needs more refining. The desired result is a hydrocarbon derived from biomass that has low oxygen content, low water solubility, and high carbon bond saturation (Smagala et al., 2013). However, the industry will have to incorporate lignocellulosic feedstocks into the fuel supply mix to keep feedstock and production costs competitive in order to further raise output volumes at a rate appropriate for shipping by sea.

While drop-in fuels are designed to comply with the same IMO requirements for marine fuel, new biofuel properties may be susceptible to aspects that are not protected by these requirements, making it more difficult to market them in full. However, with more fuel testing and production, drop-in biofuels are an alternative to traditional marine fuels, which are currently used.

49.3.3 Advanced Biofuel Production Technologies

49.3.3.1 Oleochemical Production

'Hydrotreated vegetable oils' (HVO), natural diesel, green diesel, or 'hydrotreated natural oils' (HRO) are also known as 'hydrotreated esters and fatty acids' (HEFA). They consist of vegetable oils or animal fats, hydrotreated and processed, typically in the presence of a catalyst. Hydrotreatment requires the injection of hydrogen in a two-stage cycle into the feedstock: the feedstock is first deoxygenated and its double bonds are saturated to form alkanes. The alkans undergo isomerization and cracking in the second step. During this step, called dewaxing, the length of the alkyl chain is reduced and the branching to hydrocarbons increased. This process primarily produces diesel fractions, with only a small portion being used for jet fuel blends.

HVO can be generated in oil refineries, since they already have hydrotreating facilities. However, some modifications can require new investments to build an HVO-only manufacturing facility. Usually, the overall manufacturing cycle is more costly than for FAME diesel, but HVO results in a drop-in fuel that can be directly integrated into distribution and refueling facilities as well as into existing diesel engines without further alteration. Given the use of renewable feedstocks as a starting material, HVO would have to be combined with traditional marine fuels in order to be feasible for its development in the long run, as the fuel volumes needed for short and long distance shipping are still too high for HVO to achieve alone. The prices of HVO feedstock often differ from source to source. Palm oil, for example, reached a ten-year high of USD1,250 per metric ton in February 2011, while used cooking oil traded at its peak in January 2013 reached USD720 per ton. Since then, prices for both commodities for palm oil and waste cooking oil have dropped to USD650 and USD400 per ton, respectively. Not only are lipid feedstock prices seasonal but labor costs and land use (sustainability) also affect them, contributing to their volatility. With the additional production costs to improve the fats and oils, HVO will support the aviation sector more economically as a high-quality fuel to reduce the feedstock and upgrade costs.

Fuels derived from algal biomass can have a high flash point, are compatible with traditional biodiesel, and are biodegradable. Algal feedstocks have a theoretical potential for transformation into fuels as the strains can have very high growth rates, which translates into high biomass yield per hectare. Such numbers should be treated with great caution, however, as they were not seen in either demonstration or commercial units. They can also act as a no-sulfur drop-in fuel when algal oils are hydrotreated. The key downside of algae-based biofuel production is processing costs. To date, due to its high capital and operational costs, no commercial algal fuel is yet available on a consistent basis.

49.3.3.2 Thermochemical Production

Thermochemical processes for biofuel production use high temperature and/or pressure, and likely homogeneous and/or heterogeneous catalysts to turn biomass into liquid fuels and chemicals, heat and electricity. Contrary to the lipid feedstocks used in the development of biodiesel and HVO, feedstocks processed by thermochemical processes are mainly based on lignocellulosics. Thermal conversion starts with the conversion of biomass (wet or dry) to intermediate fluid (gas or oil) and then catalytically converted or hydroprocessed to hydrocarbon fuels.

Pyrolysis oil can be used as a component in emulsion biofuels for marine fuel applications to increase the thermal efficiency and minimize particulate emissions when used in diesel engines (Chiaramonti et al., 2003).

49.3.3.3 Biofuel Blending

Mixing traditional fossil fuels with biofuels is an alternative to reducing the use of fossil fuel while adding compatible drop-in fuels into the fuel mix. However, renewable fuel blending requirements are still not in effect in the case of marine fuels, and biodiesel blends are not yet available in the shipping fuel industry, while production and distribution are available through contracts set. Testing of biofuel mixtures on board ocean-going vessels was recorded for 7 to 100% biofuel mixtures, the results of which showed that biofuels can be combined effectively with standard fuel without any drastic effects on the engine (Maersk, 2011).

49.3.3.4 Bioethanol

Bioethanol processing includes the microbial fermentation of sucrose, starch, or cellulose (feedstocks dependent on glucose) to ethanol. Bioethanol is currently the most consumed transport biofuel, produced mostly by Brazil and the USA (Konur, 2018). Commercial bioethanol production is nearly twice as much as biodiesel, with much of the fuel used for motor vehicle transport. Bioethanol has a greater volume of cetane and a lower energy content than biodiesel, but its usage decreases shipping emissions and carbon footprint. Advancing new multifuel diesel engine technologies will theoretically open up the bioethanol marine fuel market, though it will be decades before such technologies can be used in a larger number of vessels.

49.3.3.5 Emulsion Biofuels

There are many ways of processing biofuels from emulsions. One relates to animal fats or vegetable oils processed into biodiesel and emulsified with water and a surfactant. Another process involves combining biodiesel with pyrolysis oils and emulsifying them in the presence of a surfactant. In certain situations, the mixture is supplemented with oxygenated compounds such as methanol, ethanol, and dimethyl carbonate to boost the ignition efficiency of the gasoline.

'Emulsion biofuels' have a low energy density, but have been tested on ocean-going vessels and the shipping companies are regarded as such. As they have a high thermal capacity and are compatible with diesel engines, they have the ability to be used for long distance shipping. There has been no commercial production of biofuels for emulsion up to now, however.

49.3.3.6 Biogas and Biomethanol

Methanol has gained interest as a shipping fuel only recently, due to its high abundance and relatively low production costs. As with ethanol, methanol in multifuel marine diesel engines is a compatible fuel. Industrially, methane is the most economical way of processing methanol. In this process, natural gas (methane) in a vapor reformer is mixed with steam, heated, and passed over a heterogeneous catalyst. The gas/steam is transformed into syngas, then pressurized and converted over a catalyst to methanol, and eventually distilled in order to remove water and impurities to produce pure methanol. Methanol fuel has some benefits over LNG as it is a liquid fuel at ambient temperature and more compliant with current infrastructure for liquid fuel.

Biogas produced from anaerobic fermentation is potentially a feedstock for the production of 'liquefied biogas' (LBG). The only technical requirement for the transformation of biogas into LBG is to purify methane by removing the CO_2 associated with fermentation. This purification process is already popular in biogas plants linked to the gas grid.

49.4 STATUS OF MARINE BIOFUEL

The support and focus available for the production of new marine fuels is related to the current price of crude oil. When oil prices are weak, there is little economic motivation to turn to renewable fuels, despite the current fossil fuel infrastructure. Nonetheless, there are environmental and regulatory opportunities to produce alternatives that supplement traditional fossil-based alternatives.

The production of biofuels compatible with marine engines is still in its infancy. Having an ample supply of feedstock and efficient processing technologies to produce competitively priced biofuel on a large scale remains a challenge. Starting with biomass feedstocks, the price and availability of agricultural residues has increased due to an increasing world population and evolving food preferences; however, the availability of markets are becoming more unpredictable due to climate extremes. Regardless of the feedstock, the sale price of biofuels is always related to the price of oil. Experts claim that biofuels produced from farm waste will not compete with conventional fuels until the price of oil falls to at least USD60 a barrel. Therefore, as the price of crude oil is weak, the market for biofuels is increasing. From the point of view of the vessel operator, up to 50% of the operating costs are mainly driven by fuel prices, which means that fossil marine fuels have a further competitive advantage. However, biofuels reduce ship pollution and improve local air quality, in addition to being ultra-low in sulfur content. As a result, as long as higher sulfur emissions are enforced, biofuels can be integrated into the shipping sector to fulfil commitments and be more cost-effective as demand rises.

With both regulators and market drivers in place, biofuels could make up between 5 and 10% of the global marine fuel mix by 2030 (Kronemeijer, 2016), equivalent to a demand of 16–33 million tons of biofuel. Estimates for global biofuel production in 2020 are approximately 115 million tons of oil equivalent. The establishment of a global 'emission control areas' (ECA) and financial incentives in ports for ships that generate less pollution will further boost the market for biofuels. Acceptance of the use of biofuels in deep-sea shipping, on the other hand, will take longer to introduce and would only take place if fuels can be produced in large quantities at a reasonable price or on a global basis. Newer fuels such as 'dimethyl ether' (DME), bio-LNG, bioethanol, and biomethanol are compatible with modern marine diesel engines, but their widespread adoption in shipping is limited by availability. There is still no infrastructure and investment available in the fuel supply chain to incorporate such new fuels, even though they can be compatible with newly built ships.

49.4.1 Marine Biofuel Production

For any new biofuel or biofuel blends made, fuel testing will be required to check their engine compatibility and verify that it meets any fuel specifications and requirements. For a fuel analysis, normally 0.2 to 3 liters of fuel are needed, which is usually performed before the fuel is placed on board a ship. Testing requirements for a new fuel are far more stringent. They need 10 to 25 liters of fuel starting with a fuel stability check. It is known at this point if the fuel has some phase separation or bacterial growth over an eight-month span; it is stored aboard a ship for the same length of time. The fuel also undergoes intense centrifugation to see if any phase separation happens and to establish the scale of particulate filters to be mounted. In addition, the fuel stability check is performed to assess the oxidation stability of the fuel and at what temperatures the fuel can be heated. In addition, a fuel pump test is carried out, 10 to 200 liters of which are required. Whether the fuel will circulate through the device without a phase separation is decided here. An engine check would need between 250 and 2,000 liters of oil, and will be ready to undergo a service check on a ship after the fuel has passed the requirements. This will need at least 100 tons of fuel, though 2,000 tons would be ideal. If the supply of fuel is ensured it is fairly straightforward for a shipping vessel to turn to a biofuel drop-in or blend. Once production facilities are operating, the infrastructure for supplying fuel must be built.

This includes bunkering and storage of biofuels, and ensuring long-term supply of these fuels. The handling of biofuels on board would also include ground workers' knowledge of rules for the safe use of fuels, as well as the installation of modern biofuel or biofuel mixing systems.

49.4.2 Biofuel Feedstock

Present feedstocks for the production of biofuels compatible with first generation diesel engines are primarily plants with high lipid content, like palm, soybean, and rapeseed (Bacovsky et al., 2016). Nonetheless, these feedstocks are typically difficult for an increased commercial production of marine biofuels to be produced cheaply and sustainably.

The availability of biomass feedstock for processing fuel depends on the geographical location. Agricultural residues such as corn stovers are commonly available in the United States and China, while wheat straw can be found in Europe and Asian and North American pockets. Sugar cane bagasse is most popular in South America, with Brazil leading the market; rice straw production is based in Asia; while palm oil residues come from Southeast Asian palm-oil-producing countries, especially Malaysia and Indonesia.

Waste lignin, a low-value commodity from the processing of second generation biofuel or pulping processes, can also be a viable raw material for biofuel production, as it has the highest energy density of all components of biomass. Algal oils are also a recognized feedstock for the production of biodiesel, which is listed as a feedstock of the third generation. Algae can produce more biomass per unit area in a year than other types of biomass under ideal conditions.

Regardless of the form of biofuel and the equipment used to produce it, the final potential of marine biofuels entering the market would eventually be constrained by the amount of feedstock that can be made available, and the price consumers are prepared to pay.

49.5 MARINE BIOFUEL REGULATIONS

Regulations concerning the usage of biodiesel in the automotive transport sector have advanced faster than in the merchant shipping sector. In 2003, the European Commission issued a mandate to establish guidelines relating to minimum criteria and biodiesel test methods: EN 14214 to standardize biodiesel fuel. The 'American Society for Testing and Materials' (ASTM) has also released the B100 biodiesel fuel specifications ASTM D-6751-02 and the distillate fuel blends. Nevertheless, the use of biofuels in ships is not reflected in IMO legislation.

Before being sold as 'Renewable Energy Directive' (RED)-compliant and counted against the EU's renewable energy goals, biofuels have to meet other sustainability requirements. These come in four categories: biofuels must achieve GHG emission reductions as compared to conventional fossil fuels to ensure carbon savings (RED Article 17.2); feedstocks cannot be collected from conservation areas to protect high natural diversity land such as primary forests and grasslands (Article 17.3); feedstocks cannot be grown in some exempt areas to protect carbon stocks and modify land use (Article 17.4), in particular high-carbon peat land areas (Article 17.5).

Many jurisdictions likewise (e.g. Canada) are evaluating whether marine fuels will be included in their list of sustainable transport fuels as it updates its 'Clean Fuel Policy' review (ECCC, 2017). The difficulty in enforcing these schemes is that market incentives for compliance with such regulations must be made to gain or impact all parties in the supply chain of biofuels, otherwise no party will turn from the conventional status quo (Florentinus et al., 2012), while international marine biofuel regulations are relevant.

49.5.1 Marine Biofuel Deployment Initiatives

In North America, much of the initial research and development work on marine biofuels was funded federally through many US departments: the U.S. Navy Department, U.S. Department of Agriculture (USDA), Transportation Department (DOT), and Energy Department (DOE). Such steps have been taken to grow the domestic biofuel industry and to reduce US reliance on imported oil.

In addition to expanding the production of marine biofuels, fuel testing aboard marine vessels was also an important part of the funding package; the key partner was the US Navy. This is also the biggest source of marine grade fuel in the USA along with the US Coast Guard.

On the other hand, the Royal Australian Navy, with its strong ties to the US Department of Defense, has also adopted the environmental policies of the US Navy and will begin purchasing oil from a pilot biodiesel plant in Queensland's Gladstone in partnership with Southern Oil (Queensland Government, 2016).

A few UK research groups have also explored the possibility of using glycerin as a marine fuel, a byproduct of FAME biodiesel processing. The 'Glycerine Fuel for Engines and Marine Sustainability' project advocates the use of glycerine as a fuel as it is derived from plant sources, is sulfur-free, can minimize NOx emissions, and can be easily retrofitted to modern marine diesel engines. With FAME production on the rise, glycerine is becoming a readily available but under-used fuel that can be used as a non-toxic and non-volatile fuel by the shipping industry in 'emission control areas' (ECAs).

49.5.2 Biofuel Deployment Challenges

One of the main issues in shipping vessels concerning the use of biofuels, or any new alternative fuel, is the lack of long-term fuel test data to guarantee the health and continued reliability of the selected fuel. The fuel price and supply guarantee along these lines are both major challenges to overcome before shipping vessels turn on a new fuel.

In diesel engines, the efficiency of regular petroleum-based fuels is very well known. At the moment, however, the same cannot be said for modern biofuels, because they are made from different feedstocks and processes. Significant quantities of research and standardization therefore need to be carried out so that the renewable fuel sector can produce proper biofuels, which are completely compliant and as efficient as an alternative to conventional fuels. While current commercially developed biofuels have been shown to be chemically compatible with the fuel structure, biological stability during transport and, in particular, long-term storage remains a concern.

49.6 CONCLUSIONS

Maritime transport is and will continue to be one of the most vital forms of freight. Maritime merchant shipping is one of the fastest growing industries within the transport industry, which plays a significant role in the economy and the climate of the world. Marine shipping is a fairly low-carbon long-distance transport mode, and seeks to develop its energy efficiency and cost-competitiveness in order to survive in a competitive market with alternatives such as road and aviation. There are also operational benefits of using maritime routes rather than road transport, as goods can arrive faster through ports while achieving a reduction of GHG emissions at the same time.

With the existing fuel volumes needed by the merchant shipping industry and the latest regulatory fuel requirements, biofuels, in particular biofuel blends, have a strong market potential. While the first-generation bioethanol and biodiesel industries have already been commercially developed, support for second-generation biofuels has been gradually evolving and shifting as of 2016. Most policies relating to renewable liquid transport fuels have been aimed at the road transport market, although legislation to encourage the use of renewables for shipping, aviation, and rail transport have lagged behind.

The use of biofuels in shipping is an opportunity to reduce GHG emissions and improve air quality, provided that feedstocks from biofuels produce very little to no sulfur. The growth of the biofuels infrastructure and supply chain is also an opportunity to create a sustainable bioeconomy. As the commercial production of marine biofuels begins, feedstocks for the production of marine biofuels can compete with other liquid transport fuels, in particular for aviation.

The environmental benefits together with current regulatory policies and government support schemes provide a strong business case for the production of biofuels. The move to biofuels or biofuel blends is likely to be driven by forward thinking shippers, big freight shipping companies, and high-end-customer-profile shipping companies such as ferry and cruise companies.

With a few exceptions, biofuels provide a net reduction in the cost of carbon, particularly those generated from feedstocks of the second generation, thereby reducing the carbon emissions created by the shipping industry. In the merchant shipping market, an overall reduction in GHG emissions will most likely be achieved by a combination of changes in ship construction, port infrastructure, and fuel technology. 'Energy Efficiency Design Index' (EEDI) and RED II will now serve as a business case for promoting advanced biofuels: EEDI enables ship owners and operators to use more energy-efficient and low-carbon technologies to power their ships. Although at the time of writing it is uncertain how RED II would cover the maritime market, it is an example of how fuel suppliers and/or bunker parties may be expected to supply biofuels to the marine market.

REFERENCES

Bacovsky, D., S. Holzleitner, and M. Enigl. 2016. Deliverable D.2.5: overview on detailed information for each of the 9 topics. In *European Biofuels Technology Platform: Support for Advanced Biofuels Stakeholders*. Gulzow, Pruzen: Fachagentur Nachwachsende Rohstoffe e.V.

Chiaramonti, D., M. Bonini, and E. Fratini et al. 2003. Development of emulsions from biomass pyrolysis liquid and diesel and their use in engines-Part 1: Emulsion production. *Biomass Bioenergy* 25: 85–99.

ECCC. 2017. *Clean Fuel Standard*. York, ON: Environment and Climate Change Canada.

EIA. 2019. *The Effects of Changes to Marine Fuel Sulfur Limits in 2020 on Energy Markets*. March, 2019. Washington, DC: Energy Information Administration.

Florentinus, A., C. Hamelinck, A. van den Bos, R. Winkel, and M. Cuijpers, 2012. *Potential of Biofuels for Shipping*. Utrecht: Ecofys Netherlands.

IEA. 2017. *Annual Report 2017: IEA Bioenergy*. Paris, Cedex: International Energy Agency.

Issa, M., H. Ibrahim, A. Ilinca, and M. Hayyani. 2019a. A review and economic analysis of different emission reduction techniques for marine diesel engines. *Open Journal of Marine Science*, 9: 148–171.

Issa, M., H. Ibrahim, R. Lepage, and A. Ilinca. 2019b A review and comparison on recent optimization methodologies for diesel engines and diesel power generators. *Journal of Power and Energy Engineering* 7: 31–56.

Kass, M. D., Z. Abdullah, and M. J. Biddy, et al., 2018. *Understanding the Opportunities of Biofuels for Marine Shipping*. Oak Ridge, TN: Oak Ridge National Laboratory.

Kołwzan, K. and M. Narewski. 2012. Alternative fuels for marine applications. *Latvian Journal of Chemistry* 51: 398–406.

Konur, O. 2018. *Bioenergy and Biofuels*. Boca Raton, FL: CRC Press.

Kronemeijer, D. 2016. *Industry Insight: Sustainable Marine Biofuel – Scaling-up for a Crucial Role in a Low Emission Future for Shipping*. Vancouver, BC: Ship & Bunker.

Maersk. 2011. *Sustainability Report 2011: Global business - Global challenges*. Copenhagen: Maersk.

MPAS. 2019. *Bunkering Statistics*. Singapore: Maritime and Port Authority of Singapore.

Noor, C. W. M., M. M. Noor, and R. Mamat. 2018. Biodiesel as alternative fuel for marine diesel engine applications: A review. *Renewable and Sustainable Energy Reviews* 94: 127–142.

Queensland Government. 2016. *Queensland to Partner US Navy in Massive Biofuel Initiative*. Queensland.

Smagala, T. G., E. Christensen, and K. M. Christison, et al. 2013. Hydrocarbon renewable and synthetic diesel fuel blendstocks: Composition and properties. *Energy & Fuels*, 27:237–246.

Wagner, I. 2020. Number of ships in the world merchant fleet as of January 1, 2019, by type. www.statista.com

Winebrake, J. J., J. J. Corbett, and P. E. Meyer. 2007. Energy use and emissions from marine vessels: A total fuel life cycle approach. *Journal of the Air & Waste Management Association* 57: 102–110.

Witherby Seamanship. 2013. *Marine Fuels and Emissions*. Livingston: Witherby Seamanship International.

50 Desulfurization of Diesel Fuels

A Review of the Research

Ozcan Konur

CONTENTS

50.1 INTRODUCTION

Crude oils have been primary sources of energy and fuels, such as petrodiesel. However, significant public concerns about the sustainability, price fluctuations, and adverse environmental impact of crude oils have emerged since the 1970s (Ahmadun et al., 2009; Atlas, 1981; Babich and Moulijn, 2003; Kilian, 2009; Perron, 1989). Thus, biooils (Bridgwater et al., 1999; Bridgwater and Peacocke, 2000; Czernik and Bridgwater, 2004) and biooil-based biodiesel fuels (Chisti, 2007; Hill et al., 2006; Hu et al., 2008) have emerged as alternatives to crude oils and crude oil-based petrodiesel fuels, respectively, in recent decades. Nowadays, although petrodiesel fuels are still used extensively, biodiesel fuels are being used increasingly in the transportation and power sectors (Konur, 2021a–ag). Therefore, there has been great public interest in petrodiesel fuels (Birch and Cary, 1996; Khalili et al., 1995; Rogge et al., 1993; Song and Ma, 2003; Song, 2003). However, it is necessary to reduce the amount of sulfur compounds in these fuels (Bosmann et al., 2001; Kim et al., 2006; Song and Ma, 2003; Song, 2003; Yang et al., 2003) due to their significant adverse health (Hedley et al., 2002; Katsouyanni et al., 1997; Sheppard et al., 1980) and environmental impacts (Carmichael et al., 2002; Doney et al., 2007; Greaver et al., 2012) of these compounds.

Furthermore, for the efficient progression of the research in this field, it is necessary to develop efficient incentive structures for the primary stakeholders and to inform these stakeholders about the research in this field (Konur, 2000, 2002a–c, 2006a–b, 2007a–b; North, 1991a–b).

Although there have been a number of reviews and book chapters in this field (Kulkarni and Afonso, 2010; Song, 2003; Song and Ma, 2003), there has been no review of the 25-most-cited articles. Thus, this chapter reviews these articles by highlighting the key findings of these most-prolific studies on the desulfurization of sulfur compounds in petrodiesel and related fuels. Then, it discusses these key findings.

50.2 MATERIALS AND METHODOLOGY

The search for the literature was carried out in the 'Web of Science' (WOS) database in February 2020. It contains the 'Science Citation Index-Expanded' (SCI-E), the 'Social Sciences Citation Index' (SSCI), the 'Book Citation Index-Science' (BCI-S), the 'Conference Proceedings Citation Index-Science' (CPCI-S), the 'Emerging Sources Citation Index' (ESCI), the 'Book Citation Index-Social Sciences and Humanities' (BCI-SSH), the 'Conference Proceedings Citation Index-Social Sciences and Humanities' (CPCI-SSH), and the 'Arts and Humanities Citation Index' (A&HCI).

The keywords for the search of the literature are collated from the screening of abstract pages for the first 1,000 highly cited papers on petrodiesel fuels. These keywords sets are provided in the Appendix of the related book chapter (Konur, 2021z).

The 25-most-cited articles are selected for this review and the key findings are presented and discussed briefly.

50.3 RESULTS

50.3.1 'Hydrodesulfurization' (HDS) of Sulfur Compounds

Ma et al. (1994) study the HDS reactivities of various sulfur compounds in diesel fuels in a paper with 359 citations. They performed the HDS of a diesel oil in a batch autoclave reactor over the temperature range 280–420°C for 0–90 min under a total pressure of 2.9 MPa, using CoMo and NiMo catalysts in both one and two stages. They examined the HDS reactivities of 'benzothiophenes' (BTs), 'dibenzothiophenes' (DBTs), and their alkylated homologes existing in the diesel fuel in detail by means of respective quantitative analyses. They observed that the sulfur compounds could be classified into four groups according to their HDS reactivities, which were described by their pseudo-first-order rate constants. DBTs carrying two alkyl substituents at the 4- and 6-positions, respectively, were the most resistant to desulfurization. H_2S produced from reactive sulfur compounds in the early stage of the reaction was one of the main inhibitors for HDS of the unreactive species. A second stage using fresh hydrogen solved this inhibition problem, with NiMo achieving deeper desulfurization.

Bataille et al. (2000) study the promoter effect of Co or Ni on the HDS activity of Mo/alumina by using DBT and 4,6-'dimethyldibenzothiophene' (DMDBT) as reactants in a paper with 327 citations. They performed the reaction at 340°C under a 4 MPa total pressure in a fixed-bed microreactor. On the Mo/alumina catalyst, they observed that both reactants had similar reactivities, DMDBT being even slightly more reactive than DBT. However, as generally observed, on the CoMo/alumina and NiMo/alumina catalysts, DBT was much more reactive (five to six times) than DMDBT. This was mainly because of a tremendous enhancement of the rate of the direct desulfurization (DDS) pathway of the HDS of DBT, whereas for DMDBT this effect was much more limited. They conclude that the main effect of the promoter on the HDS of DBT-type molecules was to increase the rate of the C-S bond cleavage, provided this reaction was not hindered by steric constraints. They attribute this effect to an enhancement by the promoter of the basicity of certain sulfur anions in its vicinity. They also show that the lower reactivity of DMDBT compared to that of DBT measured on the promoted catalysts could not be attributed to differences in the adsorption strength of the reactants. Assuming that C-S bond cleavage occurred through a B-elimination process, they propose several other explanations for the low reactivity of DMDBT: (a) steric hindrance of the adsorption of the dihydrointermediates by the methyl groups; (b) steric hindrance by the methyl groups of the C-S bond cleavage; (c) the fact that only one H atom is available for the C-S bond cleavage; (d) an effect of the methyl group on the acidity of the H atom involved in the elimination step. They also make proposals concerning the catalytic centers involved in the hydrogenation steps and in the C-S bond cleavage steps.

Houalla et al. (1980) study the HDS of methyl-substituted DBTs catalyzed by sulfided Co-Mo-γ-Al_2O_3 in a paper with 276 citations.

Kabe et al. (1992) performed deep HDS of 'polyaromatic sulfur-containing compounds' (PASC) in light oil by using Co-MO/Al_2O_3 under experimental conditions representative of industrial practice in a paper with 256 citations. They observed that alkyl-substituted DBTs among sulfur-containing compounds in light oil were most difficult to desulfurize. Dialkyl-substituted DBTs, especially 4,6-DMDBT, remained until the final stage of the reaction (390°C) while alkylbenzothiophenes were completely desulfurized at 350°C.

Sawhill et al. (2005) prepared silica and alumina-supported nickel phosphide (Ni_xP_y) catalysts, characterized them by bulk and surface sensitive techniques, and evaluated them for the HDS of thiophene in a paper with 249 citations. They prepared a series of 30 wt% Ni_xP_y/SiO_2 and 20 wt% Ni_xP_y/Al_2O_3 catalysts from oxidic precursors with a range of P/Ni molar ratios by temperature-programmed reduction (TPR) in flowing H_2. They observed that oxidic precursors with molar ratios of P/Ni = 0.8 and 2.0 yielded catalysts containing phase-pure Ni_2P on the silica and alumina supports, respectively. At lower P/Ni ratios, significant $Ni_{12}P_5$ impurities were present in the Ni_xP_y/SiO_2 and Ni_xP_y/Al_2O_3 catalysts. The HDS activities of these catalysts depended strongly on the P/Ni molar ratio of the oxidic precursors with optimal activities obtained for catalysts containing phase-pure Ni_2P and minimal excess P. After 48 h on stream, a Ni_2P/SiO_2 catalyst was 20 and 3.3 times more active than sulfided Ni/SiO_2 and Ni-Mo/SiO_2 catalysts, respectively. A Ni_2P/Al_2O_3 catalyst was 2.7 times more active than a sulfided Ni/Al_2O_3 catalyst but only about half as active as a sulfided Ni-Mo/Al_2O_3 catalyst.

Egorova and Prins (2004) study the HDS of DBT and 4, 6-DMDBT over sulfided NiMo/γ-Al_2O_3, CoMo/γ-Al_2O_3, and Mo/γ-Al_2O_3 catalysts in a paper with 236 citations. They observed that the Ni and Co promoters strongly enhanced the activity of the Mo catalyst in the direct desulfurization pathway of the HDS of DBT and 4,6-DMDBT and in the final sulfur-removal step in the hydrogenation pathway, while the hydrogenation was moderately promoted. H_2S had a negative effect on the HDS of DBT and 4,6-DMDBT, which was strongest for the NiMo catalyst and stronger for the direct desulfurization pathway than for the hydrogenation pathway. Because the direct desulfurization pathway was less important for the HDS of 4,6-DMDBT than the hydrogenation pathway, the conversion of 4,6-DMDBT was less affected by H_2S than the conversion of DBT. The sulfur removal via the direct desulfurization pathway and the ultimate sulfur removal in the hydrogenation pathway were affected by H_2S to the same extent over all the catalysts. The removal of sulfur from tetrahydrodibenzothiophenes takes place by hydrogenolysis, like the direct desulfurization of DBT to biphenyl. The CoMo catalyst performed better than the NiMo catalyst in the final desulfurization via the hydrogenation pathway in the HDS of 4,6-DMDBT at all partial pressures of H_2S.

50.3.2 Adsorptive Desulfurization of Sulfur Compounds

Yang et al. (2003) study the desulfurization of transportation fuels with zeolites under ambient conditions in a paper with 701 citations. They showed that Cu^+ and Ag^+ zeolite Y can adsorb sulfur compounds from commercial fuels selectively and with high sulfur capacities (by π complexation) at an ambient temperature and pressure. Thus, they reduced the sulfur content from 430 to < 0.2 parts per million by weight in a commercial diesel at a sorbent capacity of 34 cubic centimeters of clean diesel produced per gram of sorbent. This sulfur selectivity and capacity are orders of magnitude higher than those obtained by previously known sorbents.

Kim et al. (2006) study the adsorptive desulfurization and denitrogenation using a model diesel fuel, which contains sulfur, nitrogen, and aromatic compounds, over three typical adsorbents (activated carbon, activated alumina, and nickel-based adsorbent) in a fixed-bed adsorption system in a paper with 396 citations. They examined the adsorptive capacity and selectivity for the various compounds and compared them on the basis of the breakthrough curves. Different adsorptive selectivities in correlation with the electronic properties of the compounds provided new insight into the fundamental understanding of the adsorption mechanism over different adsorbents. For the supported nickel adsorbent, they observed that the direct interaction between the heteroatom in the adsorbates and the surface nickel played an important role. The adsorption selectivity on the activated alumina depended dominantly on the molecular electrostatic potential and the acidic–basic interaction. The activated carbon showed higher adsorptive capacity and selectivity for both sulfur and nitrogen compounds, especially for the sulfur compounds with methyl substituents, such as 4,6-'methyldibenzothiophene' (MDBT). The hydrogen bond interaction might play an important role in adsorptive desulfurization and denitrogenation over the activated carbon and different adsorbents might be suitable for separating different sulfur compounds from different hydrocarbon streams.

Ma et al. (2002) explore a new desulfurization process by 'selective adsorption for removing sulfur' (SARS) in a paper with 357 citations. They developed an adsorbent and used it for the adsorption desulfurization of diesel fuel, gasoline, and jet fuel at room temperature. They observed that this transition metal-based adsorbent was effective for selectively adsorbing the sulfur compounds, even the refractory sulfur compounds in diesel fuels. The SARS process could effectively remove sulfur compounds in the liquid hydrocarbon fuels at ambient temperature under atmospheric pressure with low investment and operating cost. They propose a novel integrated process for deep desulfurization of the liquid hydrocarbon fuels in a refinery, which combines a selective adsorption (SARS) of the sulfur compounds and a 'hydrodesulfurization process of the concentrated sulfur fraction' (HDSCS). The SARS concept might be used for on-site or on-board removal of sulfur from fuels for fuel cell systems.

Velu et al. (2003) synthesize and evaluate adsorbents based on transition metal ion-exchanged Y zeolites (with Cu, Ni, Zn, Pd, and Ce ions) for the adsorptive desulfurization of a 'model jet fuel' (MJF) and a real jet fuel (JP-8) in a paper with 319 citations. Among the adsorbents tested, they observed that Ce-exchanged Y zeolites exhibited a better adsorption capacity of about 10 mg of sulfur/g of adsorbent at 80°C with an MJF containing 510 ppmw sulfur. The same adsorbent exhibited a sulfur adsorption capacity of about 4.5 mg/g for the real JP-8 jet fuel containing about 750 ppmw sulfur. Desulfurization of MJF under flow conditions at 80°C showed a breakthrough capacity of about 2.3 mg/g of adsorbent. Ce-exchanged zeolites exhibited higher selectivity for sulfur compounds as compared to the selectivity of aromatics, for which a comparative study indicated that the sulfur compounds are adsorbed over Ce-exchanged Y zeolites via direct sulfur–adsorbent (S-M) interaction rather than via π-complexation. While the selectivity for 2-'methylbenzothiophene' (MBT) was higher in the static adsorption studies, the adsorption selectivity decreased in the order 5-MBT > BT > 2-MBT under dynamic conditions. This trend was correlated to the electron density on sulfur atoms derived from computer-aided molecular orbital calculations.

Hernandez-Maldonado et al. (2005) study new π-complexation-based sorbents for the desulfurization of diesel, gasoline, and jet fuels in a paper with 277 citations. They obtained these sorbents by ion exchanging faujasite type zeolites with Cu^+, Ni^{2+}, or Zn^{2+} cations using different techniques, including 'liquid phase ion exchange' (LPIE). They avoided cation hydrolysis limitation when they used 'vapor phase ion exchange' (VPIE) and 'solid-state ion exchange' (SSIE) techniques. They performed the deep-desulfurization (sulfur levels of < 1ppmw) tests using fixed-bed adsorption techniques. They characterized the treated and untreated fuels, which are capable of eliminating quenching effects. After proper optimization and calibration of the detector non-linear response, it was possible to detect sulfur concentration as low as 20 ppbw S. They verified the data using total sulfur analyzers. They observed that the π-complexation sorbents' desulfurization performance decreased as follows: Cu(I)-Y(VPIE) > Ni(II)-Y(SSIE) > Ni(II)-X(LPIE) > Zn(II)-X(LPIE) > Zn(II)-Y(LPIE). These sorbents' performance decreased as follows: $Cu^+ > Ni^{2+} > Zn^{2+}$. The best sorbent, Cu(I)-Y(VPIE), had breakthrough adsorption capacities of 0.395 and 0.278 mmol S/g of sorbent for commercial jet fuel (364.1 ppmw S) and diesel (297.2 ppmw S), respectively.

Hernandez-Maldonado and Yang (2003) study fixed-bed adsorption using different π-complexation adsorbents for the desulfurization of liquid fuels in a paper with 259 citations. They used Cu(l)-Y (autoreduced Cu(II)-Y), Ag-Y, H-Y, and Na-Y zeolites to separate low-concentration thiophene from mixtures including benzene and/or n-octane, all at room temperature and atmospheric pressure. They obtained sulfur-free (i.e. below the detection limit of 4 ppmw sulfur) fuels with Cu(I)-Y, Ag-Y, and H-Y but not Na-Y. Breakthrough and saturation adsorption capacities obtained for an influent concentration of 760 ppmw sulfur (or 2,000 ppmw thiophene) in n-octane followed the order Cu(l)-Y > Ag-Y > H-Y > Na-Y and Cu(I)-Y > H-Y > Na-Y > Ag-Y, respectively. Cu(l)-Y zeolite adsorbed 5.50 and 7.54 wt% sulfur at breakthrough and saturation, respectively, for an influent concentration of 760 ppmw sulfur in n-octane. For the case of 190 ppmw sulfur in mixtures containing both benzene and n-octane, Cu(I)-Y adsorbed 0.70 and 1.40 wt% sulfur at breakthrough and saturation, respectively. They regenerated the adsorbent by using air at 350°C, followed by reactivation in helium at 450°C. The observed adsorption behavior, in general, agreed well with previous studies performed for pure-component vapor-phase adsorption of thiophene and benzene with the same adsorbents.

Hernandez-Maldonado and Yang (2004) study the desulfurization of a commercial diesel fuel by VPIE copper(I) faujasite zeolites in a fixed-bed adsorber operated at ambient temperature and pressure in a paper with 252 citations. They observed that this zeolite adsorbed approximately five thiophenic molecules per unit cell. After treating 18 cm^3 of fuel, the cumulative average sulfur concentration detected was 0.032 ppmw-S. The π-complexation sorbents selectively adsorbed highly substituted thiophenes, BTs, and DBTs from diesel, which was not possible by using conventional HDS reactors. They assert that the high sulfur selectivity and high sulfur capacity of the VPIE Cu(II)-zeolites were due to π-complexation.

Lauritsen et al. (2004) synthesized single-layer MoS_2 nanoclusters on an Au substrate as a model system for the hydrotreating catalyst and studied them by atomically resolved scanning tunneling microscopy (STM) in order to achieve atomic-scale insight into the interactions with hydrogen and thiophene (C_4H_4S) in a paper with 238 citations. Surprisingly, they observed that thiophene molecules could adsorb and react on the fully sulfided edges of triangular single-layer MoS_2 nanoclusters. They associate this unusual behavior with the presence of special brim sites exhibiting a metallic character. These sites existed only at the regions immediately adjacent to the edges of the MoS_2 nanoclusters, and from density-functional theory such sites were associated with one-dimensional electronic edge states. The full sulfur-saturated sites were capable of adsorbing thiophene, and when thiophene and hydrogen reactants were coadsorbed here, a reaction path was revealed which led to partial hydrogenation of the thiophene followed by C-S bond activation and ring opening of thiophene molecules. These might be regarded as important first steps in the HDS of thiophene. The metallic brim sites were important for other hydrotreating reactions over MoS_2-based catalysts, and the properties of the brim sites directly explain why hydrogenation reactions of aromatics were not severely inhibited by H_2S. The presence of brim sites in MoS nanoclusters also explains previous structure-activity relations and observations regarding steric effects and the influence of stacking of MoS_2 on the reactivity and selectivity.

50.3.3 Oxidative Desulfurization of Sulfur Compounds

Otsuki et al. (2000) study the oxidative desulfurization of light gas oil and vacuum gas oils by oxidation and solvent extraction in a paper with 639 citations. They performed the oxidation of model sulfur compounds (thiophene derivatives, BT derivatives, and DBT derivatives), straight run-light gas oil (SR-LGO, S: 1.35 wt%), and vacuum gas oil (VGO, S: 2.17 wt%) with a mixture of hydrogen peroxide and formic acid. They observed that the thiophene derivatives with 5.696 to 5.716 electron densities on the sulfur atoms could not be oxidized at 50°C whilst benzo[b]thiophene with a 5.739 electron density and other BT and DBTs with higher electron densities could be oxidized. The sulfur compounds in SR-LGO and VGO were oxidized to detectable levels (ca. 0.01 wt% S) where sulfones were formed by oxidation. The removal of sulfur compounds by extraction became more effective for the oxidized samples than for the original samples, whilst lighter sulfur compounds were preferentially extracted. The extraction efficiencies of solvents, i.e. 'N,N'-dimethylformamide' (DMF), acetonitrile (ACN), and methanol, varied greatly. The most effective solvent for the removal of sulfur compounds was DMF. The recovery of oil was, however, lowest with DMF.

Wang et al. (2003) study the oxidation of sulfur compounds in kerosene with 'tert-butyl hydroperoxide' (t-BuOOH) in the presence of various catalysts in a paper with 260 citations. They estimated the oxidation activities of DBT in kerosene for a series of Mo catalysts supported on Al_2O_3 with various Mo contents. They observed that the oxidation activity of DBT increased with increasing Mo content up to about 16 wt% and decreased when Mo content was beyond this value. They also performed the oxidation of BT, DBT, 4-MDBT, and 4,6-DMDBT dissolved in decalin on a 16 wt% Mo/Al_2O3 catalyst with t-BuOOH to investigate the oxidation reactivities of these sulfur compounds. They further observed that the oxidation reactivities of these sulfur compounds decreased in the order of DBT > 4-MDBT > 4,6-DMDBT, and much greater than BT. They assert that the oxidative reaction of each sulfur compound could be treated as a first-order reaction.

Lu et al. (2006) performed the oxidation of sulfur-containing compounds (BT, DBT, and their derivatives) in diesel in an emulsion oxidative system (water in oil (W/O)) composed of diesel 30 wt% hydrogen peroxide and an amphiphilic catalyst $[C_{18}H_{37}N(CH_3)_3]_4[H_2NaPW_{10}O_{36}]$ under mild conditions in a paper with 234 citations. They observed that the amphiphilic catalyst in the w/o emulsion system exhibited very high catalytic activity such that all sulfur-containing compounds in either model or actual diesel could be selectively oxidized into their corresponding sulfones using hydrogen peroxide (o/s <= 3) as an oxidant. The catalytic oxidation reactivity of sulfur-containing compounds was in the following order: BT < 5-MBT < DBT < 4,6-DMDBT. Although BT was relatively difficult to oxidize, it could be efficiently oxidized in the emulsion system, whilst the sulfones could be readily separated from diesel by an extractant. The sulfur level of a prehydrotreated diesel could be lowered from 500 to 0.1 ppm after oxidation and then extraction, whereas the sulfur level of a straight-run diesel could be decreased from 6,000 to 30 ppm after oxidation and extraction.

Li et al. (2004) study the ultra-deep desulfurization of diesel in a paper with 227 citations. They observed that a $[(C_{18}H_{37})_2N^+(CH_3)_2]_3[PW_{12}O_{40}]$ catalyst, assembled in an emulsion in diesel, could selectively oxidize the sulfur-containing molecules present in diesel into their corresponding sulfones by using H_2O_2 as the oxidant under mild conditions. The sulfones could be readily separated from the diesel using an extractant, and the sulfur level of the desulfurized diesel could be lowered from about 500 ppm to 0.1 ppm without changing the properties of the diesel. This catalyst showed high performance ($\geq$ 96% efficiency of H_2O_2, was easily recycled, and ~100% selectivity to sulfones). Metastable emulsion droplets (water in oil) acted like a homogeneous catalyst and were formed when the catalyst (as the surfactant) and H_2O_2 (30%) were mixed in the diesel. However, the catalyst could be separated from the diesel after demulsification.

Hulea et al. (2001) applied Ti-containing molecular sieves as catalysts for the oxidation of aromatic sulfur compounds with hydrogen peroxide under mild conditions in a paper with 222 citations. They determined the catalytic activity of TS-1 (titanium silicalite), Ti-β, and Ti-HMS (hexagonal mesoporous silica) in the oxidation of thiophene derivatives with hydrogen peroxide. They observed that while TS-1 had low activity due to restricted access of reactants into the porosity, Ti-β and Ti-HMS were effective under various reaction conditions, both in two-phase, solid–liquid, and in three-phase, solid–liquid–liquid, systems. They also obtained information on the applicability of the oxidation method for the removal of organic sulfur compounds from fuels. They observed that the sulfoxidation reaction could be used for decreasing the sulfur content of kerosene without consumption of hydrogen and the use of high pressure equipment. In this new method, the polyaromatic sulfur compounds (BT and DBT derivatives) were oxidized into their corresponding sulfoxides and sulfones, which were then removed by simple liquid–liquid separation.

50.3.4 Extractive Desulfurization of Sulfur Compounds

Bosmann et al. (2001) develop a new approach for the deep desulfurization of diesel fuels by extraction with 'ionic liquids' (ILs) in a paper with 471 citations.

Wu et al. (2004) study the desulfurization of flue gas through SO_2 absorption by an IL in a paper with 408 citations. They observed that the IL '1,1,3,3-tetramethylguanidinium lactate', could absorb SO_2 from simulated flue gas effectively under ambient conditions. Absorbed SO_2 could be desorbed under vacuum or by heating, and the IL could be reused. This absorption method might be used for cleaning gases that contain SO_2.

Zhang et al. (2004) study the 'extractive desulfurization' (EDS) and denitrogenation of fuels using ILs in a paper with 347 citations. They showed that two types of ILs, '1-alkyl-3-methylimidazolium' (AMIM) tetrafluoroborate and hexafluorophosphate and 'trimethylamine hydrochloride' ($AlCl_3$-TMAC), were potentially applicable for sulfur removal from transportation fuels. $EMIMBF_4$ (E=ethyl), $BMIMPF_6$ (B=butyl), BMIMBF4, and heavier $AMIMPF_6$ showed high selectivity, particularly toward aromatic sulfur and nitrogen compounds, for EDS and denitrogenation. The used ILs were readily regenerated either by distillation or by water displacement of absorbed molecules. The absorbed aromatic S-containing compounds were quantitatively recovered. Organic compounds with higher aromatic π-electron density were favorably absorbed. Alkyl substitution on the aromatic rings significantly reduced the absorption capacity, because of a steric effect. The cation and anion structure and size in the ILs were important parameters affecting the absorption capacity for aromatic compounds. At low concentrations, the N- and S-containing compounds were extracted from fuels without mutual hindrance. $AlCl_3$-TMAC ILs had remarkably high absorption capacities for aromatics.

Nie et al. (2006) study the EDS of gasoline using imidazolium-based phosphoric ILs in a paper with 231 citations. They determined sulfur partition coefficients for sulfur compounds 3-'methylthiophene' (3-MT), BT, and DBT between phosphoric ILs (ILs), namely, 'N- methyl- N- methylimidazolium dimethyl phosphate' ([MMIM][DMP]), 'N-ethyl-N-methylimidazolium diethyl phosphate' ([EMIM][DEP]), and 'N-butyl-N-methylimidazolium dibutyl phosphate' (BMIM][DBP]), and gasoline experimentally at 298.15 K over a wide range of sulfur content and compared with other IL extractants. They also measured the solubility of sulfur (as DBT, BT) in an IL aqueous solution at 298.15 K and varying water content. They observed that the desulfurization ability of the ILs for each sulfur component (DBT, BT, 3-MT) followed the order [BMIM][DBP] > [EMIM][DEP]. [MMIM][DMP], and the sulfur removal selectivity for a specified IL, followed the order DBT > BT > 3-MT. The phosphoric ILs were insoluble in gasoline while the fuel solubility in ILs was noticeable and followed the order [BMIM][DBP] >> [EMIM][DEP] > [MMIM][DMP]. Considering the relatively high sulfur removal ability, low fuel dissolvability, and small influence on the gasoline treated, [EMIM][DEP] might be used as a promising solvent for the desulfurization of gasoline via an EDS process. The sulfur components in the spent ILs could be conveniently separated via a water dilution process.

Zhang and Zhang (2002) study the properties of ILs in selective sulfur removal from fuels at room temperature in a paper with 220 citations. They observed that the ILs, '1-ethyl-3-methylimidazolium tetrafluoroborate', '1-butyl-3-methylimidazolium hexafluorophosphate', and '1-butyl-3-methylimidazolium tetrafluoroborate' were effective for the selective removal of sulfur-containing compounds from transportation fuels at room temperature. S-containing compounds with a C-5 aromatic ring were favorably absorbed over C-6 aromatics, while S-containing non-aromatic compounds were poorly absorbed. The ILs were regenerated from absorbed S-containing compounds by distillation or by dissolution in water. These ILs were air and moisture-stable at low temperature and non-corrosive. Therefore, the ILs could be used in multiple cycles for the removal of S-containing compounds from fuels. The absorption capacity of the ILs for S-containing compounds was sensitive to the structure of both the anion and cation of the ILs, which was manifested by the significant inhibiting effect of methyl group substitution on the aromatic ring.

50.3.5 Biodesulfurization of Sulfur Compounds

Gray et al. (1996) describe a unique sulfur acquisition system that *Rhodococcus* uses to obtain sulfur from very stable heterocyclic molecules in a paper with 257 citations. DBT is representative of a broad range of sulfur heterocycles found in petroleum that are recalcitrant to desulfurization via HDS. *Rhodococcus* sp. strain IGTS8 has the ability to convert DBT to 2-hydroxybiphenyl (HBP) with the release of inorganic sulfur. They observe that the conversion of DBT to HBP was catalyzed by a multienzyme pathway consisting of two monooxygenases and a desulfinase. The final reaction catalyzed by the desulfinase was the rate limiting step in the pathway. Each of the enzymes was purified to homogeneity and their kinetic and physical properties were studied. Neither monooxygenase had a tightly bound cofactor and each required an NADH-FMN oxidoreductase for activity. An NADH-FMN oxidoreductase was purified from *Rhodococcos* and was a protein of approximately 25,000 molecular weight with no apparent sequence homology to any other protein in the databases.

50.4 DISCUSSION

Table 50.1 provides information on the research fronts in this field. As this table shows the primary research fronts of 'HDS of sulfur compounds', 'adsorptive desulfurization of sulfur compounds', 'oxidative desulfurization of sulfur compounds', and 'EDS of sulfur compounds' comprise 24, 32, 20, and 20% of these papers, respectively. Furthermore, 'biodesulfurization of sulfur compounds' comprises 4% of these papers.

50.4.1 HDS of Sulfur Compounds

Ma et al. (1994) study the HDS reactivities of various sulfur compounds in diesel fuels in a paper with 359 citations. Bataille et al. (2000) study the promoter effect of Co or Ni on the HDS activity of Mo/alumina by using DBT and 4,6-DMDBT as reactants in a paper with 327 citations. Houalla et al. (1980) study the HDS of

TABLE 50.1
Research Fronts

	Research Front	Papers (%)
1	HDS of sulfur compounds	24
2	Adsorptive desulfurization of sulfur compounds	32
3	Oxidative desulfurization of sulfur compounds	20
4	Extractive desulfurization (EDS) of sulfur compounds	20
5	Biodesulfurization of sulfur compounds	4

methyl-substituted DBTs catalyzed by sulfided Co-Mo-γ-Al_2O_3 in a paper with 276 citations.

Kabe et al. (1992) perform deep HDS of 'polyaromatic sulfur-containing compounds' (PASC) in light oil by using Co-MO/Al_2O_3 under experimental conditions representative of industrial practice in a paper with 256 citations. Sawhill et al. (2005) prepare silica- and alumina-supported nickel phosphide (Ni_xP_y) catalysts, characterize them by bulk and surface sensitive techniques, and evaluate them for the HDS of thiophene in a paper with 249 citations. Egorova and Prins (2004) study the HDS of DBT and 4, 6-DMDBT over sulfided NiMo/γ-Al_2O_3, CoMo/γ-Al_2O_3, and Mo/γ-Al_2O_3 catalysts in a paper with 236 citations.

These prolific studies highlight the key findings of the research in this field.

50.4.2 Adsorptive Desulfurization of Sulfur Compounds

Yang et al. (2003) study the desulfurization of transportation fuels with zeolites under ambient conditions in a paper with 701 citations. Kim et al. (2006) study the adsorptive desulfurization and denitrogenation using a model diesel fuel, which contains sulfur, nitrogen, and aromatic compounds, over three typical adsorbents (activated carbon, activated alumina, and nickel-based adsorbent) in a fixed-bed adsorption system in a paper with 396 citations. Ma et al. (2002) explore a new desulfurization process by SARS in a paper with 357 citations.

Velu et al. (2003) synthesize and evaluate adsorbents based on transition metal ion-exchanged Y zeolites (with Cu, Ni, Zn, Pd, and Ce ions) for the adsorptive desulfurization of a model jet fuel and a real jet fuel (JP-8) in a paper with 319 citations. Hernandez-Maldonado et al. (2005) study new π-complexation-based sorbents for desulfurization of diesel, gasoline, and jet fuels in a paper with 277 citations. Hernandez-Maldonado and Yang (2003) study the fixed-bed adsorption using different π-complexation adsorbents for desulfurization of liquid fuels in a paper with 259 citations.

Hernandez-Maldonado and Yang (2004) study the desulfurization of a commercial diesel fuel by VPIE copper(I) faujasite zeolites in a fixed-bed adsorber operated at ambient temperature and pressure in a paper with 252 citations. Lauritsen et al. (2004) synthesize single-layer MoS_2 nanoclusters on an Au substrate as a model system for the hydrotreating catalyst and study them by atomically resolved 'scanning tunneling microscopy' (STM) in order to achieve atomic-scale insight into the interactions with hydrogen and thiophene (C_4H_4S) in a paper with 238 citations.

These prolific studies highlight the key findings of the research as an alternative of HDS of sulfur compounds in petrodiesel and related fuels.

50.4.3 Oxidative Desulfurization of Sulfur Compounds

Otsuki et al. (2000) study the oxidative desulfurization of light gas oils and vacuum gas oils by oxidation and solvent extraction in a paper with 639 citations. Wang et al. (2003) study the oxidation of sulfur compounds in kerosene with t-BuOOH in the presence of various catalysts in a paper with 260 citations. Lu et al. (2006) perform the oxidation of sulfur-containing compounds (BT, DBT, and their derivatives) in diesel in an emulsion oxidative system (water in oil (W/O)) composed of diesel 30 wt% hydrogen peroxide, and an ail amphiphilic catalyst $[C_{18}H_{37}N(CH_3)_3]_4[H_2NaPW_{10}O_{36}]$ under mild conditions in a paper with 234 citations.

Li et al. (2004) study the ultra-deep desulfurization of diesel in a paper with 227 citations. Hulea et al. (2001) apply Ti-containing molecular sieves as catalysts for the oxidation of aromatic sulfur compounds with hydrogen peroxide under mild conditions in a paper with 222 citations.

These prolific studies highlight the key findings of the research as an alternative of HDS of sulfur compounds in petrodiesel and related fuels.

50.4.4 Extractive Desulfurization of Sulfur Compounds

Bosmann et al. (2001) develop a new approach for the deep desulfurization of diesel fuels by extraction with ILs in a paper with 471 citations. Wu et al. (2004) study the desulfurization of flue gas through SO_2 absorption by an IL in a paper with 408 citations. Zhang et al. (2004) study the EDS and denitrogenation of fuels using ILs in a paper with 347 citations.

Nie et al. (2006) study the EDS of gasoline using imidazolium-based phosphoric ILs in a paper with 231 citations. Zhang and Zhang (2002) study the properties of ILs in selective sulfur removal from fuels at room temperature in a paper with 220 citations.

These prolific studies highlight the key findings of the research as an alternative of HDS of sulfur compounds in petrodiesel and related fuels.

50.4.5 Biodesulfurization of Sulfur Compounds

Gray et al. (1996) describe a unique sulfur acquisition system that *Rhodococcus* uses to obtain sulfur from very stable heterocyclic molecules in a paper with 257 citations.

This prolific study highlights the key findings of the research as an alternative of HDS of sulfur compounds in petrodiesel and related fuels.

50.4.6 Societal Implications of Desulfurization of Sulfur Compounds

The significant emissions of 'sulfur dioxide' (SO_2) have been well established with important contributions from China and other Asian countries as well as from shipping (Lu et al., 2010, 2011; Smith et al., 2011). The adverse ecological impacts of these emissions, such as global warming and ocean acidification, have also been well documented (Carmichael et al., 2002; Doney et al., 2007; Greaver et al., 2012). Similarly, the adverse impact of these emissions on human health and mortality have been studied extensively (Hedley et al., 2002; Katsouyanni et al., 1997; Sheppard et al., 1980).

Thus, the efficient desulfurization of sulfur compounds in petrodiesel and related fuels would be helpful in reducing SO_2 emissions, especially in shipping, transport, and the power sectors, by reducing the adverse impact of these emissions on ecology and human health, meeting the great public concern in this area.

Thus, biooils (Bridgwater et al., 1999; Bridgwater and Peacocke, 2000; Czernik and Bridgwater, 2004) and biooil-based biodiesel fuels (Chisti, 2007; Hill et al., 2006; Hu et al., 2008) have emerged as alternatives to crude oils and crude oil-based petrodiesel fuels, respectively, in recent decades in line with the public concerns about the sustainability, price fluctuations, and adverse environmental impact of crude oils that have emerged since the 1970s (Ahmadun et al., 2009; Atlas, 1981; Babich and Moulijn, 2003; Kilian, 2009; Perron, 1989) as well as the adverse health and ecological impacts of SO_2 emissions from petrodiesel fuels.

50.5 CONCLUSION

This chapter has presented the key findings of the 25-most-cited article papers in this field. Table 50.1 provides information on the research fronts in this field. As this table shows the primary research fronts of 'HDS of sulfur compounds', 'adsorptive desulfurization of sulfur compounds', 'oxidative desulfurization of sulfur compounds', and 'EDS of sulfur compounds', comprise 24, 32, 20, and 20% of these papers, respectively. Furthermore, 'biodesulfurization of sulfur compounds' comprises 4% of these papers.

These prolific studies in five complementary research fronts provide valuable evidence on the desulfurization of sulfur compounds in petrodiesel fuels and other related fuels.

The significant emissions of SO_2 have been well established with important contributions from China and other Asian countries as well as from shipping (Lu et al., 2010, 2011; Smith et al., 2011). The adverse ecological impact of these emissions, such as global warming and ocean acidification, have also been well documented (Carmichael et al., 2002; Doney et al., 2007; Greaver et al., 2012). Similarly, the adverse impact of these emissions on human health and mortality have also been studied extensively (Hedley et al., 2002; Katsouyanni et al., 1997; Sheppard et al., 1980).

Thus, the efficient desulfurization of sulfur compounds in petrodiesel and related fuels would be helpful in reducing SO_2 emissions, especially in shipping, transport and, the power sectors, by reducing the adverse impact of these emissions on ecology and human health, meeting the great public concern in this area.

Thus, biooils (Bridgwater et al., 1999; Bridgwater and Peacocke, 2000; Czernik and Bridgwater, 2004) and biooil-based biodiesel fuels (Chisti, 2007; Hill et al., 2006; Hu et al., 2008) have emerged as alternatives to crude oils and crude oil-based petrodiesel fuels, respectively, in recent decades in line with the public concerns about the sustainability, price fluctuations, and adverse environmental impact of crude oils that have emerged since the 1970s (Ahmadun et al., 2009; Atlas, 1981; Babich and Moulijn, 2003; Kilian, 2009; Perron, 1989) as well as the adverse impact on human health and ecology of SO_2 emissions from petrodiesel fuels.

ACKNOWLEDGMENTS

The contribution of the highly cited researchers in this field is greatly acknowledged.

REFERENCES

Ahmadun, F. R., A. Pendashteh, and L. C. Abdullah, et al. 2009. Review of technologies for oil and gas produced water treatment. *Journal of Hazardous Materials* 170: 530–551.

Atlas, R. M.. 1981. Microbial degradation of petroleum hydrocarbons: An environmental perspective. *Microbiological Reviews* 45: 180–209.

Babich, I. V. and J. A. Moulijn. 2003. Science and technology of novel processes for deep desulfurization of oil refinery streams: A review. *Fuel* 82: 607–631.

Bataille, F., J. L. Lemberton, and P. Michaud, et al. 2000. Alkyldibenzothiophenes hydrodesulfurization: Promoter effect, reactivity, and reaction mechanism. *Journal of Catalysis* 191: 409–422.

Birch, M. E. and R. A. Cary. 1996. Elemental carbon-based method for monitoring occupational exposures to particulate diesel exhaust. *Aerosol Science and Technology* 25: 221–241.

Bosmann, A., L. Datsevich, and A. Jess, et al. 2001. Deep desulfurization of diesel fuel by extraction with ionic liquids. *Chemical Communications* 23: 2494–2495.

Bridgwater, A. V., D. Meier, and D. Radlein. 1999. An overview of fast pyrolysis of biomass. *Organic Geochemistry* 30: 1479–1493.

Bridgwater, A. V. and G. V. C. Peacocke. 2000. Fast pyrolysis processes for biomass. *Renewable & Sustainable Energy Reviews* 4: 1–73.

Carmichael, G. R., D. G. Streets, and G. Calori, et al. 2002. Changing trends in sulfur emissions in Asia: Implications for acid deposition, air pollution, and climate. *Environmental Science & Technology* 36: 4707–4713.

Chisti, Y. 2007. Biodiesel from microalgae. *Biotechnology Advances* 25: 294–306.

Czernik, S. and A. V. Bridgwater. 2004. Overview of applications of biomass fast pyrolysis oil. *Energy & Fuels* 18: 590–598.

Doney, S. C., N. Mahowald, and I. Lima, et al. 2007. Impact of anthropogenic atmospheric nitrogen and sulfur deposition on ocean acidification and the inorganic carbon system. *Proceedings of the National Academy of Sciences of the United States of America* 104: 14580–14585.

Egorova, M. and R. Prins. 2004. Hydrodesulfurization of dibenzothiophene and 4, 6-dimethyldibenzothiophene over sulfided NiMo/γ-Al_2O_3, CoMo/γ-Al_2O_3, and Mo/γ-Al_2O_3 catalysts. *Journal of Catalysis* 225: 417–427.

Gray, K. A., O. S. Pogrebinsky, and G. T. Mrachko, et al. 1996. Molecular mechanisms of biocatalytic desulfurization of fossil fuels. *Nature Biotechnology* 14: 1705–1709.

Greaver, T. L., T. J. Sullivan, and J. D. Herrick, et al. 2012. Ecological effects of nitrogen and sulfur air pollution in the US: What do we know? *Frontiers in Ecology and the Environment* 10: 365–372.

Hedley, A. J., C. M. Wong, and T. Q. Thach, et al. 2002. Cardiorespiratory and all-cause mortality after restrictions on sulphur content of fuel in Hong Kong: An intervention study. *Lancet* 360: 1646–1652.

Hernandez-Maldonado, A. J., F. H. Yang, G. Qi, and R. T. Yang. 2005. Desulfurization of transportation fuels by π-complexation sorbents: Cu(I)-, Ni(II)-, and Zn(II)-zeolites. *Applied Catalysis B-Environmental* 56: 111–126.

Hernandez-Maldonado, A. J. and R. T. Yang. 2003. Desulfurization of liquid fuels by adsorption via pi complexation with Cu(I)-Y and Ag-Y zeolites. *Industrial & Engineering Chemistry Research* 42: 123–129.

Hernandez-Maldonado, A. J. and R. T. Yang. 2004. Desulfurization of diesel fuels by adsorption via π-complexation with vapor-phase exchanged Cu(l)-Y zeolites. *Journal of the American Chemical Society* 126: 992–993.

Hill, J., E. Nelson, D. Tilman, S. Polasky, and D. Tiffany. 2006. Environmental, economic, and energetic costs and benefits of biodiesel and ethanol biofuels. *Proceedings of the National Academy of Sciences of the United States of America* 103: 11206–11210.

Houalla, M., D. H. Broderick, and A. V. Sapre, et al. 1980. Hydrodesulfurization of methyl-substituted dibenzothiophenes catalyzed by sulfided Co-Mo-γ-Al_2O_3. *Journal of Catalysis* 61: 523–527.

Hu, Q., M. Sommerfeld, and E. Jarvis, et al. 2008. Microalgal triacylglycerols as feedstocks for biofuel production: Perspectives and advances. *Plant Journal* 54: 621–639.

Hulea, V., F. Fajula, and J. Bousquet. 2001. Mild oxidation with H_2O_2 over Ti-containing molecular sieves: A very efficient method for removing aromatic sulfur compounds from fuels. *Journal of Catalysis* 198: 179–186.

Kabe, T., A. Ishihara, and H. Tajima. 1992. Hydrodesulfurization of sulfur-containing polyaromatic compounds in light oil. *Industrial & Engineering Chemistry Research* 31: 1577–1580.

Katsouyanni, K., G. Touloumi, and C. Spix, et al. 1997. Short term effects of ambient sulphur dioxide and particulate matter on mortality in 12 European cities: Results from time series data from the APHEA project. *BMJ-British Medical Journal* 314: 1658–1663.

Khalili, N. R., P. A. Scheff, and T. M. Holsen. 1995. PAH source fingerprints for coke ovens, diesel and gasoline-engines, highway tunnels, and wood combustion emissions. *Atmospheric Environment* 29: 533–542.

Kilian, L. 2009. Not all oil price shocks are alike: Disentangling demand and supply shocks in the crude oil market. *American Economic Review* 99: 1053–1069.

Kim, J. H., X. L. Ma, A. N. Zhou, and C. S. Song. 2006. Ultra-deep desulfurization and denitrogenation of diesel fuel by selective adsorption over three different adsorbents: A study on adsorptive selectivity and mechanism. *Catalysis Today* 111: 74–83.

Konur, O. 2000. Creating enforceable civil rights for disabled students in higher education: An institutional theory perspective. *Disability & Society* 15: 1041–1063.

Konur, O. 2002a. Access to Nursing Education by disabled students: Rights and duties of nursing programs. *Nurse Education Today* 22: 364–374.

Konur, O. 2002b. Assessment of disabled students in higher education: Current public policy issues. *Assessment and Evaluation in Higher Education* 27: 131–152.

Konur, O. 2002c. Access to employment by disabled people in the UK: Is the Disability Discrimination Act working? *International Journal of Discrimination and the Law* 5: 247–279.

Konur, O. 2006a. Participation of children with dyslexia in compulsory education: Current public policy issues. *Dyslexia* 12: 51–67.

Konur, O. 2006b. Teaching disabled students in Higher Education. *Teaching in Higher Education* 11: 351–363.

Konur, O. 2007a. A judicial outcome analysis of the Disability Discrimination Act: A windfall for the employers? *Disability & Society* 22: 187–204.

Konur, O. 2007b. Computer-assisted teaching and assessment of disabled students in higher education: The interface between academic standards and disability rights. *Journal of Computer Assisted Learning* 23: 207–219.

Konur, O., ed. 2021a. *Handbook of Biodiesel and Petrodiesel Fuels: Science, Technology, Health, and Environment.* Boca Raton, FL: CRC Press.

Konur, O., ed. 2021b. *Handbook of Biodiesel and Petrodiesel Fuels: Science, Technology, Health, and Environment. Volume 1. Biodiesel Fuels: Science, Technology, Health, and Environment.* Boca Raton, FL: CRC Press.

Konur, O., ed. 2021c. *Handbook of Biodiesel and Petrodiesel Fuels: Science, Technology, Health, and Environment. Volume 2. Biodiesel Fuels based on the Edible and Nonedible Feedstocks, Wastes, and Algae: Science, Technology, Health, and Environment.* Boca Raton, FL: CRC Press.

Konur, O., ed. 2021d. *Handbook of Biodiesel and Petrodiesel Fuels: Science, Technology, Health, and Environment. Volume 3. Petrodiesel Fuels: Science, Technology, Health, and Environment.* Boca Raton, FL: CRC Press.

Konur, O. 2021e. Biodiesel and petrodiesel fuels: Science, technology, health, and environment. In *Handbook of Biodiesel and Petrodiesel Fuels: Science, Technology, Health, and Environment. Volume 1. Biodiesel Fuels: Science, Technology, Health, and Environment*, ed. O. Konur. Boca Raton, FL: CRC Press.

Konur, O. 2021f. Biodiesel and petrodiesel fuels: A scientometric review of the research. In *Handbook of Biodiesel and Petrodiesel Fuels: Science, Technology, Health, and Environment. Volume 1. Biodiesel Fuels: Science, Technology, Health, and Environment*, ed. O. Konur. Boca Raton, FL: CRC Press.

Konur, O. 2021g. Biodiesel and petrodiesel fuels: A review of the research. In *Handbook of Biodiesel and Petrodiesel Fuels: Science, Technology, Health, and Environment. Volume 1. Biodiesel Fuels: Science, Technology, Health, and Environment*, ed. O. Konur. Boca Raton, FL: CRC Press.

Konur, O. 2021h Nanotechnology applications in the diesel fuels and the related research fields: A review of the research. In *Handbook of Biodiesel and Petrodiesel Fuels: Science, Technology, Health, and Environment. Volume 1. Biodiesel Fuels: Science, Technology, Health, and Environment*, ed. O. Konur. Boca Raton, FL: CRC Press.

Konur, O. 2021i. Biooils: A scientometric review of the research. In *Handbook of Biodiesel and Petrodiesel Fuels: Science, Technology, Health, and Environment. Volume 1. Biodiesel Fuels: Science, Technology, Health, and Environment*, ed. O. Konur. Boca Raton, FL: CRC Press.

Konur, O. 2021j. Characterization and properties of biooils: A review of the research. In *Handbook of Biodiesel and Petrodiesel Fuels: Science, Technology, Health, and Environment. Volume 1. Biodiesel Fuels: Science, Technology, Health, and Environment*, ed. O. Konur. Boca Raton, FL: CRC Press.

Konur, O. 2021k. Biomass pyrolysis and pyrolysis oils: A review of the research. In *Handbook of Biodiesel and Petrodiesel Fuels: Science, Technology, Health, and Environment. Volume 1. Biodiesel Fuels: Science, Technology, Health, and Environment*, ed. O. Konur. Boca Raton, FL: CRC Press.

Konur, O. 2021l. Biodiesel fuels: A scientometric review of the research. In *Handbook of Biodiesel and Petrodiesel Fuels: Science, Technology, Health, and Environment. Volume 1. Biodiesel Fuels: Science, Technology, Health, and Environment*, ed. O. Konur. Boca Raton, FL: CRC Press.

Konur, O. 2021m. Glycerol: A scientometric review of the research. In *Handbook of Biodiesel and Petrodiesel Fuels: Science, Technology, Health, and Environment. Volume 1. Biodiesel Fuels: Science, Technology, Health, and Environment*, ed. O. Konur. Boca Raton, FL: CRC Press.

Konur, O. 2021n. Propanediol production from glycerol: A review of the research. In *Handbook of Biodiesel and Petrodiesel Fuels: Science, Technology, Health, and Environment. Volume 1. Biodiesel Fuels: Science, Technology, Health, and Environment*, ed. O. Konur. Boca Raton, FL: CRC Press.

Konur, O. 2021o. Edible oil-based biodiesel fuels: A scientometric review of the research. In *Handbook of Biodiesel and Petrodiesel Fuels: Science, Technology, Health, and Environment. Volume 2. Biodiesel Fuels based on the Edible and Nonedible Feedstocks, Wastes, and Algae: Science, Technology, Health, and Environment*, ed. O. Konur. Boca Raton, FL: CRC Press.

Konur, O. 2021p. Palm oil-based biodiesel fuels: A review of the research. In *Handbook of Biodiesel and Petrodiesel Fuels: Science, Technology, Health, and Environment. Volume 2. Biodiesel Fuels based on the Edible and Nonedible Feedstocks, Wastes, and Algae*, ed. O. Konur. Boca Raton, FL: CRC Press.

Konur, O. 2021q. Rapeseed oil-based biodiesel fuels: A review of the research. In *Handbook of Biodiesel and Petrodiesel Fuels: Science, Technology, Health, and Environment.*

Volume 2. Biodiesel Fuels based on the Edible and Nonedible Feedstocks, Wastes, and Algae, ed. O. Konur. Boca Raton, FL: CRC Press.

Konur, O. 2021r. Nonedible oil-based biodiesel fuels: A scientometric review of the research. In *Handbook of Biodiesel and Petrodiesel Fuels: Science, Technology, Health, and Environment. Volume 2. Biodiesel Fuels based on the Edible and Nonedible Feedstocks, Wastes, and Algae: Science, Technology, Health, and Environment*, ed. O. Konur. Boca Raton, FL: CRC Press.

Konur, O. 2021s. Waste oil-based biodiesel fuels: A scientometric review of the research. In *Handbook of Biodiesel and Petrodiesel Fuels: Science, Technology, Health, and Environment. Volume 2. Biodiesel Fuels based on the Edible and Nonedible Feedstocks, Wastes, and Algae: Science, Technology, Health, and Environment*, ed. O. Konur. Boca Raton, FL: CRC Press.

Konur, O. 2021t. Algal biodiesel fuels: A scientometric review of the research. In *Handbook of Biodiesel and Petrodiesel Fuels: Science, Technology, Health, and Environment. Volume 2. Biodiesel Fuels based on the Edible and Nonedible Feedstocks, Wastes, and Algae: Science, Technology, Health, and Environment*, ed. O. Konur. Boca Raton, FL: CRC Press.

Konur, O. 2021u. Algal biomass production for biodiesel production: A review of the research. In *Handbook of Biodiesel and Petrodiesel Fuels: Science, Technology, Health, and Environment. Volume 2. Biodiesel Fuels based on the Edible and Nonedible Feedstocks, Wastes, and Algae*, ed. O. Konur. Boca Raton, FL: CRC Press.

Konur, O. 2021v. Algal biomass production in wastewaters for biodiesel production: A review of the research. In *Handbook of Biodiesel and Petrodiesel Fuels: Science, Technology, Health, and Environment. Volume 2. Biodiesel Fuels based on the Edible and Nonedible Feedstocks, Wastes, and Algae*, ed. O. Konur. Boca Raton, FL: CRC Press.

Konur, O. 2021x. Algal lipid production for biodiesel production: A review of the research. In *Handbook of Biodiesel and Petrodiesel Fuels: Science, Technology, Health, and Environment. Volume 2. Biodiesel Fuels based on the Edible and Nonedible Feedstocks, Wastes, and Algae*, ed. O. Konur. Boca Raton, FL: CRC Press.

Konur, O. 2021y. Crude oils: A scientometric review of the research. In *Handbook of Biodiesel and Petrodiesel Fuels: Science, Technology, Health, and Environment. Volume 3. Petrodiesel Fuels: Science, Technology, Health, and Environment*, ed. O. Konur. Boca Raton, FL: CRC Press.

Konur, O. 2021z. Petrodiesel fuels: A scientometric review of the research. In *Handbook of Biodiesel and Petrodiesel Fuels: Science, Technology, Health, and Environment. Volume 3. Petrodiesel Fuels: Science, Technology, Health, and Environment*, ed. O. Konur. Boca Raton, FL: CRC Press.

Konur, O. 2021aa. Bioremediation of petroleum hydrocarbons in the contaminated soils: A review of the research. In *Handbook of Biodiesel and Petrodiesel Fuels: Science, Technology, Health, and Environment. Volume 3. Petrodiesel Fuels: Science, Technology, Health, and Environment*, ed. O. Konur. Boca Raton, FL: CRC Press.

Konur, O. 2021ab. Desulfurization of diesel fuels: A review of the research. In *Handbook of Biodiesel and Petrodiesel Fuels: Science, Technology, Health, and Environment. Volume 3. Petrodiesel Fuels: Science, Technology, Health, and Environment*, ed. O. Konur. Boca Raton, FL: CRC Press.

Konur, O. 2021ac. Diesel fuel exhaust emissions: A scientometric review of the research. In *Handbook of Biodiesel and Petrodiesel Fuels: Science, Technology, Health, and Environment. Volume 3. Petrodiesel Fuels: Science, Technology, Health, and Environment*, ed. O. Konur. Boca Raton, FL: CRC Press.

Konur, O. 2021ad. The adverse health and safety impact of diesel fuels: A scientometric review of the research. In *Handbook of Biodiesel and Petrodiesel Fuels: Science,*

Technology, Health, and Environment. Volume 3. Petrodiesel Fuels: Science, Technology, Health, and Environment, ed. O. Konur. Boca Raton, FL: CRC Press.

Konur, O. 2021ae. Respiratory illnesses caused by the diesel fuel exhaust emissions: A review of the research. In *Handbook of Biodiesel and Petrodiesel Fuels: Science, Technology, Health, and Environment. Volume 3. Petrodiesel Fuels: Science, Technology, Health, and Environment*, ed. O. Konur. Boca Raton, FL: CRC Press.

Konur, O. 2021af. Cancer caused by the diesel fuel exhaust emissions: A review of the research. In *Handbook of Biodiesel and Petrodiesel Fuels: Science, Technology, Health, and Environment. Volume 3. Petrodiesel Fuels: Science, Technology, Health, and Environment*, ed. O. Konur. Boca Raton, FL: CRC Press

Konur, O. 2021ag. Cardiovascular and other illnesses caused by the diesel fuel exhaust emissions: A review of the research. In *Handbook of Biodiesel and Petrodiesel Fuels: Science, Technology, Health, and Environment. Volume 3. Petrodiesel Fuels: Science, Technology, Health, and Environment*, ed. O. Konur. Boca Raton, FL: CRC Press.

Kulkarni, P. S. and C. A. M. Afonso. 2010. Deep desulfurization of diesel fuel using ionic liquids: Current status and future challenges. *Green Chemistry* 12: 1139–1149.

Lauritsen, J. V., M. Nyberg, and J. K. Norskov, et al. 2004. Hydrodesulfurization reaction pathways on MoS_2 nanoclusters revealed by scanning tunneling microscopy. *Journal of Catalysis* 224: 94–106.

Li, C., Z. X. Jiang, and J. B. Gao, et al. 2004. Ultra-deep desulfurization of diesel: Oxidation with a recoverable catalyst assembled in emulsion. *Chemistry-A European Journal* 10: 2277–2280.

Lu, H. Y., J. B. Gao, and Z. X. Jiang, et al. 2006. Ultra-deep desulfurization of diesel by selective oxidation with $[C_{18}H_{37}N(CH_3)_3]_4[H_2NaPW_{10}O_{36}]$ catalyst assembled in emulsion droplets. *Journal of Catalysis* 239: 369–375.

Lu, Z., D. G. Streets, and Q. Zhang, et al. 2010. Sulfur dioxide emissions in China and sulfur trends in East Asia since 2000. *Atmospheric Chemistry and Physics* 10: 6311–6331.

Lu, Z., Q. Zhang, and D. G. Streets. 2011. Sulfur dioxide and primary carbonaceous aerosol emissions in China and India, 1996–2010. *Atmospheric Chemistry and Physics* 11: 9839–9864.

Ma, X. L., K. Y. Sakanishi, and I. Mochida. 1994. Hydrodesulfurization reactivities of various sulfur-compounds in diesel fuel. *Industrial & Engineering Chemistry Research* 33: 218–222.

Ma, X. L., L. Sun, and C. Song. 2002. A new approach to deep desulfurization of gasoline, diesel fuel and jet fuel by selective adsorption for ultra-clean fuels and for fuel cell applications. *Catalysis Today* 77: 107–116.

Nie, Y., C. X. Li, A. J. Sun, H. Meng, and Z. H. Wang. 2006. Extractive desulfurization of gasoline using imidazolium-based phosphoric ionic liquids. *Energy & Fuels* 20: 2083–2087.

North, D. C. 1991a. *Institutions, Institutional Change and Economic Performance*. Cambridge, Mass: Cambridge University Press.

North, D. C. 1991b. Institutions. *Journal of Economic Perspectives* 5: 97–112.

Otsuki, S., T. Nonaka, and N. Takashima, et al. 2000. Oxidative desulfurization of light gas oil and vacuum gas oil by oxidation and solvent extraction. *Energy & Fuels* 14: 1232–1239.

Perron, P. 1989. The great crash, the oil price shock, and the unit root hypothesis. *Econometrica: Journal of the Econometric Society* 57: 1361–1401.

Rogge, W. F., L. M. Hildemann, M. A. Mazurek, G. R. Cass, and B. R. T. Simoneit. 1993. Sources of fine organic aerosol. 2. Noncatalyst and catalyst-equipped automobiles and heavy-duty diesel trucks. *Environmental Science & Technology* 27: 636–651.

Sawhill, S. J., K. A. Layman, and D. R. van Wyk, et al. 2005. Thiophene hydrodesulfurization over nickel phosphide catalysts: Effect of the precursor composition and support. *Journal of Catalysis* 231: 300–313.

Sheppard, D., W. S. Wong, C. F. Uehara, J. A. Nadel, and H. A. Boushey. 1980. Lower threshold and greater bronchomotor responsiveness of asthmatic subjects to sulfur dioxide. *American Review of Respiratory Disease* 122: 873–878.

Smith, S. J., J. van Aardenne, and Z. Klimont, et al. 2011. Anthropogenic sulfur dioxide emissions: 1850–2005. *Atmospheric Chemistry and Physics* 11: 1101–1116.

Song, C. and X. L. Ma. 2003. New design approaches to ultra-clean diesel fuels by deep desulfurization and deep dearomatization. *Applied Catalysis B-Environmental* 41: 207–238.

Song, C. S. 2003. An overview of new approaches to deep desulfurization for ultra-clean gasoline, diesel fuel and jet fuel. *Catalysis Today* 86: 211–263.

Velu, S., X. L. Ma, and C. S. Song. 2003. Selective adsorption for removing sulfur from jet fuel over zeolite-based adsorbents. *Industrial & Engineering Chemistry Research* 42: 5293–5304.

Wang, D. H., E. W. H. Qian, and H. Amano, et al., 2003. Oxidative desulfurization of fuel oil: Part I. Oxidation of dibenzothiophenes using tert-butyl hydroperoxide. *Applied Catalysis A-General* 253: 91–99.

Wu, W. Z., B. X. Han, and H. X. Gao, et al. 2004. Desulfurization of flue gas: SO_2 absorption by an ionic liquid. *Angewandte Chemie-International Edition* 43: 2415–2417.

Yang, R. T., A. J. Hernandez-Maldonado, and F. H. Yang. 2003. Desulfurization of transportation fuels with zeolites under ambient conditions. *Science* 301: 79–81.

Zhang, S. G., Q. L. Zhang, and Z. C. Zhang. 2004. Extractive desulfurization and denitrogenation of fuels using ionic liquids. *Industrial & Engineering Chemistry Research* 43: 614–622.

Zhang, S. G. and Z. C. Zhang. 2002. Novel properties of ionic liquids in selective sulfur removal from fuels at room temperature. *Green Chemistry* 4: 376–379.

Part XI

Emissions of Petrodiesel Fuels

51 Diesel Fuel Exhaust Emissions

A Scientometric Review of the Research

Ozcan Konur

CONTENTS

51.1 INTRODUCTION

Biodiesel and petrodiesel fuels have been primary sources of energy and fuels (Chisti, 2007, 2008; Konur, 2012g, 2015; Lapuerta et al., 2008; Marchetti et al., 2007; Srivastava and Prasad, 2000; van Gerpen, 2005). Nowadays, both biodiesel and diesel fuels are being used extensively (Konur, 2021a–ag).

However, the emissions (Birch and Cary, 1996; Busca et al., 1998; Lapuerta et al., 2008; Robinson et al., 2007; Rogge et al., 1993) from these fuels have raised significant concerns about human and environmental health and safety (Birch and Cary, 1996; Diaz-Sanchez et al., 1997; McCreanor et al., 2007; Mills et al., 2007; Salvi et al., 1999). Thus, the adverse health and safety impacts of diesel fuel exhaust emissions are of utmost public importance as studies on the science and technology of these emissions have emerged as a distinct research field over time.

Furthermore, for the efficient progression of the research in this field, it is necessary to develop efficient incentive structures for the primary stakeholders and to inform these stakeholders about the research (Konur, 2000, 2002a–c, 2006a–b, 2007a–b; North, 1991a–b).

Scientometric analysis offers ways to evaluate the research in a respective field (Garfield, 1955, 1972; Konur, 2011, 2012a–n, 2015, 2016a–f, 2017a–f, 2018a–b, 2019a–b). However, there has been no scientometric study of this field.

This chapter presents a study on the scientometric evaluation of the research in this field using two datasets. The first dataset includes the 100-most-cited papers (n = 100 sample papers) whilst the second set includes population papers (n = over 10,450 population papers) published between 1980 and 2019. The papers on the health and safety impact of the emissions from diesel fuel exhaust are considered in complementary chapters (Konur, 2021ad–ag).

The data on the indices, document types, authors, institutions, funding bodies, source titles, 'Web of Science' subject categories, keywords, research fronts, and citation impact are presented and discussed.

51.2 MATERIALS AND METHODOLOGY

The search for the literature was carried out in the 'Web of Science' database in January 2020. It contains the 'Science Citation Index-Expanded' (SCI-E), the 'Social Sciences Citation Index' (SSCI), the 'Book Citation Index-Science' (BCI-S), the 'Conference Proceedings Citation Index-Science' (CPCI-S), the 'Emerging Sources Citation Index' (ESCI), the 'Book Citation Index-Social Sciences and Humanities' (BCI-SSH), the 'Conference Proceedings Citation Index-Social Sciences and Humanities' (CPCI-SSH), and the 'Arts and Humanities Citation Index' (A&HCI).

The keywords for the search of the literature are collated from the screening of abstract pages for the first 500 highly cited papers. This keyword set is provided in the Appendix.

Two datasets are used for this study. The highly cited 100 papers comprise the first dataset (sample dataset, n = 100 papers) whilst all the papers form the second dataset (population dataset, n = over 10,450 papers). The studies on the adverse health and safety influence of diesel fuel exhaust emissions are considered in complementary chapters (Konur, 2021ad–ag). Therefore, they are not included in these two datasets.

The data on the indices, document types, publication years, institutions, funding bodies, source titles, countries, 'Web of Science' subject categories, citation impact, keywords, and research fronts are collected from these datasets. The key findings are provided in the relevant tables and figure, supplemented with explanatory notes in the text. The findings are discussed, a number of conclusions are drawn, and a number of recommendations for further study are made.

51.3 RESULTS

51.3.1 Indices and Documents

There are over 15,000 papers related to diesel fuel exhaust emissions in the 'Web of Science' as of January 2020. This original population dataset is refined for the document type (article, review, book chapter, book, editorial material, note, and letter) and language (English), resulting in over 10,450 papers comprising over 69.3% of the original population dataset.

The primary index is the SCI-E for both the sample and population papers. About 96.2% of the population papers are indexed by this database. Additionally 7.5 and 3.8% of these papers are indexed by the CPCI-S and ESCI databases. The papers on the social and humanitarian aspects of this field are relatively negligible with 0.8 and 0.0% of the population papers indexed by the SSCI and A&HCI, respectively.

Brief information on the document types for both datasets is provided in Table 51.1. The key finding is that article types of documents are the primary documents for both datasets and that reviews are over-represented by 11.7% in the sample papers.

51.3.2 Authors

Brief information about the most-prolific 19 authors with at least three sample papers each is provided in Table 51.2. Around 260 and 18,700 authors contribute to sample and population papers, respectively.

The most-prolific author is 'Ralph T. Yang' of the 'University of Michigan' with eight sample papers and a 7.5% publication surplus, working primarily on 'nitrogen oxide reduction'. The other prolific researchers are 'Hong He', 'Gongshin Qi', 'Guido Busca', 'Junhua Li', and 'Gianguido Ramis' with at least nine sample papers and 4.1% publication surpluses each, respectively.

TABLE 51.1
Document Types

	Document Type	Sample Dataset (%)	Population Dataset (%)	Difference (%)
1	Article	85	96.8	−11.8
2	Review	14	2.3	11.7
3	Book chapter	0	0.4	−0.4
4	Proceeding paper	5	7.4	−2.4
5	Editorial material	1	0.5	0.5
6	Letter	0	0.2	−0.2
7	Book	0	0.1	−0.1
8	Note	0	0.1	−0.1

TABLE 51.2
Authors

	Author	Sample Papers (%)	Population Papers (%)	Surplus (%)	Institution	Country	Research Front
1	Yang, Ralph T.	8	0.5	7.5	Univ. Michigan	USA	Nitrogen oxide reduction
2	He, Hong	6	0.9	5.1	Chinese Acad. Sci.	China	Nitrogen oxide reduction
3	Qi, Gongshin	6	0.2	5.8	General Motors	USA	Nitrogen oxide reduction
4	Busca, Guido	5	0.2	4.8	Univ. Genova	Italy	Nitrogen oxide reduction
5	Li, Junhua**	5	0.9	4.1	Tsinghua Univ.	China	Nitrogen oxide reduction
6	Ramis, Gianguido	5	0.2	4.8	Univ. Genova	Italy	Nitrogen oxide reduction
7	Moulijn, Jacob A.	4	0.3	3.7	Delft Univ. Technol.	Netherlands	Particle emissions
8	Zhang, Changbin	4	0.2	3.8	Chinese Acad. Sci.	China	Nitrogen oxide reduction
9	Elsener, Martin	3	0.3	2.7	Paul Scherrer Inst.	Switzerland	Nitrogen oxide reduction
10	Ge, Maofa	3	0.2	2.8	Chinese Acad. Sci.	China	Nitrogen oxide reduction
11	Kittelson, David B.	3	0.2	2.8	Univ. Minnesota	USA	Particle emissions
12	Koebel, Manfred	3	0.2	2.8	Paul Scherrer Inst.	Switzerland	Nitrogen oxide reduction
13	Liu, Fudong	3	0.4	2.6	Chinese Acad. Sci.	China	Nitrogen oxide reduction
14	Makkee, Michiel	3	0.3	2.7	Delft Univ. Technol.	Netherlands	Particle emissions
15	McMurry Peter A.	3	0.1	2.9	Univ. Minnesota	USA	Particle emissions
16	Smirniotis, Panagiotis G.	3	0.1	2.9	Univ. Cincinnati	USA	Nitrogen oxide reduction
17	Topsoe, Non-Yu	3	0.1	2.9	Haldor Topsoe Res. Labs.	Denmark	Nitrogen oxide reduction
18	Yi, Li	3	0.1	2.9	Univ. Genova	Italy	Nitrogen oxide reduction
19	Chen, Liang (L/LA)*	3	0.1	2.9	Tsinghua Univ.	China	Nitrogen oxide reduction

*Different spellings in parentheses;
**highly cited researchers in 2019.

The most-prolific institution for these top authors is the ‘Chinese Academy of Sciences’ of China with four authors, followed by the ‘University of Genoa’ of Italy with three authors. The other prolific institutions are ‘Delft University of Technology’ of the Netherlands, the ‘Paul Scherrer Institute’ of Switzerland, ‘Tsinghua University’ of China, and the ‘University of Minnesota’ of the USA with two authors each. In total, ten institutions house these top authors.

It is notable that only one of these top researchers are listed in the ‘Highly Cited Researchers’ (HCR) in 2019 (Clarivate Analytics, 2019; Docampo and Cram, 2019).

The most-prolific country for these top authors is China with six, followed by the USA with five. The other prolific countries are Italy, the Netherlands, and Switzerland with three, three, and two authors, respectively. In total, six countries contribute to these top papers.

There are two key research fronts for these top researchers. The top research front is ‘nitrogen oxide reduction’ with 15 authors. The other prolific research front is ‘particle emissions’ with four authors.

It is further notable that there is a significant gender deficit among these top authors as only one researcher is female (Lariviere et al., 2013; Xie and Shauman, 1998).

The authors with the most impact are ‘Ralph T. Yang’, ‘Gongshin Qi’, ‘Hong He’, ‘Guido Busca’, and ‘Gianguido Ramis’ with at least a 4.8% publication surplus each. On the other hand, the authors with the least impact are ‘Maofa Ge’, ‘David B. Kittelson’, ‘Manfred Koebel’, ‘Martin Elsener’, ‘Michiel Makkee’, and ‘Fudong Liu’ with at least 2.6% publication surpluses.

51.3.3 Publication Years

Information about the publication years for both datasets is provided in Figure 51.1.

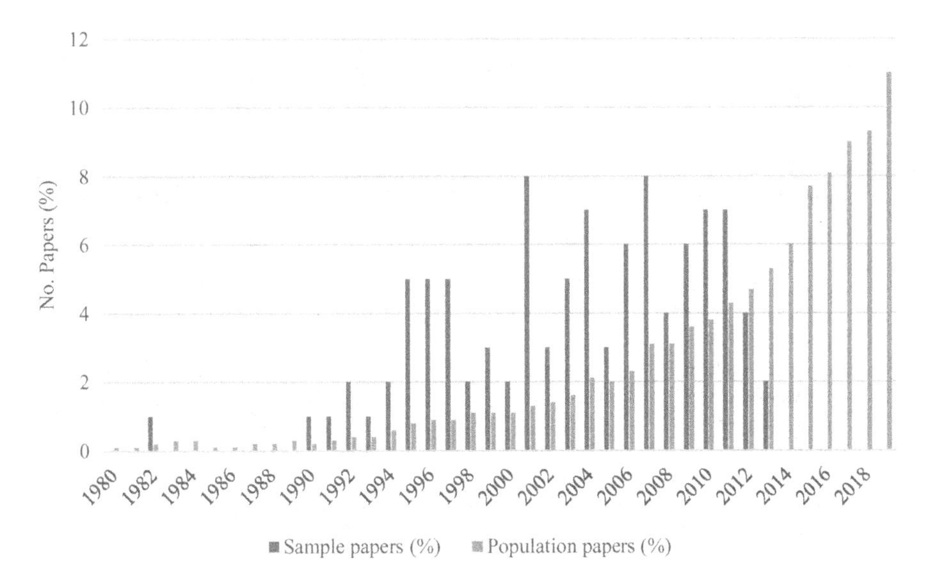

FIGURE 51.1 Research output between 1980 and 2019.

This figure shows that 27, 52, and 20% of the sample papers and 6.7, 21.6, and 69.2% of the population papers were published in the 1990s, 2000s, and 2010s, respectively.

Similarly, the most-prolific publication years for the sample dataset are 2001, 2004, 2007, 2010, and 2011 with at least seven papers each. On the other hand, the most-prolific publication years for the population dataset are 2015, 2016, 2017, 2018, and 2019 with at least 7.7% of the population papers each. It is notable that there is a sharply rising trend for the population papers.

51.3.4 Institutions

Brief information on the top 13 institutions with at least 3% of the sample papers each is provided in Table 51.3. In total, over 100 and 4,000 institutions contribute to the sample and population papers, respectively.

These top institutions publish 59.0 and 12.7% of the sample and population papers, respectively. The top institution is the 'Chinese Academy of Sciences' of China with nine sample papers and a 6.4% publication surplus. The other top institutions are the 'University of Michigan', 'Tsinghua University', the 'Paul Scherrer Institute', and the 'University of Genoa' with at least 5% of the sample papers.

The most-prolific country for these top institutions is the USA with five. The other prolific countries are China and Italy with two institutions each. The other countries are Denmark and India.

The institutions with the most impact are the 'Chinese Academy of Sciences', the 'University of Michigan', the 'University of Genoa', and the 'Paul Scherrer Institute' with at least a 4.3% publication surplus each. On the other hand, the institutions with the least impact are 'Haldor Topsoe', the 'Polytechnic University of Milan', and the 'Indian Institute of Technology' with at least a 1.2% publication surplus each.

TABLE 51.3
Institutions

	Institution	Country	No. of Sample Papers (%)	No. of Population Papers (%)	Difference (%)
1	Chinese Acad. Sci.	China	9	2.6	6.4
2	Univ. Michigan	USA	7	1.1	5.9
3	Tsinghua Univ.	China	6	2.2	3.8
4	Paul Scherrer Inst.	USA	5	0.7	4.3
5	Univ. Genoa	Italy	5	0.3	4.7
6	Delft Univ. Technol.	Netherlands	4	0.5	3.5
7	Ford Moto Co.	USA	4	1.1	2.9
8	Univ. Cincinnati	USA	4	0.3	3.7
9	Haldor Topsoe	Denmark	3	0.6	2.4
10	Indian Inst. Technol.	India	3	1.8	1.2
11	Natl. Renew. Energ. Lab.	USA	3	0.3	2.7
12	Polytech. Univ. Milan	Italy	3	0.9	2.1
13	Univ. Amsterdam	Netherlands	3	0.3	2.7

51.3.5 Funding Bodies

Brief information about the top five funding bodies with at least 2% of the sample papers each is provided in Table 51.4. It is significant that only 19.0 and 49.1% of the sample and population papers declare any funding, respectively.

The top funding body is the 'National High Technology Research and Development Program of China' of China, funding 8.0 and 1.8% of the sample and population papers, respectively, with a 6.2% publication surplus. This top funding body is followed by the 'National Natural Science Foundation of China' with 7.0 and 14.6% of the sample and population papers, respectively and with a –7.6% publication deficit.

It is notable that some top funding agencies for the population studies do not enter this top funding body list. Some of them are the 'Fundamental Research Funds for the Central Universities' (1.6%), the 'National Basic Research Program of China' (1.3%), the 'European Union' (1.0%), the 'Engineering Physical Sciences Research Council' of the UK (0.9%), the 'National Science Foundation' of the USA (0.9%), the 'China Postdoctoral Science Foundation' (0.9%), and the 'National Key R&D Program of China' (0.8%).

It is notable that the most-prolific country for these top funding bodies is China with four funding bodies.

51.3.6 Source Titles

Brief information about the top 12 source titles with at least three sample papers each is provided in Table 51.5. In total, 38 and 1,008 source titles publish the sample and population papers, respectively. On the other hand, these top 12 journals publish 65.0 and 24.7% of the sample and population papers, respectively.

The top journal is the 'Journal of Catalysis' publishing 13 sample papers with an 11.4% publication surplus. This top journal is closely followed by 'Applied Catalysis B Environmental' with 12 sample papers and a 7.6% publication surplus. The other top journals are 'Environmental Science Technology', 'Catalysis Today', 'Atmospheric Environment', and 'Energy Conversion and Management' with at least four sample papers each.

TABLE 51.4
Funding Bodies

	Institution	Country	No. of Sample Papers (%)	No. of Population Papers (%)	Difference (%)
1	National High Technology Research and Development Program of China	China	8	1.8	6.2
2	National Natural Science Foundation of China	China	7	14.6	–7.6
3	US Department of Energy	USA	3	1.7	1.3
4	National Basic Research Program of China	China	2	1.3	0.7
5	Ministry of Education China	China	2	0.5	1.5

TABLE 51.5
Source Titles

	Source Title	Wos Subject Category	No. of Sample Papers (%)	No. of Population Papers (%)	Difference (%)
1	Journal of Catalysis	Chem. Phys., Eng. Chem.	13	1.6	11.4
2	Applied Catalysis B Environmental	Chem. Phys., Eng. Env., Eng. Chem.	12	4.4	7.6
3	Environmental Science Technology	Eng. Env., Env. Sci.	8	2.8	5.2
4	Catalysis Today	Chem. Appl., Chem. Phys., Eng. Chem.	6	2.7	3.3
5	Atmospheric Environment	Env. Sci., Meteor. Atmosph. Sci.	4	1.9	2.1
6	Energy Conversion and Management	Therm., Energy Fuels, Mechs.	4	2.3	1.7
7	Combustion and Flame	Therm., Energy Fuels, Eng. Mult., Eng. Chem., Eng. Mech.	3	0.7	2.3
8	Fuel	Energy Fuel., Eng. Chem.	3	5.8	–2.8
9	Fuel Processing Technology	Chem. Appl., Energy Fuels, Eng. Chem.	3	0.8	2.2
10	Journal of Aerosol Science	Eng. Chem., Eng, Mech., Env. Sci., Meteor. Atmosph. Sci.	3	0.8	2.2
11	Progress in Energy and Combustion Science	Therm., Energy Fuels, Eng. Chem., Eng. Mech.	3	0.2	2.8
12	Renewable Sustainable Energy Reviews	Green Sust. Sci. Technol., Energy Fuels	3	0.7	2.3
			65	24.7	40.3

Although these journals are indexed by 12 subject categories, the top category is 'Engineering Chemical' with eight journals. This top category is followed by 'Energy Fuels' with five journals. The other prolific subjects with three journals are 'Chemistry Physical', 'Environmental Sciences', and 'Thermodynamics'. Additionally, subjects with two journals are 'Engineering Mechanical', 'Engineering Environmental', 'Chemistry Applied', and 'Meteorology Atmospheric Sciences'.

The journals with the most impact are the 'Journal of Catalysis', 'Applied Catalysis B Environmental', 'Environmental Science Technology', 'Catalysis Today', and 'Progress in Energy and Combustion Science' with at least 2.8% publication surpluses. On the other hand, the journals with the least impact are 'Atmospheric Environment', 'Energy Conversion and Management', and 'Fuel' with at least –2.8 publication deficits.

51.3.7 Countries

Brief information about the top 13 countries with at least two sample papers each is provided in Table 51.6. In total, 20 and over 210 countries contribute to the sample and population papers, respectively.

The top country is the USA publishing 47.0 and 17.1% of the sample and population papers, respectively. On the other hand, China follows the USA with 16 and 27% of the sample and population papers, respectively. The other prolific countries are Italy, the Netherlands, Switzerland, and India publishing seven, six, six, and five sample papers, respectively.

The European and Asian countries represented in this table publish altogether 33 and 25% of the sample papers and 22.6 and 41.5% of the population papers, respectively.

It is notable that the publication surplus for the USA and these European and Asian countries is 29.9, 10.4, and –16.5%, respectively.

51.3.8 Web of Science Subject Categories

Brief information about the top 17 'Web of Science' subject categories with at least three sample papers each is provided in Table 51.7. The sample and population papers are indexed by 21 and 132 subject categories, respectively.

For the sample papers, the top subjects are 'Engineering Chemical', 'Chemistry Physical', and 'Energy Fuels' with 51, 43, and 30 papers, respectively. The other prolific subjects are 'Engineering Environmental', 'Environmental Sciences', Thermodynamics', Engineering Mechanical', and 'Chemistry Applied' with over ten papers each.

TABLE 51.6
Countries

	Country	No. of Sample Papers (%)	No. of Population Papers (%)	Difference (%)
1	USA	47	17.1	29.9
2	China	16	23.7	–7.7
3	Italy	7	3.9	3.1
4	Netherlands	6	1.2	4.8
5	Switzerland	6	1.9	4.1
6	India	5	12.2	7.2
7	Denmark	3	0.9	2.1
8	UK	3	4.8	–1.8
9	Germany	3	4.0	–1.0
10	Spain	3	4.0	–1.0
11	Bangladesh	2	0.3	1.7
12	Greece	2	1.9	0.1
13	Japan	2	5.3	–3.3
	Europe-8	33	22.6	10.4
	Asia-4	25	41.5	–16.5

TABLE 51.7
Web of Science Subject Categories

	Subjecs	No. of Sample Papers (%)	No. of Population Papers (%)	Difference (%)
1	Engineering Chemical	51	33.1	17.9
2	Chemistry Physical	43	22.2	20.8
3	Energy Fuels	30	30.7	–0.7
4	Engineering Environmental	22	12.0	10.0
5	Environmental Sciences	20	1.7	18.3
6	Thermodynamics	13	12.8	0.2
7	Engineering Mechanical	12	1.5	10.5
8	Chemistry Applied	10	6.5	3.5
9	Meteorology Atmospheric Sciences	9	4.4	4.6
10	Mechanics	6	4.7	1.3
11	Green Sustainable Science Technology	5	3.8	1.2
12	Chemistry Multidisciplinary	4	4.6	–0.6
13	Agricultural Engineering	3	0.8	2.2
14	Biotechnology Applied Microbiology	3	0.7	2.3
15	Engineering Multidisciplinary	3	3.3	–0.3
16	Materials Science Multidisciplinary	3	3.3	–0.3
17	Nanoscience Nanotechnology	3	1.4	1.6

It is notable that the publication surplus is most significant for 'Chemistry Physical', 'Environmental Sciences', and 'Engineering Chemical' with 20.8, 18.3, and 17.9% surpluses, respectively. The other high-impact subjects are 'Engineering Mechanical' and 'Engineering Environmental' with 10.5 and 10.0% publication surpluses, respectively.

On the other hand, the subjects with least impact are 'Engineering Multidisciplinary', 'Materials Science Multidisciplinary', 'Chemistry Multidisciplinary', and 'Energy Fuels' with –0.3, –0.3, –0.6, and –0.7% publication deficits, respectively.

51.3.9 Citation Impact

These sample papers received about 36,000 citations as of January 2020. Thus, the average number of citations per paper is 360.

51.3.10 Keywords

Although a number of keywords are listed in the Appendix for the datasets related to both diesel fuels and diesel fuel emissions in this field, some of them are more significant for the sample papers. The most-prolific keywords is '*diesel' with 64 occurrences. The other prolific keywords are 'emission*', 'selective catalytic reduction', 'NH3', 'NOx', 'ammonia', 'exhaust*', and 'SCR' with 39, 38, 26, 27, 114, 11, and 11 citations, respectively. Further keywords are 'particl*' and 'soot*' with eight and seven occurrences, respectively.

51.3.11 Research Fronts

Brief information about the key research fronts is provided in Table 51.8. There are three primary research fronts for these sample papers: 'diesel fuel exhaust emissions in general', 'diesel fuel exhaust particle emissions', and 'diesel fuel exhaust NO_x emissions'.

The most-prolific research front is 'diesel fuel exhaust NO_x emissions' with 49 sample papers. The other prolific research fronts are 'diesel fuel exhaust emissions in general' and 'diesel fuel exhaust particle emissions' with 30 and 21 sample papers, respectively.

It is notable that 22 and 5 of these papers are related to biodiesel fuels and other biofuels, respectively. The remaining papers are all related to petrodiesel fuels.

51.4 DISCUSSION

The size of the research on diesel fuel exhaust has increased to over 10,450 papers as of January 2020. It is expected that the number of the population papers in this field will exceed 30,000 papers by the end of the 2020s.

The research has developed more in the technological aspects of this field, rather than the social and humanitarian pathways as evidenced by the negligible number of population papers in the indices of the 'Web of Science', SSCI, and A&HCI.

The article types of documents are the primary documents for both datasets and reviews are over-represented by 11.7% in the sample papers (Table 51.1). Thus, the contribution of reviews by 14% of the sample papers in this field is highly exceptional (cf. Konur, 2011, 2012a–n, 2015, 2016a–f, 2017a–f, 2018a–b, 2019a–b).

Nineteen authors from ten institutions have at least three sample papers each (Table 51.2). Six and five of these authors are from China and the USA, respectively and the remaining ones are from Italy, the Netherlands, and Switzerland.

These authors focus on 'nitrogen oxide reduction' and 'particle emissions'. It is significant that there is ample 'gender deficit' among these top authors as only one researcher is female (Lariviere et al., 2013; Xie and Shauman, 1998).

TABLE 51.8
Research Fronts

	Research Front	No. of Sample Papers (%)
1	Diesel fuel exhaust emissions in general	30
1.1	Biodiesel fuel emissions	21
1.2	Petrodiesel fuel missions	4
1.3	Other biofuel emissions	5
2	Diesel fuel exhaust particle emissions	21
2.1	Petrodiesel fuel exhaust particle emissions	20
2.2	Biodiesel fuel exhaust particle emissions	1
3	Diesel fuel exhaust NO_x emissions	49
3.1	Selective catalytic reduction (SCR) of NO_x	48
3.2	Exhaust gas recirculation (EGR)	1

Of the sample papers 27, 52, and 20% and 6.7, 21.6, and 69.2% of the population papers were published in the 1990s, 2000s, and 2010s, respectively (Figure 51.1). This finding suggests that the population papers have built on the sample papers, primarily published in the 1980s, 1990s, and 2000s, in the 2000s and 2010s.

The engagement of the institutions in this field at the global scale is significant as over 100 and 4,000 institutions contribute to the sample and population papers, respectively.

Thirteen top institutions publish 59.0 and 12.7% of the sample and population papers, respectively (Table 51.3). The top institution is the 'Chinese Academy of Sciences' of China with nine sample papers and a 6.4% publication surplus. The other top institutions are the 'University of Michigan', 'Tsinghua University', the 'Paul Scherrer Institute', and the 'University of Genoa' with at least 5% of the sample papers. As in the case of the top authors, the most-prolific countries for these top institutions are the USA, China, and Italy with six, four, and three sample papers, respectively.

It is significant that only 19.0 and 49.1% of the sample and population papers declare any funding, respectively. The most-prolific country for these top funding bodies is China with four (Table 51.4). It is evident that the level of the research funding is relatively low (cf. Konur, 2011, 2012a–n, 2015, 2016a–f, 2017a–f, 2018a–b, 2019a–b).

Four Chinese funding bodies dominate this top funding table. This finding is in line with the studies showing the heavy research funding in China, where the NSFC is the primary funding agency (Wang et al., 2012).

The sample and population papers are published by 38 and over 1,008 journals, respectively. It is significant that the top 12 journals publish 65.0 and 24.7% of the sample and population papers, respectively (Table 51.5).

The top journals, the 'Journal of Catalysis' and 'Applied Catalysis B Environmental', publish together 18 sample papers with 19% publication surpluses.

The top subject categories for these top journals are 'Engineering Chemical', 'Energy Fuels', 'Chemistry Physical', 'Environmental Sciences', and 'Thermodynamics'.

In total, 20 and over 210 countries contribute to the sample and population papers, respectively. The top country is the USA publishing 47.0 and 17.1% of the sample and population papers, respectively with a 29.9% publication surplus (Table 51.6). This finding is in line with the studies arguing that the USA is not losing ground in science and technology (Leydesdorff and Wagner, 2009).

The other prolific countries are China, Italy, the Netherlands, Switzerland, and India, publishing 16, 7, 6, 6, and 5 sample papers, respectively. These findings are in line with the studies showing that European countries have superior publication performance in science and technology (Franceschet and Costantini, 2011; Rinia et al., 1998; Schreiber and Kindler, 2005; Youtie et al., 2008).

The European and Asian countries represented in this table publish altogether 33 and 25% of the sample papers and 22.6 and 41.5% of the population papers, respectively.

It is notable that the publication surplus for the USA and these European and Asian countries is 29.9, 10.4, and –16.5%, respectively. It is further notable that

China has a significant publication deficit (–7.4%). This finding is in contrast with China's efforts to be a leading nation in science and technology (Zhou and Leydesdorff, 2006) but it is in line with the findings of Guan and Ma (2007) and Youtie et al. (2008) relating to China's performance in nanotechnology.

It is further notable that, like China, India also has a significant publication deficit (–7.2%). This finding is also in contrast with India's efforts to be a leading nation in science and technology (Anuradha and Urs, 2007).

Similarly, South Korea, Turkey, France, Malaysia, and Iran have no place in this top country table, although they have significant contributions to the population papers each (Aytac, 2010; Park et al., 2016; Tahira et al., 2013; van Raan et al., 2011; Waast and Rossi, 2010).

The sample and population papers are indexed by 21 and 132 subject categories, respectively. For the sample papers, the top subjects are 'Engineering Chemical', 'Chemistry Physical', and 'Energy Fuels' with 51, 43, and 30 papers, respectively (Table 51.7). The other prolific subjects are 'Engineering Environmental', 'Environmental Sciences', 'Thermodynamics', 'Engineering Mechanical', and 'Chemistry Applied' with over ten papers each.

It is notable that the publication surplus is most significant for 'Chemistry Physical', 'Environmental Sciences', and 'Engineering Chemical' with 20.8, 18.3, and 17.9% surpluses, respectively. The other high-impact subjects are 'Engineering Mechanical' and 'Engineering Environmental' with 10.5 and 10.0% publication surpluses, respectively.

On the other hand, the subjects with least impact are 'Engineering Multidisciplinary', 'Materials Science Multidisciplinary', 'Chemistry Multidisciplinary', and Energy Fuels' with –0.3, –0.3, –0.6, and –0.7% publication deficits, respectively.

These sample papers received about 36.000 citations as of January 2020. Thus, the average number of citations per paper is 360. Thus the citation impact of the top 100 papers in this field has been significant.

Although a number of keywords are listed in the Appendix for the dataset in this field, the most-prolific keyword is '*diesel' with 64 occurrences. The other prolific keywords are 'emission*', 'selective catalytic reduction', 'NH3','NOx', 'ammonia', 'exhaust*', and 'scr' with 39, 38, 26, 27, 114, 11, and 11 citations. Further keywords are 'particl*' and 'soot*' with eight and seven occurrences, respectively. As expected, these keywords provide valuable information about the pathways of the research in this field.

Three research fronts emerge from the examination of the sample papers: 'diesel fuel exhaust emissions in general', 'diesel fuel exhaust particle emissions', and 'diesel fuel exhaust NO_x emissions' (Table 51.8).

The most-prolific research front is 'diesel fuel exhaust NO_x emissions' with 49 sample papers. The other prolific research fronts are 'diesel fuel exhaust emissions in general' and 'diesel fuel exhaust particle emissions' with 30 and 21 sample papers, respectively.

It is notable that the research in this field is a subfield of the research on the health and environmental impact of particles in general (Brauer et al., 2001; Heyder et al., 1986; Peters et al., 1997) as well as pollutant emissions in general (Jaramillo et al.,

2007; Olivier et al., 1998; Zhao et al., 2008). Therefore, many findings related to the relationships between the structure and properties of these particle and air emissions are also relevant for the research in this field.

The key emphasis in these research fronts is the exploration of the structure–processing–property relationships of diesel fuel exhausts (Cheng and Ma, 2011; Konur and Matthews, 1989; Rogers and Hopfinger, 1994; Scherf and List, 2002).

51.5 CONCLUSION

This chapter has mapped the research on diesel fuel exhaust emissions using a scientometric method, complementing the book chapters on the adverse health and safety impact of these emissions (Konur, 2021ad–ag).

The size of over 10,450 population papers shows the public importance of this interdisciplinary research field. However, it is significant that the research has developed more in the technological aspects in this field, rather than the social and humanitarian pathways.

Articles and reviews dominate the sample papers, primarily published in the 1990s, 2000s, and 2010s. The population papers, primarily published in the 2000s and 2010s, build on these sample papers.

The data presented in the tables and figure show that a small number of authors, institutions, funding bodies, journals, keywords, research fronts, subject categories, and countries have shaped the research in this field.

It is notable that the authors, institutions, and funding bodies from the USA, China, Italy, and the Netherlands dominate the research in this field. Furthermore, China, India, Turkey, Japan, and South Korea are under-represented significantly in the sample papers.

These findings show the importance of the progression of efficient incentive structures for the development of the research in this field as in other fields. It seems that the USA and European countries (such as Italy and the Netherlands) have efficient incentive structures for the development of the research in this field, contrary to China, India, Turkey, and South Korea.

It further seems that although the research funding is a significant element of these incentive structures, it might not be a sole solution for increasing the incentives for the research in this field as in the case of China, India, and South Korea.

On the other hand, it seems there is more to do to reduce the significant gender deficit in this field as in other fields of science and technology (Lariviere et al., 2013; Xie and Shauman, 1998).

The data on the research fronts, keywords, source titles, and subject categories provides valuable evidence for the interdisciplinary (Lariviere and Gingras, 2010; Morillo et al., 2001) nature of the research in this field. These findings are in line with studies arguing for the interdisciplinarity of the research in nanomaterials, relevant for diesel fuel exhaust nanoparticles (Meyer and Persson, 1998; Schummer, 2004).

There is ample justification for the broad search strategy employed in this study due to the interdisciplinary nature of this research field as evidenced by the top subject categories. The search strategy employed in this study is in line with the search strategies employed for related and other research fields (Konur, 2011, 2012a–n, 2015, 2016a–f, 2017a–f, 2018a–b, 2019a–b).

Three research fronts emerge from the examination of the sample papers: 'diesel fuel exhaust emissions in general', 'diesel fuel exhaust particle emissions', and 'diesel fuel exhaust NOx emissions'.

It is recommended that further scientometric studies are carried out for each of these research fronts building on the pioneering studies in these fields.

ACKNOWLEDGMENTS

The contribution of the highly cited researchers in the field of diesel fuel exhaust emissions is greatly acknowledged.

51.A APPENDIX

The keyword set: (TI = (*diesel* or "compression ignition" or "Ci engine*" or trans-ester* or "trans-ester*" or diesohol or "methyl ester*") and TI = (particular or particulate* or soot* or particle* or pm or nox or denox or "de-nox" or "*no(x)" or "nitr* *oxide" or "no* oxidation" or no2 or scr or "selective catalytic reduction" or exhaust* or emission* or *smoke or aerosol or "Exhaust gas recirculation" or egr)) or TI = (("selective catalytic reduction" or scr or "exhaust gas recirculation" or egr) and (nox or denox or "de-nox" or "*no(x)" or "nitr* *oxide" or "no* oxidation" or no2 or nh3 or ammonia)

REFERENCES

Anuradha, K. and S. Urs. 2007. Bibliometric indicators of Indian research collaboration patterns: A correspondence analysis. *Scientometrics* 71:179–189.

Aytac, S. 2010. Scientific international collaboration of Turkey, Greece, Poland, and Portugal: A bibliometric analysis. *Proceedings of the American Society for Information Science and Technology* 47:1–3.

Birch, M. E. and R. A. Cary. 1996. Elemental carbon-based method for monitoring occupational exposures to particulate diesel exhaust. *Aerosol Science and Technology* 25:221–241.

Brauer, M. S. T., Ebelt, T. V., Fisher, J., et al. 2001. Exposure of chronic obstructive pulmonary disease patients to particles: Respiratory and cardiovascular health effects. *Journal of Exposure Science and Environmental Epidemiology* 11:490–500.

Busca, G., L. Lietti, G. Ramis, G. and F. Berti. 1998. Chemical and mechanistic aspects of the selective catalytic reduction of NO_x by ammonia over oxide catalysts: A review. *Applied Catalysis B-Environmental* 18:1–36.

Cheng, Y. Q. and E. Ma. 2011. Atomic-level structure and structure–property relationship in metallic glasses. *Progress in Materials Science* 56:379–473.

Chisti, Y. 2007. Biodiesel from microalgae. *Biotechnology Advances* 25:294–306.

Chisti, Y. 2008. Biodiesel from microalgae beats bioethanol. *Trends in Biotechnology* 26:126–131.

Clarivate Analytics. 2019. *Highly Cited Researchers: 2019 Recipients*. Philadelphia, PA: Clarivate Analytics. https://recognition.webofsciencegroup.com/awards/highly-cited/2019/ (accessed January, 3, 2020).

Diaz-Sanchez, D., A. Tsien, J. Fleming, and A. Saxon. 1997. Combined diesel exhaust particulate and ragweed allergen challenge markedly enhances human in vivo nasal ragweed-specific IgE and skews cytokine production to a T helper cell 2-type pattern. *Journal of Immunology* 158:2406–2413.

Docampo, D. and L. Cram. 2019. Highly cited researchers: A moving target. *Scientometrics* 118:1011–1025.

Franceschet, M. and A. Costantini. 2011. The first Italian research assessment exercise: A bibliometric perspective. *Journal of Informetrics* 5:275–291.

Garfield, E. 1955. Citation indexes for science. *Science* 122:108–111.

Garfield, E. 1972. Citation analysis as a tool in journal evaluation. *Science* 178:471–479.

Guan, J. C. and N. Ma. 2007. China's emerging presence in nanoscience and nanotechnology: A comparative bibliometric study of several nanoscience 'giants'. *Research Policy* 36:880–886.

Heyder, J. J. Gebhart, G. Rudolf, C. F. Schiller, and W. Stahlhofen. 1986. Deposition of particles in the human respiratory tract in the size range 0.005–15 µm. *Journal of Aerosol Science* 17:811–825.

Jaramillo, P., W. M. Griffin, and H. S. Matthews. 2007. Comparative life-cycle air emissions of coal, domestic natural gas, LNG, and SNG for electricity generation. *Environmental Science & Technology* 41:6290–6296.

Konur, O. 2000. Creating enforceable civil rights for disabled students in higher education: An institutional theory perspective. *Disability & Society* 15:1041–1063.

Konur, O. 2002a. Access to Nursing Education by disabled students: Rights and duties of nursing programs. *Nurse Education Today* 22:364–374.

Konur, O. 2002b. Assessment of disabled students in higher education: Current public policy issues. *Assessment and Evaluation in Higher Education* 27:131–152.

Konur, O. 2002c. Access to employment by disabled people in the UK: Is the Disability Discrimination Act working? *International Journal of Discrimination and the Law* 5:247–279.

Konur, O. 2006a. Participation of children with dyslexia in compulsory education: Current public policy issues. *Dyslexia* 12:51–67.

Konur, O. 2006b. Teaching disabled students in Higher Education. *Teaching in Higher Education* 11:351–363.

Konur, O. 2007a. A judicial outcome analysis of the Disability Discrimination Act: A windfall for the employers? *Disability & Society* 22:187–204.

Konur, O. 2007b. Computer-assisted teaching and assessment of disabled students in Higher Education: The interface between academic standards and disability rights. *Journal of Computer Assisted Learning* 23:207–219.

Konur, O. 2011. The scientometric evaluation of the research on the algae and bio-energy. *Applied Energy* 88:3532–3540.

Konur, O. 2012a. Evaluation of the research on the social sciences in Turkey: A scientometric approach. *Energy Education Science and Technology Part B: Social and Educational Studies* 4:1893–1908.

Konur, O. 2012b. Prof. Dr. Ayhan Demirbas' scientometric biography. *Energy Education Science and Technology Part A: Energy Science and Research* 28:727–738.

Konur, O. 2012c. The evaluation of the biogas research: A scientometric approach. *Energy Education Science and Technology Part A: Energy Science and Research* 29:1277–1292.

Konur, O. 2012d. The evaluation of the educational research: A scientometric approach. *Energy Education Science and Technology Part B: Social and Educational Studies* 4:1935–1948.

Konur, O. 2012e. The evaluation of the global energy and fuels research: A scientometric approach. *Energy Education Science and Technology Part A: Energy Science and Research* 30:613–628.

Konur, O. 2012f. The evaluation of the research on the Arts and Humanities in Turkey: A scientometric approach. *Energy Education Science and Technology Part B: Social and Educational Studies* 4:1603–1618.

Konur, O. 2012g. The evaluation of the research on the biodiesel: A scientometric approach. *Energy Education Science and Technology Part A: Energy Science and Research* 28:1003–1014.

Konur, O. 2012h. The evaluation of the research on the bioethanol: A scientometric approach. *Energy Education Science and Technology Part A: Energy Science and Research* 28:1051–1064.

Konur, O. 2012i. The evaluation of the research on the biofuels: A scientometric approach. *Energy Education Science and Technology Part A: Energy Science and Research* 28:903–916.

Konur, O. 2012j. The evaluation of the research on the biohydrogen: A scientometric approach. *Energy Education Science and Technology Part A: Energy Science and Research* 29:323–338.

Konur, O. 2012k. The evaluation of the research on the microbial fuel cells: A scientometric approach. *Energy Education Science and Technology Part A: Energy Science and Research* 29:309–322.

Konur, O. 2012l. The scientometric evaluation of the research on the production of bioenergy from biomass. *Biomass and Bioenergy* 47:504–515.

Konur, O. 2012m. The scientometric evaluation of the research on the deaf students in higher education. *Energy Education Science and Technology Part B: Social and Educational Studies* 4:1573–1588.

Konur, O. 2012n. The scientometric evaluation of the research on the students with ADHD in higher education. *Energy Education Science and Technology Part B: Social and Educational Studies* 4:1547–1562.

Konur, O. 2015. Current state of research on algal biodiesel. In *Marine Bioenergy: Trends and Developments*, ed. S. K. Kim, and C. G. Lee, 487–512. Boca Raton, FL: CRC Press.

Konur, O. 2016a. Scientometric overview in nanobiodrugs. In *Nanoarchitectonics for Smart Delivery and Drug Targeting*, A. M. Holban, A.M. Grumezescu, ed., 405–428. Amsterdam: Elsevier.

Konur, O. 2016b. Scientometric overview regarding nanoemulsions used in the food industry. In *Emulsions: Nanotechnology in the Agri-Food Industry*, A. M. Grumezescu, ed., 689–711. Amsterdam: Elsevier.

Konur, O. 2016c. Scientometric overview regarding the nanobiomaterials in antimicrobial therapy. In *Nanobiomaterials in Antimicrobial Therapy*, A. M. Grumezescu, ed., 511–535. Amsterdam: Elsevier.

Konur, O. 2016d. Scientometric overview regarding the nanobiomaterials in dentistry. In *Nanobiomaterials in Dentistry*, A. M. Grumezescu, ed., 425–453. Amsterdam: Elsevier.

Konur, O. 2016e. Scientometric overview regarding the surface chemistry of nanobiomaterials. In *Surface Chemistry of Nanobiomaterials*, A. M. Grumezescu, ed., 463–486. Amsterdam: Elsevier.

Konur, O. 2016f. The scientometric overview in cancer targeting. In *Nanoarchitectonics for Smart Delivery and Drug Targeting*, A. M. Holban, A. Grumezescu, ed., 871–895. Amsterdam; Elsevier.

Konur, O. 2017a. Recent citation classics in antimicrobial nanobiomaterials. In *Nanostructures for Antimicrobial Therapy*, A. Ficai and A. M. Grumezescu, ed., 669–685. Amsterdam: Elsevier.

Konur, O. 2017b. Scientometric overview in nanopesticides. In *New Pesticides and Soil Sensors*, A. M. Grumezescu, ed. 719–744. Amsterdam: Elsevier.

Konur, O. 2017c. Scientometric overview regarding oral cancer nanomedicine. In *Nanostructures for Oral Medicine*, E. Andronescu, A. M. Grumezescu, ed., 939–962. Amsterdam: Elsevier;

Konur, O. 2017d. Scientometric overview regarding water nanopurification. In *Water Purification*, A. M. Grumezescu, ed., 693–716. Amsterdam: Elsevier.

Konur, O. 2017e. Scientometric overview in food nanopreservation. In *Food Preservation*, A. M. Grumezescu, ed., 703–729. Amsterdam: Elsevier.

Konur, O. 2017f. The top citation classics in alginates for biomedicine. In *Seaweed Polysaccharides: Isolation, Biological and Biomedical Applications*, J. Venkatesan, S. Anil, S. K. Kim, ed., 223–249. Amsterdam: Elsevier.

Konur, O. 2018a. Scientometric evaluation of the global research in spine: An update on the pioneering study by Wei et al. *European Spine Journal* 27:525–529.

Konur, O. 2018b. Bioenergy and biofuels science and technology: scientometric overview and citation classics. In *Bioenergy and Biofuels*, O. Konur, ed., 3–63. Boca Raton: CRC Press.

Konur, O. 2019a. Cyanobacterial bioenergy and biofuels science and technology: A scientometric overview. In *Cyanobacteria: From Basic Science to Applications*, ed. A. K. Mishra, D. N. Tiwari and A. N. Rai, 419–442. Amsterdam: Elsevier.

Konur, O. 2019b. Nanotechnology applications in food: A scientometric overview. In *Nanoscience for Sustainable Agriculture*, R. N. Pudake, N. Chauhan, and C. Kole, ed., 683–711. Cham: Springer.

Konur, O., ed. 2021a. *Handbook of Biodiesel and Petrodiesel Fuels: Science, Technology, Health, and Environment.* Boca Raton, FL: CRC Press.

Konur, O., ed. 2021b. *Handbook of Biodiesel and Petrodiesel Fuels: Science, Technology, Health, and Environment. Volume 1. Biodiesel Fuels: Science, Technology, Health, and Environment.* Boca Raton, FL: CRC Press.

Konur, O., ed. 2021c. *Handbook of Biodiesel and Petrodiesel Fuels: Science, Technology, Health, and Environment. Volume 2. Biodiesel Fuels based on the Edible and Nonedible Feedstocks, Wastes, and Algae: Science, Technology, Health, and Environment.* Boca Raton, FL: CRC Press.

Konur, O., ed. 2021d. *Handbook of Biodiesel and Petrodiesel Fuels: Science, Technology, Health, and Environment. Volume 3. Petrodiesel Fuels: Science, Technology, Health, and Environment.* Boca Raton, FL: CRC Press.

Konur, O. 2021e. Biodiesel and petrodiesel fuels: Science, technology, health, and environment. In *Handbook of Biodiesel and Petrodiesel Fuels: Science, Technology, Health, and Environment. Volume 1. Biodiesel Fuels: Science, Technology, Health, and Environment*, ed. O. Konur. Boca Raton, FL: CRC Press.

Konur, O. 2021f. Biodiesel and petrodiesel fuels: A scientometric review of the research. In *Handbook of Biodiesel and Petrodiesel Fuels: Science, Technology, Health, and Environment. Volume 1. Biodiesel Fuels: Science, Technology, Health, and Environment*, ed. O. Konur. Boca Raton, FL: CRC Press.

Konur, O. 2021g. Biodiesel and petrodiesel fuels: A review of the research. In *Handbook of Biodiesel and Petrodiesel Fuels: Science, Technology, Health, and Environment. Volume 1. Biodiesel Fuels: Science, Technology, Health, and Environment*, ed. O. Konur. Boca Raton, FL: CRC Press.

Konur, O. 2021h Nanotechnology applications in the diesel fuels and the related research fields: A review of the research. In *Handbook of Biodiesel and Petrodiesel Fuels: Science, Technology, Health, and Environment. Volume 1. Biodiesel Fuels: Science, Technology, Health, and Environment*, ed. O. Konur. Boca Raton, FL: CRC Press.

Konur, O. 2021i. Biooils: A scientometric review of the research. In *Handbook of Biodiesel and Petrodiesel Fuels: Science, Technology, Health, and Environment. Volume 1. Biodiesel Fuels: Science, Technology, Health, and Environment*, ed. O. Konur. Boca Raton, FL: CRC Press.

Konur, O. 2021j. Characterization and properties of biooils: A review of the research. In *Handbook of Biodiesel and Petrodiesel Fuels: Science, Technology, Health, and*

Environment. Volume 1. Biodiesel Fuels: Science, Technology, Health, and Environment, ed. O. Konur. Boca Raton, FL: CRC Press.
Konur, O. 2021k. Biomass pyrolysis and pyrolysis oils: A review of the research. In *Handbook of Biodiesel and Petrodiesel Fuels: Science, Technology, Health, and Environment. Volume 1. Biodiesel Fuels: Science, Technology, Health, and Environment*, ed. O. Konur. Boca Raton, FL: CRC Press.
Konur, O. 2021l. Biodiesel fuels: A scientometric review of the research. In *Handbook of Biodiesel and Petrodiesel Fuels: Science, Technology, Health, and Environment. Volume 1. Biodiesel Fuels: Science, Technology, Health, and Environment*, ed. O. Konur. Boca Raton, FL: CRC Press.
Konur, O. 2021m. Glycerol: A scientometric review of the research. In *Handbook of Biodiesel and Petrodiesel Fuels: Science, Technology, Health, and Environment. Volume 1. Biodiesel Fuels: Science, Technology, Health, and Environment*, ed. O. Konur. Boca Raton, FL: CRC Press.
Konur, O. 2021n. Propanediol production from glycerol: A review of the research. In *Handbook of Biodiesel and Petrodiesel Fuels: Science, Technology, Health, and Environment. Volume 1. Biodiesel Fuels: Science, Technology, Health, and Environment*, ed. O. Konur. Boca Raton, FL: CRC Press.
Konur, O. 2021o. Edible oil-based biodiesel fuels: A scientometric review of the research. *In Handbook of Biodiesel and Petrodiesel Fuels: Science, Technology, Health, and Environment. Volume 2. Biodiesel Fuels based on the Edible and Nonedible Feedstocks, Wastes, and Algae: Science, Technology, Health, and Environment*, ed. O. Konur. Boca Raton, FL: CRC Press.
Konur, O. 2021p. Palm oil-based biodiesel fuels: A review of the research. In *Handbook of Biodiesel and Petrodiesel Fuels: Science, Technology, Health, and Environment. Volume 2. Biodiesel Fuels based on the Edible and Nonedible Feedstocks, Wastes, and Algae*, ed. O. Konur. Boca Raton, FL: CRC Press.
Konur, O. 2021q. Rapeseed oil-based biodiesel fuels: A review of the research. In *Handbook of Biodiesel and Petrodiesel Fuels: Science, Technology, Health, and Environment. Volume 2. Biodiesel Fuels based on the Edible and Nonedible Feedstocks, Wastes, and Algae*, ed. O. Konur. Boca Raton, FL: CRC Press.
Konur, O. 2021r. Nonedible oil-based biodiesel fuels: A scientometric review of the research. In *Handbook of Biodiesel and Petrodiesel Fuels: Science, Technology, Health, and Environment. Volume 2. Biodiesel Fuels based on the Edible and Nonedible Feedstocks, Wastes, and Algae: Science, Technology, Health, and Environment*, ed. O. Konur. Boca Raton, FL: CRC Press.
Konur, O. 2021s. Waste oil-based biodiesel fuels: A scientometric review of the research. In *Handbook of Biodiesel and Petrodiesel Fuels: Science, Technology, Health, and Environment. Volume 2. Biodiesel Fuels based on the Edible and Nonedible Feedstocks, Wastes, and Algae: Science, Technology, Health, and Environment*, ed. O. Konur. Boca Raton, FL: CRC Press.
Konur, O. 2021t. Algal biodiesel fuels: A scientometric review of the research. In *Handbook of Biodiesel and Petrodiesel Fuels: Science, Technology, Health, and Environment. Volume 2. Biodiesel Fuels based on the Edible and Nonedible Feedstocks, Wastes, and Algae: Science, Technology, Health, and Environment*, ed. O. Konur. Boca Raton, FL: CRC Press.
Konur, O. 2021u. Algal biomass production for biodiesel production: A review of the research. In *Handbook of Biodiesel and Petrodiesel Fuels: Science, Technology, Health, and Environment. Volume 2. Biodiesel Fuels based on the Edible and Nonedible Feedstocks, Wastes, and Algae*, ed. O. Konur. Boca Raton, FL: CRC Press.
Konur, O. 2021v. Algal biomass production in wastewaters for biodiesel production: A review of the research. In *Handbook of Biodiesel and Petrodiesel Fuels: Science, Technology,*

Health, and Environment. Volume 2. Biodiesel Fuels based on the Edible and Nonedible Feedstocks, Wastes, and Algae, ed. O. Konur. Boca Raton, FL: CRC Press.

Konur, O. 2021x. Algal lipid production for biodiesel production: A review of the research. In *Handbook of Biodiesel and Petrodiesel Fuels: Science, Technology, Health, and Environment. Volume 2. Biodiesel Fuels based on the Edible and Nonedible Feedstocks, Wastes, and Algae*, ed. O. Konur. Boca Raton, FL: CRC Press.

Konur, O. 2021y. Crude oils: A scientometric review of the research. In *Handbook of Biodiesel and Petrodiesel Fuels: Science, Technology, Health, and Environment. Volume 3. Petrodiesel Fuels: Science, Technology, Health, and Environment*, ed. O. Konur. Boca Raton, FL: CRC Press.

Konur, O. 2021z. Petrodiesel fuels: A scientometric review of the research. In *Handbook of Biodiesel and Petrodiesel Fuels: Science, Technology, Health, and Environment. Volume 3. Petrodiesel Fuels: Science, Technology, Health, and Environment*, ed. O. Konur. Boca Raton, FL: CRC Press.

Konur, O. 2021aa. Bioremediation of petroleum hydrocarbons in the contaminated soils: A review of the research. In *Handbook of Biodiesel and Petrodiesel Fuels: Science, Technology, Health, and Environment. Volume 3. Petrodiesel Fuels: Science, Technology, Health, and Environment*, ed. O. Konur. Boca Raton, FL: CRC Press.

Konur, O. 2021ab. Desulfurization of diesel fuels: A review of the research. In *Handbook of Biodiesel and Petrodiesel Fuels: Science, Technology, Health, and Environment. Volume 3. Petrodiesel Fuels: Science, Technology, Health, and Environment*, ed. O. Konur. Boca Raton, FL: CRC Press.

Konur, O. 2021ac. Diesel fuel exhaust emissions: A scientometric review of the research. In *Handbook of Biodiesel and Petrodiesel Fuels: Science, Technology, Health, and Environment. Volume 3. Petrodiesel Fuels: Science, Technology, Health, and Environment*, ed. O. Konur. Boca Raton, FL: CRC Press.

Konur, O. 2021ad. The adverse health and safety impact of diesel fuels: A scientometric review of the research. In *Handbook of Biodiesel and Petrodiesel Fuels: Science, Technology, Health, and Environment. Volume 3. Petrodiesel Fuels: Science, Technology, Health, and Environment*, ed. O. Konur. Boca Raton, FL: CRC Press.

Konur, O. 2021ae. Respiratory illnesses caused by the diesel fuel exhaust emissions: A review of the research. In *Handbook of Biodiesel and Petrodiesel Fuels: Science, Technology, Health, and Environment. Volume 3. Petrodiesel Fuels: Science, Technology, Health, and Environment*, ed. O. Konur. Boca Raton, FL: CRC Press.

Konur, O. 2021af. Cancer caused by the diesel fuel exhaust emissions: A review of the research. In *Handbook of Biodiesel and Petrodiesel Fuels: Science, Technology, Health, and Environment. Volume 3. Petrodiesel Fuels: Science, Technology, Health, and Environment*, ed. O. Konur. Boca Raton, FL: CRC Press.

Konur, O. 2021ag. Cardiovascular and other illnesses caused by the diesel fuel exhaust emissions: A review of the research. In *Handbook of Biodiesel and Petrodiesel Fuels: Science, Technology, Health, and Environment. Volume 3. Petrodiesel Fuels: Science, Technology, Health, and Environment*, ed. O. Konur. Boca Raton, FL: CRC Press.

Konur, O. and F. L. Matthews. 1989. Effect of the properties of the constituents on the fatigue performance of composites: A review. *Composites* 20:317–328.

Lapuerta, M., O. Armas and J. Rodriguez-Fernandez. 2008. Effect of biodiesel fuels on diesel engine emissions. *Progress in Energy and Combustion Science* 34:198–223.

Lariviere, V. and Y. Gingras. 2010. On the relationship between interdisciplinarity and scientific impact. *Journal of the American Society for Information Science and Technology* 61:126–131.

Lariviere, V., C. Ni, Y. Gingras, B. Cronin, and C.R. Sugimoto. 2013. Bibliometrics: Global gender disparities in science. *Nature News* 504:211–213.

Leydesdorff, L. and Wagner, C. 2009. Is the United States losing ground in science? A global perspective on the world science system. *Scientometrics*, 78:23–36.

Marchetti, J. M., V. U. Miguel and A. F. Errazu. 2007. Possible methods for biodiesel production. *Renewable and Sustainable Energy Reviews* 11:1300–1311.

McCreanor, J, P. Cullinan, and M. J. Nieuwenhuijsen, et al. 2007. Respiratory effects of exposure to diesel traffic in persons with asthma. *New England Journal of Medicine* 357:2348–2358.

Meyer, M. and O. Persson. 1998. Nanotechnology - Interdisciplinarity, patterns of collaboration and differences in application. *Scientometrics* 42:195–205.

Mills, N. L., H. Tornqvist, and M. C. Gonzalez. 2007. Ischemic and thrombotic effects of dilute diesel-exhaust inhalation in men with coronary heart disease. *New England Journal of Medicine* 357:1075–1082.

Morillo, F., M. Bordons and I. Gomez. 2001. An approach to interdisciplinarity through bibliometric indicators. *Scientometrics* 51:203–222.

North, D. C. 1991a. *Institutions, Institutional Change and Economic Performance*. Cambridge, Mass.: Cambridge University Press.

North, D.C. 1991b. Institutions. *Journal of Economic Perspectives* 5:97–112.

Olivier, J. G. J., A. F. Bouwman, K. W. van der Hoek, and J. J. M. Berdowski. 1998. Global air emission inventories for anthropogenic sources of NO_x, NH_3 and N_2O in 1990. *Environmental Pollution* 102:135–148.

Park, H. W., J. Yoon and L. Leydesdorff. 2016. The normalization of co-authorship networks in the bibliometric evaluation: The government stimulation programs of China and Korea. *Scientometrics* 109:1017–1036.

Peters, A., H. E. Wichmann, T. Tuch, J. Heinrich, and J. Heyder. 1997. Respiratory effects are associated with the number of ultrafine particles. *American Journal of Respiratory and Critical Care Medicine* 155:1376–1383.

Rinia, E. J., T. N. van Leeuwen, H. G. van Vuren, and A. F. Van Raan. 1998. Comparative analysis of a set of bibliometric indicators and central peer review criteria: Evaluation of condensed matter physics in the Netherlands. *Research Policy* 27:95–107.

Robinson, A. L., N. M. Donahue, and M. K. Shrivastava, et al. 2007. Rethinking organic aerosols: Semivolatile emissions and photochemical aging. *Science* 315:1259–1262.

Rogers, D., and A. J. Hopfinger. 1994. Application of genetic function approximation to quantitative structure-activity relationships and quantitative structure-property relationships. *Journal of Chemical Information and Computer Sciences* 34:854–866.

Rogge, W. F., L. M. Hildemann, M. A. Mazurek, G. R. Cass, and B. R. T. Simoneit. 1993. Sources of fine organic aerosol. 2. Noncatalyst and catalyst-equipped automobiles and heavy-duty diesel trucks. *Environmental Science & Technology* 27:636–651.

Salvi, S., A. Blomberg, and B. Rudell, B, et al. 1999. Acute inflammatory responses in the airways and peripheral blood after short-term exposure to diesel exhaust in healthy human volunteers. *American Journal of Respiratory and Critical Care Medicine* 159:702–709.

Scherf, U. and E. J. List. 2002. Semiconducting polyfluorenes—towards reliable structure–property relationships. *Advanced Materials* 14:477–487.

Schreiber, K. and C. H. Kindler. 2005. Bibliometric analysis of anaesthetic molecular biology research in Germany, Austria and Switzerland. *Anaesthesist* 54:1094–1099.

Schummer, J. 2004. Multidisciplinarity, interdisciplinarity, and patterns of research collaboration in nanoscience and nanotechnology. *Scientometrics* 59:425–465.

Srivastava, A. and R. Prasad. 2000. Triglycerides-based diesel fuels. *Renewable and Sustainable Energy Reviews* 4:111–133.

Tahira, M., R. A. Alias and A. Bakri, 2013. Scientometric assessment of engineering in Malaysians universities. *Scientometrics* 96:865–879.

Van Gerpen, J. 2005. Biodiesel processing and production. *Fuel Processing Technology* 86:1097–1107.

Van Raan, A. F., T. N. van Leeuwen and M. S. Visser. 2011. Severe language effect in university rankings: Particularly Germany and France are wronged in citation-based rankings. *Scientometrics* 88:495–498.

Waast, R. and P. L. Rossi. 2010. Scientific production in Arab countries: A bibliometric perspective. *Science, Technology and Society* 15:339–370.

Wang, X., D. Liu, K. Ding, K. and X. Wang. 2012. Science funding and research output: A study on 10 countries. *Scientometrics* 91:591–599.

Xie, Y. and K. A. Shauman. 1998. Sex differences in research productivity: New evidence about an old puzzle. *American Sociological Review* 847–870.

Youtie, J, P. Shapira, and A. L. Porter. 2008. Nanotechnology publications and citations by leading countries and blocs. *Journal of Nanoparticle Research* 10:981–986.

Zhao, Y., S. Wang, and L. Duan, et al. 2008. Primary air pollutant emissions of coal-fired power plants in China: Current status and future prediction. *Atmospheric Environment* 42:8442–8452.

Zhou, P. and L. Leydesdorff. 2006. The emergence of China as a leading nation in science. *Research Policy* 35:83–104.

52 Diesel Emissions and Approaches to their Mitigation

I. M. Rizwanul Fattah
Hwai Chyuan Ong
T. M. Indra Mahlia
M. Mofijur

CONTENTS

52.1 INTRODUCTION

Throughout history, the expansion of human civilization has been supported by the steady growth in our use of high-quality exosomatic energy. This growth has been driven by an increasing population and increasing levels of activity. As we learned to harness the energy sources around us we progressed from horse-drawn plows, hand forges, and wood fires to our present level of mechanization with its wide variety of high-density energy sources. As industrialization has progressed around the world, the amount of energy each one of us uses has also increased, with the global average per capita consumption of all forms of energy rising by 50% in the last 40 years alone (Fattah et al., 2013). Historically, fossil fuels have played a vital role in global energy demand. The diesel engine, named after its inventor Rudolf Diesel, was patented in 1892 and has catered for this energy demand ever since (Cummins, 1976). Being the powerhouse of heavy-duty and commercial transport vehicles, both on land and at sea, has been the most important use of diesel engines; and their importance is increasing consistently. The diesel engine is the most efficient type of internal combustion engine, offering good fuel economy and low 'carbon dioxide' (CO_2) emissions (Imtenan et al., 2014). It converts the chemical energy contained in the fuel into mechanical power. High-pressure diesel fuel is injected into the engine cylinder where it mixes with air and combustion occurs due to the lower 'self-ignition temperature' (SIT) of the fuel. While diesel engines are arguably superior to any other power-production device for the transportation sector in terms of efficiency, torque, and overall drivability, they suffer from inferior performance in terms of emissions. Due to their mode of operation, involving stratification and the presence of fuel-rich, soot-prone regions and simultaneously lean 'nitrogen oxide' (NO_x), prone regions in the flame vicinity, a number of challenges are posed, necessitating further research effort into pollutant reduction technologies.

As far as emissions are concerned, diesel engines, especially heavy-duty vehicle engines, have been subjected to progressively stringent emission control standards. The adoption of emission standards was initially launched by the USA, followed by the European Union and Japan, and, subsequently, by other countries, such as Australia, Brazil, China, and India. These standards have led to the current strict emission limits, imposed by US EPA 2010 and Euro VI and other variants, and provided huge emission reductions of criteria emissions, i.e. 'hydrocarbons' (HC), 'carbon monoxide' (CO), NO_x, 'particulate matter' (PM), and smoke. Nevertheless, due to climate change and air quality concerns, more stringent emission control is still a regulatory demand. One prominent example of emission control regulation is the US 2010 standard which required PM and NO_x emission reductions by at least 90%, compared with initial standards (Barbosa, 2020).

This chapter first discusses the diesel combustion process which results in emissions in gaseous and solid forms. The criteria emissions, i.e. NO_x, soot/PM, HC, and CO, are discussed next in detail so that the reader has a clear idea of their characterization. After that, some of the salient pre-combustion and post-combustion technologies are explored, though they are in no way a complete list of the available technologies. The simultaneous reduction of NO_x and PM is also highlighted here. Finally, a quick review of the strategies and the available technological pathways for emissions reduction is discussed.

52.2 DIESEL COMBUSTION

The diesel engine is the most fuel-efficient commercial combustion engine ever built for transportation purposes. These engines operate on the principle of 'compression ignition' (CI), unlike 'spark-ignited (SI) engines' in which a spark plug initiates combustion. These engines rely on compression in the cylinder to raise the air temperature and pressure such that upon injecting the fuel, the resulting combustible mixture auto-ignites. In conventional diesel combustion, spray droplets are formed from each hole of the fuel injector, when liquid fuel is pushed through multiple holes at high pressure. The spray penetrates the combustion chamber entraining in-cylinder or ambient gases into the jet, forming a roughly conical jet with its volume increasing with downstream distance from the injector. The increasing entrainment downstream causes the 'equivalence ratio' (φ), defined as the 'ratio of the local fuel-ambient charge mass ratio to the stoichiometric fuel-ambient mass ratio' (Musculus et al., 2013), to vary along the jet axis, approximately inversely with the downstream distance.

The liquid fuel is vaporized by the entrained gases and the vaporized fuel-ambient mixture is then carried downstream of the liquid length by momentum and continues to entrain more ambient gases. First-stage ignition reactions then commence (Dec and Espey, 1995, 1998), which is followed by a highly exothermic second-stage reaction that leads to the premixed burn phase of diesel combustion. As the temperature rises, soot-precursor species like 'polycyclic aromatic hydrocarbon' (PAH) quickly form in the hot (~1600 K), fuel-rich combustion products of the premixed burn (Dec, 1997; Siebers and Higgins, 2001). Soot is formed afterwards, which fills the entire downstream jet cross-section (Pickett and Siebers, 2004). Any of this soot that is not oxidized in the later phases of combustion becomes engine-out PM. The diffusion flame is formed on the periphery of the fuel-rich, high-temperature downstream region (in excess of 2600 K) of the jet where NO_x are formed. Thus, traditional mixing-controlled diesel combustion contributes to undesirable engine-out NO_x and PM emissions because of its inherent nature.

52.3 DIESEL EXHAUST EMISSIONS

Diesel engine exhaust contains a wide range of gaseous and particulate phased organic and inorganic compounds with higher amounts of aromatics and sulfur compared to gasoline engines. The particles have hundreds of chemicals adsorbed onto their surfaces, comprising many recognized or suspected mutagens and carcinogens. The gaseous phase also contains many toxic chemicals and irritants. These have a serious adverse effect on human health and an environmental impact (Ackerman et al., 2000; Reitmayer et al., 2019).

52.3.1 DIESEL GASEOUS EMISSIONS

Diesel gaseous emissions primarily comprise, but are not limited to, NO_x, HC, CO, CO_2, and sulfur dioxide (SO_2). Out of these, as discussed previously, NO_x, HC, and CO are regulated by most emission standards.

52.3.1.1 NO_x

NO_x is the generalized term for NO and NO_2. NO_x is the most harmful gaseous emission of diesel engines; the reduction of it is always a target for engine researchers and manufacturers. As discussed, during typical mixing-controlled diesel combustion, a diffusion flame is formed at near stoichiometric fuel–air mixtures where NO_x formation is high due to the very high temperatures in typical diesel operating conditions (exceeding 2600 K) (Ahmad and Plee, 1983; Dec and Canaan, 1998). NO_x, primarily NO, is typically formed by three main mechanisms, i.e. thermal (Zeldovich), prompt (Fenimore), and fuel-bound nitrogen mechanisms (Sun et al., 2010).

The thermal mechanism is based on the Zeldovich mechanism, represented by Equations 52.1 through 52.3, which involves the breakdown of the strong triple bond of atmospheric (molecular) nitrogen and which occurs during combustion or shortly thereafter in the post-flame gas region (Bowman, 1975). The residence time and concentration of nitrogen and oxygen also play a vital role in the production of thermal NO (Ban-Weiss et al., 2007).

$$O + N_2 \leftrightarrow NO + N \tag{52.1}$$

$$N + O_2 + NO + O \tag{52.2}$$

$$N + OH \leftrightarrow NO + H \tag{52.3}$$

The presence of a second mechanism leading to NO formation was first identified by Fenimore and termed 'prompt NO' (Fenimore, 1971). In this mechanism, the formation of free radicals in the flame front of the hydrocarbon flames leads to rapid production of NO_x. Prompt NO_x is prevalent in some combustion environments, such as in low-temperature, fuel-rich conditions and where residence time is short (Fenimore, 1971).

The fuel NO_x is formed by the combustion of nitrogen which is chemically associated with the fuel (apart from the molecular nitrogen) (Cofala and Syri, 1998). The production process is complex because this includes in the order of 50 intermediate species and several hundred reversible reactions; the true values of the rate constants are still unknown. However, diesel fuel has very low nitrogen levels. The addition of additives containing nitrogen atoms, e.g. pyridine and pyrrole, may lead to fuel NO_x formation.

52.3.1.2 Hydrocarbons (HC)

Diesel engines run on an overall fuel-lean φ ($\varphi < 1$), whereas the gasoline engine tends to run on a near stoichiometric ratio ($\varphi \approx 1$). As such, diesel engines generally produce about one-fifth of the HC emissions of a gasoline engine (Ganesan, 1996). In general, a CI engine has a combustion efficiency of 98%, which means only about 2% of the HC fuel is emitted after combustion. The prevalent reasons of HC emission include: over-mixing of fuel with air, which leads to over-leaning, thereby making it difficult to support combustion, especially at low temperature, low load, and idle; under-mixing of fuel and air, causing an over-rich mixture which is difficult to ignite; and flame quenching at low-temperature walls which causes partial burning (Alozie and Ganippa, 2019). In addition, other causes, including low velocity and late fuel injections such as post-injection, can also be significant (Fattah et al., 2013).

52.3.1.3 CO

CO is a poisonous gas and, when inhaled, replaces the oxygen in the bloodstream, which causes the metabolism to function improperly. Small amounts of CO concentration slow down physical and mental activities and produce headaches, while large amounts can kill. CO is generally formed due to the incomplete combustion of diesel fuel in the absence of a sufficient air supply. However, combustion in most diesel engines is practically lean beyond the stoichiometric, therefore CO emission is normally low except for during transient operations. In general, CO emissions in the exhaust represent lost chemical energy that is not fully utilized in the engine (Ozsezen and Canakci, 2010).

52.3.2 Solid-Phase Emissions

The solid phase emissions, known as particulates in broader terms, are the combination of soot and other liquid or solid-phase materials that are collected when product (exhaust) gases pass through a filter (Agarwal et al., 2011).

52.3.2.1 Soot

Particulates are often separated into a soluble and an insoluble or dry fraction. The fraction of particulate, which is soot, is often estimated by finding the insoluble portion of the particulates (Lee et al., 1998; Ullman, 1989). Soot is not a clearly defined substance. In general, it is a solid substance consisting of roughly eight parts carbon and one part hydrogen (Tree and Svensson, 2007). It is formed from unburned fuel, which nucleates from the vapor phase to a solid phase in fuel-rich regions at elevated temperature without sufficient oxygen concentration. In engine-out emission, soot primarily consists of fractal-like carbonaceous agglomerates and adsorbed material in a size span of 30–500 nm (Kittelson et al., 2006). Liquid phase materials and hydrocarbons are adsorbed on the surface of soot, depending on engine operating conditions. The fraction of soot in particulates from diesel exhaust varies but is typically higher than 50% (Tree and Svensson, 2007). Soot is of particular interest to researchers studying spray combustion because it can be quantified using advanced combustion diagnostics such as optical and/or laser diagnostics (Chan et al., 2011; Fattah et al., 2019). Soot is often deposited on combustion chamber walls in diesel engines. Smoke opacity is an indirect indicator of soot content in the exhaust gases of a diesel engine (Agudelo et al., 2010). As such, this parameter can be correlated with the fuel's tendency to form soot during engine operation.

52.3.2.2 PM

As mentioned previously, particulates are often separated into a soluble and an insoluble or dry fraction. The soluble portion is generally termed 'PM', which is a mixture of solid and liquid particles that differ in surface area, solubility, number, size, shape, chemical composition, and origin (Pope and Dockery, 2006; Sakurai et al.,

2003). The soluble fraction present in particulates mainly consists of aldehydes, alkanes, alkenes, aliphatic hydrocarbons, PAHs, and its derivatives (Mohankumar and Senthilkumar, 2017). Other PM constituents include un/partially burned fuel/lubricant oil, bound water, wear metals, and fuel-derived sulfates (Lee et al., 1998; Ullman, 1989). Researchers studying engine combustion are predominantly interested in PM because it can be quantified using exhaust gas measurement devices (Khalife et al., 2017). The aerodynamic diameter has been recognized as a simple means of defining the particle size of PM (Bhat and Kumar, 2012), as these particles exist in different shapes and densities in the air. These are: PM_{10} – inhalable particles, with diameters that are generally 10 micrometers and smaller; and $PM_{2.5}$ – fine inhalable particles, with diameters that are generally 2.5 micrometers and smaller.

There have been deep concerns about public health because of particulate emissions in the atmosphere. This is mainly because of the nature of particles in emissions being very small (more than 90% are less than 1 mm by mass), which makes them readily breathable.

52.4 METHODS OF MITIGATION

As discussed previously, traditional mixing-controlled diesel combustion contributes to undesirable engine-out NO_x and PM emissions, the reduction of which is often the primary objective of many studies.

52.4.1 NO_x Emission Mitigation

This section summarizes the developments regarding NO_x mitigation in CI engines. NO_x can be mitigated through several techniques depending on when it is implemented. Various pre-combustion and after-treatment technologies are employed to control NO_x emission. The complete layout of NO_x control techniques is shown in Figure 52.1.

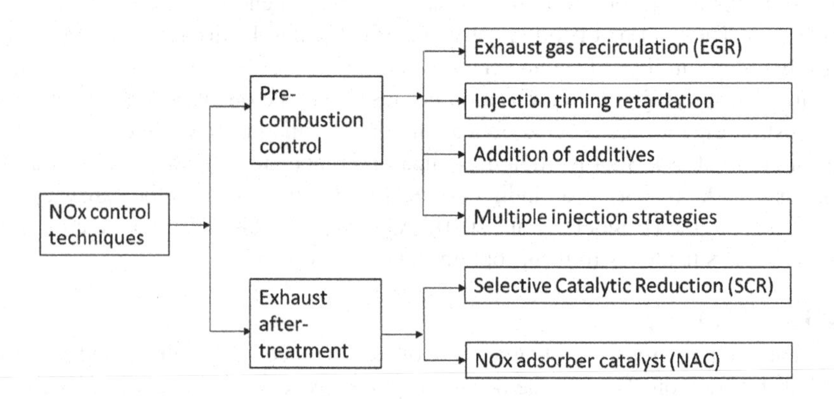

FIGURE 52.1 Schematic layout of NO_x control techniques.

52.4.1.1 Pre-Combustion Techniques

'Exhaust gas recirculation' (EGR) is an engine-out (i.e. prior to exhaust after-treatment) emission control technology, allowing significant NO_x emission reductions from most types of diesel engines: from light-duty engines through medium and heavy-duty engine applications. EGR can be achieved by diluting the intake stream with combustion products, thereby reducing both the intake-oxygen concentration and the flame temperature (Alriksson et al., 2005; Ladommatos et al., 2000). However, this benefit of EGR comes at a cost, i.e. increase in fuel consumption, higher emission of PM, HC, and CO, higher engine wear, and reduction in engine durability (Abd-Alla, 2002). In particular, EGR exacerbates the trade-off between NO_x and particulate emissions at high loads. The percentage reduction of NO_x depends on the EGR level with intake oxygen concentrations in the range of 10–15%, engine speed and load condition, intake air condition, etc. (Musculus et al., 2013; Pandurangi et al., 2014).

Different types of additives – such as metal-based ones, i.e. ferric chloride ($FeCl_3$) and magnesium oxide (MgO); nanoparticles, i.e. cerium oxide (CeO_2), alumina nanoparticles, and carbon nanotubes (CNT); oxygenated additives, i.e. dimethyl ether, ethanol, and methanol; antioxidants, i.e. cetane number improver 2-Ethylhexyl nitrate (EHN), di-t-butyl peroxide (DTBP), etc. – in conjunction with biodiesel are generally used to reduce the NO_x emissions (Fattah et al., 2014; Hosseinzadeh-Bandbafha et al., 2018; Soudagar et al., 2018). Biodiesel is an alternative fuel commonly described as fatty acid methyl ester derived from vegetable oils and animal fats (Ong et al., 2014, 2019; Silitonga et al. 2020). It is renewable, biodegradable, and an oxygenated fuel consisting of triglycerides of long-chain saturated and unsaturated fatty acids that can be converted into monoglycerides by the transesterification process.

Retarding the main injection timing is always found to reduce NO_x emissions due to the fact that it reduces the combustion temperature in the cylinder, as well as the residence time of high-temperature-burned gas in the combustion chamber where NO_x is actively formed (Qi et al., 2011).

The multiple injections, i.e. pilot-injections, split-injections, and post-injections, are the three most commonly used multiple injection strategies (O'Connor and Musculus, 2013). Among these two strategies, pilot-injection and split-injections are used to mitigate NO_x emissions. The pilot injection strategy is initialized with a pilot injection that precedes the main injection, whereby the quantity of fuel injected in the pilot is typically less than that of the main injection (Tow et al., 1994). Pilot combustion raises the in-cylinder temperature prior to the main combustion event, which allows for a reduced ignition delay, limiting the amount of premixed combustion and lowering the rate of combustion (Sahoo et al., 2013). As a result, the peak heat release rate is reduced and, consequently, so is the NO_x produced. Split-injection, on the other hand, is often used for splitting the overall heat release which alleviates the wall-wetting of long injection pulses achieved through splitting that into a few pulses thereby interrupting their maximum stabilized length (Pickett et al., 2009) and reducing emissions. The number of pulses into which the single injection is split is vital for emission reduction. Ehleskog et al. (2007) found that splitting the main injection into three and four pulses was found to result in a reduction in NO_x emissions.

52.4.1.2 Exhaust After-Treatment Techniques

There are two main after-treatment technologies available to reduce NO_x emissions from diesel engines (Hoekman and Robbins, 2012). The first one involves the use of a 'selective catalytic reduction' (SCR) catalyst. In the SCR system, a reductant (e.g. ammonia in the form of aqueous urea) is injected into the engine exhaust line ahead of the SCR catalyst, and a different catalyst bed is used to directly reduce NO_x to N_2. The SCR technique has a high NO_x conversion efficiency of about 90%. SCR with hydrocarbon reductant is known as a 'lean-NO_x trap' (LNT) (Alozie and Ganippa, 2019). The second one involves the use of a 'NO_x adsorber catalyst' (NAC). An NAC system utilizes a catalyst containing a basic oxide, i.e. $Pt/Ba/Al_2O_3$ catalysts to convert all NO to NO_2, which is then trapped in an adsorbent bed, i.e. a Ba surface. Once the adsorber is saturated with NO_2, the exhaust stream is forced into a fuel-rich condition and the trapped NO_2 is reduced to N_2 on the Pt surface, which is exhausted (Tripathi et al., 2018). An NAC is also called a 'NO_x storage reduction catalyst' (NSRC/NSC) in the literature (Liu and Gao, 2011). NAC has up to 80% reduction efficiency.

52.4.2 Particulates Emission Mitigation

This section summarizes the developments regarding soot/PM mitigation in 'CI engines'. Soot can be mitigated through several techniques depending on when it is implemented. Various pre-combustion and post-combustion technologies are employed to control particulates. The complete layout of the particulates control technique is shown in Figure 52.2.

52.4.2.1 Pre-Combustion Techniques

The addition of oxygenated additives is a well-documented technique for soot/PM mitigation. These chemicals have fuel-bound oxygen in their structure and typically

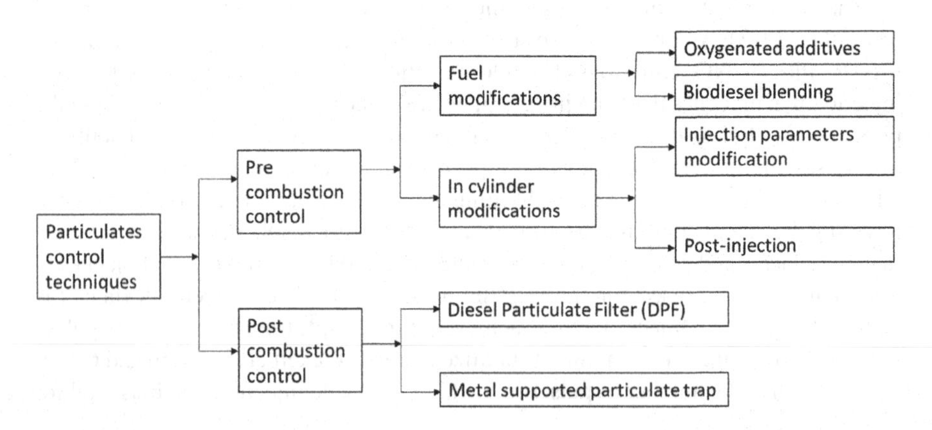

FIGURE 52.2 Schematic layout of particulate control techniques.

possess a very high cetane number which enhances combustion and the ignition quality of the fuel, thereby increasing the likelihood of oxidation of particulates (Mohankumar and Senthilkumar, 2017). Some of the oxygenated additives used in previous studies include methanol, ethanol, butanol, diethyl ether, diphenyl ether, diethylene glycol, dimethyl ether, nitromethane, and dimethyl carbonate (Gorski et al., 2013; Mueller and Martin, 2002; Wang et al. 2009).

Biodiesel blending with diesel at various proportions (~10–50%) has been reported to reduce particulate emissions significantly. The presence of long-chain saturated and unsaturated fatty acids, low sulfur and aromatic content has been advantageous in the reduction of PM (Mahlia et al., 2020; Silitonga et al., 2013).

Increasing the fuel injection pressure greatly reduces particulate emission by shortening the ignition delay thereby improving spray atomization and enhancing the fuel–air mixing rate (Fang et al., 2010; Gao et al., 2007). In addition, the reduction of the injector nozzle diameter also produces the same effect (Kuti et al., 2013). The shortened ignition delay advances the combustion and also the soot formation processes, giving ample time for the oxidation of soot. As such, increasing the injection pressure and reducing the nozzle hole diameter to a micro-hole size helps in reducing particulate emission.

Post-injection is where a quantity of fuel (usually up to 20%) is allocated into separate portions in a way that the second injection duration is much shorter than that of the main injection. As discussed previously, a high level of EGR can cause an increase in the engine-out soot emissions when used in conventional diesel engines due to the suppression of soot oxidation (Idicheria and Pickett 2005; Tree and Svensson, 2007). This unwanted increase can be countered by post-injection which is introduced early enough to interact with the main combustion (Fattah et al., 2018; Yip et al., 2019). Another advantage of post-injection is that it increases exhaust temperature.

52.4.2.2 Post-Combustion Techniques

'Diesel particulate filter' (DPF) is a type of monolithic filter which is used to trap particles of micron and sub-micron sizes carried by the exhaust gases (Salvat et al., 2000). The most commonly used one is wall-flow DPF which has the cells alternately plugged at each end where the exhaust gas permeates through the walls of the filter while the particles are trapped. Due to the continuous use of DPF, there is a chance of clogging, which requires periodic regeneration to oxidize the accumulated soot in the filter and keep its pressure drop at a reasonable level to ensure fuel economy and the proper working of the engine. In general, DPF should possess high filtration efficiency, less pressure drop, high soot-storing capacity, compatibility, and stability (mechanical, thermal, and chemical) with regeneration methods.

DPF technology cannot be applied to all applications since the filter regeneration is limited by the engine-out NO_x to PM ratio as well as exhaust temperature. Thus, these cannot be applied to older diesel engines like heavy-duty trucks with high soot/PM emissions. To overcome this difficulty, a passive PM control technology, known as metal-supported flow through diesel filters employing continuously the principle of the regenerating trap, has been developed for this type of engine (Jacobs et al., 2006; Pace et al., 2005). This technology is referred to as partial filter technology (PFT) which can reduce PM by up to 77%.

52.4.3 Simultaneous NO_x and Particulate Reduction

Simultaneous reduction of PM and NO_x emissions from diesel exhaust is the prime focus of research and development in modern diesel engines.

52.4.3.1 'Low-Temperature Combustion' (LTC)

LTC is one of the methods to achieve a simultaneous in-cylinder reduction of NO_x and soot/PM reduction. LTC takes place at temperatures below the formation regime of NO_x and at local φ below the formation regime of diesel soot (Imtenan et al., 2014). During LTC, a large fraction of the fuel burns in the premixed combustion phase and combustion of the homogeneous lean mixture takes place throughout the combustion chamber. This unique feature of LTC offers ultra-low NO_x (due to the low in-cylinder temperature) and soot emissions (due to premixed combustion) simultaneously. LTC is generally achieved through two pathways, i.e. 'homogeneous charge compression ignition' (HCCI) combustion and 'reactivity controlled compression ignition (RCCI). HCCI utilizes volumetric autoignition and combustion in a lean or diluted mixture which is achieved by different methods including premixed, early direct injection, and late direct-injection (Juttu et al., 2007). The main problem is that the charge ignites too rapidly at higher loads which are addressed with RCCI, wherein stratification is introduced into the charge by using two fuels of differing reactivity (Splitter et al., 2011). In-cylinder fuel blending using port fuel-injection of a low reactivity fuel and optimized direct injection of higher reactivity fuel is usually used to control combustion phasing and duration. The combustion is spread out over more crank angle degrees maintaining low NO_x and low PM.

52.4.3.2 Water Injection

Water injection is one of the methods for introducing water to the diesel combustion chamber which helps in reducing the emissions of NO_x and particulates simultaneously. There are various ways of injecting water into a combustion chamber, directly into the combustion chamber and by emulsifying fuel (Tauzia et al., 2010). The emulsion is the more practical way of injecting water. The emulsion is created when water is dispersed throughout the fuel, usually in the form of spherical droplets. With the use of emulsions, an improvement in the mixing process is usually observed due to two confirmed phenomena. The first one is added momentum in the jet's behavior due to water addition (Andrews et al., 1988). The second one is the internal droplet micro-explosions of water induced by the volatility difference between the water and the fuel. This violent disintegration disperses the fine droplets, producing a secondary atomization and consequently enhancing the fuel–air mixing in the combustion chamber (Kadota and Yamasaki, 2002). This increases the premixed combustion duration as well as the ignition delay period, which in turn allows more time for fuel–air mixing, leading to a reduction in PM formation. In addition, water droplet evaporation also reduces peak cycle temperature, which leads to the reduction of NO_x emissions.

52.4.3.3 After-Treatment Technologies

Simultaneous particulate and NO_x reduction represent the next step in the reduction of diesel emissions. To achieve this target, the combination of two technologies already in use is utilized. There are a few combinations that have been reported in the literature. For example, combined SCR and DPF (Conway et al., 2005) and combined NSC and DPF (Ranalli et al., 2004). Another stand-alone technology which has been developed is known as a 'diesel particulate-NO_x reduction' (DPNR) system (Mizuno and Suzuki, 2004; Nakatani et al., 2002). This is a type of catalytic converter consisting of a fine porous ceramic having a monolithic honeycomb structure, coated with a NO_x storage reduction catalyst.

52.4.4 HC and CO Emission Mitigation

52.4.4.1 Use of a Turbocharger

A simple way to reduce some of the potentially harmful emissions is to use a turbocharger which increases the mass of air inducted into the cylinder. This also allows for a proportional increase in fuel injection to create additional power output compared to a naturally aspirated engine which helps in downsizing it (Tigelaar et al., 2017). Since the air velocity is higher through the intake, air–fuel mixing is enhanced. This leads to better fuel oxidation and reduction in the emissions of HC, soot/PM, and CO. However, it can increase NO_x formation.

52.4.4.2 'Diesel Oxidation Catalyst' (DOC)

DOC assists in the oxidation of HC, CO, soot particles, and PM to CO_2. To promote this oxidation, DOC consists of precious metals, i.e. platinum (Pt) and palladium (Pd) as a catalyst. This catalyst is covered in a wash-coat material of alumina (Al_2O_3) or silica (Si_2O_3). The layer of wash coat and catalyst is spread on a DOC substrate which can be either ceramic or metallic in a honeycomb structure (Hora et al., 2018). DOC is highly durable in operation and can achieve a high conversion efficiency of up to 90% at a sufficiently high exhaust temperature. It can also remove diesel odor. The DOC is also used to generate the necessary exothermal energy for the regeneration of the wall-flow filters.

52.5 CONCLUSIONS

The diesel combustion process is a complex phenomenon, the understanding of which is required to comprehend the emission formation process. Engine systems design, i.e. air delivery, fuel delivery, mixing design; the interplay of the relevant integrated technologies; and the transient nature of the combustion event due to load variation play key parts in this regard, albeit the efficiency of the in-cylinder combustion is central to the quality of emission occurring at the tailpipe. The efforts in reducing this emission have resulted in a series of continuously evolving emission standards over the last 30 years. The emission control focus has been on the so-called 'criteria pollutants' (HC, CO, PM, and NO_x). Heavy-duty diesel engines have been scrutinized heavily due to their high NO_x and PM emission potential, which is intrinsic to their combustion process. Several pre-combustion and post-combustion technologies are available for emission mitigation, such as EGR, additive addition, SCR, NAC, multiple injections, LTC, and DOC. These days, it is customary to incorporate a series of after-treatment devices, especially for automotive applications.

REFERENCES

Abd-Alla, G. H. 2002. Using exhaust gas recirculation in internal combustion engines: A review. *Energy Conversion and Management* 43:1027–1042.

Ackerman, A. S., O. B. Toon, and D. E. Stevens, et al. 2000. Reduction of tropical cloudiness by soot. *Science* 288:1042–1047.

Agarwal, A. K., T. Gupta, and A. Kothari. 2011. Particulate emissions from biodiesel vs diesel fuelled compression ignition engine. *Renewable and Sustainable Energy Reviews* 15:3278–3300.

Agudelo, J., P. Benjumea, and A. P. Villegas. 2010. Evaluation of nitrogen oxide emissions and smoke opacity in a HSDI diesel engine fuelled with palm oil biodiesel. *Revista Facultad de Ingenieria Universidad de Antioquia* 2010:62–71.

Ahmad, T. and S. L. Plee. 1983. Application of flame temperature correlations to emissions from a direct-injection diesel engine. *SAE Technical Paper* 831734.

Alozie, N. S., and L. C. Ganippa. 2019. Diesel exhaust emissions and mitigations. In *Introduction to Diesel Emissions*, ed. R. Viskup, 2019:85248. London: IntechOpen.

Alriksson, M., T. Rente, and I. Denbratt. 2005. Low soot, low NO_x in a heavy duty diesel engine using high levels of EGR. *SAE Technical Paper* 2005-01-3836.

Andrews, G. E., K. D. Bartle, S. W. Pang, A. M. Nurein, and P. T. Williams. 1988. The reduction in diesel particulate emissions using emulsified fuels. *SAE Technical Paper* 880348.

Ban-Weiss, G. A., J. Y. Chen, B. A. Buchholz, and R. W. Dibble. 2007. A numerical investigation into the anomalous slight NO_x increase when burning biodiesel: A new (old) theory. *Fuel Processing Technology* 88:659–667.

Barbosa, F. C. 2020. Heavy duty diesel emission standards regulation evolution review: Current outcomes and future perspectives. *SAE Technical Paper* 2019-36-0174.

Bhat, A. and A. Kumar. 2012. Particulate characteristics and emission rates during the injection of class B biosolids into an agricultural field. *Science of the Total Environment* 414:328–334.

Bowman, C. T. 1975. Kinetics of pollutant formation and destruction in combustion. *Progress in Energy and Combustion Science* 1:33–45.

Chan, Q. N., P. R. Medwell, and P. A. M. Kalt, et al. 2011. Simultaneous imaging of temperature and soot volume fraction. *Proceedings of the Combustion Institute* 33:791–798.

Cofala, J. and S. Syri. 1998. Nitrogen oxides emission mechanisms. In *Nitrogen Oxides Emissions, Abatement Technologies and Related Costs for Europe in the RAINS Model Database*, IR-98-088. Laxenburg, Austria: International Institute for Applied Systems Analysis.

Conway, R., S. Chatterjee, and A. Beavan, et al. 2005. Combined SCR and DPF technology for heavy duty diesel retrofit. *SAE Technical Paper* 2005-01-3548.

Cummins, C. L. 1976. Early IC and automotive engines. *SAE Transactions* 85:1960–1971.

Dec, J. E. 1997. A conceptual model of DI diesel combustion based on laser-sheet imaging. *SAE Technical Paper* 970873.

Dec, J. E. and R. E. Canaan. 1998. PLIF imaging of NO formation in a DI diesel engine. *SAE Technical Paper* 980147.

Dec, J. E. and C. Espey. 1995. Ignition and early soot formation in a DI diesel engine using multiple 2-D imaging diagnostics. *SAE Technical Paper* 950456.

Dec, J. E. and C. Espey. 1998. Chemiluminescence imaging of autoignition in a DI diesel engine. *SAE Technical Paper* 982685.

Ehleskog, R., R. L. Ochoterena, and S. Andersson. 2007. Effects of multiple injections on engine-out emission levels including particulate mass from an HSDI diesel engine. *SAE Technical Paper* 2007-01-0910.

Fang, T., C. F. Lee, R. Coverdill, and R. White. 2010. Effects of injection pressure on low-sooting combustion in an optical HSDI diesel engine using a narrow angle injector. *SAE Technical Paper* 2010-01-0339.

Fattah, I. M. R., H. H. Masjuki, and M. A. Kalam, et al. 2014. Effect of antioxidants on oxidation stability of biodiesel derived from vegetable and animal based feedstocks. *Renewable and Sustainable Energy Reviews* 30:356–370.

Fattah, I. M. R., H. H. Masjuki, and A. M. Liaquat, et al. 2013. Impact of various biodiesel fuels obtained from edible and non-edible oils on engine exhaust gas and noise emissions. *Renewable and Sustainable Energy Reviews* 18:552–567.

Fattah, I. M. R., C. Ming, and Q. N. Chan, et al. 2018. Spray and combustion investigation of post injections under low-temperature combustion conditions with biodiesel. *Energy & Fuels* 32:8727–8742.

Fattah, I. M. R., H. L. Yip, and Z. Jiang, et al. 2019. Effects of flame-plane wall impingement on diesel combustion and soot processes. *Fuel* 255:115726.

Fenimore, C. P. 1971. Formation of nitric oxide in premixed hydrocarbon flames. *Symposium (International) on Combustion* 13:373–380.

Ganesan, V. 1996. *Internal Combustion Engines*. New York: McGraw-Hill.

Gao, J., Y. Matsumoto, and K. Nishida. 2007. Effects of group-hole nozzle specifications on fuel atomization and evaporation of direct injection diesel sprays. *SAE Technical Paper* 2007-01-1889.

Gorski, K., A. K. Sen, V. Lotko, and M. Swat. 2013. Effects of ethyl-tert-butyl ether (ETBE) addition on the physicochemical properties of diesel oil and particulate matter and smoke emissions from diesel engines. *Fuel* 103:1138–1143.

Hoekman, S. K. and C. Robbins. 2012. Review of the effects of biodiesel on NO_x emissions. *Fuel Processing Technology* 96:237–249.

Hora, T. S., A. P. Singh, and A. K. Agarwal. 2018. Future mobility solutions of Indian automotive industry: BS-VI, hybrid, and electric vehicles. In *Advances in Internal Combustion Engine Research*, ed. D. K. Srivastava, A. K. Agarwal, A. Datta and R. K. Maurya, 309–345. Singapore: Springer Singapore.

Hosseinzadeh-Bandbafha, H., M. Tabatabaei, M. Aghbashlo, M. Khanali, and A. Demirbas. 2018. A comprehensive review on the environmental impacts of diesel/biodiesel additives. *Energy Conversion and Management 174*:579–614.

Idicheria, C. A. and L. M. Pickett. 2005. Soot formation in diesel combustion under high-EGR conditions. *SAE Technical Paper* 2005-01-3834.

Imtenan, S., M. Varman, and H. H. Masjuki, et al. 2014. Impact of low temperature combustion attaining strategies on diesel engine emissions for diesel and biodiesels: A review. *Energy Conversion and Management* 80:329–356.

Jacobs, T., S. Chatterjee, and R. Conway, et al. 2006. Development of partial filter technology for HDD retrofit. *SAE Technical Paper* 2006-01-0213.

Juttu, S., S. S. Thipse, N. V. Marathe, and M. K. G. Babu. 2007. Homogeneous charge compression ignition (HCCI): A new concept for near zero NO_x and particulate matter (PM) from diesel engine combustion. *SAE Technical Paper* 2007-26-020.

Kadota, T. and H. Yamasaki. 2002. Recent advances in the combustion of water fuel emulsion. *Progress in Energy and Combustion Science* 28:385–404.

Khalife, E., M. Tabatabaei, A. Demirbas, and M. Aghbashlo. 2017. Impacts of additives on performance and emission characteristics of diesel engines during steady state operation. *Progress in Energy and Combustion Science* 59:32–78.

Kittelson, D. B., W. F. Watts, and J. P. Johnson. 2006. On-road and laboratory evaluation of combustion aerosols-Part1: Summary of diesel engine results. *Journal of Aerosol Science* 37:913–930.

Kuti, O. A., J. Zhu, K. Nishida, X. Wang, and Z. Huang. 2013. Characterization of spray and combustion processes of biodiesel fuel injected by diesel engine common rail system. *Fuel* 104:838–846.

Ladommatos, N., S. Abdelhalim, and H. Zhao. 2000. The effects of exhaust gas recirculation on diesel combustion and emissions. *International Journal of Engine Research* 1:107–126.

Lee, R., J. Pedley, and C. Hobbs. 1998. Fuel quality impact on heavy duty diesel emissions: A literature review. *SAE Technical Paper* 982649.

Liu, G. and P.-X. Gao. 2011. A review of NO_x storage/reduction catalysts: Mechanism, materials and degradation studies. *Catalysis Science & Technology* 1:552–568.

Mahlia, T. M. I., Z. Syazmi, and M. Mofijur, et al. 2020. Patent landscape review on biodiesel production: Technology updates. *Renewable & Sustainable Energy Reviews* 118:109526.

Mizuno, T. and J. Suzuki. 2004. Development of a new DPNR Catalyst. *SAE Technical Paper* 2004-01-0578.

Mohankumar, S. and P. Senthilkumar. 2017. Particulate matter formation and its control methodologies for diesel engine: A comprehensive review. *Renewable and Sustainable Energy Reviews* 80:1227–1238.

Mueller, C. J. and G. C. Martin. 2002. Effects of oxygenated compounds on combustion and soot evolution in a DI diesel engine: Broadband natural luminosity imaging. *SAE Technical Paper* 2002-01-1631.

Musculus, M. P. B., P. C. Miles, and L. M. Pickett. 2013. Conceptual models for partially premixed low-temperature diesel combustion. *Progress in Energy and Combustion Science* 39:246–283.

Nakatani, K., S. Hirota, and S. Takeshima, et al. 2002. Simultaneous PM and NO_x reduction system for diesel engines. *SAE Technical Paper* 2002-01-0957.

O'Connor, J. and M. Musculus. 2013. Post injections for soot reduction in diesel engines: A review of current understanding. *SAE Technical Paper* 2013-01-0917.

Ong, H. C., H. H. Masjuki, and T. M. I. Mahlia, et al. 2014. Engine performance and emissions using *Jatropha curcas*, *Ceiba pentandra* and *Calophyllum inophyllum* biodiesel in a CI diesel engine. *Energy* 69:427–445.

Ong, H. C., J. Milano, and A. S. Silitonga, et al. 2019. Biodiesel production from *Calophyllum inophyllum-Ceiba pentandra* oil mixture: Optimization and characterization. *Journal of Cleaner Production* 219:183–198.

Ozsezen, A. N. and M. Canakci. 2010. The emission analysis of an IDI diesel engine fueled with methyl ester of waste frying palm oil and its blends. *Biomass and Bioenergy* 34:1870–1878.

Pace, L., R. Konieczny, and M. Presti. 2005. Metal supported particulate matter-cat, a low impact and cost effective solution for a 1.3 Euro IV diesel engine. *SAE Technical Paper* 2005-01-0471.

Pandurangi, S. S., N. Frapolli, M. Bolla, K. Boulouchos, and Y. M. Wright. 2014. Influence of EGR on post-injection effectiveness in a heavy-duty diesel engine fuelled with n-heptane. *SAE International Journal of Engines* 7:1851–1862.

Pickett, L. M., S. Kook, and T. C. Williams. 2009. Visualization of diesel spray penetration, cool-flame, ignition, high-temperature combustion, and soot formation using high-speed imaging. *SAE International Journal of Engines* 2:439–459.

Pickett, L. M. and D. L. Siebers. 2004. Soot in diesel fuel jets: Effects of ambient temperature, ambient density, and injection pressure. *Combustion and Flame* 138:114–135.

Pope, C. A. and D. W. Dockery. 2006. Health effects of fine particulate air pollution: Lines that connect. *Journal of the Air & Waste Management Association* 56:709–742.

Qi, D., M. Leick, Y. Liu, and C.-F. F. Lee. 2011. Effect of EGR and injection timing on combustion and emission characteristics of split injection strategy DI-diesel engine fueled with biodiesel. *Fuel* 90:1884–1891.

Ranalli, M., S. Schmidt, and L. Watts. 2004. NO_x-particulate filter (NPF): Evaluation of an after-treatment concept to meet future diesel emission standards. *SAE Technical Paper* 2004-01-0577.

Reitmayer, C. M., J. M. W. Ryalls, and E. Farthing, et al. 2019. Acute exposure to diesel exhaust induces central nervous system stress and altered learning and memory in honey bees. *Scientific Reports* 9:5793.

Sahoo, D., P. C. Miles, J. Trost, and A. Leipertz. 2013. The impact of fuel mass, injection pressure, ambient temperature, and swirl ratio on the mixture preparation of a pilot injection. *SAE International Journal of Engines* 6:1716–1730.

Sakurai, H., K. Park, and P. H. McMurry, et al. 2003. Size-dependent mixing characteristics of volatile and nonvolatile components in diesel exhaust aerosols. *Environmental Science & Technology* 37:5487–5495.

Salvat, O., P. Marez, and G. Belot. 2000. Passenger car serial application of a particulate filter system on a common rail direct injection diesel engine. *SAE Technical Paper* 2000-01-0473.

Siebers, D. L. and B. Higgins. 2001. Flame lift-off on direct-injection diesel sprays under quiescent conditions. *SAE Technical Paper* 2001-01-0530.

Silitonga, A. S., H. H. Masjuki, and T. M. I. Mahlia, et al. 2013. Overview properties of biodiesel diesel blends from edible and non-edible feedstock. *Renewable and Sustainable Energy Reviews* 22:346–360.

Silitonga, A. S., A. H. Shamsuddin, and T. M. I. Mahlia, et al. 2020. Biodiesel synthesis from *Ceiba pentandra* oil by microwave irradiation-assisted transesterification: ELM modeling and optimization. *Renewable Energy* 146:1278–1291.

Soudagar, M. E. M., N.-N. Nik-Ghazali, and M. A. Kalam, et al. 2018. The effect of nano-additives in diesel-biodiesel fuel blends: A comprehensive review on stability, engine performance and emission characteristics. *Energy Conversion and Management* 178:146–177.

Splitter, D., R. Hanson, S. Kokjohn, and R. D. Reitz. 2011. Reactivity controlled compression ignition (RCCI) heavy-duty engine operation at mid-and high-loads with conventional and alternative fuels. *SAE Technical Paper* 2011-01-0363.

Sun, J., J. A. Caton, and T. J. Jacobs. 2010. Oxides of nitrogen emissions from biodiesel-fuelled diesel engines. *Progress in Energy and Combustion Science* 36:677–695.

Tauzia, X., A. Maiboom, and S. R. Shah. 2010. Experimental study of inlet manifold water injection on combustion and emissions of an automotive direct injection diesel engine. *Energy* 35:3628–3639.

Tigelaar, J., K. Jaquet, D. Cox, and A. Peter. 2017. Utilization of turbocharger speed data to increase engine power and improve air path control strategy and diagnostics. *SAE Technical Paper* 201-01-1068.

Tow, T. C., D. A. Pierpont, and R. D. Reitz. 1994. Reducing particulate and NO_x emissions by using multiple injections in a heavy duty D.I. diesel engine. *SAE Technical Paper* 940897.

Tree, D. R. and K. I. Svensson. 2007. Soot processes in compression ignition engines. *Progress in Energy and Combustion Science* 33:272–309.

Tripathi, G., A. Dhar, and A. Sadiki. 2018. Recent advancements in after-treatment technology for internal combustion engines: An overview. In *Advances in Internal Combustion Engine Research*, ed. D. K. Srivastava, A. K. Agarwal, A. Datta and R. K. Maurya, 159–179. Singapore: Springer Singapore.

Ullman, T. L. 1989. Investigation of the effects of fuel composition on heavy-duty diesel engine emissions. *SAE Technical Paper* 982072.

Wang, X., F. Wu, J. Xiao, and S. Shuai. 2009. Oxygenated blend design and its effects on reducing diesel particulate emissions. *Fuel* 88:2037–2045.

Yip, H. L., I. M. R. Fattah, and A. C. Y. Yuen, et al. 2019. Flame-wall interaction effects on diesel post-injection combustion and soot formation processes. *Energy & Fuels* 33:7759–7769.

53 Particles from Compression and Spark Ignition Engines

Jianbing Gao
Ye Liu

CONTENTS

53.1 INTRODUCTION

'Internal combustion (IC) engines' (IC engines) powered vehicles, as one of the main sources of traffic related particles, are attracting much attention. The adverse impact on environmental and human health related to 'particulate matter' (PM) from IC engines have gained increasing attention in recent years (Khan and Gillies, 2019; Sgro et al., 2012). Nowadays, engine technologies for vehicles can progressively reduce the mass concentration of particles in diesel exhaust (Shi et al., 2000). However, particle number emissions are attracting much attention, and emission regulations limit them. Smaller particles are more easily absorbed into the respiratory system, where they do more harm to human health, especially when their diameter is smaller than 23 nm, which is beyond the scope of the Euro 6 emission regulations. So, it is necessary for IC engines to be equipped appropriately with after-treatment devices to reduce effectively 'diesel exhaust particles' (DEPs) (Khobragade et al., 2019). Meanwhile, the physical-chemical properties (including chemical compositions, particle diameters, and nanostructures) of particles are closely related to the design of the after-treatment devices, a 'gasoline particulate filter' (GPF), and their regeneration.

Due to the differences of in-cylinder combustion and particle formation mechanisms between 'compression ignition (CI) engines' (CI engines) and 'spark ignition (SI) engines' (SI engines), particles from these two types of engines are significantly different. This chapter introduces the physical properties of particles emitted by CI engines and SI engines individually. Particles from CI engines are introduced firstly, followed by particles from SI engines.

53.2 PARTICLES FROM CI ENGINES

In this section, the particle number and size distribution, chemical composition, and particle control approaches are introduced.

53.2.1 Number and Size Distribution of Particles from CI Engines

Current diesel engine emission regulations are based on a gravimetric method for PM measurement; the particle number in diesel exhaust is controlled in emission regulations. Figure 53.1 shows the representative size distribution, both in number and mass weightings, of diesel engines. The distributions shown are trimodal and lognormal in form. The concentration of particles in any size range is proportional to the area under the corresponding curve in that range. In general, there are three typical particle modes of diesel engines (Kittelson 1998): nucleation, accumulation, and coarse modes.

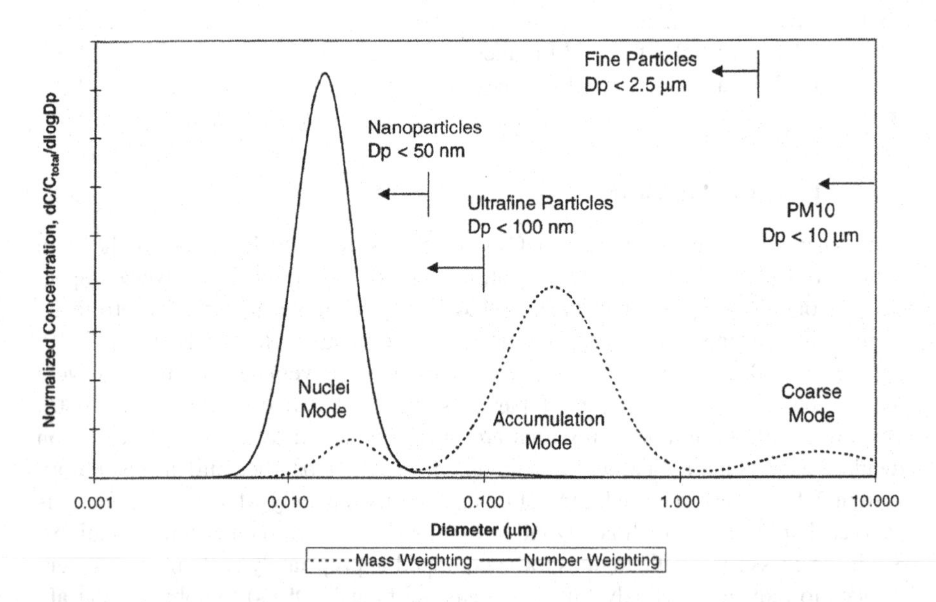

FIGURE 53.1 Typical size distribution of both mass and number weightings of PM.

Source: Reprinted from Kittelson (1998).

Nucleation mode particles typically range in diameter from 0.005 to 0.05 nm and typically consist of volatile organic and sulfur compounds that form during exhaust dilution and cooling, and may also contain solid carbon and metal compounds. For diesel engine aerosols, the nuclei mode typically contains 1–20% of the particle mass and more than 90% of the particle number. The number of nucleation mode particles below 50 nm continuously increases with the biodiesel blend ratios increasing from 0 to 100% (Tan et al., 2009). This phenomenon is ascribed to the following factors (Yamane et al., 2001):

1. Compared to diesel fuel, the evaporation and air mixing in the combustion chamber deteriorate due to the increase in viscosity and lower volatility of biodiesel, which may result in an increase in 'volatile organic compounds' (VOCs).
2. Biodiesel reduces the number of accumulation mode particles, thus providing less soot surface that can be used for condensation and adsorption of volatile compounds. This situation may lead to the formation of more new nuclei particles.
3. The increase in oxygen content of biodiesel fuel causes carbonaceous particles to change from a fine size to an ultrafine size or even a nanoparticle size, which in turn results in an increase of nucleation mode particles.

The accumulation mode ranges in size from roughly 0.05 to 1.00 μm. Most of the particle mass, primarily carbonaceous agglomerates and adsorbed materials, is found here. The number of accumulation mode particles decreases as the biodiesel blend ratio increases (Jung et al., 2006; Ristovski et al., 2006; Tsolakis, 2006). The oxygen content of biodiesel improves combustion in fuel-rich diffusion flame regions in the combustion chamber, and promotes the oxidation of the already formed soot (Rakopoulos et al., 2008). In addition, a less aromatic hydrocarbon of biodiesel fuel also reduces soot precursors (Tan et al., 2009). The above-mentioned factors can well interpret the phenomenon of the reduction in accumulation mode particles.

The coarse mode consists of particles larger than 1 μm and contains 5–20% of the particle mass. It consists of accumulation mode particles that have been deposited on cylinder and exhaust system surfaces and later reentrained. Some definitions of size for atmospheric particles are also observed in Figure 53.1: PM_{10}, D (diameter) < 10 μm; fine particles, $D < 2.5$ μm; ultrafine particles, $D < 0.10$ μm; and nanoparticles, $D < 0.05$ μm or 50 nm. PM_{10} and fine particles are actually defined by standard sampling systems in which the sampling probability falls to 50% at the designated aerodynamic diameter. Exact definitions of ultrafine and nanoparticles have not been agreed upon.

It is worthwhile noting that particle size significantly influences the behavior of particles within the engine itself and in the environment (Kittelson, 1998). The difference in size distribution of PM would dramatically affect the performance of after-treatment devices of diesel engines (Khobragade et al., 2019). In general, a 'diesel particulate filter' (DPF) and 'diesel oxidation catalyst' (DOC) are used to reduce diesel particles. DPF is most effective at removing solid accumulation mode particles, while DOC removes mainly organic compounds including the volatile organic fraction and nuclei mode particle.

The size of engine exhaust particles influences the environment in several ways: the atmospheric residence time of the particles, the optical properties of the particles, the particle surface area, and the ability to participate in atmospheric chemistry. PM ranging from 0.1 to 10.0 μm possesses the longest residence time of about one week at atmosphere (Wei et al., 2001). Larger particles are easier to remove from the atmosphere by diffusion and coagulation with smaller ones. A typical residence time for 10 nm particles is only about 15 min. The main mechanism for removal of these nanoparticles is coagulation with particles in the accumulation mode. Thus, although they lose their nature as individual particles, they remain in the atmosphere for the same time as the larger accumulation mode particles.

PM causes not only a change in atmospheric visibility but also the soiling of buildings. The extent of influence, however, depends upon particle size, shape, and composition (Jaiprakash, 2017). PM interact with light by absorption and scattering. For PM, absorption is much stronger than scattering and is relatively independent of particle size for light in the visible range. The absorption is due to the presence of the carbon content of PM. Light scattering is strongly dependent on particle size and shape and is typically maximum for particles that are a few tenths of a micron in diameter. The scattering is due mainly to particles in the accumulation mode particles. Ultrafine and nanoparticles scatter quite weakly. The specific surface area of PM is typically in the range of 100 m^2g^1, corresponding to the specific surface area of about a 0.03 μm carbon sphere (Kittelson, 1998; Ouf et al., 2019). This means that the surface area of PM is probably more a function of the size of the individual nuclei in the agglomerates rather than the agglomerate size. Meanwhile, the surface area is available for atmospheric reactions, which various constituents of the atmosphere will compete with diesel exhaust constituents for this surface area.

Currently, fine and ultrafine particles of PM receive the greatest attention since they harm human health (Zhang and Balasubramanian, 2014). The number of particles and particle surface area per unit mass increases with decreasing particle size. The efficiency of deposition in the human respiratory tract is dependent upon particle size. In particular, pulmonary deposition increases with decreasing particle size.

53.2.2 Chemical Composition of Particles from CI Engines

Complete combustion of a fuel containing hydrocarbons in diesel engines yields only CO_2 and H_2O as combustion products, while the small fraction of unburned fuel and lubricating oil in diesel engines yields a great number of PM that affect urban air quality and human health (Majewski and Khair, 2006; Ristovski et al., 2012). PM is not only a complex, multipollutant mixture of solid and liquid particles suspended in a gas but also a very dynamic physical and chemical system that exhibits very strong spatial and temporal dependency in terms of its composition (Eastwood, 2008; Wilt, 2007). The formation of PM depends on many factors including the engine operating conditions (e.g. speed/load, injection timing, and strategy), the application of after-treatment devices (such as a DPF), the maintenance status of the engine, and the type of fuel and lubricants used (Davies, 2002; Konstandopoulos et al., 2007). PM are mainly composed of non-volatile (insoluble) and volatile (soluble) fractions. Volatile compounds are constituted by organic carbon, sulfate, and nitrate compounds while the non-volatile fraction consists of carbonaceous (soot) fraction and ash content (Khobragade et al., 2019).

Figures 53.2 and 53.3 present typical compositions of PM. A diesel particle consists of many primary carbonaceous particles that agglomerate together to produce a complex, fractal-like morphology (Eastwood, 2008; Ristovski et al., 2012). The carbonaceous component of PM provides a surface for other compounds such as organic compounds, sulfates, and metal oxides to adsorb or condense on. The organic compounds that present in PM are derived from heavy hydrocarbons with a high boiling point, originating from unburnt fuel and lubricating oil. Whether the organic component of PM is the gas phase or particle phases primarily depends upon the level of dilution and cooling employed during diesel particle sampling (Robinson et al., 2007). Meanwhile, these are considered as imparting toxicity, including potential carcinogenicity. A sulfate like sulfuric acid (H_2SO_4) and the ammonium sulfate ($(NH_3)_2SO_4)$) component of PM originates from sulfur present in the fuel and lubricating oil. Metallic ash like zinc oxide (ZnO) and iron oxide (Fe_2O_3) can also adsorb to the diesel particle surface, with lubricating oil providing a metallic source during combustion (Mayer et al., 2010). An exhaust sample also includes secondary particles that form during the sampling process in the particle sampling system. Some losses of primary particles may occur during the particulate collection.

FIGURE 53.2 The physicochemical composition of PM.

Source: Reprinted from Maricq (2007).

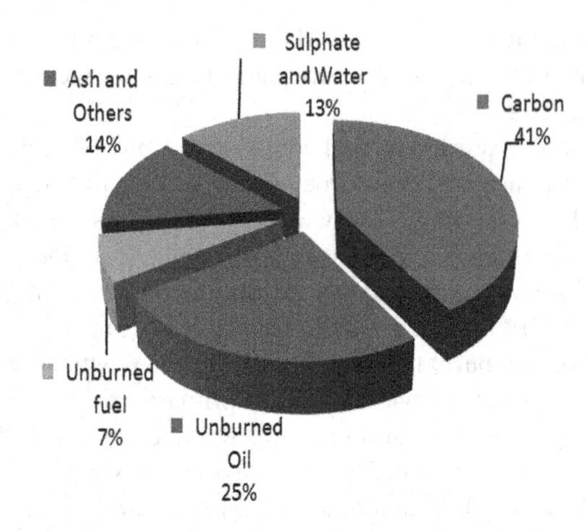

FIGURE 53.3 Typical particle composition for a heavy-duty diesel engine tested in a heavy-duty transient cycle.

Source: Reprinted from Kittelson (1998).

53.2.3 Methods for CI Particle Control

With the increasingly rigorous emission control regulations, efficient after-treatment technologies are essential measures to reduce effectively DEPs, especially with the aging of diesel engines. DOC and DPF are generally used to deal with diesel exhaust emissions. The DOC and DPF assembly is recognized as the best after-exhaust treatment option and a schematic diagram of their assembly, shown in Figure 53.4, makes an important contribution to diesel exhaust emission control. It is an open monolith non-filter system that utilizes a catalytic reaction process to oxidize pollutants in diesel exhaust streams, turning them into less harmful components (Cooper and Roth, 1991). DOC is generally constituted by the assembly of the following

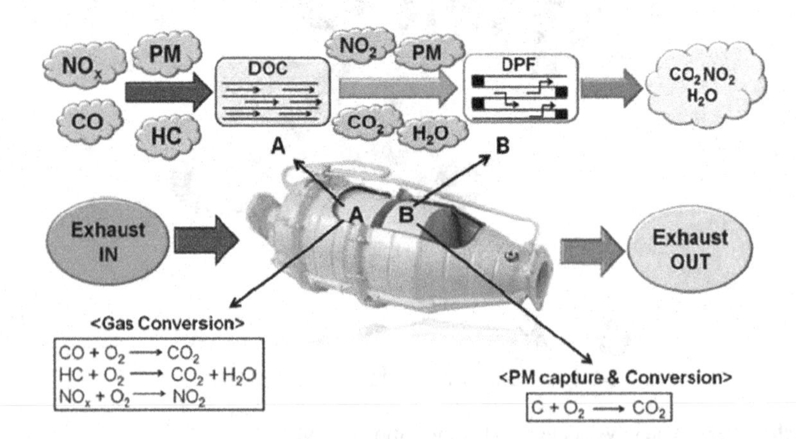

FIGURE 53.4 Schematic diagram of DOC and DPF assembly.

Source: Reprinted from Khobragade et al. (2019).

components: substrate (catalyst support), alumina wash coat (about 20–40 μm thick), and impregnated catalyst followed by canning. Cordierite honeycomb monolith and corrugated metallic types of structures are the most commonly used substrates (Labhsetwar et al., 2006). Pt/Pd is commercially the most popular oxidation catalyst used in DOCs (Prasad and Bella, 2010; Stratakis, 2004).

Diesel particulate emissions are accompanied by CO, NO_x, and unburned hydrocarbons (UHCs) in gas streams with a high concentration of O_2, CO_2, and H_2O. The emission temperature is typically below 500–550°C, and this situation is well suitable for the oxidation of CO and HC but not for NO_x and diesel particles (Hernandez-Gimenez et al., 2014). Therefore, DPF is used mainly to physically capture diesel particles from the exhaust stream in order to prevent diesel particles entering the atmosphere, which shows significant filtration efficiencies of more than 90%, as well as good mechanical and thermal durability. As shown in Figure 53.4, the filter is a monolith-based honeycomb structure that is partially blocked at each entry and exit end, providing entrapment for diesel particles. The porous walls of the filter allow gas to flow through, capturing diesel particles, depending upon the porosity of material used in DPF (Twigg, 2007), while DPF may have limited influence in controlling the nonsolid fraction of diesel particulate emissions – the volatile organic fraction and sulfate particles. These volatile organic fraction emissions can be minimized or removed by typical oxidation catalysts, while sulfate particulates can be minimized or avoided by use of ultra-low sulfur fuels (Song et al., 2002).

DPF accumulate a large amount of soot because soot particles have a low density (~0.1 g/cm³). Removal of these particulates, called filter regeneration, is necessary since filter clogging leads to a high exhaust gas pressure drop in the filter that negatively affects engine operation. Filter regeneration can be done either periodically, after a predetermined quantity of diesel particles has been accumulated, or continuously, during regular operation of the filter. The filters where the exhaust stream itself

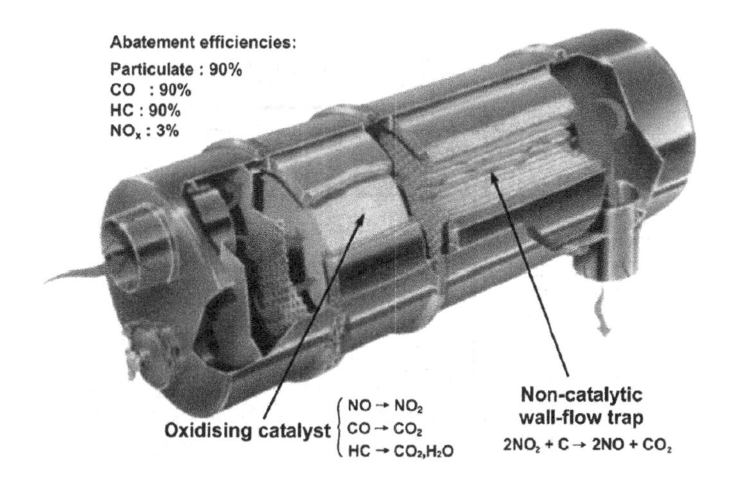

FIGURE 53.5 The continuously regenerable trap system.

Source: Reprinted from Fino and Specchia (2008).

provides the temperature required for particulate oxidation at the sufficient rate are called passive filters or continuous regenerating traps, as shown in Figure 53.5.

Alternatively, there are filters that need active strategies for increasing the filter temperature. These active strategies aim at increasing the exhaust temperature by either late cycle injection of additional fuel quantities which is an in-cylinder engine management method, or injection and combustion of fuel in the exhaust gas. In addition, electrical heating is also used in many vehicle configurations. Such heaters can be placed either upstream of the filter substrate or incorporated into the filters or else electrically conductive media can be used, which can act as both the filter and the heater (Kim et al., 2010). The third category is the combination of passive and active regeneration. Figure 53.6 summarizes the current diesel filter regeneration strategies.

53.3 PARTICLES FROM SI ENGINES

SI engines are classified as 'port fuel injection (PFI) engines' (PFI engines) and 'direct fuel injection (DI) engines' (DI engines) based on the fuel injection methods (Saliba et al., 2017). Particle emission is less for PFI engines than CI engines, due to the preformation of a stoichiometric air/fuel mixture in the intake manifolds. The in-cylinder charging efficiency of SI engines is low under part-load conditions due to the serious throttle loss (Wang et al., 2014b). Compared to PFI engines, DI engines have higher charging efficiency since fuel is directly injected into the cylinders, which leads to more fresh air in them. In the meantime, it has a cooling effect on the in-cylinder air, which contributes to a decrease in NO_x emissions. However, more particles are formed in cylinders due to the existence of the rich combustion regions. This subsection is focused on particle number and mass distributions and nanostructures.

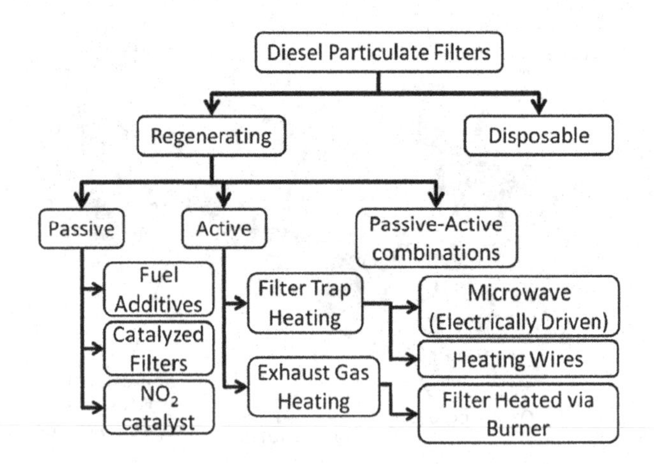

FIGURE 53.6 Possible routes for a controlled regeneration of a DPF.

Source: Reprinted from Khobragade et al. (2019).

53.3.1 Particles from PFI Engines

As for PFI engines, the stoichiometric air/fuel mixture is sucked into the cylinders. Meanwhile, the quantity of the air/fuel mixture changes with engine operation conditions significantly, e.g. engine torque and rotation speed (Amann et al., 2011). Figure 53.7 shows particle number distributions as the series of particle diameters over various engine torques and speeds. Peak values of the particle concentration are in the range of 4–36 × 10^6 particles/cm^3 for the given scenarios (3,500 rpm and 0–40 Nm in Figure 53.7a. Particle number concentration is low under small engine torque situations, diameters corresponding to the peak particle number concentrations are shifted to big values as well. This means that more nucleation mode particles are transferred to accumulative mode over high engine load conditions. Due to the stoichiometric air/fuel mixture in the cylinders, VOCs account for a high percentage of particles from PFI engines (Gupta et al., 2010). Particles from IC engines significantly depend on the concentrations of hydrocarbon, which will condense on soot in the liquid

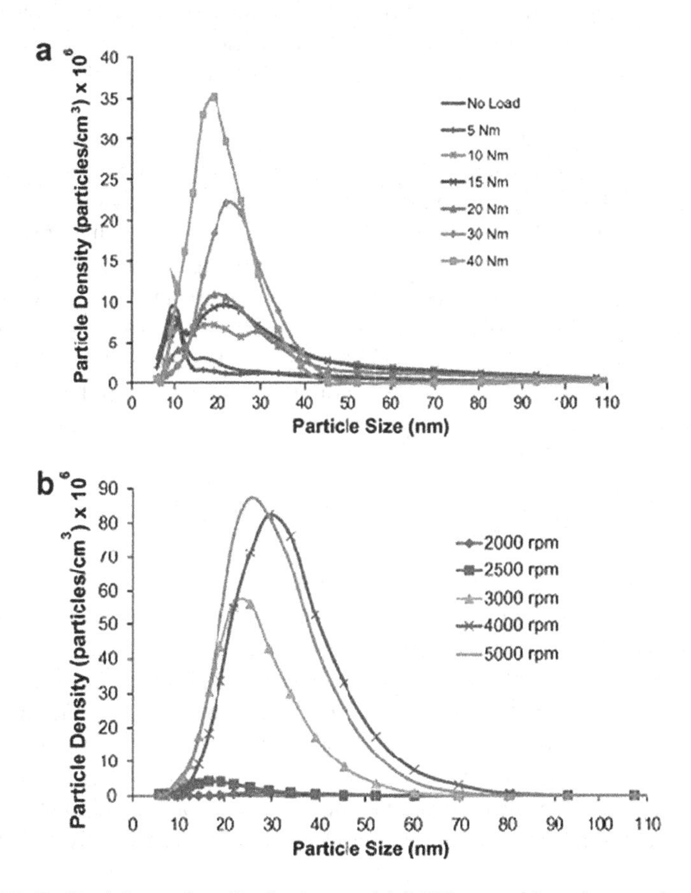

FIGURE 53.7 Particle number distributions at (a) 3,500 rpm with various engine loads and (b) a 50% rated load with speed variations.

Source: Reprinted from Gupta et al. (2010).

phase. Figure 53.7b presents the particle number distributions over a 50% rated load with speed variations. Particle numbers are quite low for a 2,000 rpm condition. The peak number concentration reaches 88 × 10^6 particles/cm^3, corresponding to a particle diameter around 30 nm under 5,000 rpm. Particle emissions both in mass and number significantly depend on the engine types which vary significantly, even when they meet the same emission levels, as shown in the work (Barrios et al., 2014a–b).

Biofuels have been widely used for vehicles due to their reproducible character and excellent combustion performance, such as methanol which is commonly used as a blend with gasoline. Figure 53.8 shows particle number concentrations under various engine operation conditions when fueled with methanol/gasoline blends. Particle number distributions show three peaks, with diameters around 6 nm, 30 nm, and 110 nm correspondingly. For this type of engine, particle number distribution changes slightly with engine torque over the given engine speed condition (2,000 rpm). However, it changes significantly with the methanol blends.

Particle surface area is related to the adsorption of organic compounds and particle oxidation activities, which influences particle filter regeneration (Gao et al., 2018, 2019a–b). Figure 53.9 shows the particle surface area and volume density distributions under different engine operation scenarios. The peak value of the surface area density is in the range of 6–46 × 10^9 nm^2/cm^3, with their corresponding diameters of 25–80 nm when the engine load changes from 0 to 40 Nm over 3,500 rpm. The surface area density is lower than 7 × 10^9 nm^2/cm^3 when the particle diameter is bigger than 150 nm, and it increases significantly with the engine speed under a 50% rated load of their corresponding engine speed conditions. The peak value of surface area density reaches 275 × 10^9 nm^2/cm^3, and its corresponding diameter is approximately 45 nm. By comparison, engine speed shows more significant effect on particle surface area density than engine load. Particle volume density distribution profiles show double peaks and that the corresponding diameter of the first peak is in the range of 20–110 nm, and more than 500 nm for that of the second peak. Similar to the particle surface area, particle volume density distribution is more sensitive to engine speed for diameters smaller than 70 nm.

53.3.2 Particles from DI Engines

Particle number and mass distributions under various fuel injection pressures and an 'indicated mean effective pressure' (IMEP) are shown in Figure 53.10. As for the particle number distributions, the profiles show a single peak under 4.5 and 6.5 bar IMEP conditions, where the distributions are slightly dependent on the fuel injection pressures (under given values of 50–172 bar). The diameter of the nucleation mode particles decreases with increasing engine load and drops from 20 to 10 nm when the IMEP decreases from 6.5 to 4.5 bar. The distributions show double peaks if the IMEP is further increased to 8.5 bar, meantime the diameters correspond to 15 and 70 nm for the nucleation mode and accumulation mode particles respectively. For an 8.5 bar

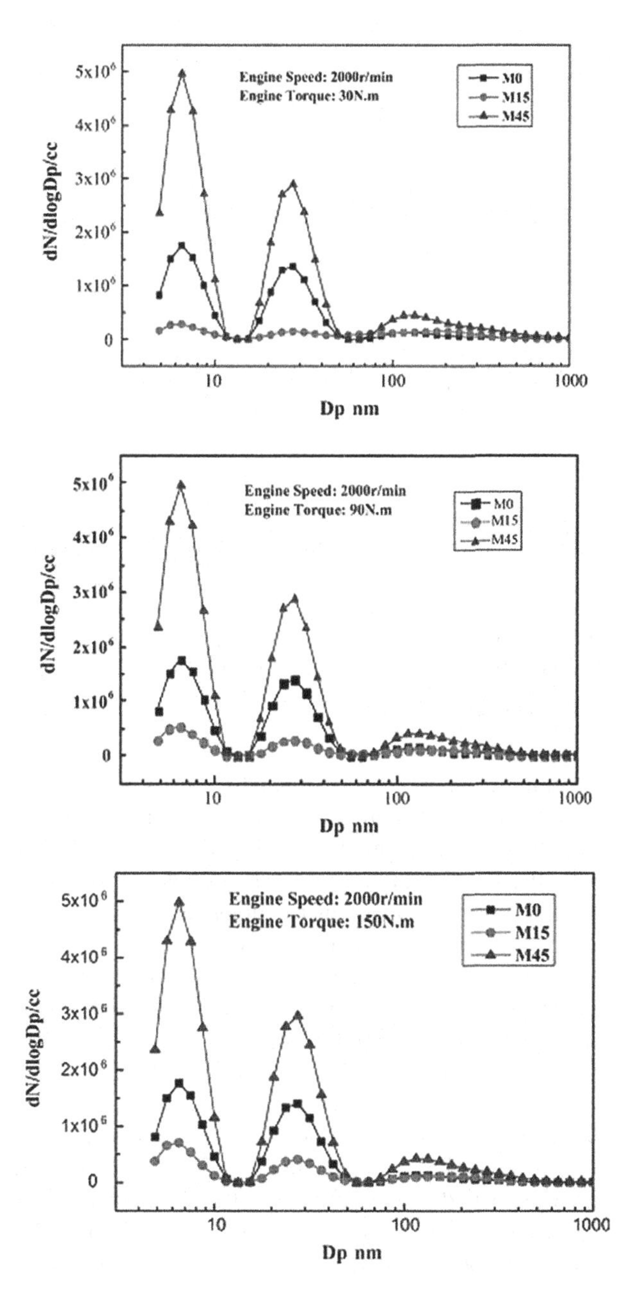

FIGURE 53.8 Effects of methanol on particulate number concentration and size distributions under various engine loads.

Source: Reprinted from Geng et al. (2015).

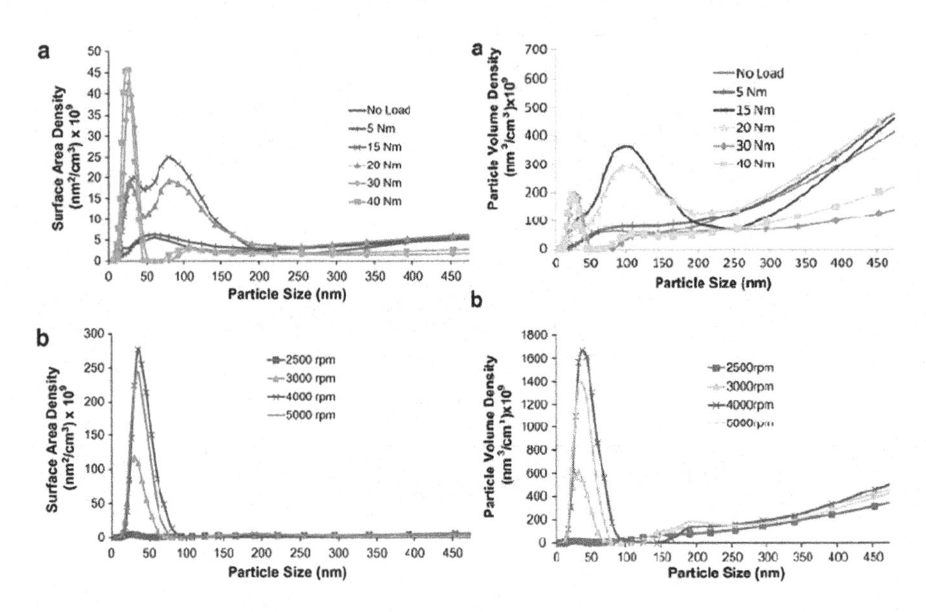

FIGURE 53.9 Particle surface area and volume density distribution at (a) 3,500 rpm with varying engine load and (b) a 50% rated load with varying speed.

Source: Reprinted from Gupta et al. (2010).

IMEP scenario, fuel injection pressure significantly affects particle number distributions such that a high injection pressure causes a high nucleation mode particle concentration and a low accumulation mode particle concentration. A scenario of a 6.5 bar IMEP presents the highest nucleation mode particle concentration among the three load situations. For particle mass distributions, they show a single peak with a diameter around 200 nm. Fuel injection pressures show a small impact on the distributions when diameters are smaller than 80, 50, and 40 nm for 4.5, 6.5, and 8.5 bar IMEP situations, respectively. Low fuel injection pressures lead to a high mass concentration over their peak value positions.

Figure 53.11 shows particle mass and number emission factors for different types of SI engines and fuels. Particles emitted by gasoline DI engines are much higher than PFI engines due to the different combustion mechanisms. The emission factors of particle mass and number reach 14 mg/km and 4.7×10^{12}/km, respectively, and it is almost the same level with diesel engines, such that the Euro 6 emission regulation limits particle mass and particle number emissions for DI engines. The addition of methanol effectively decreases particle emissions both in mass and numbers for DI and PFI engines.

Both particle number and mass emissions under different cold conditions for DI and PFI engines are presented in Figure 53.12. Cold conditions not only increase CO and HC emissions, but also cause more particle emissions due to the more serious condensation of VOC. The particle emission factor is almost four times higher for DI engine fueling with gasoline than that of E10 fuel, and it is almost 50 times for PFI engines. The adoption of E10 significantly decreases particle emissions under warm conditions for DI engines; however, it changes slightly for PFI engines.

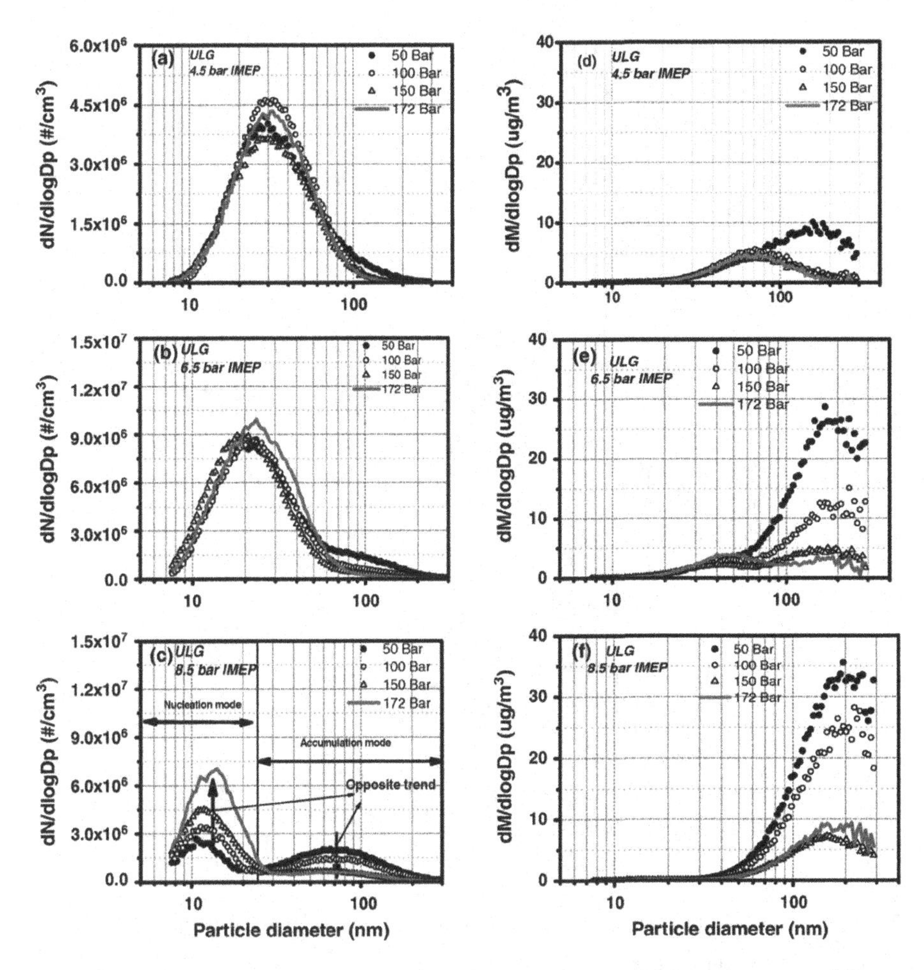

FIGURE 53.10 Particle number and mass distributions of DI engine over different conditions.

Source: Reprinted from Wang et al. (2014a).

Nanostructures of particles emitted by DI and PFI engines are shown in Figure 53.13. Particles show branch-like shapes under low resolution images (100 nm scale). Particles from DI engines are more seriously overlapped than those from PFI engines. More void cores are shown in PFI engine particles than DI engines. Crystallites are more densely and more orderly arranged for DI engine particles. The primary diameter is in the range of 15 to 125 nm for DI engine particles; however, it is 15–115 nm for PFI engine particles. Meanwhile, the primary diameter corresponding to peak number values is 85 and 40 nm respectively for particles emitted by DI and PFI engines.

53.4 CONCLUSIONS

In the first section, the mass and number distributions, chemical compositions, and control methods of particles emitted by CI engines were introduced; in the

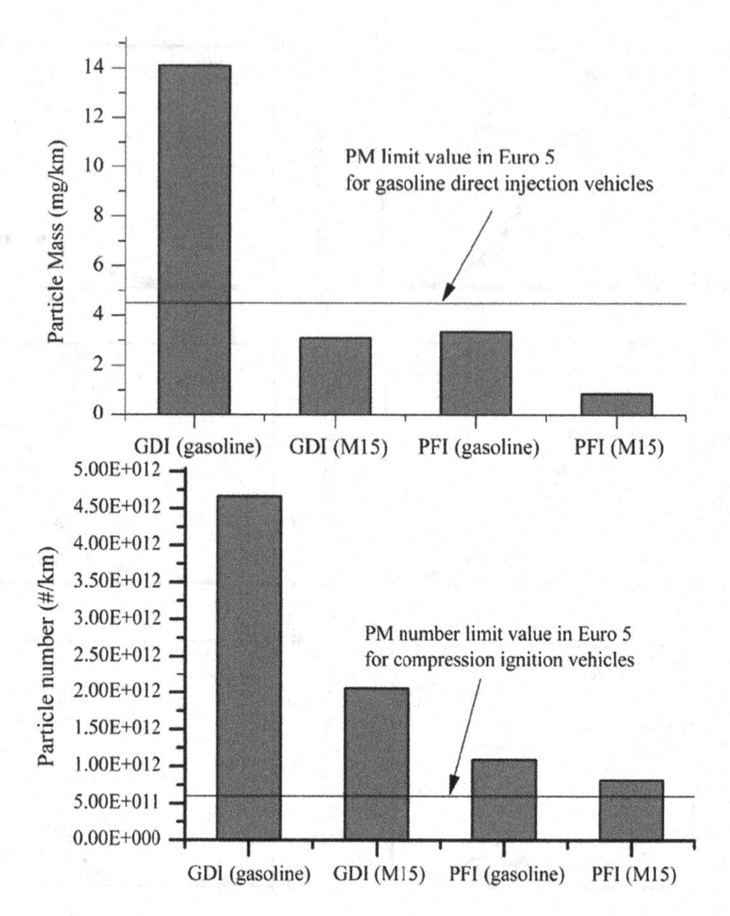

FIGURE 53.11 Particle mass and number emissions for both DI and FPI engines.

Source: Reprinted from Liang et al. (2013).

second section, the physical properties of particles from PFI and DI engines were presented.

The main conclusions are based on the categories of the engines.

53.4.1 Particles from CI Engines

There are three typical particle modes of diesel engines: nucleation, accumulation, and coarse modes. Nucleation mode particles typically range in diameter from 0.005 to 0.05 μm and consist of VOCs, sulfate, metallic, and carbonaceous compounds. The accumulation mode ranges in size from roughly 0.05 to 1.00 μm. Most of the particle mass, primarily carbonaceous agglomerates, and adsorbed materials are found here. For diesel engine aerosols, the nuclei mode typically contains 1–20% of the particle mass and more than 90% of the particle numbers. The coarse mode particles consist of particles larger than 1 μm and contain 5–20 % of the particle mass.

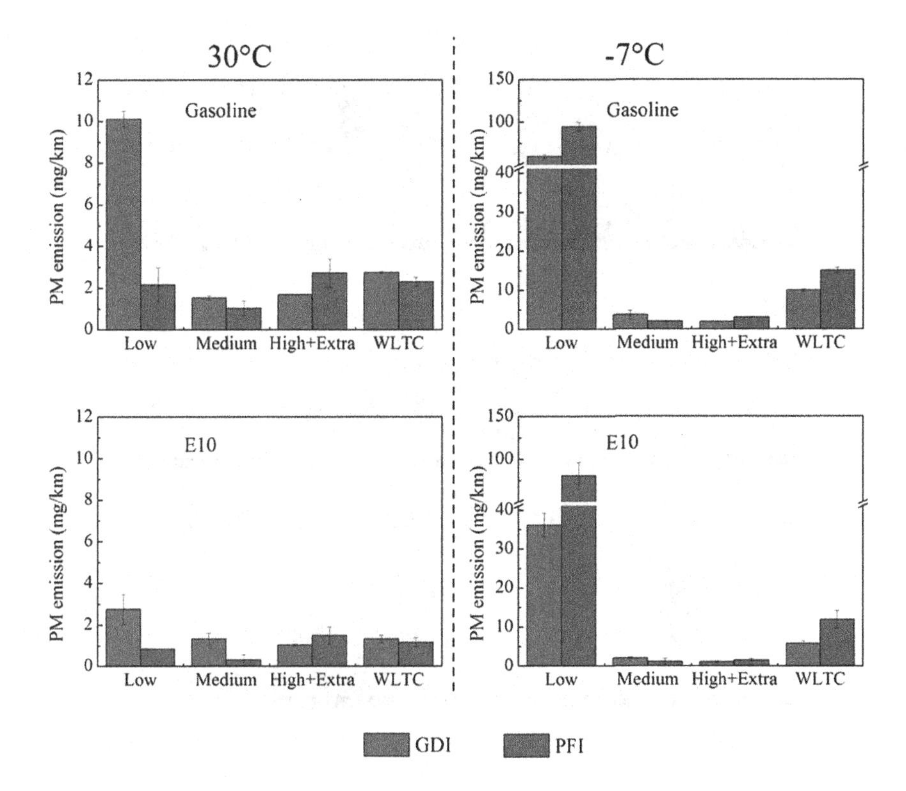

FIGURE 53.12 PM emission factors of DI and PFI vehicles fueling with gasoline and E10 during each phase of the 'world-wide harmonized light duty test cycle' (WLTC) in cold-start tests.

Source: Reprinted from Zhu et al. (2016).

PM is mainly composed of non-volatile (insoluble) and volatile (soluble) fractions. Volatile compounds are constituted by organic carbon, sulfate, and nitrate compounds, while the non-volatile fraction consists of a carbonaceous (soot) fraction and ash content.

In order to minimize or remove diesel exhaust emissions, the after-treatment devices are essential measures. DOC makes an important contribution to diesel exhaust emission control. It is an open monolith non-filter system that utilizes a catalytic reaction process to oxidize pollutants in the diesel exhaust stream, turning them into less harmful components. On the other hand, DPF is used mainly to physically capture diesel particles from the exhaust stream in order to prevent diesel particles entering the atmosphere.

53.4.2 Particles from SI Engines

The particle number concentration, mass, surface area density, and volume density distributions of SI (PFI and DI) engines present multi-peaks, and the diameters

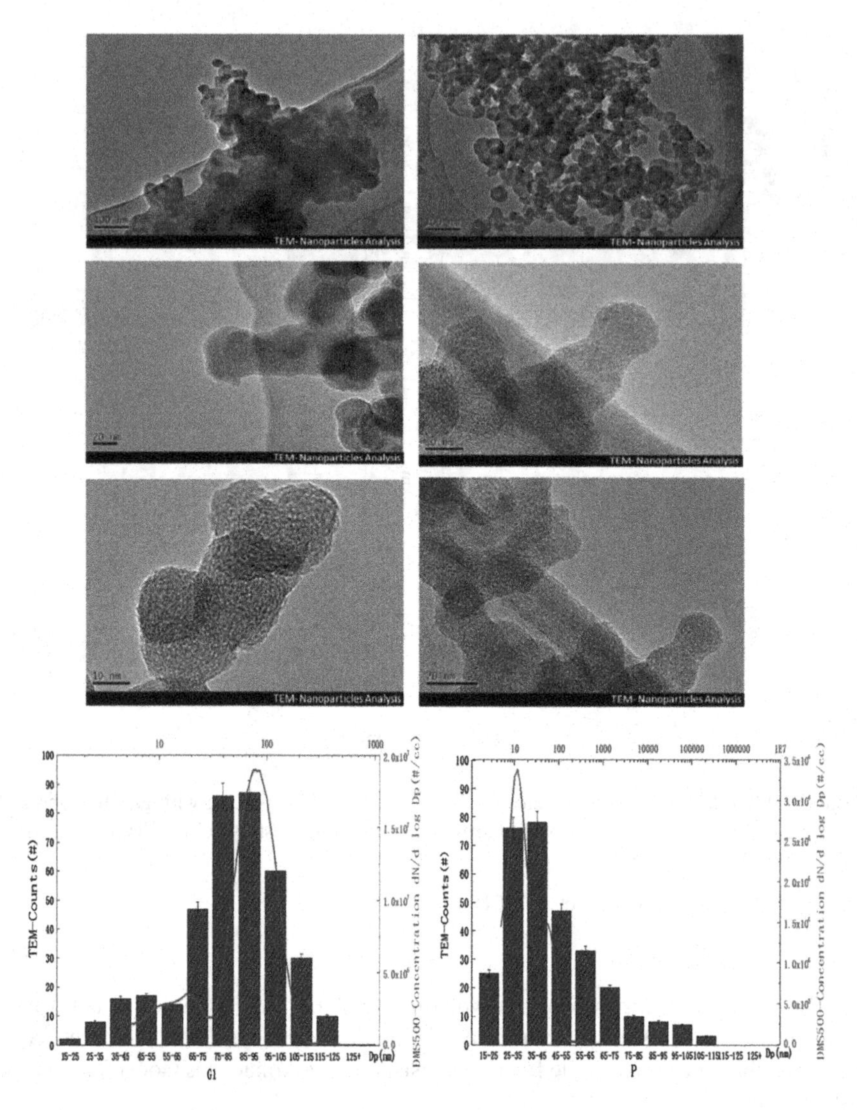

FIGURE 53.13 Upper image: The aggregate particles in TEM images of DI vehicle (left) and PFI vehicle (right) in three different scales (10, 20, and 100 nm).

Source: Reprinted from Chen et al. (2017).

corresponding to the peak values vary with engine types, fuel types, fuel injection pressure, engine torque, engine speed, and so on.

Particles from SI engines show chain-like structures and the crystallites are orderly and densely arranged; meanwhile, the particles from PFI engines are more seriously overlapped than that from DI engines.

Biofuels contribute to the decrease in particle number and mass emissions from SI engines; additionally, the cold start conditions present a significant effect on particle emissions, especially for DI engines.

REFERENCES

Amann, M., D. Mehta, and T. Alger. 2011. Engine operating condition and gasoline fuel composition effects on low-speed pre-ignition in high-performance spark ignited gasoline engines. *SAE International Journal of Engines* 4:274–285.

Barrios, C. C., A. Dominguez-Saez, C. Martin, and P. Alvarez. 2014a. Effects of animal fat based biodiesel on a TDI diesel engine performance, combustion characteristics and particle number and size distribution emissions. *Fuel* 117:618–623.

Barrios, C. C., C. Martin, and A. Dominguez-Saez, et al. 2014b. Effects of the addition of oxygenated fuels as additives on combustion characteristics and particle number and size distribution emissions of a TDI diesel engine. *Fuel* 132:93–100.

Chen, L., Z. Liang, X. Zhang, and S. Shuai. 2017. Characterizing particulate matter emissions from GDI and PFI vehicles under transient and cold start conditions. *Fuel* 189:131–140.

Cooper, B. J. and S. A. Roth. 1991. Flow-through catalysts for diesel engine emissions control. *Platinum Metals Review* 35:178.

Davies, B. 2002. Diesel particulate control strategies at some Australian underground coal mines. *AIHAJ-American Industrial Hygiene Association* 63:554–558.

Eastwood, P. 2008. *Particulate Emissions from Vehicles*. Chichester: John Wiley & Sons.

Fino, D. and V. Specchia. 2008. Open issues in oxidative catalysis for diesel particulate abatement. *Powder Technology* 180:64–73

Gao, J., C. Ma, S. Xing, L. Sun, and L. Huang. 2018. A review of fundamental factors affecting diesel PM oxidation behaviors. *Science China Technological Sciences* 61:330–345.

Gao, J., H. Chen, and J. Chen, et al. 2019a. Explorations on the continuous oxidation kinetics of diesel PM from heavy-duty vehicles using a single ramp rate method. *Fuel* 248, 254–257.

Gao, J., H. Chen, G. Tian, C. Ma, and F. Zhu. 2019b. Oxidation kinetic analysis of diesel particulate matter using single-and multistage methods. *Energy & Fuels* 33:6809–6816.

Geng, P., H. Zhang, and S. Yang. 2015. Experimental investigation on the combustion and particulate matter (PM) emissions from a port-fuel injection (PFI) gasoline engine fueled with methanol–ultralow sulfur gasoline blends. *Fuel* 145:221–227.

Gupta, T., A. Kothari, D. K. Srivastava, and A. K. Agarwal. 2010. Measurement of number and size distribution of particles emitted from a mid-sized transportation multipoint port fuel injection gasoline engine. *Fuel* 89:2230–2233.

Jaiprakash, G. H. 2017. Chemical and optical properties of $PM_{2.5}$ from on-road operation of light duty vehicles in Delhi city. *Science of the Total Environment* 586:900–916.

Hernandez-Gimenez, A. M., D. L. Castello, and A. Bueno-Lopez. 2014. Diesel soot combustion catalysts: Review of active phases. *Chemical Papers* 68:1154–1168.

Jung, H., D. B. Kittelson, and M. R. Zachariah. 2006. Characteristics of SME biodiesel-fueled diesel particle emissions and the kinetics of oxidation. *Environmental Science & Technology* 40:4949–4955.

Khan, M. U. and S. Gillies. 2019. Diesel particulate matter: Monitoring and control improves safety and air quality. In *Advances in Productive, Safe, and Responsible Coal Mining*, ed. J. Hirschi, 199–213. Cambridge: Woodhead Publishing.

Khobragade, R., S. K. Singh, and P. C. Shukla, et al. 2019. Chemical composition of diesel particulate matter and its control *Catalysis Reviews: Science and Engineering* 61:447–515.

Kim, H. J., B. Han, and W. S. Hong, et al., 2010. Development of electrostatic diesel particulate matter filtration systems combined with a metallic flow-through filter and electrostatic methods. *International Journal of Automotive Technology* 11:447–453.

Kittelson, D. B. 1998. Engines and nanoparticles: A review. *Journal of Aerosol Science* 29:575–588.

Konstandopoulos, A., D. Zarvalis, and I. Dolios. 2007. Multi-instrumental assessment of diesel particulate filters. *SAE Technical Paper* 2007-01-0313.

Labhsetwar, N., R. B. Biniwale, R. Kumar, S. Rayalu, and S. Devotta. 2006. Application of supported perovskite-type catalysts for vehicular emission control. *Catalysis Surveys from Asia* 10:55–64.

Liang, B., Y. Ge, and J. Tan, et al. 2013. Comparison of PM emissions from a gasoline direct injected (GDI) vehicle and a port fuel injected (PFI) vehicle measured by electrical low pressure impactor (ELPI) with two fuels: Gasoline and M15 methanol gasoline. *Journal of Aerosol Science* 57:22–31.

Majewski, W. A. and M. K. Khair. 2006. *Diesel Emissions and their Control.* Warrendale, PA: SAE International.

Maricq, M. M. 2007. Chemical characterization of particulate emissions from diesel engines: A review. *Journal of Aerosol Science* 38:1079–1118.

Mayer, A. C., A. Ulrich, J. Czerwinski, and J. J. Mooney. 2010: Metal-oxide particles in combustion engine exhaust. *SAE Technical Paper* 2010-01-0792.

Ouf, F.-X., S. Bourrous, C. Vallieres, J. Yon, and L. Lintis. 2019. Specific surface area of combustion emitted particles: Impact of primary particle diameter and organic content. *Journal of Aerosol Science* 137:105436.

Prasad, R. and V. R. Bella. 2010. A review on diesel soot emission, its effect and control. *Bulletin of Chemical Reaction Engineering and Catalysis* 5:69–86.

Rakopoulos, C. D., D. C. Rakopoulos, D. T. Hountalas, E. G. Giakoumis, and E. C. Andritsakis. 2008. Performance and emissions of bus engine using blends of diesel fuel with biodiesel of sunflower or cottonseed oils derived from Greek feedstock. *Fuel* 87:147–157.

Ristovski, Z. D., E. R. Jayaratne, M. Lim, G. A. Ayoko, and L. Morawska. 2006. Influence of diesel fuel sulfur on nanoparticle emissions from city buses. *Environmental Science & Technology* 40:1314–1320.

Ristovski, Z. D., B. Miljevic, and N. C. Surawski, et al. 2012. Respiratory health effects of diesel particulate matter. *Respirology* 17:201–212.

Robinson, A. L., N. M. Donahue, and M. K. Shrivastava, et al. 2007. Rethinking organic aerosols: Semivolatile emissions and photochemical aging. *Science* 315:1259–1262.

Saliba, G., R. Saleh, and Y. Zhao, et al. 2017. Comparison of gasoline direct-injection (GDI) and port fuel injection (PFI) vehicle emissions: Emission certification standards, cold-start, secondary organic aerosol formation potential, and potential climate impacts. *Environmental Science & Technology* 51:6542–6552.

Sgro, L. A., A. D'Anna, and P. Minutolo. 2012. On the characterization of nanoparticles emitted from combustion sources related to understanding their effects on health and climate. *Journal of Hazardous Materials* 211–212:420–426.

Shi, J. P., D. Mark, and R. M. Harrison. 2000. Characterization of particles from a current technology heavy-duty diesel engine. *Environmental Science & Technology* 34:748–755.

Song, J., M. Alam, and V. Zello, et al. 2002: Fuel sulfur effect on membrane coated diesel particulate filter. *SAE Technical Paper* 2002-01-2788.

Stratakis, G. A. 2004. *Experimental Investigation of Catalytic Soot Oxidation and Pressure Drop Characteristics in Wall-Flow Diesel Particulate Filters.* Thessaly: University of Thessaly.

Tan, P., Z. Hu, D. Lou, and B. Li. 2009. Particle number and size distribution from a diesel engine with Jatropha biodiesel fuel. *SAE Technical Paper* 2009-01-2726.

Tsolakis, A. 2006. Effects on particle size distribution from the diesel engine operating on RME-biodiesel with EGR. *Energy & Fuels* 20:1418–1424.

Twigg, M. V. 2007. Progress and future challenges in controlling automotive exhaust gas emissions. *Applied Catalysis B: Environmental* 70:2–15.

Wang, C., H. Xu, J. M. Herreros, J. Wang, and R. Cracknell. 2014a. Impact of fuel and injection system on particle emissions from a GDI engine. *Applied Energy* 132:178–191.

Wang, S., C. Ji, B. Zhang, and X. Liu. 2014b. Lean burn performance of a hydrogen-blended gasoline engine at the wide open throttle condition. *Applied Energy* 136:43–50.

Wei, Q., D. B. Kittelson, and W. F. Watts. 2001. Single-stage dilution tunnel performance. *SAE Technical Paper* 2001-01-0201.

Wilt, G. A. 2007. *Growth of Diesel Exhaust Particulate Matter in a Ventilated Mine Tunnel.* Morgantown, WV: West Virginia University.

Yamane, K., A. Ueta, and Y. Shimamoto. 2001. Influence of physical and chemical properties of biodiesel fuels on injection, combustion and exhaust emission characteristics in a direct injection compression ignition engine. *International Journal of Engine Research* 2:249–261.

Zhang, Z. H. and R. Balasubramanian. 2014. Physicochemical and toxicological characteristics of particulate matter emitted from a non-road diesel engine: Comparative evaluation of biodiesel-diesel and butanol-diesel blends. *Journal of Hazardous Materials* 264:395–402.

Zhu, R., J. Hu, and X. Bao, et al. 2016. Tailpipe emissions from gasoline direct injection (GDI) and port fuel injection (PFI) vehicles at both low and high ambient temperatures. *Environmental Pollution* 216:223–234.

54 Selective Catalytic Reduction of NO_x Emissions

R. J. G. Nuguid
F. Buttignol
A. Marberger
O. Kröcher

CONTENTS

54.1 INTRODUCTION

'Selective catalytic reduction' (SCR) is currently the most efficient technology to curb 'nitrogen oxide' (NO_x) emissions from diesel-powered vehicles and thermal power plants. First introduced in Japan in the 1970s (Ando et al., 1976; Takagi et al., 1977), SCR quickly gained global recognition as an integral part of stationary 'exhaust gas after-treatment'. In the late 1980s, the first experiments were conducted to transfer the SCR process to mobile applications as well, for which it has been established as the best option for NO_x abatement in engines that operate under excess oxygen.

As the term implies, SCR requires a reductant that will preferentially react with NO_x to form environmentally benign products. During the early stages of SCR development, several chemical substances were extensively investigated as potential reducing agents. The most studied ones are hydrocarbons, H_2, and NH_3. Hydrocarbon-SCR is particularly convenient because the fuel itself and/or the unburned hydrocarbons in the exhaust can serve as the reducing agent directly (Mrad et al., 2015).

However, the overall efficiency of the process was found to be low because most of the hydrocarbons are actually oxidized to CO_2 instead of selectively reacting with NO_x. Side reactions can also form HCN, an extremely poisonous gas, at concentrations above the threshold limit value (Radtke et al., 1995). H_2-SCR was proposed as a much greener alternative, but it requires a large excess of H_2 to achieve sufficient conversions and expensive Pt-based catalysts (Costa et al., 2007). Due to these serious limitations, hydrocarbon-SCR and H_2-SCR did not achieve widespread use. Indeed, most modern SCR installations use ammonia (NH_3) as a reducing agent for the conversion of NO_x. While this technology is more expensive to implement, its efficiency and selectivity make up for the added cost. To solve the problem of toxicity and storage of NH_3, urea was proposed as a safe NH_3 precursor compound. Under SCR-relevant conditions, urea easily decomposes *in situ* yielding two equivalents of NH_3. With the introduction of the SCR system in diesel vehicles, 32.5 wt% urea solution (trade name AdBlue®) became the most used NH_3 precursor worldwide (Bowers, 1988; Gabrielsson, 2004).

There are three main types of NH_3-SCR reactions that lead to the reduction of NO_x, depending on the NO/NO_2 ratio in the feed and the reaction temperature. Under normal conditions, NO accounts for nearly 90% of the total NO_x in the exhaust (Koebel et al., 2001), and the chemical equation corresponding to its reduction is known as the standard SCR reaction:

$$4NO + 4NH_3 + O_2 \rightarrow 4N_2 + 6H_2O \qquad (54.1)$$

It should be noted that the reaction also requires oxygen, which is available in excess in the exhaust gas of diesel engines. From the perspective of the active metal center, the SCR process can be thought of as a redox cycle. The first half involves the reduction of the metal ions in the presence of NO and NH_3 and in the second half the catalytic cycle is closed by oxygen that regenerates the active sites through reoxidation. The presence of NO_2 can accelerate the NO_x conversion through a different reaction pathway, known as the fast SCR reaction:

$$2NO + 2NO_2 + 4NH_3 \rightarrow 4N_2 + 6H_2O \qquad (54.2)$$

The rate enhancement is thought to occur from the faster reoxidation of the active sites in the second half of the catalytic cycle. This is because NO_2 is a much stronger oxidizing agent than O_2 (Koebel et al., 2002a–b). For this reason, diesel oxidation catalysts are usually installed upstream of the SCR system in order to partially oxidize NO to NO_2 and induce the fast SCR reaction. However, too much NO_2 in the gas feed is not beneficial for SCR. When NO_2 is present above the equimolar amount of NO, the SCR process is hampered through the occurrence of the following reaction:

$$6NO_2 + 8NH_3 \rightarrow 7N_2 + 12H_2O \qquad (54.3)$$

This equation is known as NO_2-SCR or slow SCR due to its sluggish kinetics (Kato et al., 1981).

Side reactions also compete with the main SCR reactions during non-optimal conditions. At high temperatures, NH_3 could be directly oxidized to N_2 (and to a lesser degree, N_2O and NO), thereby limiting the amount of reductant that can selectively react with NO_x (Madia et al., 2002). Some catalyst formulations are also susceptible to forming significant amounts of N_2O, which has approximately 300 times higher greenhouse warming potential than CO_2. In the presence of S(VI) gas in the exhaust, $(NH_4)_2SO_4$ can precipitate at low temperatures and cause catalyst fouling (Koebel et al., 2002a). Preventing the occurrence of these undesirable reactions is a major objective of ongoing studies.

The last four decades of SCR development yielded different classes of catalysts with diverse formulations. Vanadium-based catalysts were the first to be commercialized for SCR, and they have remained the preferred materials for stationary and marine applications (Amiridis et al., 1996). On the other hand, copper and iron-exchanged zeolites have emerged as a commercial solution for mobile SCR, for which a wider operating window is required. Although the composition of SCR catalysts can be completely different, they all have the following two key characteristics: (1) the presence of acidic sites that adsorb NH_3 from the gas phase; and (2) a redox site where the actual reaction between NH_3 and NO occurs.

The variety of SCR catalysts today reflects the diversity of diesel and gas engines used in the field. Indeed, engines for on-road applications (e.g. passenger car, trucks, or buses) and non-road applications (e.g. mining trucks, harvesters, trains, ships, or power-generating applications) differ in engine size, working speed, quality of fuel, and field of application (Nova and Tronconi, 2014). Ultimately, the final choice of the SCR catalyst to be used depends on the specific application.

In the following sections, the three major SCR catalysts (V-, Cu-, and Fe-based materials) will be described in detail, with emphasis on their molecular structure, key advantages and disadvantages, as well as reaction mechanisms.

54.2 VANADIUM-BASED SCR CATALYSTS

V-based catalysts typically consist of 1–3 wt% V_2O_5 as the redox-active species and TiO_2 in the anatase phase as the support. Commercial V_2O_5/TiO_2 catalysts are promoted with around 3–10 wt% WO_3 for improved activity and stability (Marberger et al., 2015).

Despite the V_2O_5 notation, vanadia is actually not present as a separate crystalline phase but as amorphous VO_x units due to the strong interaction with TiO_2. This has profound structural implications as vanadia exists as octahedral VO_6 units in bulk but as tetrahedral VO_4 units on the surface. Depending on the loading, VO_x can be present as a monomeric, dimeric, and/or polymeric species (Went et al., 1992). Only when the monolayer coverage of 7–8 VO_x units per nm^2 (Wachs, 1996) is reached, V_2O_5 domains are formed. The formal oxidation state of supported vanadium oxide is either V^{5+} or V^{4+} and the active phase cycles between both states during the reaction. VO_x sites can exhibit Lewis and Bronsted acidity, and the ratio between the two acid sites can vary depending on the reaction conditions. For instance, the presence of water can increase the proportion of Bronsted sites relative to Lewis sites (Ramis et al., 1990).

TiO_2 proved to be the best suited support for V-based SCR catalysts because of its advantageous chemical interaction with the active phase, resulting in a homogeneous dispersion of vanadium oxide and high SCR activity (Forzatti and Lietti, 1996). Furthermore, TiO_2 plays a role in substrate adsorption as it interacts strongly with NH_3 (Nuguid et al., 2019b). The anatase modification is used because it provides a high surface area and thermodynamic stability up to 700°C. Above this temperature, anatase irreversibly transforms into rutile (Marberger et al., 2019), which is undesirable, as the latter has a significantly lower surface area. To delay the onset of reutilization, WO_3 is added as a promoter that induces high levels of TiO_2 lattice distortions that limit sintering and particle growth (Cristiani et al., 1993). Another important effect of WO_3 is the increase of surface acidity (Yamaguchi et al., 1980), which guarantees the NH_3 supply for the SCR reaction. Through the addition of WO_3, the weak basic sites of TiO_2 disappear and the surface becomes much more acidic. The presence of WO_x species also tends to crowd VO_x sites together, forming oligomeric VO_x units that have a higher turnover frequency than isolated species (Jaegers et al., 2019). Additional promotional effects poison resistance to alkali metal and arsenious oxides, and lower NH_3 and SO_2 oxidation at elevated reaction temperature (Alemany et al., 1995; Chen and Yang, 1992; Ramis et al., 1992).

Since the 1980s, the reaction mechanism on V-based catalysts has been actively investigated. It is now generally accepted that the standard SCR reaction proceeds according to the stoichiometry of Equation 54.1. The first popular proposal was an Eley-Rideal type mechanism, where gaseous NO reacted with adsorbed NH_4^+ to form N_2 and H_2O (Inomata et al., 1980). The ammonium ion is thereby adsorbed on a V-OH Bronsted acid site and an adjacent V^{5+}=O species acts as the redox active site. The subsequent studies confirmed this reaction scheme with slight modifications and alterations (Janssen et al., 1987a–b).

In the early 1990s, the amide-nitrosamide mechanism was proposed, which involves a Lewis acid site for the adsorption of NH_3 instead of a Bronsted acid site (Busca et al., 1998; Ramis et al., 1990, 1996). This mechanism was more detailed especially because of the introduction of the nitrosamide (NH_2NO) intermediate, which readily decomposes into N_2 and H_2O.

A combination of the adjacent V species and the amide-nitrosamide mechanisms were also proposed (Topsoe, 1994, 1995). In this combined mechanism, NH_3 adsorbs on a Bronsted acid site and is activated by an adjacent V^{5+}=O site acting as the redox center.

Recent time-resolved studies have advanced the understanding of the SCR reaction over V_2O_5/TiO_2 at the molecular level. Through transient IR spectroscopy, Marberger et al. (2016) showed that the active sites for SCR are Lewis acid sites, on which NH_3 binds and react instantaneously with NO. The introduced NH_3 adsorbed on Lewis and Bronsted acid sites as NH_3 and NH_4^+, respectively. Upon NO addition, the SCR reaction started and adsorbed NH_3 molecules were consumed preferentially over NH_4^+ species. The evolution of the water by-product of SCR also coincided with the consumption of Lewis-bound NH_3. Only when the NH_3 supply from Lewis acid sites was almost fully depleted, did the NH_4^+ species from Bronsted acid sites start to decrease. Parallel measurements using UV-visible spectroscopy revealed the reduction of the VO_x sites upon NO and NH_3 introduction and its subsequent reoxidation once all of the adsorbed NH_3 was consumed. In the same experiment, Marberger et al. (2016) verified the formation of the NH_2NO intermediate, thereby confirming for the first time under reaction conditions the SCR intermediate postulated almost three decades earlier. The importance of the Lewis acid sites in SCR was further confirmed by time-resolved Raman spectroscopy. Under reaction conditions, vanadyl species (VO_x) adopt various states of coordination, as demonstrated by the significant widening of the VO_x peak in the time-resolved Raman spectra. However, only the coordinatively unsaturated species, which correspond to Lewis acid sites, were found to be responsible for the SCR activity as their response to repeated NH_3 pulses caused their characteristic signal to appear in the phase-resolved spectra (Nuguid et al., 2019a). The involvement of the coordinatively unsaturated VO_x sites was verified by Arrhenius-type relationships, wherein the apparent activation energy (E_a) obtained from activity measurements correlated with the E_a calculated from the Raman intensity changes.

Although there are still open questions for the SCR mechanism over V_2O_5/TiO_2, some important reaction steps are generally accepted (Figure 54.1).

In the first step, NH_3 coordinates with VO_x sites through an acid-base interaction. Adsorbed NH_3 then reacts with NO, most probably in the gas phase, to form nitrosamide (NH_2NO) as intermediate. This step is also accompanied by the reduction of the V center from +5 to the +4 state. The intermediate nitrosamide is unstable and decomposes easily into N_2 and H_2O. In the last step, molecular O_2 oxidizes the V center back to the +5 state, releasing H_2O and closing the catalytic cycle.

The commercial success of V_2O_5/TiO_2 catalysts derives from their moderate cost and superior sulfur resistance. However, this catalyst type possesses two major drawbacks that limit their application in vehicles. First, they have a rather poor activity at low temperatures, which poses a problem during cold start-ups and short-distance driving, where the exhaust gas does not reach the required temperature for sufficient NO_x conversion. Second, 'diesel particulate filters' (DPF), located upstream of SCR catalysts, need to be periodically regenerated at temperatures that are high enough to release volatile vanadium species and cause the undesirable transformation of TiO_2, from anatase to the low-surface-area rutile. Nonetheless, they still find applications

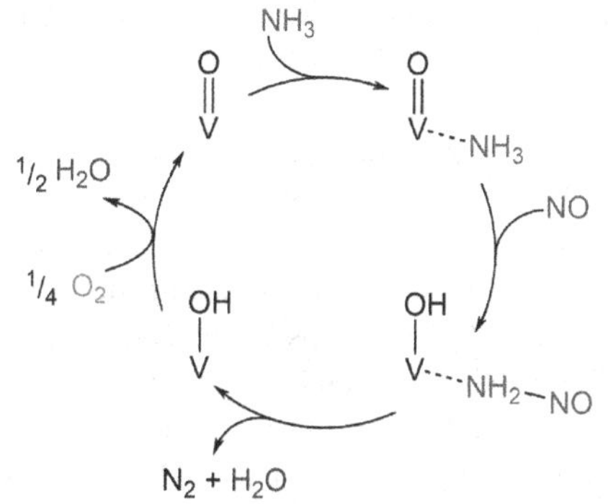

FIGURE 54.1 Proposed mechanism of SCR over V_2O_5/TiO_2.

Source: Adapted with minor revisions from Marberger et al. (2016).

in stationary power plants, locomotives, and ships, which require only a relatively narrow temperature window.

54.3 COPPER-BASED SCR CATALYSTS

Cu-based catalysts have been known for a long time as catalysts for NO_x abatement. As early as the 1980s, researchers had found that Cu-exchanged zeolites could decompose NO into N_2 and O_2 directly (Iwamoto et al., 1981). Further research revealed that the NO_x conversion can be strongly improved when a selective reductant was introduced (Held et al., 1990). Since then, Cu-based catalysts have remained in the focus of SCR research.

The chemical nature of Cu-exchanged zeolites is well studied in the scientific literature. Cu can coordinate with negatively charged Al sites in the zeolite framework to form isolated Cu^{2+} or Cu-hydroxo (Cu^{+}–OH) sites. CuO clusters typically do not form, unless all of the Al sites are already occupied. The Cu metal center serves as the active site of the catalyst and possesses both redox and acidic functionalities. During the reaction cycle, Cu shuttles between two oxidation states: +1 in the presence of both NO and NH_3 and +2 after the reoxidation with O_2 or a stronger oxidizing agent such as NO_2 (Nuguid et al., 2019a; Negri et al., 2019). Likewise, Cu ions, as opposed to Cu metal, contain vacant *d* orbitals and can therefore function as Lewis acid sites, where reactant molecules can adsorb. The zeolite support is also involved in the catalytic process: aside from ensuring a high dispersion of the active site, its unique pore structure allows the permeation of reactants and provides spatial confinement for the reaction to occur. Furthermore, hydroxyl bridges (i.e. Al sites that were not ion-exchanged) act as Bronsted acid sites, which facilitate reactant adsorption.

For many years, the large-pore zeolites ZSM-5 and beta have been the standard carriers for Cu-based catalysts used in NH_3-SCR. While Cu-ZSM-5 was slightly more active than Cu-beta, the latter was more durable and enjoyed a wider acceptance in the industry. However, neither possessed sufficient hydrothermal stability for long-term use. In one study, both catalyst samples lost more than 25% of their initial turnover rate after aging under 10 vol% H_2O at 800°C (Kwak et al., 2012). The deactivation was reported to have started already at 550°C in some cases (Palella et al., 2003). Several mechanisms were proposed for the observed activity loss and there is consensus that deactivation occurs when Al sites irreversibly detach from their original positions in the framework. Although this phenomenon, known as dealumination, is well documented for all zeolite materials working in the presence of steam, the situation is prohibitive for NH_3-SCR because large amounts of water are generated during combustion. Higher water contents favor the conditions for dealumination since water molecules can cooperate through proton shuttling to stabilize the transition state in the subsequent hydrolysis steps (Nielsen et al., 2019). This process eventually leads to loss of Bronsted acidity and collapse of the porous zeolite structure in favor of denser, more stable crystalline phases. Without the Al sites for anchoring, the high dispersion of Cu is lost and SCR-inactive CuO phases are formed.

The most straightforward way to deal with this problem is to partially decrease the number of Al sites on the catalyst so that water has less sites to attack (Silaghi et al., 2014). This can be achieved by controlled dealumination of the zeolite prior to the ion-exchange procedure. Although this strategy has been implemented in some applications, it comes with the serious drawback of obtaining a material with a sub-optimal Cu loading. Alternatively, a more thermostable support can be used, since the major constraint on catalyst stability is related to the support rather than the active phase. Recently, two support materials have been identified that allowed the preparation of stable Cu-zeolites: SSZ-13 and SAPO-34. Both supports contain smaller pores than ZSM-5 or beta. Indeed, the enhancement of the hydrothermal stability stems from the fact that the pores of these materials are so small (< 4 Å) that the dealumination product $Al(OH)_3$ (~5 Å) cannot exit, thus preserving the integrity of the zeolite structure even under hydrothermal conditions (Fickel et al., 2011; Wang et al., 2014). Consequently, both Cu-SAPO-34 and Cu-SSZ-13 have been commercialized and marketed as SCR catalysts.

SAPO-34 is a silicoaluminophosphate molecular sieve that shares its three-dimensional microporous framework with the chabazite-type zeolites, to which SSZ-13 belongs. It has excellent hydrothermal stability as evidenced by the fact that it is able to keep its structure intact at 600°C under 20 vol% H_2O (Lok et al., 1984). Interestingly, prolonged hydrothermal aging even improved the activity of Cu-SAPO-34 relative to the fresh state (Wang et al., 2013). The rate enhancement was attributed to the migration of CuO_x clusters from the external surface of SAPO-34 to the Al exchange sites, where they could be stabilized as single Cu^{2+} sites. The demonstrated hydrothermal stability of Cu-SAPO-34 at high temperature is unprecedented, even surpassing that of Cu-SSZ-13 at 800°C (Wang et al., 2015). But surprisingly, Cu-SAPO-34 was shown to undergo irreversible deactivation upon hydrothermal treatment at temperatures below 100°C (Leistner and Olsson, 2015). The deactivated samples still have the preserved zeolite framework and Bronsted acidity, suggesting that the mechanism of activity loss is not due to dealumination as it was with Cu-ZSM-5 but more likely due to the transformation of the Cu species. Recently, the molecular basis of the deactivation was revealed to originate from the reaction of $Cu(OH)_2$ with terminal Al species, forming SCR-inactive Cu-aluminate species (Wang et al., 2019). Condensed water molecules facilitate this transformation, thus explaining the low-temperature nature of the deactivation. Even Cu-SAPO-34 samples stored under ambient conditions for a prolonged period are prone to this type of deactivation and exhibit transformation of the active Cu^{2+} sites into Cu-aluminates. Because of this unexpected but serious limitation, the industrial acceptance of Cu-SAPO-34 has started to decline.

Nowadays, Cu-SSZ-13 is the reference catalyst material for mobile SCR applications. It solves the pitfalls of vanadia-based and other Cu-exchanged zeolites, and offers both superior activity and hydrothermal stability. This comprises sufficient activity at temperatures below 200°C as well as stability at high temperatures when the DPF has to be regenerated. Due to its commercial success, it has been the subject of many spectroscopic and mechanistic studies. Atomically dispersed Cu^{2+} species are believed to be the active sites for NH_3-SCR. The molecular details of the reaction mechanism are still being intensively debated, but there are some unifying features in all of the reports, which are summarized in Figure 54.2.

Due to the basicity and polarity of NH_3, it coordinates more strongly with the Cu^{2+} sites than NO or its derivatives. Nonetheless, most reported mechanistic cycles

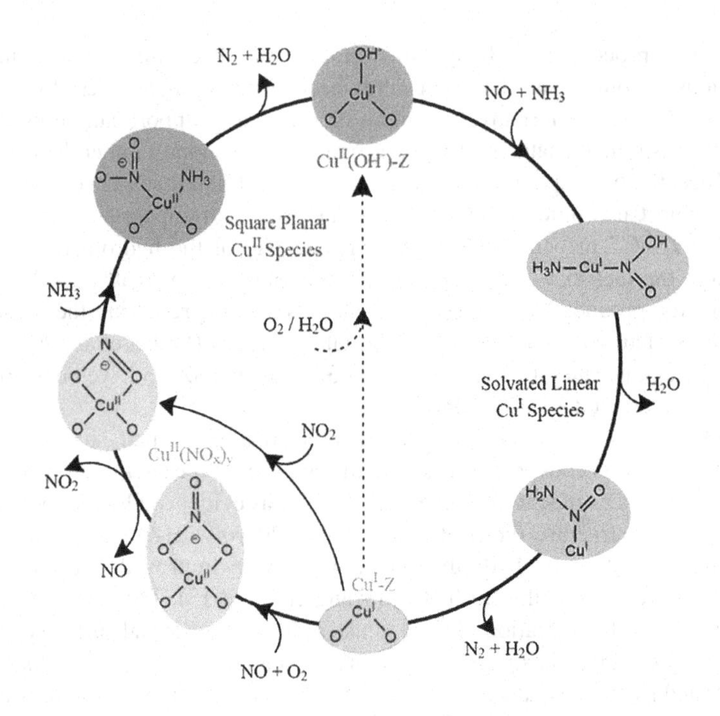

FIGURE 54.2 Proposed mechanism of SCR over Cu-SSZ-13.

Source: Clark et al. (2020).

indicate the co-adsorption of NH_3 and NO on the Cu^{2+} site to form a solvated linear Cu^{+} complex as the first step (Janssens et al., 2015; Marberger et al., 2018). The Cu site only reduces in the presence of both NH_3 and NO. The formed intermediate is quite unstable and decomposes readily into the SCR products, N_2 and H_2O. This generates a free Cu^{+} species, which is the final product of the reducing half-cycle of the mechanism. The second half-cycle features the reoxidation of the Cu sites, which can be accomplished in two different ways. First, Cu^{+} can be reoxidized directly to Cu^{2+} by the oxygen in the gas feed, closing the catalytic cycle immediately (Clark et al., 2020). Alternatively, Cu^{+} can form $Cu(NO_x)_y$ upon the adsorption of NO and oxygen. This is also accompanied by the change of oxidation state of Cu to +2. Then, NH_3 can coordinate with $Cu(NO_x)_y$ to yield a square planar Cu^{2+} complex that decomposes to produce N_2 and H_2O and regenerate the starting Cu^{2+} species. The rate-limiting step is thought to be the oxidation of Cu^{+} back to Cu^{2+}, but recent data suggest that at high temperatures, when parasitic NH_3 oxidation becomes prevalent, it is the reduction of Cu^{2+} that determines the reaction rate (Clark et al., 2020; Marberger et al., 2018).

Cu-SSZ-13 is currently the best option for mobile SCR, providing unsurpassed low-temperature activity and high-temperature stability. However, it is not without drawbacks. Besides its much higher costs compared to vanadium-based systems, it is susceptible to N_2O formation, especially at high temperatures due to unselective NH_3 oxidation (Zhang and Yang, 2018) and it deactivates in the presence of sulfur. The overall NO_x reduction also tends to decrease above 400°C, although the catalyst remains stable. These points must be addressed in the future development of Cu-based catalysts for SCR.

54.4 IRON-BASED SCR CATALYSTS

Fe-exchanged zeolites can serve as an alternative to Cu-based materials mostly for high-temperature applications. Although Fe catalysts suffer from low activity below 350°C, they already surpass Cu-SSZ-13 in terms of NO_x conversion above 450°C, and even vanadia-based catalysts above 550°C. Moreover, thanks to the capability of Fe in catalyzing N_2O decomposition and reduction, negligible N_2O emissions are generally detected at elevated temperatures (Krocher et al., 2006).

Analogous with Cu-exchanged zeolites, the Fe ions in the zeolite framework act as Lewis acid sites where NH_3 is activated to react with NO. The zeolite framework also provides Bronsted acidity in the form of protonated Al-O-Si bonds, where NH_3 can also adsorb. The active Fe center changes from the +3 (oxidized) to the +2 state (reduced) and vice versa during the catalytic cycle. Fe-exchanged zeolites generally have a lower NH_3 storage capacity than their Cu counterpart, which may be explained by the inability of Fe to form a strong complex with NH_3 (Colombo et al., 2010). While Cu usually maintains a high degree of dispersion across the zeolite framework, Fe is more prone to form clusters or even particles.

An Fe-based catalyst can have: (1) monomeric Fe sites with different coordination and framework environments; (2) oligomeric $Fe_xO_y(OH)_z$ species possessing various degrees of agglomeration; and (3) bulk Fe_xO_y particles of varying sizes and crystallinity. The ratio between these species primarily depends on the Fe loading, the synthesis procedure, and the zeolite structure (Brandenberger et al. 2010; Capek et al., 2005). Moreover, these species may convert from one form to another under SCR conditions (Kumar et al., 2004). This diverse and dynamic speciation makes the molecular understanding of the SCR process over Fe-based materials particularly arduous.

Brandenberger et al. (2010) proposed a direct relationship between the reaction temperature and the activity of Fe species of varying nuclearity. They suggested that at temperatures below 300°C, monomeric Fe sites are the only active species that determine the SCR activity. As the reaction temperature increases, the dimeric, oligomeric, and large Fe_xO_y sites also start to become active in reverse order of their degree of nuclearity (i.e. the large Fe_xO_y clusters are the last to become active at higher temperatures). SCR selectivity is also influenced by the degree of agglomeration, with oligomeric and bulk Fe_xO_y species being widely recognized as the main contributors to the undesired NH_3 oxidation reaction. Hence, a high dispersion of Fe species is preferred. These findings were corroborated by Hoj et al. (2009), who reported a direct correlation between the NO_x conversion and degree of isolated Fe sites in the catalyst. They additionally proposed that not all monomeric sites have equal activity, thus suggesting that the zeolite environment has an impact on the activity as well.

The earliest Fe-exchanged zeolites studied had not only limited activity at low temperatures but also poor stability under hydrothermal conditions. Upon aging at high temperature in the presence of steam, Fe species detach from the exchange sites and agglomerate, which has a negative impact on selectivity (Brandenberger et al., 2008). Fe-based catalysts are also prone to poisoning by SO_2 and the formation of coke (Ma et al., 2012).

To improve the low-temperature activity of Fe-based catalysts, researchers adopted a synergistic approach where Fe is coupled with another material. It is well established that Fe is more active at high temperatures whereas Cu is at low temperatures. Hence, it is intuitive to combine both materials in one system to have a catalyst that can operate in a wider temperature range. This strategy was exploited to prepare Fe/Cu-exchanged zeolite catalysts (Boron et al., 2015), sequential dual-catalytic beds (Krocher and Elsener, 2008), and dual-layer Fe/Cu catalysts (Metkar et al., 2012). More recently, Liu et al. (2017) prepared core-shell MoFe/Beta@CeO_2 materials, which improved the low-temperature activity by partially oxidizing NO to NO_2, thereby allowing the fast SCR reaction to proceed at low temperature.

Large pore-BEA and medium pore-ZSM-5 zeolites have been the most investigated supports for Fe-based SCR catalysts. Although satisfactory NO_x reduction activities are generally achieved, they deactivate quite readily at high temperatures in the presence of steam (Iwasaki et al., 2011). Because of this, small-pore chabazite-type zeolites such as SSZ-13 have received growing attention. Gao et al. (2015) synthesized a Fe-SSZ-13 catalyst with a remarkably low temperature activity that can be retained up to 550°C in the fresh state. Upon aging, the catalyst exhibited lower activity but still achieved 80% NO conversion at 450°C (GHSV = 200,000 h^{-1}). The hydrocarbon tolerance can be improved by careful selection of the zeolite framework. Ma et al. (2012) studied the effect of propene poisoning on the activity of Fe-BEA, Fe-ZSM-5, and Fe-MOR. While Fe-BEA and Fe-ZSM-5 deactivated significantly, Fe-MOR retained high NO_x conversion even after propene coking at 350°C. The exceptional stability of Fe-MOR could be traced back to the one-dimensional pore structure of the zeolite framework, which could limit hydrocarbon diffusion in the first place. Furthermore, hydrocarbon activation is hindered in MOR because it is not as acidic as BEA and ZSM-5.

As with the two previously discussed catalysts, the mechanism of SCR over Fe-exchanged zeolites is still under debate. Nonetheless, the general features of the reaction cycle can be described without ambiguity (Figure 54.3).

Under reaction conditions, NH_3 populates the catalyst surface adsorbing on both Bronsted (protonated Al-O-Si groups) and Lewis acid sites (Fe centers). The first step of the reaction is the so-called NO oxidative activation, in which NO from the gas or weakly adsorbed state oxidizes on Fe^{3+} sites. This step is coupled with the reduction of the transition metal center to Fe^{2+}. While the occurrence of this step is widely accepted, the nature of the oxidized intermediate is still a matter of debate. Historically, NO_2 and its derivative N_2O_3 were proposed as intermediates (Brandenberger et al., 2008), but recent evidence pointed toward a chemical species possessing an N atom in the 3+ oxidation state such as HONO or nitrites (Ruggeri et al., 2014). Nonetheless, HONO and nitrites are ultimately produced by either NO_2

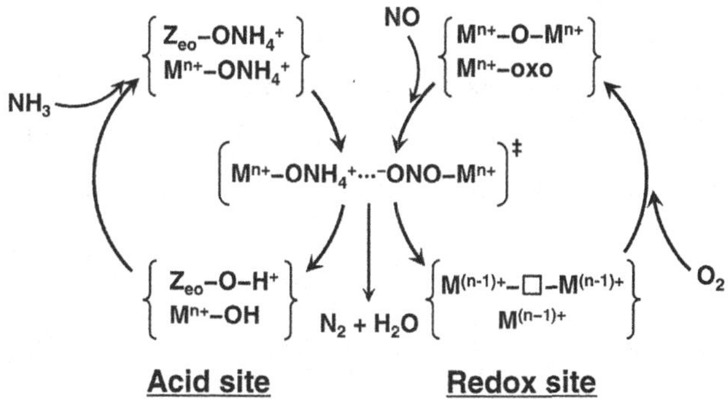

FIGURE 54.3 Proposed mechanism of SCR over Fe-exchanged zeolites.

Notes: M^{n+} corresponds to Fe^{3+} while $M^{(n-1)+}$ corresponds to Fe^{2+}.
Source: Nova and Tronconi (2014).

dimerization and disproportionation or N_2O_3 disproportionation. The successive reaction between HONO or nitrites species and adsorbed NH_3 possibly lead to the formation of the intermediate ammonium nitrite that decomposes into N_2 and H_2O. This step automatically regenerates the Bronsted and Lewis acid sites, which are once again ready to adsorb NH_3 molecules. On the other hand, the reduced Fe^{2+} is oxidized back to Fe^{3+} by the action of molecular oxygen. While this step is considered rate-limiting by Ruggeri et al. (2014), Metkar et al. (2012) suggested NO oxidation to surface-bound NO_2 as the rate-determining step.

Currently, commercial Fe SCR catalysts rely on BEA and certain chabazite structures such as SAPO-34 and CHA (Toops et al., 2014). Future research should focus on achieving Fe-exchanged zeolites characterized by high Fe loading with remarkable monomeric dispersion and enhanced stability under hydrothermal conditions.

54.5 SUMMARY

SCR is an indispensable component of diesel exhaust after-treatment, where it is used to lower the NO_x emissions to acceptable values. Although other reductants were employed in the past, only NH_3 achieved commercial success and widespread use due to its high efficiency and selectivity. Three families of materials have emerged as the leading SCR catalysts. Each offers a unique set of advantages and disadvantages that must be considered upon catalyst selection (Table 54.1).

Time-resolved studies have advanced the molecular view of the SCR process, and new spectroscopic evidence may rewrite the SCR mechanism that is generally accepted today. Ultimately, this molecular-level information will be crucial for designing more active and more robust SCR catalysts in a rational manner.

TABLE 54.1
Comparison between the Three Most Industrially Relevant SCR Catalysts

Type of Catalyst	V-based Catalysts	Cu-based Catalysts	Fe-based Catalysts
Main advantages	High resistance to SO_2 poisoning; Moderate cost	Unparalleled activity at low temperature; High thermal stability (SSZ-13)	High SCR activity at high temperatures; Able to decompose and reduce N_2O
Main disadvantages	Potential release of toxic vanadium oxides; Poor thermal stability due to sintering/reutilization	Release of N_2O; Decrease in NO_x activity at high temperature	Prone to form clusters which lower the activity; Poor resistance against coking and steam (BEA and ZSM-5)
Typical applications	Stationary/marine systems	Mobile systems	

REFERENCES

Alemany, L. J., L. Lietti, and N. Ferlazzo, et al. 1995. Reactivity and physicochemical characterization of V_2O_5-WO_3/TiO_2 de-NO_x catalysts. *Journal of Catalysis* 155:117–130.

Amiridis, M. D., I. E. Wachs, D. Deo, J. M. Jehng, and D. S. Kim. 1996. Reactivity of V_2O_5 catalysts for the selective catalytic reduction of NO by NH_3: Influence of vanadia loading, H_2O, and SO_2. *Journal of Catalysis* 161:247–253.

Ando, J., H. Tohata, and G. A. Isaacs. 1976. *NOx Abatement for Stationary Sources in Japan.* Research Triangle Park, NC: Environmental Protection Agency.

Boron, P., L. Chmielarz, and S. Dzwigaj. 2015. Influence of Cu on the catalytic activity of FeBEA zeolites in SCR of NO with NH_3. *Applied Catalysis B: Environmental* 168–169:377–384.

Bowers, W. E. 1988. Reduction of nitrogen-based pollutants through the use of urea solutions containing oxygenated hydrocarbon solvents. *Atmospheric Environment* 22:II. US 4719092

Brandenberger, S., O. Krocher, A. Tissler, and R. Althoff. 2008. The state of the art in selective catalytic reduction of NO_x by ammonia using metal-exchanged zeolite catalysts. *Catalysis Reviews Science and Engineering* 50:492–531.

Brandenberger, S., O. Krocher, A. Tissler, and R. Althoff. 2010. Estimation of the fractions of different nuclear iron species in uniformly metal-exchanged Fe-ZSM-5 samples based on a Poisson distribution. *Applied Catalysis A: General* 373:168–175.

Busca, G., L. Lietti, G. Ramis, and F. Berti. 1998. Chemical and mechanistic aspects of the selective catalytic reduction of NO_x by ammonia over oxide catalysts: A review. *Applied Catalysis B: Environmental* 18:1–36.

Capek, L., V. Kreibich, and J. Dedecek, et al. 2005. Analysis of Fe species in zeolites by UV-VIS-NIR, IR spectra and voltammetry. Effect of preparation, Fe loading and zeolite type. *Microporous and Mesoporous Materials* 80:279–289.

Chen, J. P. and R. T. Yang. 1992. Role of WO_3 in mixed V_2O_5-WO_3/TiO_2 catalysts for selective catalytic reduction of nitric oxide with ammonia. *Applied Catalysis A General* 80:135–148.

Clark, A. H., R. J. G. Nuguid, and P. Steiger, et al. 2020. Selective catalytic reduction of NO with NH_3 on Cu-SSZ-13: Deciphering the low and high-temperature rate-limiting steps by transient XAS experiments. *ChemCatChem* 12: 1429–1435.

Colombo, M., I. Nova, and E. Tronconi. 2010. A comparative study of the NH_3-SCR reactions over a Cu-zeolite and a Fe-zeolite catalyst. *Catalysis Today* 151:223–230.

Costa, C. N., P. G. Savva, J. L. G. Fierro, and A. M. Efstathiou. 2007. Industrial H_2-SCR of NO on a novel Pt/MgO-CeO_2 catalyst. *Applied Catalysis B: Environmental* 75:147–156.

Cristiani, C., M. Bellotto, P. Forzatti, and F. Bregani. 1993. On the morphological properties of tungsta-titania de-NO_xing catalysts. *Journal of Materials Research* 8:2019–2025.

Fickel, D. W., E. D'Addio, J. A. Lauterbach, and R. F. Lobo. 2011. The ammonia selective catalytic reduction activity of copper-exchanged small-pore zeolites. *Applied Catalysis B: Environmental* 102:441–448.

Forzatti, P. and L. Lietti. 1996. Recent advances in de-NOxing catalysis for stationary applications. *Heterogeneous Chemistry Reviews* 3:33–51.

Gabrielsson, P. L. T. 2004. Urea-SCR in automotive applications. *Topics in Catalysis* 28:177–184.

Gao, F., M. Kollar, and R. K. Kukkadapu, et al. 2015. Fe/SSZ-13 as an NH_3-SCR catalyst: A reaction kinetics and FTIR/Mossbauer spectroscopic study. *Applied Catalysis B: Environmental* 164:407–419.

Held, W., A. Konig, T. Richter, and L. Puppe. 1990. Catalytic NO_x reduction in net oxidizing exhaust gas. *SAE Technical Paper* 900496.

Hoj, M., M. J. Beier, J. D. Grunwaldt, and S. Dahl. 2009. The role of monomeric iron during the selective catalytic reduction of NO_x by NH_3 over Fe-BEA zeolite catalysts. *Applied Catalysis B: Environmental* 93:166–176.

Inomata, M., A. Miyamoto, and Y. Murakami. 1980. Mechanism of the reaction of NO and NH_3 on vanadium oxide catalyst in the presence of oxygen under the dilute gas condition. *Journal of Catalysis* 62:140–148.

Iwamoto, M., S. Yokoo, K. Sakai, and S. Kagawa. 1981. Catalytic decomposition of nitric oxide over copper(II)-exchanged, Y-type zeolites. *Journal of the Chemical Society, Faraday Transactions 1: Physical Chemistry in Condensed Phases* 77:1629–1638

Iwasaki, M., K. Yamazaki, and H. Shinjoh. 2011. NO_x reduction performance of fresh and aged Fe-zeolites prepared by CVD: Effects of zeolite structure and Si/Al_2 ratio. *Applied Catalysis B: Environmental 102*:302–309.

Jaegers, N. R., J.-K. Lai, and Y. He, et al. 2019. Mechanism by which tungsten oxide promotes the activity of supported V_2O_5/TiO_2 catalysts for NO_X abatement: Structural effects revealed by ^{51}V MAS NMR Spectroscopy. *Angewandte Chemie International Edition* 58:12609–12616.

Janssen, F. J. J. G., F. M. G. van den Kerkhof, H. Bosch, and J. R. H. Ross. 1987a. Mechanism of the reaction of nitric oxide, ammonia, and oxygen over vanadia catalysts I. The role of oxygen studied by way of isotopic transients under dilute conditions. *Journal of Physical Chemistry* 91:5921–5927.

Janssen, F. J. J. G., F. M. G. van den Kerkhof, H. Bosch, and J. R. H. Ross. 1987b. Mechanism of the reaction of nitric oxide, ammonia, and oxygen over vanadia catalysts. 2. Isotopic transient studies with oxygen-18 and nitrogen-15. *Journal of Physical Chemistry* 91:6633–6638.

Janssens, T. V. W., H. Falsig, and L. F. Lundegaard, et al. 2015. A consistent reaction scheme for the selective catalytic reduction of nitrogen oxides with ammonia. *ACS Catalysis* 5:2832–2845.

Kato, A., S. Matsuda, and T. Kamo, et al. 1981. Reaction between nitrogen oxide (NO_x) and ammonia on iron oxide-titanium oxide catalyst. *Journal of Physical Chemistry* 85:4099–4102.

Koebel, M., M. Elsener, and G. Madia. 2001. Reaction pathways in the selective catalytic reduction process with NO and NO_2 at low temperatures. *Industrial & Engineering Chemistry Research* 40:52–59.

Koebel, M., G. Madia, and M. Elsener. 2002a. Selective catalytic reduction of NO and NO_2 at low temperatures. *Catalysis Today* 73:239–247.

Koebel, M., G. Madia, F. Raimondi, and A. Wokaun. 2002b. Enhanced reoxidation of vanadia by NO_2 in the fast SCR Reaction. *Journal of Catalysis* 209:159–165.

Krocher, O., M. Devadas, and M. Elsener, et al. 2006. Investigation of the selective catalytic reduction of NO by NH_3 on Fe-ZSM5 monolith catalysts. *Applied Catalysis B: Environmental* 66:208–216.

Krocher, O. and M. Elsener. 2008. Combination of V_2O_5/WO_3-TiO_2, Fe-ZSM5, and Cu-ZSM5 catalysts for the selective catalytic reduction of nitric oxide with ammonia. *Industrial and Engineering Chemistry Research* 47:8588–8593.

Kumar, M. S., M. Schwidder, W. Grunert, and A. Bruckner. 2004. On the nature of different iron sites and their catalytic role in Fe-ZSM-5 deNO_x catalysts: New insights by a combined EPR and UV/VIS spectroscopic approach. *Journal of Catalysis* 227:384–397.

Kwak, J. H., D. Tran, and S. D. Burton, et al. 2012. Effects of hydrothermal aging on NH_3-SCR reaction over Cu/zeolites. *Journal of Catalysis* 287:203–209.

Leistner, K. and L. Olsson. 2015. Deactivation of Cu/SAPO-34 during low-temperature NH_3-SCR. *Applied Catalysis B: Environmental* 165:192–199.

Liu, J., Y. Du, and J. Liu, et al. 2017. Design of MoFe/Beta@CeO_2 catalysts with a core-shell structure and their catalytic performances for the selective catalytic reduction of NO with NH_3. *Applied Catalysis B: Environmental* 203:704–714.

Lok, B. M., C. A. Messina, and R. L. Patton, et al. 1984. Silicoaluminophosphate molecular sieves: Another new class of microporous crystalline inorganic solids. *Journal of the American Chemical Society* 106:6092–6093.

Ma, L., J. Li, and Y. Cheng, et al. 2012. Propene poisoning on three typical Fe-zeolites for SCR of NO_x with NH_3: From mechanism study to coating modified architecture. *Environmental Science & Technology* 46:1747–1754.

Madia, G., M. Koebel, M. Elsener, and A. Wokaun. 2002. Side reactions in the selective catalytic reduction of NO_x with various NO_2 fractions. *Industrial & Engineering Chemistry Research* 41:4008–4015.

Marberger, A., M. Elsener, D. Ferri, and O. Krocher. 2015. VO_x surface coverage optimization of V_2O_5/WO_3-TiO_2 SCR catalysts by variation of the V loading and by aging. *Catalysts* 5:1704–1720.

Marberger, A., M. Elsener, R. J. G. Nuguid, D. Ferri, and O. Krocher. 2019. Thermal activation and aging of a V_2O_5/WO_3-TiO_2 catalyst for the selective catalytic reduction of NO with NH_3. *Applied Catalysis A: General* 573:64–72.

Marberger, A., D. Ferri, M. Elsener, and O. Krocher. 2016. The significance of Lewis acid sites for the selective catalytic reduction of nitric oxide on vanadium-based catalysts. *Angewandte Chemie International Edition* 55:11989–11994.

Marberger, A., A. W. Petrov, and P. Steiger, et al. 2018. Time-resolved copper speciation during selective catalytic reduction of NO on Cu-SSZ-13. *Nature Catalysis* 1:221–227.

Metkar, P. S., M. P. Harold, and V. Balakotaiah. 2012. Selective catalytic reduction of NO_x on combined Fe- and Cu-zeolite monolithic catalysts: Sequential and dual layer configurations. *Applied Catalysis B: Environmental* 111–112:67–80.

Mrad, R., A. Aissat, R. Cousin, D. Courcot, and S. Siffert. 2015. Catalysts for NO_x selective catalytic reduction by hydrocarbons (HC-SCR). *Applied Catalysis A: General* 504:542–548.

Negri, C., M. Signorile, and N. G. Porcaro, et al. 2019. Dynamic Cu^{II}/Cu^{I} speciation in Cu-CHA catalysts by *in situ* diffuse reflectance UV-vis-NIR spectroscopy. *Applied Catalysis A: General* 578:1–9.

Nielsen, M., A. Hafreager, and R. Y. Brogaard, et al. 2019. Collective action of water molecules in zeolite dealumination. *Catalysis Science & Technology* 9:3721–3725.

Nova, I. and E. Tronconi. (Eds.) 2014. *Urea-SCR Technology for deNOx after Treatment of Diesel Exhausts*. New York, NY: Springer.

Nuguid, R. J. G., D. Ferri, and O. Krocher. 2019a. Design of a reactor cell for modulated excitation Raman and diffuse reflectance studies of selective catalytic reduction catalysts. *Emission Control Science and Technology* 5:307–316.
Nuguid, R. J. G., D. Ferri, A. Marberger, M. Nachtegaal, and O. Krocher. 2019b. Modulated excitation Raman spectroscopy of V_2O_5/TiO_2: Mechanistic insights into the selective catalytic reduction of NO with NH_3. *ACS Catalysis* 9:6814–6820.
Palella, B. I., M. Cadoni, and A. Frache, et al. 2003. On the hydrothermal stability of CuAPSO-34 microporous catalysts for N_2O decomposition: A comparison with CuZSM-5. *Journal of Catalysis* 217:100–106.
Radtke, F., R. A. Koeppel, and A. Baiker. 1995. Formation of hydrogen cyanide over Cu/ZSM-5 catalyst used for the removal of nitrogen oxides from exhausts of lean-burn engines. *Environmental Science & Technology* 29:2703–2705.
Ramis, G., G. Busca, F. Bregani, and P. Forzatti. 1990. Fourier transform-infrared study of the adsorption and coadsorption of nitric oxide, nitrogen dioxide and ammonia on vanadia-titania and mechanism of selective catalytic reduction. *Applied Catalysis* 64:259–278.
Ramis, G., G. Busca, and C. Cristiani, et al. 1992. Characterization of tungsta-titania catalysts. *Langmuir* 8:1744–1749.
Ramis, G., L. Yi, and G. Busca. 1996. Ammonia activation over catalysts for the selective catalytic reduction of NO_x and the selective catalytic oxidation of NH_3. An FT-IR study. *Catalysis Today* 28:373–380.
Ruggeri, M. P., T. Selleri, M. Colombo, I. Nova, and E. Tronconi. 2014. Identification of nitrites/HONO as primary products of NO oxidation over Fe-ZSM-5 and their role in the standard SCR mechanism: A chemical trapping study. *Journal of Catalysis* 311:266–270.
Silaghi, M.-C., C. Chizallet, and P. Raybaud. 2014. Challenges on molecular aspects of dealumination and desilication of zeolites. *Microporous and Mesoporous Materials* 191:82–96.
Takagi, M., T. Kawai, M. Soma, T. Onishi, and K. Tamaru. 1977. The mechanism of the reaction between NO_x and NH_3 on V_2O_5 in the presence of oxygen. *Journal of Catalysis* 50:441–446.
Toops, T. J., J. A. Pihl, and W. P. Partridge. 2014. Fe-zeolite functionality, durability, and deactivation mechanisms in the selective catalytic reduction (SCR) of NO_x with ammonia. In *Urea-SCR Technology for deNOx After Treatment of Diesel Exhausts. Fundamental and Applied Catalysis* ed. I. Nova and E. Tronconi, 97–121. New York, NY: Springer.
Topsoe, N. Y. 1994. Mechanism of the selective catalytic reduction of nitric oxide by ammonia elucidated by *in situ* on-line Fourier transform infrared spectroscopy. *Science* 265:1217–1219.
Topsoe, N. Y., H. Topsoe, and J. A. Dumesic. 1995. Vanadia/titania catalysts for selective catalytic reduction (SCR) of nitric oxide by ammonia: I. Combined temperature programmed *in situ* FTIR and on-line mass spectroscopy studies. *Journal of Catalysis* 151:226–240.
Wachs, I. E. 1996. Raman and IR studies of surface metal oxide species on oxide supports: Supported metal oxide catalysts. *Catalysis Today* 27:437–455.
Wang, A., Y. Chen, and E. D. Walter, et al. 2019. Unraveling the mysterious failure of Cu/SAPO-34 selective catalytic reduction catalysts. *Nature Communications* 10:1137.
Wang, D., Y. Jangjou, and Y. Liu, et al. 2015. A comparison of hydrothermal aging effects on NH_3-SCR of NO_x over Cu-SSZ-13 and Cu-SAPO-34 catalysts. *Applied Catalysis B: Environmental* 165:438–445.
Wang, D., L. Zhang, J. Li, K. Kamasamudram, and W. S. Epling. 2014. NH_3-SCR over Cu/SAPO-34: Zeolite acidity and Cu structure changes as a function of Cu loading. *Catalysis Today* 231:64–74.
Wang, L., J. R. Gaudet, W. Li, and D. Weng. 2013. Migration of Cu species in Cu/SAPO-34 during hydrothermal aging. *Journal of Catalysis* 306:68–77.

Went, G. T., L.-J. Leu, R. R. Rosin, and A. T. Bell. 1992. The effects of structure on the catalytic activity and selectivity of V_2O_5/TiO_2 for the reduction of NO by NH_3. *Journal of Catalysis* 134:492–505.

Yamaguchi, T., Y. Tanaka, and K. Tanabe. 1980. Isomerization and disproportionation of olefins over tungsten oxides supported on various oxides. *Journal of Catalysis* 65:442–447.

Zhang, D. and R. T. Yang. 2018. N_2O formation pathways over zeolite-supported Cu and Fe catalysts in NH_3-SCR. *Energy & Fuels* 32:2170–2182.

Part XII

The Health Impact of Petrodiesel Fuel Emissions

55 The Adverse Health and Safety Impact of Diesel Fuels

A Scientometric Review of the Research

Ozcan Konur

CONTENTS

55.1 INTRODUCTION

Biodiesel and petrodiesel fuels have been primary sources of energy and fuels (Chisti, 2007, 2008; Konur, 2012g, 2015; Lapuerta et al., 2008; Marchetti et al., 2007;

Srivastava and Prasad, 2000van Gerpen, 2005). Nowadays both fuels are being used extensively (Konur, 2021a–ag).

However, the emissions (Busca et al., 1998; Birch and Cary, 1996; Lapuerta et al., 2008; Rogge et al., 1993; Robinson et al., 2007) from these fuels have raised significant concerns about human and environmental health and safety (Birch and Cary, 1996; Diaz-Sanchez et al., 1997; McCreanor et al., 2007; Mills et al., 2007; Salvi et al., 1999). Thus, the adverse health and safety impact of diesel fuel exhaust emissions have utmost public importance.

Furthermore, for the efficient progression of the research in this field, it is necessary to develop efficient incentive structures for the primary stakeholders and to inform these stakeholders about the research (Konur, 2000, 2002a–c, 2006a–b, 2007a–b; North, 1991a–b).

Scientometric analysis offers ways to evaluate the research in a respective field (Garfield, 1955, 1972; Konur, 2011, 2012a–n, 2015, 2016a–f, 2017a–f, 2018a–b, 2019a–b). However, there has been no scientometric study of this field.

This chapter presents a study of the scientometric evaluation of the research in this field using two datasets. The first dataset includes the 100-most-cited papers (n = 100 sample papers) whilst the second set includes population papers (n = over 2,150 population papers) published between 1980 and 2019.

The data on the indices, document types, authors, institutions, funding bodies, source titles, 'Web of Science' subject categories, keywords, research fronts, and citation impact are presented and discussed.

55.2 MATERIALS AND METHODOLOGY

The search for the literature was carried out in the 'Web of Science' database in January 2020. It contains the 'Science Citation Index-Expanded' (SCI-E), the 'Social Sciences Citation Index' (SSCI), the 'Book Citation Index-Science' (BCI-S), the 'Conference Proceedings Citation Index-Science' (CPCI-S), the 'Emerging Sources Citation Index' (ESCI), the 'Book Citation Index-Social Sciences and Humanities' (BCI-SSH), the 'Conference Proceedings Citation Index-Social Sciences and Humanities' (CPCI-SSH), and the 'Arts and Humanities Citation Index' (A&HCI).

The keywords for the search of the literature are collated from the screening of abstract pages for the first 500 highly cited papers. This keyword set is provided in the Appendix.

Two datasets are used for this study. The highly cited 100 papers comprise the first dataset (sample dataset, n = 100 papers) whilst all the papers form the second dataset (population dataset, n = over 2,150 papers).

The data on the indices, document types, publication years, institutions, funding bodies, source titles, countries, 'Web of Science' subject categories, citation impact, keywords, and research fronts are collated from these datasets. The key findings are provided in the relevant tables and figure, supplemented with explanatory notes in the text. The findings are discussed and a number of conclusions are drawn and a number of recommendations for further study are made.

55.3 RESULTS

55.3.1 INDICES AND DOCUMENTS

There are over 3,000 papers related to the health and safety impact of diesel fuel exhaust emissions in the 'Web of Science' as of January 2020. This original population dataset was refined for the document type (article, review, book chapter, book, editorial material, note, and letter) and language (English), resulting in over 2,150 papers comprising 71.7% of the original population dataset.

The primary index is the SCI-E for both the sample and population papers. The papers on the social and humanitarian aspects of this field are relatively negligible, with only a few papers from the SSCI and A&HCI.

Brief information on the document types for both datasets is provided in Table 55.1. The key finding is that article types of documents are the primary documents for both datasets and that reviews are over-represented by 3.6% in the sample papers.

55.3.2 AUTHORS

Brief information about the 19 most-prolific authors with at least four sample papers each is provided in Table 55.2. Around 450 and 6,500 authors contribute to the sample and population papers, respectively.

The most-prolific author is 'David Diaz-Sanchez' of the 'University of California Los Angeles' with 13 sample papers and an 11.9% publication surplus, working primarily on the 'respiratory illnesses caused by diesel exhaust particles'. The other prolific researchers are 'Thomas Sandstrom', 'Anders Blomberg', 'Andrew Saxon', 'Andre E. Nel', and 'Masaru Sagai' with at least nine sample papers and a 5.9% publication surplus each.

The most-prolific institution for these top authors is the 'University of California Los Angeles' of the USA with six authors. The other prolific institutions are the 'University of Edinburgh' of the UK, the 'National Institute for Environmental Studies' of Japan, and 'Umea University' of Sweden with three sample papers each,

TABLE 55.1
Document Types

	Document Type	Sample Dataset (%)	Population Dataset (%)	Difference (%)
1	Article	91	90.2	0.8
2	Review	7	3.4	3.6
3	Book chapter	0	0.1	−0.1
4	Proceeding paper	6	4.1	1.9
5	Editorial material	1	2.0	−1.0
6	Letter	0	3.6	−3.6
7	Book	0	0.0	0.0
8	Note	0	0.7	−0.7

TABLE 55.2
Authors

	Author	Sample Papers (%)	Population Papers (%)	Surplus (%)	Institution	Country	Research Front
1	Diaz-Sanchez, David (DiazSanchez)*	13	1.1	11.9	Univ. Calif. L.A.	USA	Respiratory illnesses
2	Sandstrom, Thomas	12	2.2	9.8	Umea Univ.	Sweden	Respiratory and vascular illnesses
3	Blomberg, Anders	11	1.9	9.1	Umea Univ.	Sweden	Respiratory and vascular illnesses
4	Saxon, Andrew	10	0.7	9.3	Univ. Calif. L.A.	USA	Respiratory illnesses
5	Nel, Andre E (A/AE)* **	9	0.5	8.5	Univ. Calif. L.A.	USA	Respiratory illnesses
6	Sagai, Masaru	9	3.1	5.9	Natl. Inst. Env. Stud.	Japan	Cancer
7	Li, Ning	8	0.6	7.4	Univ. Calif. L.A.	USA	Respiratory illnesses
8	Ichinose, Takamichi	6	1.9	4.1	Natl. Inst. Env. Stud.	Japan	Cancer
9	Kelly, Frank J. (F/FJ)*	6	0.5	5.5	King's Coll.	UK	Respiratory illnesses
10	Wang, Meiying (M/MY)*	6	0.4	5.6	Univ. Calif. L.A.	USA	Respiratory illnesses
11	Donaldson, Kenneth	5	0.8	4.2	Univ. Edinburgh	UK	Respiratory and vascular illnesses
12	Frew, Anthony J. (A/AJ)*	5	0.6	4.4	Southampton Gen. Hosp.	UK	Respiratory illnesses
13	Block, Michelle L.	4	0.4	3.6	Virginia Commonwealth Univ.	USA	Neurologic illnesses
14	Mills, Nicholas L.	4	0.8	3.2	Univ. Edinburgh	UK	Vascular diseases
15	Mudway, Ian S. (I/IS)*	4	0.4	3.6	King's Coll.	UK	Respiratory illnesses
16	Newby, David E. **	4	1.0	3.0	Univ. Edinburgh	UK	Respiratory and vascular illnesses
17	Stenfors, Nikolai	4	0.4	3.6	Umea Univ.	Sweden	Respiratory illnesses
18	Tsien, Albert	4	0.4	3.6	Univ. Calif. L.A.	USA	Respiratory illnesses
19	Yoshikawa, Toshikazu	4	1.4	2.6	Natl. Inst. Env. Stud.	Japan	Cancer

*Different spellings in parentheses
**highly cited researchers in 2019 (Clarivate Analytics, 2019).

and 'King's College' of the UK with two sample papers. In total, seven institutions house these top authors.

It is notable that two of these top researchers are listed in the 'Highly Cited Researchers' (HCR) in 2019 (Clarivate Analytics, 2019; Docampo and Cram, 2019).

The most-prolific country for these top authors is the USA with eight. The other prolific countries are the United Kingdom, Sweden, and Japan with five, three, and three authors, respectively.

There are three key research fronts for these top researchers. The top research front is 'respiratory illnesses caused by diesel fuel exhaust emissions' with 14 authors. The other prolific research fronts are 'vascular illnesses' and 'cancer caused by diesel fuel exhaust emissions' with five and three authors, respectively.

It is further notable that there is a significant gender deficit among these top authors as only two of them are female (Lariviere et al., 2013; Xie and Shauman, 1998).

55.3.3 Publication Years

Information about the publication years for both datasets is provided in Figure 55.1.

This figure shows that 36 and 44% of the sample papers were published in the 1990s and 2000s whilst 30.0 and 48.1% of the population papers were published in the 2000s and 2010s, respectively. It is notable that 13.0 and 9.5% of the sample and population papers were published in the 1980s, respectively. Furthermore, 12.9% of the population papers were published in the 1990s.

Similarly, the most-prolific publication years for the sample dataset are 1997, 1998, and 1999 with nine, eight, and eight sample papers, respectively. On the other

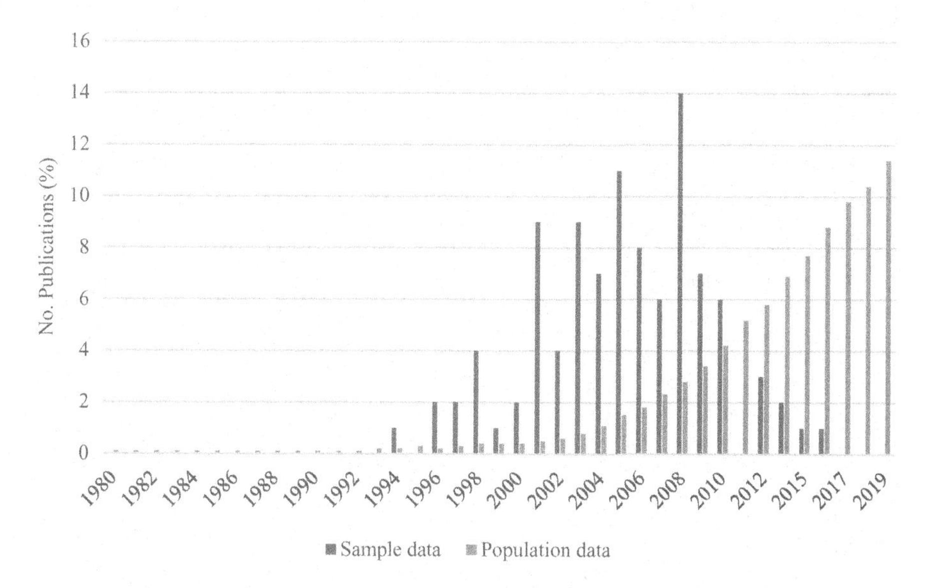

FIGURE 55.1 Research output between 1980 and 2019.

hand, the most-prolific publication years for the population dataset are 2015, 2012, 2019, and 2016 with over 5% of the population papers each.

55.3.4 Institutions

Brief information on the top 15 institutions with at least 4% of the sample papers each is provided in Table 55.3. In total, over 140 and 1,800 institutions contribute to the sample and population papers, respectively.

These top institutions publish 109 and 39.9% of the sample and population papers. The top institution is the 'University of California Los Angeles' of the USA with 28 sample papers and a 25.6% publication surplus. The other top institutions are 'Umea University' of Sweden and the 'National Institute for Environmental Studies' of Japan with 12% of the sample papers each and 9.4 and 4.2% publication surpluses, respectively.

The most-prolific country for these top institutions is the USA with six institutions. The other prolific countries are Japan and the UK with four and three sample papers, respectively. The other countries are France and Sweden.

55.3.5 Funding Bodies

Brief information about the top six funding bodies with at least 3% of the sample papers each is provided in Table 55.4. It is significant that only 38 and 46% of the sample and population papers declare any funding, respectively.

The top funding body is the 'National Institute of Allergy Infectious Diseases' of the USA, funding 18.0 and 1.6% of the sample and population papers, respectively, with a 16.4% publication surplus. The 'National Institute of Environmental Health

TABLE 55.3
Institutions

	Institution	Country	No. of Sample Papers (%)	No. of Population Papers (%)	Difference (%)
1	Univ. Calif. L.A.	USA	28	2.4	25.6
2	Natl. Inst. Env. Stud.	Japan	12	7.8	4.2
3	Umea Univ.	Sweden	12	2.6	9.4
4	King's Coll. London	UK	6	1.1	4.9
5	Univ. Tokyo	Japan	6	1.5	4.5
6	Ctr. Dis. Cont. Prev.	USA	5	4.4	0.6
7	Loveplace Resp. Res. Inst.	USA	5	3.3	1.7
8	Natl. Inst. Occ. Hlth. Saf.	USA	5	4.4	0.6
9	Univ. Edinburgh	UK	5	1.5	3.5
10	Univ. Southampton	UK	5	1.0	4.0
11	Natl. Inst. Hlth. Med. Res.	France	4	1.7	2.3
12	Kyoto Pref. Univ. Med.	Japan	4	1.3	2.7
13	Env. Prot. Agcy.	USA	4	5.5	–1.5
14	Univ. Calif. S. F.	USA	4	0.4	3.6
15	Univ. Tsukuba	Japan	4	1.0	3.0

TABLE 55.4
Funding Bodies

	Institution	Country	No. of Sample Papers (%)	No. of Population Papers (%)	Difference (%)
1	Natl. Inst. Allerg. Inf. Dis.	USA	18	1.6	16.4
2	Natl. Inst. Env. Hlth. Serv.	USA	14	5.2	8.8
3	Med. Res. Counc.	UK	5	1.0	4.0
4	Brit. Heart Found.	UK	3	1.3	1.7
5	Natl. Cancer Inst.	USA	3	2.2	0.8
6	Natl. Heart Lung Blood Inst.	USA	3	1.6	1.4

Sciences' of Japan closely follows the top funding agency with 14.0 and 5.2% of the sample and population studies, respectively, with an 8.8% publication surplus.

It is notable that some top funding agencies for the population studies do not enter this list. Some of them are the 'National Council for Scientific and Technological Development' of Brazil (2.6%), the 'Environmental Protection Agency' of the USA (2.3%), the 'National Natural Science Foundation of China' (2.1%), the 'Ministry of Education Culture Sports Science and Technology' (2.0%), and the 'Japan Society for the Promotion of Science' (1.4%) of Japan.

The most-prolific country for these top funding bodies is the USA with four bodies, followed by the UK with two.

55.3.6 Source Titles

Brief information about the top 11 source titles with at least three sample papers each is provided in Table 55.5. In total, 46 and 545 source titles publish the sample and population papers, respectively. On the other hand, the top 11 journals publish 55 and 18% of the sample and population papers, respectively.

The top journal is the 'Journal of Allergy and Clinical Immunology' publishing ten sample papers with an 8.7% publication surplus. This top journal is closely followed by the 'Journal of Immunology' with eight sample papers and a 7.2% publication surplus. The other top journals are the 'American Journal of Respiratory and Critical Care Medicine', 'Environmental Health Perspectives', and 'Free Radical Biology and Medicine' with at least five sample papers each.

It is significant that nearly all top journals are related to biomedicine. Although these journals are indexed by 12 subject categories, the top category is 'Immunology' with four journals. The other prolific subjects are 'Respiratory System', 'Toxicology', 'Allergy', 'Environmental Sciences', 'Biochemistry Molecular Biology', and 'Biology'.

55.3.7 Countries

Brief information about the top eight countries with at least two sample papers each is provided in Table 55.6. In total, 13 and over 120 countries contribute to the sample and population papers, respectively.

TABLE 55.5
Source Titles

	Source Title	WOS Subject Category	No. of Sample Papers (%)	No. of Population Papers (%)	Difference (%)
1	Journal of Allergy and Clinical Immunology	Allerg., Immunol.	10	1.3	8.7
2	Journal of Immunology	Immunol.	8	0.8	7.2
3	American Journal of Respiratory and Critical Care Medicine	Crit. Care Med., Resp. Syst.	6	2.3	3.7
4	Environmental Health Perspectives	Env. Sci., Pub. Env. Occ. Hlth, Toxic.	6	2.6	3.4
5	Free Radical Biology and Medicine	Bioch. Mol. Biol., Biol., Endocr. Metabol.	5	0.7	4.3
6	European Respiratory Journal	Respir. Syst.	4	0.6	3.4
7	Inhalation Toxicology	Toxic.	4	4.7	–0.7
8	American Journal of Respiratory Cell and Molecular Biology	Bioch. Mol. Biol., Biol., Respir. Syst.	3	0.6	2.4
9	Environmental Science Technology	Eng. Env., Env. Sci.	3	1.7	1.3
10	Mutation Research	Gen. Hered., Toxic.	3	1.1	1.9
11	Toxicology	Toxic	3	1.6	1.4
			55	18.0	37

TABLE 55.6
Countries

	Country	No. of Sample Papers (%)	No. of Population Papers (%)	Difference (%)
1	USA	54	38.3	15.7
2	Japan	21	19.0	2.0
3	UK	16	14.9	1.1
4	Sweden	13	4.3	8.7
5	France	5	3.7	1.3
6	Netherlands	5	3.5	1.5
7	Germany	3	4.7	–1.7
8	Australia	2	2.5	–0.5
	Europe-5	42	31.1	10.9
	Asia-1	21	19.0	2.0

The top country is the USA, publishing 54.0 and 38.3% of the sample and population papers, respectively. The other prolific countries are Japan, the UK, and Sweden, publishing 21, 16, and 13 sample papers, respectively.

The European and Asian countries represented in this table publish altogether 42 and 21% of the sample papers and 31.1 and 19.0% of the population papers, respectively.

It is notable that the publication surplus for the USA and European and Asian countries is 15.7, 10.9, and 2.0%, respectively.

55.3.8 Web of Science Subject Categories

Brief information about the top 17 'Web of Science' subject categories with at least three sample papers each is provided in Table 55.7. The sample and population papers are indexed by 11 and 120 subject categories, respectively.

For the sample papers, the top subjects are 'Toxicology', 'Immunology', and 'Respiratory System' with 25, 24, and 19 papers, respectively. The other prolific subjects are 'Allergy', 'Environmental Sciences', 'Biochemistry Molecular Biology', and 'Public Environmental Occupational Health' with 13, 13, 12, and 11 papers, respectively.

It is notable that the publication surplus is most significant for 'Immunology' and 'Respiratory System' with 16.1 and 13.9% surpluses, respectively. The other high-impact subjects are 'Allergy', 'Biochemistry Molecular Biology', and 'Critical Care Medicine' with 9.8, 8.1, and 4.7% publication surpluses, respectively.

On the other hand, the subjects with least impact are 'Public Environmental Occupational Health', 'Environmental Sciences', and 'Toxicology' with –8.4, –14.2, and –16.1% publication deficits, respectively.

55.3.9 Citation Impact

These sample and population papers received over 20,500 and 62,500 citations, respectively, as of January 2020. Thus, the average number of citations per paper for the sample and population papers are 205 and 29, respectively.

TABLE 55.7
Web of Science Subject Categories

	Subject	No. of Sample Papers (%)	No. of Population Papers (%)	Difference (%)
1	Toxicology	25	41.1	–16.1
2	Immunology	22	5.9	16.1
3	Respiratory System	19	5.1	13.9
4	Allergy	13	3.2	9.8
5	Environmental Sciences	13	27.2	–14.2
6	Biochemistry Molecular Biology	12	3.9	8.1
7	Public Environmental Occupational Health	11	19.4	–8.4
8	Critical Care Medicine	6	1.3	4.7
9	Pharmacology Pharmacy	6	7.6	–1.6
10	Cell Biology	5	1.6	3.4
11	Endocrinology Metabolism	5	1.1	3.9
12	Oncology	5	2.7	2.3
13	Cardiac Cardiovascular Systems	3	1.2	1.8
14	Chemistry Analytical	3	1.4	1.6
15	Engineering Chemical	3	2.6	0.4
16	Engineering Environmental	3	3.2	–0.2
17	Genetics Heredity	3	3.2	–0.2

55.3.10 Keywords

Although a number of keywords are listed in the Appendix for the dataset of health and safety, some of them are more significant for the sample papers. For this dataset, the most-prolific keywords are 'respir*', 'cell*', 'human*', '*inflam*', 'health', 'lung*', and 'airway*' at the first level with 30, 24, 21, 17, 16, 18, and 15 occurrences. The other prolific keywords are 'asthma*', 'ige', 'disease*', 'in vivo*', 'nasal', 'inhalation', 'cytokine*', 'mutagen*', 'cancer', 'neur*', 'bronch*', 'in vitro', 'mice', 'genes', 'carcin*', 'allergen*', 'antioxidant*', and 'macrophage*'. On the other hand, there is only one word for the dataset on diesel fuels: '*diesel'.

55.3.11 Research Fronts

Brief information about the key research fronts is provided in Table 55.8. There are four research fronts for these sample papers: 'respiratory illnesses', 'cancer', and 'other illnesses' caused by diesel fuel exhausts, and 'diesel fuel exhaust particle characterization'.

The most-prolific research front is 'respiratory illnesses' caused by diesel fuel exhausts with 65 sample papers. The other prolific research fronts are 'cancer' and 'other illnesses' caused by diesel fuel exhausts with 35 and 18 papers, respectively. There are also two sample papers on the 'characterization of diesel exhaust particles' relevant to the determination of the health effect of diesel fuels.

55.4 DISCUSSION

The size of the research on the health and safety impact of diesel fuel exhausts has increased to over 2,150 papers as of January 2020. The research has developed more in the technological aspects of this field, rather than the social and humanitarian pathways as evidenced by the negligible number of population papers in the indices of the 'Web of Science', SSCI, and A&HCI.

The article types of documents are the primary documents for both datasets and reviews are over-represented by 3.6% in the sample papers (Table 55.1). Thus, the

TABLE 55.8
Research Fronts

	Research Front	No. of Sample Papers (%)
1	**Respiratory illnesses**	62
1.1	Exhaust in general	9
1.2	Particles	53
2	**Cancer**	35
2.1	Exhaust in general	14
2.2	Particles	21
3	**Other illnesses**	18
3.1	Exhaust in general	4
3.2	Particles	14
4	**Particle characterization**	2

contribution of reviews by 7% of the sample papers in this field is not highly exceptional (cf. Konur, 2011, 2012a–n, 2015, 2016a–f, 2017a–f, 2018a–b, 2019a–b).

Nineteen authors from 7seveninstitutions have at least four sample papers each (Table 55.2). Eight of these authors are from the USA and the remaining ones are from the United Kingdom, Sweden, and Japan.

These authors focus on 'respiratory illnesses', 'vascular illnesses', and 'cancer', caused by diesel fuel exhaust. It is significant that there is ample 'gender deficit' among these top authors as only two of them are female (Lariviere et al., 2013; Xie and Shauman, 1998).

Whilst 36 and 44% of the sample papers are published in the 1990s and 2000s, 30.0 and 48.1% of the population papers were published in the 2000s and 2010s, respectively (Figure 55.1). This finding suggests that the population papers have built on the sample papers, which were primarily published in the 1990s and 2000s, in the 2000s and 2010s.

The engagement of the institutions in this field at the global scale is significant as 140 and over 1,800 institutions contribute to the sample and population papers, respectively.

Fifteen top institutions publish 109 and 39.9% of the sample and population papers, respectively (Table 55.3). The top institution is the 'University of California Los Angeles' of the USA with 28 sample papers and a 25.6% publication surplus. The other top institutions are 'Umea University' of Sweden and the 'National Institute for Environmental Studies' of Japan. As in the case of the top authors, most-prolific countries for these top institutions are the USA, Japan, and the UK with six, four, and three sample papers, respectively.

It is significant that only 38 and 46% of the sample and population papers declare any funding, respectively. The most-prolific country for these top funding bodies is the USA with four bodies followed by the UK with two (Table 55.4).

The three institutes of the 'National Health Institute' of the USA dominate this top funding table. On the other hand, there is no Chinese funding body in this table, although they are over-represented in the population papers. For example, the 'National Natural Science Foundation of China' (NSFC) funds 2.1% of the population papers. These findings are in line with the studies showing heavy research funding in China, where the NSFC is the primary funding agency (Wang et al., 2012).

The sample and population papers are published by 46 and over 545 journals, respectively. It is significant that the top 11 journals publish 55 and 18% of the sample and population papers, respectively (Table 55.5).

The top journals, the 'Journal of Allergy and Clinical Immunology' and the 'Journal of Immunology' publish together 18 sample papers with a 15.9% publication surplus. It is significant that nearly all top journals are related to biomedicine. The top subject categories are 'Immunology', 'Respiratory System', 'Toxicology', 'Allergy', 'Environmental Sciences', 'Biochemistry Molecular Biology', and 'Biology'.

In total, 13 and 120 countries contribute to the sample and population papers, respectively. The top country is the USA, publishing 54.0 and 38.3% of the sample and population papers, respectively, with a 15.7% publication surplus (Table 55.6). This finding is in line with studies arguing that the USA is not losing ground in science and technology (Leydesdorff and Wagner, 2009).

The other prolific countries are Japan, the UK, and Sweden, publishing 21, 16, and 13 sample papers, respectively. These findings are in line with studies showing that European countries and Japan have a superior publication performance in science and technology (Carpenter et al., 1988; Glanzel, 2000; Hayashi, 2003; Youtie et al., 2008).

The European and Asian countries represented in this table publish altogether 42 and 21% of the sample papers and 31.1 and 19.0% of the population papers, respectively.

It is notable that the publication surplus for the USA and European and Asian countries is 15.7, 10.9, and 2.0%, respectively. It is further notable that China has no place in this top country table as it publishes 2.0 and 4.9% of the sample and population papers, respectively. This finding is in contrast with China's efforts to be a leading nation in science and technology (Zhou and Leydesdorff, 2006) but it is in line with the findings of Guan and Ma (2007) and Youtie et al. (2008) relating to China's performance in nanotechnology. Similarly, Brazil, Canada, India, and South Korea have no place in this top country table, although they each make significant contributions to the population papers (Anuradha and Urs, 2007; Fu and Ho, 2015; Glanzel et al., 2006; Park et al., 2016).

The sample and population papers are indexed by 11 and 120 subject categories, respectively. For the sample papers, the top subjects are 'Toxicology', 'Immunology', and 'Respiratory System' with 25, 24, and 19 papers, respectively (Table 55.7). The other prolific subjects are 'Allergy', 'Environmental Sciences', 'Biochemistry Molecular Biology', and 'Public Environmental Occupational Health'.

It is notable that the publication surplus is most significant for 'Immunology' and 'Respiratory System' with 16.1 and 13.9% surpluses, respectively. The other high-impact subjects are 'Allergy', 'Biochemistry Molecular Biology', and 'Critical Care Medicine' with 9.8, 8.1, and 4.7% publication surpluses, respectively.

On the other hand, the subjects with least impact are 'Public Environmental Occupational Health', 'Environmental Sciences', and 'Toxicology' with –8.4, –14.2, and –16.1% publication deficits, respectively.

These sample and population papers receive over 20,500 and 62,500 citations, respectively, as of January 2020. Thus, the average number of citations per paper for the sample and population papers are 205 and 29, respectively. Thus, the citation impact of the research in this field has been relatively significant (cf. Konur, 2011, 2012a–n, 2015, 2016a–f, 2017a–f, 2018a–b, 2019a–b).

Although a number of keywords are listed in the Appendix for the dataset of health and safety, some of them are more significant. For this dataset, the most-prolific keywords are 'respir*', 'cell*', 'human*', '*inflam*', 'health', 'lung*', and 'airway*' at the first level with 30, 24, 21, 17, 16, 18, and 15 occurrences. The other prolific keywords are 'asthma*', 'ige', 'disease*', 'in vivo*', 'nasal', 'inhalation', 'cytokine*', 'mutagen*', 'cancer', 'neur*', 'bronch*', 'in vitro', 'mice', 'genes', 'carcin*', 'allergen*', 'antioxidant*', and 'macrophage*'.

On the other hand, there is only one word for the dataset of diesel fuels: '*diesel'. As expected, these keywords provide valuable information about the pathways of the research in this field.

Four research fronts emerge from the examination of the sample papers: 'respiratory illnesses', 'cancer', and 'other illnesses' caused by diesel fuel exhausts, and 'diesel fuel exhaust particle characterization' (Table 55.8).

The most-prolific research front is 'respiratory illnesses' caused by diesel fuel exhaust with 65 sample papers. The other prolific research fronts are 'cancer' and other illnesses' caused by diesel fuel exhaust with 35 and 18 papers, respectively. There are also two sample papers on the 'characterization of diesel exhaust particles' relevant to the determination of the health effect of diesel fuels.

It is notable that the research on the health and safety impact of diesel fuel exhaust has been a subfield of the research on the health and safety impact of particles in general (Brauer et al., 2001; Heyder et al., 1986; Peters et al., 1997). Therefore, many findings related to the relationships between the structure and health and safety impact of these particles are primarily also relevant for the research on the health and safety impact of diesel fuel exhaust particles.

The key emphasis in these research fronts is the exploration of the structure–processing–property relationships between petrodiesel fuel exhausts and their significant impact on humans (Cheng and Ma, 2011; Konur and Matthews, 1989; Rogers and Hopfinger, 1994; Scherf and List, 2002).

55.5 CONCLUSION

This chapter has mapped the research on the health and safety impact of diesel fuel exhausts using a scientometric method.

The size of over 2,150 population papers shows the public importance of this interdisciplinary research field. However, it is significant that the research has developed more in the technological aspects in this field, rather than the social and humanitarian pathways.

Articles and reviews dominate the sample papers, primarily published in the 1990s and 2000s. The population papers, primarily published in the 2000s and 2010s, build on these sample papers.

The data presented in the tables and figure show that a small number of authors, institutions, funding bodies, journals, keywords, research fronts, subject categories, and countries have shaped the research in this field.

It is notable that the authors, institutions, and funding bodies from the USA, the UK, Japan, and Sweden dominate the research in this field. Furthermore, China, Brazil, and South Korea are under-represented significantly in the sample papers.

These findings show the importance of the progression of efficient incentive structures for the development of the research in this field as in other fields. It seems that the US, the UK, Sweden, and Japan have efficient incentive structures for the development of the research, contrary to China, Brazil, Australia, and South Korea.

It further seems that although research funding is a significant element of these incentive structures, it might not be a sole solution for increasing the incentives for the research in this field as in the case of China, Brazil, Australia, and South Korea.

On the other hand, it seems there is more to do to reduce the significant gender deficit in this field as in other fields of biomedical research (Lariviere et al., 2013; Xie and Shauman, 1998).

The data on the research fronts, keywords, source titles, and subject categories provide valuable evidence for the interdisciplinary nature of the research in this field. These findings are in line with studies arguing for the interdisciplinarity of research in nanomaterials, relevant to diesel fuel exhaust nanoparticles (Meyer and Persson, 1998; Schummer, 2004).

There is ample justification for the broad search strategy employed in this study due to the interdisciplinary nature of this research field as evidenced by the top subject categories. The search strategy employed in this study is in line with the search strategies employed for related and other research fields (Konur, 2011, 2012a–n, 2015, 2016a–f, 2017a–f, 2018a–b, 2019a–b).

Four research fronts emerge from the examination of the sample papers. The most-prolific research fronts are 'respiratory illnesses', 'cancer' and 'other illnesses' caused by diesel fuel exhausts, and 'characterization of diesel exhaust particles' relevant to the determination of the health effect of diesel fuels.

It is recommended that further scientometric studies are carried out for each of these research fronts, building on the pioneering studies in these fields.

ACKNOWLEDGMENTS

The contribution of the highly cited researchers in the field of the health impact of diesel fuel emissions is greatly acknowledged.

55.A APPENDIX

(TI = (*diesel*) and (TI = (cancer or lung* or respir* or asthma* or *inflam* or ige or *allerg* or inhal* or disease* or health* or *toxic* or cardio* or *vascul* or airway* or dysfunction* or macrophage* or bronch* or *carcin* or mutagen* or *nasal or *dna or "biological eff*" or *pulmon* or occupat* or tumor* or cytokine* or worker* or lymph* or reproduct* or immun* or endothel* or miner* or myocardial or infarction or *thrombosis or nrf2* or disorder* or "oxidative stress*" or "reactive oxygen" or microglia* or *interleukin* or nitroarene* or *thromb* or eosinophil* or neutrophil* or apopto* or adjuvant or ovalbumin or blood* or tnf or histamine or "benzo(a)pyrene*" or heart or coronary or ischemic or ischaemic or fibrinolysis or antigen or mucos* or chronic* or mice or rats or rat or hamster or cytokine or *vitro or *vivo or *trach* or *alveol* or nitrophenol* or *estrogen* or *androgen* or vaso* or hypertens* or hyper-respons* or endocr* or spermato* or serum or testosterone or hormon* or testis or testicular or epithel* or *nitrobenzanthrone or syndrome or autis* or *sclerosis or *sclerotic* or leukemia or nerve* or infect* or rodent* or fibrosis or hepatic or "adverse effect*" or listeria or superoxide or chemokine* or atopic or monkey* or dentritic or *lipoprotein* or murine or phagocyt* or mrna* or aort* or cardiac or arrhythmia* or mitochond* or hyperlipidemia or lavage or glutathione or *quinone* or cholesterol) OR AU = ("sagai m" or "takano h" or "sandstrom t" or "ichinose t" or "blomberg a" or "mauderly jl" or "mcclellan ro" or "silverman dt" or "yanagisawa r" or "diaz-sanchez d" or "diazsanchez d" or "carlsten c" or "gilmour mi" or "loft s" or "vermeulen r" or "yoshikawa t" or "moller p" or "inoue k" or "nemmar a") OR WC = (oncol* or respir* or med* or cardiac* or clin* or toxic* or "public env*" or pharm* or immun* or allergy or periph* or "critical care*" or endoc* or path* or "chemistry med*" or reproduct* or hemat* or surg* or gastro*))) NOT TI = (nontox* or ultrason* or bmi)

REFERENCES

Anuradha, K. and S. Urs. 2007. Bibliometric indicators of Indian research collaboration patterns: A correspondence analysis. *Scientometrics* 71:179–189.

Birch, M. E. and R. A. Cary. 1996. Elemental carbon-based method for monitoring occupational exposures to particulate diesel exhaust. *Aerosol Science and Technology* 25:221–241.

Brauer, M., S. T. Ebelt, T. V. Fisher, et al. 2001. Exposure of chronic obstructive pulmonary disease patients to particles: Respiratory and cardiovascular health effects. *Journal of Exposure Science and Environmental Epidemiology* 11:490–500.

Busca, G., L. Lietti, G. Ramis and F. Berti. 1998. Chemical and mechanistic aspects of the selective catalytic reduction of NO_x by ammonia over oxide catalysts: A review. *Applied Catalysis B-Environmental* 18:1–36.

Carpenter, M., F. Gibb, M. Harris, J. Irvine, B. Martin, F. Narin. 1988. Bibliometric profiles for British academic institutions: An experiment to develop research output indicators. *Scientometrics* 14:213–233.

Cheng, Y. Q. and E. Ma. 2011. Atomic-level structure and structure–property relationship in metallic glasses. *Progress in Materials Science* 56:379–473.

Chisti, Y. 2007. Biodiesel from microalgae. *Biotechnology Advances* 25:294–306.

Chisti, Y. 2008. Biodiesel from microalgae beats bioethanol. *Trends in Biotechnology* 26:126–131.

Clarivate Analytics. 2019. *Highly cited researchers: 2019 Recipients*. Philadelphia, PA: Clarivate Analytics. https://recognition.webofsciencegroup.com/awards/highly-cited/2019/ (accessed January 3, 2020).

Diaz-Sanchez, D., A. Tsien, J. Fleming, and A. Saxon. 1997. Combined diesel exhaust particulate and ragweed allergen challenge markedly enhances human in vivo nasal ragweed-specific IgE and skews cytokine production to a T helper cell 2-type pattern. *Journal of Immunology* 158:2406–2413.

Docampo, D. and L. Cram. 2019. Highly cited researchers: A moving target. *Scientometrics* 118:1011–1025.

Fu, H. Z. and Y. S. Ho. 2015. Highly cited Canada articles in Science Citation Index Expanded: A bibliometric analysis. *Canadian Social Science* 11:50.

Garfield, E. 1955. Citation indexes for science. *Science* 122:108–111.

Garfield, E. 1972. Citation analysis as a tool in journal evaluation. *Science* 178:471–479.

Glanzel, W. 2000. Science in Scandinavia: A bibliometric approach. *Scientometrics* 48:121–150.

Glanzel, W., J. Leta and B. Thijs. 2006. Science in Brazil. Part 1: A macro-level comparative study. *Scientometrics* 67:67–86.

Guan, J. C. and N. Ma. 2007. China's emerging presence in nanoscience and nanotechnology – A comparative bibliometric study of several nanoscience 'giants'. *Research Policy* 36:880–886.

Hayashi, T. 2003. Bibliometric analysis on additionality of Japanese R&D programmes. *Scientometrics* 56:301–316.

Heyder, J. J. Gebhart, G. Rudolf, C. F. Schiller, and W. Stahlhofen. 1986. Deposition of particles in the human respiratory tract in the size range 0.005–15 μm. *Journal of Aerosol Science* 17:811–825.

Konur, O. 2000. Creating enforceable civil rights for disabled students in higher education: An institutional theory perspective. *Disability & Society* 15:1041–1063.

Konur, O. 2002a. Access to Nursing Education by disabled students: Rights and duties of nursing programs. *Nurse Education Today* 22:364–374.

Konur, O. 2002b. Assessment of disabled students in higher education: Current public policy issues. *Assessment and Evaluation in Higher Education* 27:131–152.

Konur, O. 2002c. Access to employment by disabled people in the UK: Is the Disability Discrimination Act working? *International Journal of Discrimination and the Law* 5:247–279.

Konur, O. 2006a. Participation of children with dyslexia in compulsory education: Current public policy issues. *Dyslexia* 12:51–67.

Konur, O. 2006b. Teaching disabled students in Higher Education. *Teaching in Higher Education* 11:351–363.

Konur, O. 2007a. A judicial outcome analysis of the Disability Discrimination Act: A windfall for the employers? *Disability & Society* 22:187–204.

Konur, O. 2007b. Computer-assisted teaching and assessment of disabled students in Higher Education: The interface between academic standards and disability rights. *Journal of Computer Assisted Learning* 23:207–219.

Konur, O. 2011. The scientometric evaluation of the research on the algae and bio-energy. *Applied Energy* 88:3532–3540.

Konur, O. 2012a. Evaluation of the research on the social sciences in Turkey: A scientometric approach. *Energy Education Science and Technology Part B: Social and Educational Studies* 4:1893–1908.

Konur, O. 2012b. Prof. Dr. Ayhan Demirbas' scientometric biography. *Energy Education Science and Technology Part A: Energy Science and Research* 28:727–738.

Konur, O. 2012c. The evaluation of the biogas research: A scientometric approach. *Energy Education Science and Technology Part A: Energy Science and Research* 29:1277–1292.

Konur, O. 2012d. The evaluation of the educational research: A scientometric approach. *Energy Education Science and Technology Part B: Social and Educational Studies* 4:1935–1948.

Konur, O. 2012e. The evaluation of the global energy and fuels research: A scientometric approach. *Energy Education Science and Technology Part A: Energy Science and Research* 30:613–628.

Konur, O. 2012f. The evaluation of the research on the Arts and Humanities in Turkey: A scientometric approach. *Energy Education Science and Technology Part B: Social and Educational Studies* 4:1603–1618.

Konur, O. 2012g. The evaluation of the research on the biodiesel: A scientometric approach. *Energy Education Science and Technology Part A: Energy Science and Research* 28:1003–1014.

Konur, O. 2012h. The evaluation of the research on the bioethanol: A scientometric approach. *Energy Education Science and Technology Part A: Energy Science and Research* 28:1051–1064.

Konur, O. 2012i. The evaluation of the research on the biofuels: A scientometric approach. *Energy Education Science and Technology Part A: Energy Science and Research* 28:903–916.

Konur, O. 2012j. The evaluation of the research on the biohydrogen: A scientometric approach. *Energy Education Science and Technology Part A: Energy Science and Research* 29:323–338.

Konur, O. 2012k. The evaluation of the research on the microbial fuel cells: A scientometric approach. *Energy Education Science and Technology Part A: Energy Science and Research* 29:309–322.

Konur, O. 2012l. The scientometric evaluation of the research on the production of bioenergy from biomass. *Biomass and Bioenergy* 47:504–515.

Konur, O. 2012m. The scientometric evaluation of the research on the deaf students in higher education. *Energy Education Science and Technology Part B: Social and Educational Studies 4*:1573–1588.

Konur, O. 2012n. The scientometric evaluation of the research on the students with ADHD in higher education. *Energy Education Science and Technology Part B: Social and Educational Studies* 4:1547–1562.

Konur, O. 2015. Current state of research on algal biodiesel. In *Marine Bioenergy: Trends and Developments*, ed. S. K. Kim, and C. G. Lee, 487–512. Boca Raton, FL: CRC Press.

Konur, O. 2016a. Scientometric overview in nanobiodrugs. In *Nanoarchitectonics for Smart Delivery and Drug Targeting*, ed. A. M. Holban, A.M. Grumezescu,405–428. Amsterdam: Elsevier.

Konur, O. 2016b. Scientometric overview regarding nanoemulsions used in the food industry. In *Emulsions: Nanotechnology in the Agri-Food Industry*, ed. A. M. Grumezescu, 689–711. Amsterdam: Elsevier.

Konur, O. 2016c. Scientometric overview regarding the nanobiomaterials in antimicrobial therapy. In *Nanobiomaterials in Antimicrobial Therapy*, ed. A. M. Grumezescu, 511–535. Amsterdam: Elsevier.

Konur, O. 2016d. Scientometric overview regarding the nanobiomaterials in dentistry. In *Nanobiomaterials in Dentistry*, ed. A. M. Grumezescu, 425–453. Amsterdam: Elsevier.

Konur, O. 2016e. Scientometric overview regarding the surface chemistry of nanobiomaterials. In *Surface Chemistry of Nanobiomaterials*, ed. A. M. Grumezescu, 463–486. Amsterdam: Elsevier.

Konur, O. 2016f. The scientometric overview in cancer targeting. In *Nanoarchitectonics for Smart Delivery and Drug Targeting*, ed. A. M. Holban, A. Grumezescu, 871–895. Amsterdam: Elsevier.

Konur, O. 2017a. Recent citation classics in antimicrobial nanobiomaterials. In *Nanostructures for Antimicrobial Therapy*, ed. A. Ficai and A. M. Grumezescu, 669–685. Amsterdam: Elsevier.

Konur, O. 2017b. Scientometric overview in nanopesticides. In *New Pesticides and Soil Sensors*, ed. A. M. Grumezescu, 719–744. Amsterdam: Elsevier.

Konur, O. 2017c. Scientometric overview regarding oral cancer nanomedicine. In *Nanostructures for Oral Medicine*, ed. E. Andronescu, A. M. Grumezescu, 939–962. Amsterdam: Elsevier.

Konur, O. 2017d. Scientometric overview regarding water nanopurification. In *Water Purification*, ed. A. M. Grumezescu, 693–716. Amsterdam: Elsevier.

Konur, O. 2017e. Scientometric overview in food nanopreservation. In *Food Preservation*, ed. A. M. Grumezescu, 703–729. Amsterdam: Elsevier.

Konur, O. 2017f. The top citation classics in alginates for biomedicine. In *Seaweed Polysaccharides: Isolation, Biological and Biomedical Applications*, ed. J. Venkatesan, S. Anil, S. K. Kim, 223–249. Amsterdam: Elsevier.

Konur, O. 2018a. Scientometric evaluation of the global research in spine: An update on the pioneering study by Wei et al. *European Spine Journal* 27:525–529.

Konur, O. 2018b. Bioenergy and biofuels science and technology: scientometric overview and citation classics. In *Bioenergy and Biofuels*, ed. O. Konur, 3–63. Boca Raton: CRC Press.

Konur, O. 2019a. Cyanobacterial bioenergy and biofuels science and technology: A scientometric overview. In *Cyanobacteria: From Basic Science to Applications*, ed. A. K. Mishra, D. N. Tiwari and A. N. Rai, 419–442. Amsterdam: Elsevier.

Konur, O. 2019b. Nanotechnology applications in food: A scientometric overview. In *Nanoscience for Sustainable Agriculture*, ed. R. N. Pudake, N. Chauhan, and C. Kole, 683–711. Cham: Springer.

Konur, O., ed. 2021a. *Handbook of Biodiesel and Petrodiesel Fuels: Science, Technology, Health, and Environment.* Boca Raton, FL: CRC Press.

Konur, O., ed. 2021b. *Handbook of Biodiesel and Petrodiesel Fuels: Science, Technology, Health, and Environment. Volume 1. Biodiesel Fuels: Science, Technology, Health, and Environment.* Boca Raton, FL: CRC Press.

Konur, O., ed. 2021c. *Handbook of Biodiesel and Petrodiesel Fuels: Science, Technology, Health, and Environment. Volume 2. Biodiesel Fuels based on the Edible and Nonedible Feedstocks, Wastes, and Algae: Science, Technology, Health, and Environment.* Boca Raton, FL: CRC Press.

Konur, O., ed. 2021d. *Handbook of Biodiesel and Petrodiesel Fuels: Science, Technology, Health, and Environment. Volume 3. Petrodiesel Fuels: Science, Technology, Health, and Environment.* Boca Raton, FL: CRC Press.

Konur, O. 2021e. Biodiesel and petrodiesel fuels: Science, technology, health, and environment. In *Handbook of Biodiesel and Petrodiesel Fuels: Science, Technology, Health, and Environment. Volume 1. Biodiesel Fuels: Science, Technology, Health, and Environment*, ed. O. Konur. Boca Raton, FL: CRC Press.

Konur, O. 2021f. Biodiesel and petrodiesel fuels: A scientometric review of the research. In *Handbook of Biodiesel and Petrodiesel Fuels: Science, Technology, Health, and Environment. Volume 1. Biodiesel Fuels: Science, Technology, Health, and Environment*, ed. O. Konur. Boca Raton, FL: CRC Press.

Konur, O. 2021g. Biodiesel and petrodiesel fuels: A review of the research. In *Handbook of Biodiesel and Petrodiesel Fuels: Science, Technology, Health, and Environment. Volume 1. Biodiesel Fuels: Science, Technology, Health, and Environment*, ed. O. Konur. Boca Raton, FL: CRC Press.

Konur, O. 2021h Nanotechnology applications in the diesel fuels and the related research fields: A review of the research. In *Handbook of Biodiesel and Petrodiesel Fuels: Science, Technology, Health, and Environment. Volume 1. Biodiesel Fuels: Science, Technology, Health, and Environment*, ed. O. Konur. Boca Raton, FL: CRC Press.

Konur, O. 2021i. Biooils: A scientometric review of the research. In *Handbook of Biodiesel and Petrodiesel Fuels: Science, Technology, Health, and Environment. Volume 1. Biodiesel Fuels: Science, Technology, Health, and Environment*, ed. O. Konur. Boca Raton, FL: CRC Press.

Konur, O. 2021j. Characterization and properties of biooils: A review of the research. In *Handbook of Biodiesel and Petrodiesel Fuels: Science, Technology, Health, and Environment. Volume 1. Biodiesel Fuels: Science, Technology, Health, and Environment*, ed. O. Konur. Boca Raton, FL: CRC Press.

Konur, O. 2021k. Biomass pyrolysis and pyrolysis oils: A review of the research. In *Handbook of Biodiesel and Petrodiesel Fuels: Science, Technology, Health, and Environment. Volume 1. Biodiesel Fuels: Science, Technology, Health, and Environment*, ed. O. Konur. Boca Raton, FL: CRC Press.

Konur, O. 2021l. Biodiesel fuels: A scientometric review of the research. In *Handbook of Biodiesel and Petrodiesel Fuels: Science, Technology, Health, and Environment. Volume 1. Biodiesel Fuels: Science, Technology, Health, and Environment*, ed. O. Konur. Boca Raton, FL: CRC Press.

Konur, O. 2021m. Glycerol: A scientometric review of the research. In *Handbook of Biodiesel and Petrodiesel Fuels: Science, Technology, Health, and Environment. Volume 1. Biodiesel Fuels: Science, Technology, Health, and Environment*, ed. O. Konur. Boca Raton, FL: CRC Press.

Konur, O. 2021n. Propanediol production from glycerol: A review of the research. In *Handbook of Biodiesel and Petrodiesel Fuels: Science, Technology, Health, and Environment. Volume 1. Biodiesel Fuels: Science, Technology, Health, and Environment*, ed. O. Konur. Boca Raton, FL: CRC Press.

Konur, O. 2021o. Edible oil-based biodiesel fuels: A scientometric review of the research. In *Handbook of Biodiesel and Petrodiesel Fuels: Science, Technology, Health, and Environment. Volume 2. Biodiesel Fuels based on the Edible and Nonedible Feedstocks, Wastes, and Algae: Science, Technology, Health, and Environment*, ed. O. Konur. Boca Raton, FL: CRC Press.

Konur, O. 2021p. Palm oil-based biodiesel fuels: A review of the research. In *Handbook of Biodiesel and Petrodiesel Fuels: Science, Technology, Health, and Environment. Volume 2. Biodiesel Fuels based on the Edible and Nonedible Feedstocks, Wastes, and Algae*, ed. O. Konur. Boca Raton, FL: CRC Press.

Konur, O. 2021q. Rapeseed oil-based biodiesel fuels: A review of the research. In *Handbook of Biodiesel and Petrodiesel Fuels: Science, Technology, Health, and Environment. Volume 2. Biodiesel Fuels based on the Edible and Nonedible Feedstocks, Wastes, and Algae*, ed. O. Konur. Boca Raton, FL: CRC Press.

Konur, O. 2021r. Nonedible oil-based biodiesel fuels: A scientometric review of the research. In *Handbook of Biodiesel and Petrodiesel Fuels: Science, Technology, Health, and Environment. Volume 2. Biodiesel Fuels based on the Edible and Nonedible Feedstocks, Wastes, and Algae: Science, Technology, Health, and Environment*, ed. O. Konur. Boca Raton, FL: CRC Press.

Konur, O. 2021s. Waste oil-based biodiesel fuels: A scientometric review of the research. In *Handbook of Biodiesel and Petrodiesel Fuels: Science, Technology, Health, and Environment. Volume 2. Biodiesel Fuels based on the Edible and Nonedible Feedstocks, Wastes, and Algae: Science, Technology, Health, and Environment*, ed. O. Konur. Boca Raton, FL: CRC Press.

Konur, O. 2021t. Algal biodiesel fuels: A scientometric review of the research. In *Handbook of Biodiesel and Petrodiesel Fuels: Science, Technology, Health, and Environment. Volume 2. Biodiesel Fuels based on the Edible and Nonedible Feedstocks, Wastes, and Algae: Science, Technology, Health, and Environment*, ed. O. Konur. Boca Raton, FL: CRC Press.

Konur, O. 2021u. Algal biomass production for biodiesel production: A review of the research. In *Handbook of Biodiesel and Petrodiesel Fuels: Science, Technology, Health, and Environment. Volume 2. Biodiesel Fuels based on the Edible and Nonedible Feedstocks, Wastes, and Algae*, ed. O. Konur. Boca Raton, FL: CRC Press.

Konur, O. 2021v. Algal biomass production in wastewaters for biodiesel production: A review of the research. In *Handbook of Biodiesel and Petrodiesel Fuels: Science, Technology, Health, and Environment. Volume 2. Biodiesel Fuels based on the Edible and Nonedible Feedstocks, Wastes, and Algae*, ed. O. Konur. Boca Raton, FL: CRC Press.

Konur, O. 2021x. Algal lipid production for biodiesel production: A review of the research. In *Handbook of Biodiesel and Petrodiesel Fuels: Science, Technology, Health, and Environment. Volume 2. Biodiesel Fuels based on the Edible and Nonedible Feedstocks, Wastes, and Algae*, ed. O. Konur. Boca Raton, FL: CRC Press.

Konur, O. 2021y. Crude oils: A scientometric review of the research. In *Handbook of Biodiesel and Petrodiesel Fuels: Science, Technology, Health, and Environment. Volume 3. Petrodiesel Fuels: Science, Technology, Health, and Environment*, ed. O. Konur. Boca Raton, FL: CRC Press.

Konur, O. 2021z. Petrodiesel fuels: A scientometric review of the research. In *Handbook of Biodiesel and Petrodiesel Fuels: Science, Technology, Health, and Environment. Volume 3. Petrodiesel Fuels: Science, Technology, Health, and Environment*, ed. O. Konur. Boca Raton, FL: CRC Press.

Konur, O. 2021aa. Bioremediation of petroleum hydrocarbons in the contaminated soils: A review of the research. In *Handbook of Biodiesel and Petrodiesel Fuels: Science, Technology, Health, and Environment. Volume 3. Petrodiesel Fuels: Science, Technology, Health, and Environment*, ed. O. Konur. Boca Raton, FL: CRC Press.

Konur, O. 2021ab. Desulfurization of diesel fuels: A review of the research. In *Handbook of Biodiesel and Petrodiesel Fuels: Science, Technology, Health, and Environment. Volume 3. Petrodiesel Fuels: Science, Technology, Health, and Environment*, ed. O. Konur. Boca Raton, FL: CRC Press.

Konur, O. 2021ac. Diesel fuel exhaust emissions: A scientometric review of the research. In *Handbook of Biodiesel and Petrodiesel Fuels: Science, Technology, Health, and Environment. Volume 3. Petrodiesel Fuels: Science, Technology, Health, and Environment*, ed. O. Konur. Boca Raton, FL: CRC Press.

Konur, O. 2021ad. The adverse health and safety impact of diesel fuels: A scientometric review of the research. In *Handbook of Biodiesel and Petrodiesel Fuels: Science, Technology, Health, and Environment. Volume 3. Petrodiesel Fuels: Science, Technology, Health, and Environment*, ed. O. Konur. Boca Raton, FL: CRC Press.

Konur, O. 2021ae. Respiratory illnesses caused by the diesel fuel exhaust emissions: A review of the research. In *Handbook of Biodiesel and Petrodiesel Fuels: Science, Technology, Health, and Environment. Volume 3. Petrodiesel Fuels: Science, Technology, Health, and Environment*, ed. O. Konur. Boca Raton, FL: CRC Press.

Konur, O. 2021af. Cancer caused by the diesel fuel exhaust emissions: A review of the research. In *Handbook of Biodiesel and Petrodiesel Fuels: Science, Technology, Health, and Environment. Volume 3. Petrodiesel Fuels: Science, Technology, Health, and Environment*, ed. O. Konur. Boca Raton, FL: CRC Press.

Konur, O. 2021ag. Cardiovascular and other illnesses caused by the diesel fuel exhaust emissions: A review of the research. In *Handbook of Biodiesel and Petrodiesel Fuels: Science, Technology, Health, and Environment. Volume 3. Petrodiesel Fuels: Science, Technology, Health, and Environment*, ed. O. Konur. Boca Raton, FL: CRC Press.

Konur, O. and F. L. Matthews. 1989. Effect of the properties of the constituents on the fatigue performance of composites: A review. *Composites* 20:317–328.

Lapuerta, M., O. Armas and J. Rodriguez-Fernandez. 2008. Effect of biodiesel fuels on diesel engine emissions. *Progress in Energy and Combustion Science* 34:198–223.

Lariviere, V., C. Ni, Y. Gingras, B. Cronin, and C. R. Sugimoto. 2013. Bibliometrics: Global gender disparities in science. *Nature News* 504:211–213.

Leydesdorff, L. and Wagner, C. 2009. Is the United States losing ground in science? A global perspective on the world science system. *Scientometrics* 78:23–36.

Marchetti, J. M., V. U. Miguel and A. F. Errazu. 2007. Possible methods for biodiesel production. *Renewable and sustainable Energy Reviews* 11:1300–1311.

McCreanor, J, P. Cullinan, and M. J. Nieuwenhuijsen, et al. 2007. Respiratory effects of exposure to diesel traffic in persons with asthma. *New England Journal of Medicine* 357:2348–2358.

Meyer, M. and O. Persson. 1998. Nanotechnology - Interdisciplinarity, patterns of collaboration and differences in application. *Scientometrics* 42:195–205.

Mills, N. L., H. Tornqvist, and M. C. Gonzalez. 2007. Ischemic and thrombotic effects of dilute diesel-exhaust inhalation in men with coronary heart disease. *New England Journal of Medicine 357*:1075–1082.

North, D. C. 1991a. *Institutions, Institutional Change and Economic Performance*. Cambridge, Mass.: Cambridge University Press.

North, D.C. 1991b. Institutions. *Journal of Economic Perspectives* 5:97–112.

Park, H. W., J. Yoon and L. Leydesdorff. 2016. The normalization of co-authorship networks in the bibliometric evaluation: The government stimulation programs of China and Korea. *Scientometrics* 109:1017–1036.

Peters, A., H. E. Wichmann, T. Tuch, J. Heinrich, and J. Heyder. 1997. Respiratory effects are associated with the number of ultrafine particles. *American Journal of Respiratory and Critical Care Medicine* 155:1376–1383.

Robinson, A. L., N. M. Donahue, and M. K. Shrivastava, et al. 2007. Rethinking organic aerosols: Semivolatile emissions and photochemical aging. *Science* 315:1259–1262.

Rogers, D., and A. J. Hopfinger. 1994. Application of genetic function approximation to quantitative structure-activity relationships and quantitative structure-property relationships. *Journal of Chemical Information and Computer Sciences* 34:854–866.

Rogge, W. F., L. M. Hildemann, M. A. Mazurek, G. R. Cass, and B. R. T. Simoneit. 1993. Sources of fine organic aerosol. 2. Noncatalyst and catalyst-equipped automobiles and heavy-duty diesel trucks. *Environmental Science & Technology* 27:636–651.

Salvi, S., A. Blomberg, and B. Rudell, et al. 1999. Acute inflammatory responses in the airways and peripheral blood after short-term exposure to diesel exhaust in healthy human volunteers. *American Journal of Respiratory and Critical Care Medicine* 159:702–709.

Scherf, U. and E. J. List. 2002. Semiconducting polyfluorenes—towards reliable structure–property relationships. *Advanced Materials* 14:477–487.

Schummer, J. 2004. Multidisciplinarity, interdisciplinarity, and patterns of research collaboration in nanoscience and nanotechnology. *Scientometrics* 59:425–465.

Srivastava, A. and R. Prasad. 2000. Triglycerides-based diesel fuels. *Renewable and Sustainable Energy Reviews* 4:111–133.

Van Gerpen, J. 2005. Biodiesel processing and production. *Fuel Processing Technology* 86:1097–1107.

Wang, X., D. Liu, K. Ding, and X. Wang. 2012. Science funding and research output: A study on 10 countries. *Scientometrics* 91:591–599.

Xie, Y. and K. A. Shauman. 1998. Sex differences in research productivity: New evidence about an old puzzle. *American Sociological Review* 847–870.

Youtie, J, P. Shapira, and A. L. Porter. 2008. Nanotechnology publications and citations by leading countries and blocs. *Journal of Nanoparticle Research* 10:981–986.

Zhou, P. and L. Leydesdorff. 2006. The emergence of China as a leading nation in science. *Research Policy* 35:83–104.

56 Respiratory Illnesses Caused by Diesel Fuel Exhaust Emissions

A Review of the Research

Ozcan Konur

CONTENTS

56.1 INTRODUCTION

Crude oils have been primary sources of energy and fuels, such as petrodiesel. However, significant public concerns about the sustainability, price fluctuations, and adverse environmental impact of crude oils have emerged since the 1970s (Ahmadun et al., 2009; Atlas, 1981; Babich and Moulijn, 2003; Kilian, 2009; Perron, 1989). Thus, biooils (Bridgwater and Peacocke, 2000; Czernik and Bridgwater, 2004) and biooil-based biodiesel fuels (Chisti, 2007; Hill et al., 2006) have emerged as alternatives to crude oils and crude oil-based petrodiesel fuels in recent decades. Nowadays, although biodiesel fuels are being used increasingly in the transportation and power sectors, petrodiesel fuels are still used extensively (Konur, 2021a–ag).

Therefore, the adverse health impact of 'diesel fuel exhaust particles' (DEPs) have been a great source of public concern (Konur 2021ac–ad). In this context, the causation of respiratory illnesses (Ashbaugh et al., 1967; Eder et al., 2006; Force et al., 2012; Galli et al., 2008; Kay, 2001; Wills-Karp et al., 1998) by DEPs in humans has been studied more extensively (Konur, 2021ac–ad).

However, for the efficient progression of the research in this field, it is necessary to develop efficient incentive structures for the primary stakeholders and to inform these stakeholders about the research (Konur, 2000, 2002a–c, 2006a–b, 2007a–b; North, 1991a–b).

Although there have been a number of reviews and book chapters in this field (McClellan, 1987; Nel et al., 1998; Pandya et al., 2002; Riedl and Diaz-Sanchez, 2005), there has been no review of the 25-most-cited articles in this field. Thus, this chapter reviews these articles by highlighting the key findings of these most-prolific studies on the causation of respiratory illnesses by DEPs. Then, it discusses the findings of the review.

56.2 MATERIALS AND METHODOLOGY

The search for the literature was carried out in the 'Web of Science' (WOS) database in February 2020. It contains the 'Science Citation Index-Expanded' (SCI-E), the 'Social Sciences Citation Index' (SSCI), the 'Book Citation Index-Science' (BCI-S), the 'Conference Proceedings Citation Index-Science' (CPCI-S), the 'Emerging Sources Citation Index' (ESCI), the 'Book Citation Index-Social Sciences and Humanities' (BCI-SSH), the 'Conference Proceedings Citation Index-Social Sciences and Humanities' (CPCI-SSH), and the 'Arts and Humanities Citation Index' (A&HCI).

The keywords for the search of the literature are collated from the screening of abstract pages for the first 500 highly cited papers on the adverse health impact of DEPs. These keywords sets are provided in the Appendix of the related book chapter (Konur, 2021ac).

The 25-most-cited articles are selected for this review and the key findings of it are discussed briefly.

56.3 RESULTS

56.3.1 Human Studies

Salvi et al. (1999) study the acute inflammatory responses in the airways and peripheral blood after short-term exposure to 'diesel exhaust' (DE) in healthy human volunteers in a paper with 608 citations. They exposed 15 healthy human volunteers to air and diluted DE under controlled conditions for 1 h with intermittent exercise. They found that while standard lung function measures did not change following DE exposure, there was a significant increase in neutrophils and B lymphocytes in airway lavage, along with increases in histamine and fibronectin. There was a significant increase in neutrophils, mast cells, CD4+, and CD8+ T lymphocytes along with upregulation of the endothelial adhesion molecules 'intercellular adhesion molecule

1' (ICAM-1) and 'vascular cell adhesion molecule 1' (VCAM-1), with increases in the numbers of 'lymphocyte function-associated antigen 1' (LFA-1) + cells in the bronchial tissue. There were also significant increases in neutrophils and platelets in peripheral blood following DE exposure. They conclude that at high ambient concentrations, acute short-term DE exposure produced a well-defined and marked systemic and pulmonary inflammatory response in healthy human volunteers.

McCreanor et al. (2007) study the respiratory effects of exposure to diesel traffic in persons with asthma in a paper with 531 citations. They found that participants had significantly higher exposure to fine particles, ultrafine particles, elemental carbon, and nitrogen dioxide in Oxford Street than in Hyde Park. Walking for two hours in Oxford Street induced asymptomatic but consistent reductions in the 'forced expiratory volume in one second' (FEV1) and 'forced vital capacity' (FVC) that were significantly larger than the reductions in FEV1 and FVC after exposure in Hyde Park. The effects were greater in subjects with moderate asthma than in those with mild asthma. These changes were accompanied by increases in biomarkers of neutrophilic inflammation and airway acidification. The changes were associated most consistently with exposures to ultrafine particles and elemental carbon.

Diaz-Sanchez et al. (1994) study the effects of DEP on localized 'immunoglobulin E' (IgE) production by performing nasal challenges with varying doses of DEP in human subjects in a paper with 300 citations. They found a significant rise in nasal IgE but not IgG, IgA, IgM, or albumin in subjects 4 d after being challenged with 0.30 mg DEP. Direct evidence for DEP-enhanced local production of IgE is that the challenge increased the number of IgE-secreting cells in lavage fluid but did not alter the number of IgA-secreting cells. There was a concomitant increase in ε mRNA production in the lavage cells. Additionally, DEP altered the relative amounts of five different ε mRNAs generated by alternative splicing, mRNAs that code for different IgE proteins. They conclude that DEP exposure *in vivo* causes both quantitative and qualitative changes in local IgE production with the implication that natural exposure to DEP may result in increased expression of respiratory allergic disease.

Diaz-Sanchez et al. (1999) study the ability of DEP exposure to lead to primary sensitization of humans by driving a de novo mucosal IgE response to a neoantigen, 'keyhole limpet hemocyanin' (KLH) in a paper with 225 citations. They gave ten atopic subjects an initial nasal immunization with 1 mg of KLH followed by two biweekly nasal challenges with 100 μg of KLH. They then performed identical nasal KLH immunization on 15 different atopic subjects; DEPs were administered 24 hours before each KLH exposure. They found that exposure to KLH alone led to the generation of an anti-KLH IgG and IgA humoral response, which was detected in nasal fluid samples. No anti-KLH IgE appeared in any subjects. In contrast, when challenged with KLH preceded by DEPs, 9 of the 15 subjects produced anti-KLH-specific IgE. KLH-specific IgG and IgA at levels similar to that seen with KLH alone could also be detected. Subjects who received DEPs and KLH had significantly increased 'interleukin-4' (IL-4), but not IFN-γ, levels in nasal lavage fluid, whereas these levels were unchanged in subjects receiving KLH alone. They conclude that DEPs can act as mucosal adjuvants to a de novo IgE response and may increase allergic sensitization.

Salvi et al. (2000) study the IL-8 and 'growth-regulated oncogene' (GRO-α) production in healthy human airways exposed to DE in a paper with 219 citations. They hypothesized that the leukocyte infiltration and the various inflammatory responses induced by DE were mediated by enhanced chemokine and cytokine production by resident cells of the airway tissue and lumen. They exposed 15 healthy human volunteers to diluted DE and air on two separate occasions for 1 h each in an exposure chamber. They show that DE enhanced gene transcription of IL-8 in the bronchial tissue and BW cells along with increases in IL-8 and GRO-α protein expression in the bronchial epithelium, and an accompanying trend toward an increase in IL-5 mRNA gene transcripts in the bronchial tissue. There were no significant changes in the gene transcript levels of IL-1β, 'tumor necrosis factor-α' (TNF-α), 'interferon γ' (IFN-γ), and 'granulocyte macrophage colony-stimulating factor' (GM-CSF), either in the bronchial tissue or BW cells after DE exposure at this time point. They propose an underlying mechanism for DE-induced airway leukocyte infiltration.

Diaz-Sanchez et al. (1996) study the enhanced nasal cytokine production in human beings exposed to DEP to determine whether DEPs can alter the production of cytokines by cells residing in the nasal mucosa in a paper with 205 citations. They exposed participants for 18 hours to intranasal saline solution or DEPs. Before the challenge, most subjects' nasal lavage cells had detectable levels of only IFN-γ , IL-2, and IL-13 mRNA. After the challenge, the cells produced readily detectable mRNA for IL-2, IL-4, IL-5, IL-6, IL-10, IL-13, and IFN-γ . In addition, the levels of all cytokine mRNA increased. Enhanced IL-4 protein was also present in the post-challenge lavage fluid. Although the cells in nasal lavage before and after the challenge do not necessarily represent the same cells in either number or type, the broad increase in cytokine production was not simply the result of an increase in T cells recovered in the lavage fluid. They conclude that an increase in nasal cytokine expression after exposure to DEPs can contribute to enhanced local IgE production and thus play a role in the increased incidence of respiratory allergic disease.

Stenfors et al. (2004) study the airway inflammatory responses in asthmatic and healthy humans exposed to DEPs in a paper with 179 citations. They hypothesize that exposure to DE would induce airway neutrophilia in healthy subjects, and that these responses would be exaggerated in subjects with mild allergic asthma, or DE would exacerbate pre-existent allergic airways. They exposed healthy and mild asthmatic subjects for 2 h to ambient levels of DE (108 microg. × m^{-3}). They find that both groups showed an increase in airway resistance of similar magnitude after DE exposure. Healthy subjects developed airway inflammation 6 h after DE exposure, with airway neutrophilia and lymphocytosis together with an increase in IL-8 protein in lavage fluid, increased IL-8 messenger RNA expression in the bronchial mucosa, and upregulation of endothelial adhesion molecules. In asthmatic subjects, DE exposure did not induce a neutrophilic response or exacerbate their pre-existing eosinophilic airway inflammation. Epithelial staining for the cytokine IL-10 was increased after DE in the asthmatic group. They observed differential effects on the airways of healthy subjects and asthmatics of particles with a 50% cut-off aerodynamic diameter of 10 microm.

Nightingale et al. (2000) study the inflammatory response to inhalation of DEPs in healthy volunteers in a paper with 180 citations. They exposed ten nonsmoking healthy volunteers for 2 h at rest to a controlled concentration of DEP (monitored at 200 μg/m^3 'particulate matter' (PM) of less than 10 μm aerodynamic diameter

(PM_{10})) or air. They found that there were no changes in cardiovascular parameters or lung function following exposure to DEP. Levels of exhaled CO were increased after exposure to DEP, and were maximal at 1 h (aic 2.9 ppm, DEP: 4.4 ppm). There was an increase *in sputum* neutrophils and 'myeloperoxidase' (MPO) at 4 h after DEP exposure as compared with 4 h after air exposure (neutrophils: 41% versus 32 %; MPO: 151 ng/ml versus 115 ng/ml), but no change in concentrations of inflammatory markers in peripheral blood. They conclude that exposure to DEPs at high ambient concentrations leads to an airway inflammatory response in healthy nonsmoking volunteers.

Baulig et al. (2003) study the metabolic pathways triggered by DEPs in human airway epithelial cells in a paper with 178 citations. DEPs induce a proinflammatory response in 'human bronchial epithelial cells' (16HBE) characterized by the release of proinflammatory cytokines after activation of transduction pathways involving 'mitogen-activated protein kinase' (MAPK) and the transcription factor 'NF-ₖB'. They detected 'reactive oxygen species' (ROS) production in bronchial and nasal epithelial cells exposed to native DEP, 'organic extracts of DEP' (OE-DEP), or several 'polycyclic aromatic hydrocarbons' (PAHs). 'Carbon black particles' mimicking the inorganic part of DEP did not increase ROS production. DEP and OE-DEP also induced the expression of genes for phase I ['cytochrome P-450 1A1' (CYP1A1)] and phase II [NADPH quinone oxidoreductase-1 (NQO-1)] xenobiotic metabolization enzymes, suggesting that DEP-adsorbed organic compounds become bioavailable, activate transcription, and are metabolized, since the CYP1A1 enzymatic activity is increased. Because NQO-1 gene induction is reduced by antioxidants, it could be related to the ROS generated by DEP, most likely through the activation of the stress-sensitive 'nuclear factor erythroid 2-related factor 2' (NRF2) transcription factor. Indeed, DEP induced the translocation of NRF2 to the nucleus and increased protein nuclear binding to the antioxidant responsive element. They conclude that DEP-organic compounds generate an oxidative stress, activate the NRF2 transcription factor, and increase the expression of genes for phase I and II metabolization enzymes.

Diaz-Sanchez (1997) studies the role of DEPs and their associated PAHs (PAH-DEP) in the induction of allergic airway disease in a paper with 172 citations. He examined whether extracts of PAH-DEP act as mucosal adjuvants to help initiate or enhance IgE production in response to common inhaled allergens. *In vitro* studies showed that PAH-DEP enhanced IgE production by tonsillar B-cells in the presence of IL-4 and CD40 monoclonal antibody, and altered the nature of the IgE produced, i.e. a decrease in the CH4'-CHe5 variant, a marker for differentiation of IgE-producing B-cells, and an increase in the M2' variant. *In vivo* nasal provocation studies using 0.30 mg DEP in saline also showed enhanced IgE production in the human upper respiratory mucosa, accompanied by a reduced CH4'-CHe5 mRNA splice variant. The effects of DEP were also isotype-specific, with no effect on IgG, IgA, IgM, or albumin, but it did produce a small increase in the IgG4 subclass. They examined the ability of DEP to act as an adjuvant to the ragweed allergen Amb a I by nasal provocation in ragweed allergic subjects using 0.3 mg DEP, Amb a I, or both. Although allergen and DEP each enhanced ragweed-specific IgE, DEP plus allergen promoted a 16-times greater antigen-specific IgE production. A nasal challenge with

DEP also influenced cytokine production. A ragweed challenge resulted in a weak response, a DEP challenge caused a strong but non-specific response, while allergen plus DEP caused a significant increase in the expression of mRNA for TH0 and TH2-type cytokines (IL-4, IL-5, IL-6, IL-10, IL-13) with a pronounced inhibitory effect on IFN-γ gene expression. He concludes that DEP can enhance B-cell differentiation, and by initiating and elevating IgE production, may play an important role in the increased incidence of allergic airway disease.

56.3.2 Animal Studies

Takano et al. (1997) study the effects of DEP inoculated intratracheally on antigen-induced airway inflammation, local expression of cytokine proteins, and antigen-specific IgE production in mice in a paper with 349 citations. They found that DEP aggravated 'ovalbumin' (OA) -induced airway inflammation, characterized by infiltration of eosinophils and lymphocytes and an increase in goblet cells in the bronchial epithelium. DEP with an antigen markedly increased IL-5 protein levers in lung tissue and bronchoalveolar ravage supernatants compared with either an antigen or DEP alone. The combination of DEP and an antigen induced significant increases in the local expression of IL-4, a GM-CSF, and IL-2, whereas the expression of IFN-γ was not affected. In addition, DEP exhibited adjuvant activity for the antigen-specific production of IgG and IgE. They conclude that DEP can enhance the manifestations of allergic asthma and that this enhancement may be mediated mainly by the increased local expression of IL-5, and by the modulated expression of IL-4 GM-CSF and IL-2.

Sagai et al. (1993) study the role of active oxygen species in pulmonary injury by DEP in mice in a paper with 308 citations. They find that DEP could produce superoxide $O_2^{\bullet-}$ and hydroxyl radical ($^{\bullet}OH$) *in vitro* without any biological activating systems. In this reaction system, $O_2^{\bullet-}$ and $^{\bullet}OH$ productions were inhibited by the addition of 'superoxide dismutase' (SOD) and dimethylsulfoxide, respectively. DEPs, which were washed with methanol, could no longer produce $O_2^{\bullet-}$ and $^{\bullet}OH$, indicating that active components were extractable with organic solvents. Furthermore, DEP instilled intratracheally to a mouse caused high mortality at low dose, although methanol-washed DEP did not kill any mouse. The cause of death is pulmonary edema mediated by endothelial cell damage. The instilled DEP markedly decreased the activities of SOD, glutathione peroxidase, and 'glutathione S-transferase' (GST) in mouse lungs. On the other hand, the death rate and lung injury were markedly prevented by 'polyethylene glycol conjugated SOD' (PEG-SOD) pretreatment prior to DEP administration. The mortality and lung injury by DEP were also suppressed by 'butylated hydroxytoluene' (BHT) pretreatment. They conclude that most parts of DEP toxicity in lungs are due to active oxygen radicals such as $O_2^{\bullet-}$ and $^{\bullet}OH$, and that the cause of death is due to pulmonary edema mediated by endothelial cell damage.

Muranaka et al. (1986) study the adjuvant activity of DEPs for the production of IgE antibody in mice in a paper with 294 citations. They found that the primary IgE antibody responses in mice immunized with an intraperitoneal injection of OA mixed with DEPs were higher than those in the animals immunized with OA alone. They also show this effect of DEP on the production of an IgE antibody in mice when they were immunized with repeated injections of dinitrophenylated-OA. In addition, they observed persistent IgE-antibody response to a major allergen of 'Japanese cedar pollen' (JCPA) in mice immunized with JCPA mixed with DEP but not in the animals immunized with JCPA alone. They conclude that the adjuvant activity of DEP is a possible cause of the associated change in the number of diesel cars and allergic rhinitis caused by pollen in Japan.

Takafuji et al. (1987) study the adjuvant activity of DEP inoculated via the intranasal route for IgE production in mice in a paper with 252 citations. In three-week interval immunization, they found that the IgE antibody responses in mice immunized with intranasal inoculation of OA mixed with DEP were higher than responses in the animals immunized with OA alone. DEP had an adjuvant activity for anti-OA IgE antibody production, even in a small dose such as 1 μg administered over a three-week interval. Also, in one-week interval immunization, they showed the enhancing effect of DEP on anti-OA IgE antibody production when mice were immunized with intranasal inoculation of OA and DEP. They conclude that DEP may exert an adjuvant activity for IgE antibody production after being inhaled into the human body and have some relation to the mechanism of the outbreak of allergic rhinitis caused by pollens in Japan.

Sagai et al. (1996) study the pathogenesis of asthma-like symptoms in mice exposed to DEPs in a paper with 172 citations. They found that DEPs instilled intratracheally and repeatedly to mice (once a week for 16 weeks) caused marked infiltration of inflammatory cells, proliferation of goblet cells, increased mucus secretion, respiratory resistance, and airway constriction. Eosinophils in the submucosa of the proximal bronchi and medium bronchioles increased eightfold following instillation. Eosinophil infiltration was significantly suppressed by pretreatment with 'polyethyleneglycol-conjugated SOD' (PEG-SOD). Bound sialic acid concentrations in bronchial alveolar lavage fluids, an index of mucus secretion, increased with DEP, but were suppressed by pretreatment with PEG-SOD. Goblet cell hyperplasia, airway narrowing, and airway constriction also were observed with DEP. Respiratory resistance in the DEP group to acetylcholine was 11 times higher than in controls, and the increased resistance was significantly suppressed by PEG-SOD pretreatment. They conclude that DEP and/or oxygen radicals derived from DEP cause bronchial asthma in mice as chronic airway inflammation, mucus hypersecretion, reversible airway constriction, and bronchial hyperresponsiveness are important pathogenic features of asthma.

56.3.3 *In vitro* Studies

Diaz-Sanchez et al. (1997) study the combined effects of an intranasal challenge with DEPs plus a ragweed allergen on local humoral immune responses in a paper with 423 citations. They collected nasal lavages from ragweed sensitized subjects at different times after a nasal challenge. As compared with the challenge with ragweed alone, they found that a challenge with both DEP and ragweed induced markedly higher ragweed-specific IgE but not total IgE levels or IgE-secreting cell numbers. Total and specific IgG4 levels also were enhanced, while total IgG levels were not. Synergy was also observed between DEP and ragweed in altering the profile of ε mRNAs generated by alternative splicing, mRNAs that code for different expressed IgE proteins. A challenge with both ragweed plus DEP resulted in decreased expression for Th1-type cytokines (IFN-γ and IL-2) but elevated expression of mRNA for other cytokines (IL-4, IL-5, IL-6, IL-10, IL-13). They conclude that this synergy between DEP and natural allergen exposure is a key feature in increasing allergen-induced respiratory allergic disease.

Li et al. (2004) study the mechanism by which redox cycling organic chemicals, prepared from DEP, induce phase II enzyme expression as a protective response in a paper with 327 citations. They showed that aromatic and polar DEP fractions, which are enriched in PAHs and quinones, respectively, induce the expression of 'heme oxygenase-1' (HO-1), GST, and other phase II enzymes in macrophages and epithelial cells. They then showed that HO-1 expression is mediated through accumulation of the bZIP transcription factor, NRF2, in the nucleus, and that NRF2 gene targeting significantly weakens this response. NRF2 accumulation and subsequent activation of the antioxidant response element is regulated by the proteasomal degradation of NRF2. This pathway is sensitive to pro-oxidative and electrophilic DEP chemicals and is also activated by ambient ultrafine particles. They propose that NRF2-mediated phase II enzyme expression protects against the proinflammatory effects of particulate pollutants in the setting of allergic inflammation and asthma.

Xiao et al. (2003) study the hierarchical oxidative stress response to DEP chemicals in a macrophage cell line by a proteomics approach in a paper with 279 citations. They showed that in the dose range 10–100 μg/ml, organic DEP extracts induce a progressive decline in the cellular 'glutathione/glutathione disulfide' (GSH/GSSG) ratio, in parallel with a linear increase in newly expressed proteins on the two-dimensional gel. They identified 32 newly induced/NAC-suppressed proteins. These include antioxidant enzymes, pro-inflammatory components, and products of intermediary metabolism that are regulated by oxidative stress. HO-1 was induced at a low extract dose and with minimal decline in the GSH/GSSG ratio, whereas MAP kinase activation required a higher chemical dose and incremental levels of oxidative stress. Moreover, at extract doses above 50 μg/ml, there was a steep decline in cellular viability. They conclude that DEPs induce a hierarchical oxidative stress response in which some of these proteins may serve as markers for oxidative stress during PM exposures.

Li et al. (2002a) study the oxidative and proinflammatory effects of organic DEP chemicals in bronchial epithelial cells and macrophages in a paper with 243 citations. They showed that organic chemicals extracted from DEP induce oxidative stress in normal and transformed bronchial epithelial cells, leading to the expression of HO-1, activation of the 'c-Jun N-terminal kinase' cascade, IL-8 production, as well as induction of cytotoxicity. Among these effects, HO-1 expression is the most sensitive marker for oxidative stress, while 'c-Jun N-terminal kinase' activation and induction of apoptosis-necrosis require incremental amounts of the organic chemicals and increased levels of oxidative stress. While a 'macrophage cell line' (THP-1) responded in similar fashion, epithelial cells produced more superoxide radicals and were more susceptible to cytotoxic effects than macrophages. Cytotoxicity is the result of mitochondrial damage, which manifests as ultramicroscopic changes in organelle morphology, a decrease in the mitochondrial membrane potential, superoxide production, and ATP depletion. Epithelial cells also differ from macrophages in not being protected by a thiol antioxidant, 'N-acetylcysteine' (NAC), which effectively protects macrophages against cytotoxic DEP chemicals. They conclude that epithelial cells exhibit a hierarchical oxidative stress response that differs from that of macrophages by a more rapid transition from cytoprotective to cytotoxic responses. Moreover, epithelial cells are not able to convert NAC to cytoprotective glutathione.

Takenaka et al. (1995) study the enhanced human IgE production from exposure to the aromatic hydrocarbons from DEPs in a paper with 223 citations. They showed that the extract of PAHs from DEPs (PAH-DEP) enhances human IgE production from purified B cells. IL-4 plus CD40 monoclonal antibody-stimulated IgE production was enhanced by 20% to 360% by the addition of PAH-DEP over a period of 10 to 14 days. This effect was increased when PAH-DEP was added two to five days after cultures were initiated. PAH-DEP itself did not induce IgE production or synergize with IL-4 alone to induce IgE from purified B cells, suggesting that it was enhancing ongoing IgE production rather than inducing germline transcription or isotype switching. The prototype nonmetabolized aromatic hydrocarbon '2,3,7,8 tetracholorodibenzo-p-dioxin', which functions solely through activation of the cytosolic aromatic hydrocarbon receptor complex, also increased IgE production. Additionally, the pattern of mRNAs coding for distinct isoforms of the ϵ chain was altered by PAH-DEP, and B-cell expression of the low-affinity IgE receptor was upregulated by PAH-DEP. They conclude that enhanced IgE production in the human airway, resulting from exposure to PAH-DEP, may be an important factor in the increase in airway allergic disease.

Li et al. (2000) study the induction of HO-1 expression in macrophages by DEP chemicals and quinones in a paper with 211 citations. They showed that a crude DEP total extract, aromatic and polar DEP fractions, a benzo(a)pyrene quinone, and a phenolic antioxidant induce HO-1 expression in RAW264.7 cells in an 'antioxidant response element' (ARE)-dependent manner. NAC and the flavonoid luteolin inhibited HO-1 protein expression. They also showed that the same stimuli induce HO-1 mRNA expression in parallel with the activation of the SX2 enhancer of that gene. Mutation of the ARE core, but not the overlapping AP-1 binding sequence, disrupted SX2 activation. Finally, they showed that biological agents, such as oxidized '1-palmitoyl-2-arachidonoyl-sn-glycero-3-phosphocholine', could also induce HO-1 expression via an ARE-dependent mechanism. Prior induction of HO-1 expression, using cobalt-protoporphyrin, protected RAW264.7 cells against DEP-induced toxicity. They conclude that HO-1 plays an important role in cytoprotection against redox-active DEP chemicals, including quinones.

Knox et al. (1997) study the binding of free grass pollen allergen molecules (Lol p 1) to DEPs in a paper with 207 citations. They used 'DE carbon particles' (DECP) derived from the exhaust of a stationary diesel engine and natural highly purified Lol p 1. They found that DECP are small carbon spheres, each 30–60 nm in diameter, forming fractal aggregates about 1–2 microns in diameter. They showed by *in vitro* experiments that the major grass pollen allergen, Lol p 1, binds to one defined class of fine particles, DECP. They conclude that DECPs are in the respirable size range, can bind to the major grass pollen allergen Lol p 1 under *in vitro* conditions, and represent a possible mechanism by which allergens can become concentrated in polluted air and thus trigger attacks of asthma.

Bayram et al. (1998) study the effect of DEPs on cell function and the release of inflammatory mediators from 16HBEs *in vitro* to investigate the mechanisms underlying DEP-induced airway disease in humans in a paper with 188 citations. They cultured 16HBE from surgically obtained bronchial explants and investigated the effects of purified DEP on the permeability and 'ciliary beat frequency' (CBF) of 16HBE, and on the release of inflammatory mediators from these cells. They found that exposure to 10–100 microg./ml DEP and a filtered solution of 50 microg./ml DEP significantly increased the electrical resistance of the cultures, reaching a maximum of 200% over the baseline after a 6 h incubation with 100 microg./ml DEP. In contrast, the movement of 14C-labeled bovine serum albumin across cell cultures was not significantly altered by incubation of 16HBE with DEP. Exposure to 50 microg./ml DEP, a filtered DEP solution, and 100 migrog./ml DEP significantly attenuated the CBF of these cells by 51, 33, and 73%, respectively, from the baseline after a 24 h incubation. Similarly, 50 microg./ml DEP, a filtered DEP solution, and 100 microg./ml DEP significantly increased the release of IL-8 from 12.9 pg/microg cellular protein to 41.6, 114.9, and 44.3 pg/microg cellular protein, respectively, after a 24 h incubation. The release of a GM-CSF and a 'soluble intercellular adhesion molecule-1' (sICAM-1) was also significantly increased after exposure for 24 h to 50 microg./ml DEP (GM-CSF) from 0.033 pg/microg. cellular protein to 0.056 pg/μg cellular protein, and sICAM-1 from 7.2 pg/microg. cellular protein to 12.5 pg/microg cellular protein). They conclude that exposure of 16HBE to DEP may lead to adverse functional changes and the release of proinflammatory mediators from these cells, and that these effects may influence the development of airway disease.

Bonvallot et al. (2001) study the critical role of organic compounds in the DEP-induced proinflammatory response in human airway epithelial cells in a paper with 184 citations. They found that OE-DEP and nDEP induce GM-CSF release, 'nuclear factor NF-ₖB activation', and MAPK phosphorylation. The carbonaceous core generally induces less intense effects. ROS are produced in 16HBE and are involved in GM-CSF release and in the stimulation of NF-ₖB DNA binding by nDEP and OE-DEP. They showed in airway epithelial cells *in vitro* that nDEP induces the expression of CYP1A1, a cytochrome P450 specifically involved in PAHs metabolism, thereby demonstrating the critical role of organic compounds in DEP-induced proinflammatory response. They conclude the DEP-induced inflammatory response in airway epithelial cells mainly involves organic compounds, such as PAH, which induce CYP1A1 gene expression.

Diaz-Sanchez (1997) studies the role of DEPs and their associated PAHs (PAH-DEP) in the induction of allergic airway disease in a paper with 172 citations. He examined whether extracts of PAH-DEP acted as mucosal adjuvants to help initiate or enhance IgE production in response to commonly inhaled allergens. *In vitro* studies show that PAH-DEP enhanced IgE production by tonsillar B-cells in the presence of IL-4 and a CD40 monoclonal antibody, and altered the nature of the IgE produced, i.e. a decrease in the CH4'-CHe5 variant, a marker for the differentiation of IgE-producing B-cells, and an increase in the M2' variant. *In vivo* nasal provocation studies using 0.30 mg DEP in saline also showed enhanced IgE production in the human upper respiratory mucosa, accompanied by a reduced CH4'-CHe5 mRNA splice variant. The effects of DEP were also isotype-specific, with no effect on IgG, IgA, IgM, or albumin, but it did produce a small increase in the IgG4 subclass. They examined the ability of DEP to act as an adjuvant to the ragweed allergen Amb a I by nasal provocation in ragweed allergic subjects using 0.3 mg DEP, Amb a I, or both. Although the allergen and DEP each enhanced ragweed-specific IgE, DEP plus an allergen promoted a 16-times greater antigen-specific IgE production. A nasal challenge with DEP also influenced cytokine production. A ragweed challenge resulted in a weak response, a DEP challenge caused a strong but non-specific response, while the allergen plus DEP caused a significant increase in the expression of mRNA for TH0 and TH2-type cytokines (IL-4, IL-5, IL-6, IL-10, IL-13) with a pronounced inhibitory effect on IFN-γ gene expression. He concludes that DEP can enhance B-cell differentiation, and by initiating and elevating IgE production, may play an important role in the increased incidence of allergic airway disease.

Li et al. (2002b) study the biological effects of ambient air PM and DEPs through a stratified oxidative stress model in a paper with 169 citations. They showed that organic DEP extracts induce a stratified oxidative stress response, leading to HO-1 expression at normal GSH/GSSG ratios, proceed to Jun kinase activation and IL-8 production at intermediary oxidative stress levels, and culminate in cellular apoptosis in parallel with a sharp decline in GSH/GSSG ratios. They next showed that PM mimics the effects of organic DEP extracts at lower oxidative stress levels. While fine PM consistently induced HO-1 expression, coarse particulates were effective at inducing that effect during fall and winter. Moreover, HO-1 expression was positively correlated to the higher 'organic carbon' (OC) and PAHs content of fine versus coarse PM, as well as the rise in PAH content that occurs in coarse PM during the winter months. Although coarse and fine PM lead to a decrease in cellular GSH/GSSG ratios, oxidative stress did not increase to cytotoxic levels. They conclude that it is possible to use the stratified oxidative stress model developed for DEP to interpret the biological effects of coarse and fine PM.

56.4 DISCUSSION

Table 56.1 provides information on the research fronts in this field. As this table shows the research fronts of 'human studies', 'animal studies', and '*in vitro* studies' comprise 40, 20, and 44% of these papers, respectively.

TABLE 56.1
Research Fronts

	Research Front	Papers (%)
1	Human studies	40
2	Animal studies	20
3	*In vitro* studies	44

56.4.1 Human Studies

In the first group of papers, Salvi et al. (1999) study the acute inflammatory responses in the airways and peripheral blood after short-term exposure to DE in healthy human volunteers in a paper with 608 citations. They conclude that at high ambient concentrations, acute short-term DE exposure produced a well-defined and marked systemic and pulmonary inflammatory response in healthy human volunteers.

McCreanor et al. (2007) study respiratory effects of exposure to diesel traffic in persons with asthma in a paper with 531 citations. They conclude that the effects were greater in subjects with moderate asthma than in those with mild asthma. These changes were accompanied by increases in biomarkers of neutrophilic inflammation and airway acidification.

Diaz-Sanchez et al. (1994) study the effects of DEP on localized IgE production by performing nasal challenges with varying doses of DEP in human subjects in a paper with 300 citations. They conclude that DEP exposure *in vivo* causes both quantitative and qualitative changes in local IgE production with the implication that natural exposure to DEP may result in increased expression of respiratory allergic disease.

Diaz-Sanchez et al. (1999) study the ability of DEP exposure to lead to primary sensitization of humans by driving a de novo mucosal IgE response to a neoantigen, a KLH, in a paper with 225 citations. They conclude that DEPs can act as mucosal adjuvants to a de novo IgE response and may increase allergic sensitization.

Salvi et al. (2000) study the IL-8 and GRO-α production in healthy human airways exposed to DE in a paper with 219 citations. They propose an underlying mechanism for DE-induced airway leukocyte infiltration.

Diaz-Sanchez et al. (1996) study the enhanced nasal cytokine production in human beings exposed to DEPs to determine whether they can alter the production of cytokines by cells residing in the nasal mucosa in a paper with 205 citations. They conclude that an increase in nasal cytokine expression after exposure to DEPs can contribute to enhanced local IgE production and thus play a role in the increased incidence of respiratory allergic disease.

Stenfors et al. (2004) study the airway inflammatory responses in asthmatic and healthy humans exposed to DEPs in a paper with 179 citations. They observe differential effects on the airways of healthy subjects and asthmatics of particles with a 50% cut-off aerodynamic diameter of 10 microm.

Nightingale et al. (2000) study the inflammatory response to inhalation of DEPs in healthy volunteers in a paper with 180 citations. They conclude that exposure to DEPs at high ambient concentrations leads to an airway inflammatory response in healthy nonsmoking volunteers.

Baulig et al. (2003) study the metabolic pathways triggered by DEPs in human airway epithelial cells in a paper with 178 citations. They conclude that DEP-organic compounds generate an oxidative stress, activate the NRF2 transcription factor, and increase the expression of genes for phase I and II metabolization enzymes.

Diaz-Sanchez (1997) studies the role of DEPs and their associated PAHs (PAH-DEP) in the induction of allergic airway disease in a paper with 172 citations. They conclude that DEPs can enhance B-cell differentiation, and by initiating and elevating IgE production, may play an important role in the increased incidence of allergic airway disease.

These prolific studies show that DEPs can cause respiratory illnesses in humans with strong public policy implications.

56.4.2 Animal Studies

Takano et al. (1997) study the effects of DEP inoculated intratracheally on antigen-induced airway inflammation, the local expression of cytokine proteins, and antigen-specific IgE production in mice caused by diesel fuel exhaust emissions in a paper with 349 citations. They conclude that DEP can enhance the manifestations of allergic asthma and that this enhancement may be mediated mainly by the increased local expression of IL-5, and also by the modulated expression of IL-4 GM-CSF and IL-2.

Sagai et al. (1993) study the role of active oxygen species in pulmonary injury by DEP in mice in a paper with 308 citations. They conclude that most parts of DEP toxicity in lungs are due to active oxygen radicals such as $O_2^{\bullet-}$ and $^{\bullet}OH$, and that the cause of death is due to pulmonary edema mediated by endothelial cell damage.

Muranaka et al. (1986) study the adjuvant activity of DEPs for the production of IgE antibody in mice in a paper with 294 citations. They conclude that the adjuvant activity of DEPs is a possible cause of the associated change in the number of diesel cars and allergic rhinitis caused by pollen in Japan.

Takafuji et al. (1987) study the adjuvant activity of DEP inoculated via the intranasal route for IgE production in mice in a paper with 252 citations. They conclude that DEP may exert an adjuvant activity for IgE antibody production after being inhaled into the human body and have some relation to the mechanism of the outbreak of allergic rhinitis caused by pollens in Japan.

Sagai et al. (1996) study the pathogenesis of asthma-like symptoms in mice exposed to DEP in a paper with 172 citations. They conclude that DEP and/or oxygen radicals derived from DEP cause bronchial asthma in mice, as chronic airway inflammation, mucus hypersecretion, reversible airway constriction, and bronchial hyperresponsiveness are important pathogenic features of asthma.

These prolific studies show that DEPs can cause respiratory illnesses in animals with strong public policy implications.

56.4.3 *In vitro* Studies

Diaz-Sanchez et al. (1997) study the combined effects of an intranasal challenge with DEPs plus a ragweed allergen on local humoral immune responses in a paper with 423 citations. They conclude that this synergy between DEP and natural allergen exposure is a key feature in increasing allergen-induced respiratory allergic disease.

Li et al. (2004) study the mechanism by which redox cycling organic chemicals, prepared from DEP, induce phase II enzyme expression as a protective response in a paper with 327 citations. They propose that NRF2-mediated phase II enzyme expression protects against the proinflammatory effects of particulate pollutants in the setting of allergic inflammation and asthma.

Xiao et al. (2003) study the hierarchical oxidative stress response to DEP chemicals in a macrophage cell line by a proteomics approach in a paper with 279 citations. They conclude that DEP induce a hierarchical oxidative stress response in which some of these proteins may serve as markers for oxidative stress during PM exposures.

Li et al. (2002a) study the oxidative and proinflammatory effects of organic DEP chemicals in bronchial epithelial cells and macrophages in a paper with 243 citations. They conclude that epithelial cells exhibit a hierarchical oxidative stress response that differs from that of macrophages by a more rapid transition from cytoprotective to cytotoxic responses. Moreover, epithelial cells are not able to convert NAC to cytoprotective glutathione.

Takenaka et al. (1995) study the enhanced human IgE production from exposure to the aromatic hydrocarbons from DEPs in a paper with 223 citations. They conclude that enhanced IgE production in the human airway, resulting from exposure to PAH-DEP, may be an important factor in the increase in airway allergic disease.

Li et al. (2000) study the induction of HO-1 expression in macrophages by DEP chemicals and quinones in a paper with 211 citations. They conclude that HO-1 plays an important role in cytoprotection against redox-active DEP chemicals, including quinones.

Knox et al. (1997) study the binding of free grass pollen allergen molecules (Lol p 1) to DEPs in a paper with 207 citations. They conclude that DECP are in the respirable size range, can bind to the major grass pollen allergen Lol p 1 under *in vitro* conditions, and represent a possible mechanism by which allergens can become concentrated in polluted air and thus trigger attacks of asthma.

Bayram et al. (1998) study the effect of DEPs on cell function and the release of inflammatory mediators from 16HBE *in vitro* to investigate the mechanisms underlying DEP-induced airway disease in humans in a paper with 188 citations. They conclude that exposure of 16HBE to DEP may lead to adverse functional changes and release of proinflammatory mediators from these cells, and that these effects may influence the development of airway disease.

Bonvallot et al. (2001) study the critical role of organic compounds in a DEP-induced proinflammatory response in human airway epithelial cells in a paper with 184 citations. They conclude the DEP-induced inflammatory response in airway epithelial cells mainly involves organic compounds such as PAH, which induce CYP1A1 gene expression.

Diaz-Sanchez (1997) studies the role of DEPs and their associated PAHs (PAH-DEP) in the induction of allergic airway disease in a paper with 172 citations. He concludes that DEP can enhance B-cell differentiation, and by initiating and elevating IgE production, may play an important role in the increased incidence of allergic airway disease.

Li et al. (2002b) study the biological effects of ambient air PM and DEPs through a stratified oxidative stress model in a paper with 169 citations. They conclude that it is possible to use the stratified oxidative stress model developed for DEP to interpret the biological effects of coarse and fine PM.

These *in vitro* prolific studies show that DEPs can cause respiratory illnesses with strong public policy implications.

56.5 CONCLUSION

This chapter has presented a review of the 25-most-cited article papers in this field. Table 56.1 provides information on the research fronts.

As this table shows the research fronts of 'human studies', 'animal studies', and '*in vitro* studies' comprise 40, 20, and 44% of these papers, respectively.

These prolific studies in three different research fronts provide valuable evidence on the causation of respiratory illnesses by DEPs with strong public policy implications.

It is recommended that similar studies are carried out for each research front as well.

ACKNOWLEDGMENTS

The contribution of the highly cited researchers in this field is greatly acknowledged.

REFERENCES

Ahmadun, F. R., A. Pendashteh, and L. C. Abdullah, et al. 2009. Review of technologies for oil and gas produced water treatment. *Journal of Hazardous Materials* 170:530–551.

Ashbaugh, D., D. B. Bigelow, T. Petty, and B. Levine. 1967. Acute respiratory distress in adults. *Lancet* 290:319–323.

Atlas, R. M. 1981. Microbial degradation of petroleum hydrocarbons: An environmental perspective. *Microbiological Reviews* 45:180–209.

Babich, I. V. and J. A. Moulijn. 2003. Science and technology of novel processes for deep desulfurization of oil refinery streams: A review. *Fuel* 82:607–631.

Baulig, A., M. Garlatti, and V. Bonvallot, et al. 2003. Involvement of reactive oxygen species in the metabolic pathways triggered by diesel exhaust particles in human airway epithelial cells. *American Journal of Physiology-Lung Cellular and Molecular Physiology* 285:L671–L6L9.

Bayram, H., J. L. Devalia, and R. J. Sapsford, et al. 1998. The effect of diesel exhaust particles on cell function and release of inflammatory mediators from human bronchial epithelial cells in vitro. *American Journal of Respiratory Cell and Molecular Biology* 18:441–448.

Bonvallot, V., A. Baeza-Squiban, and A. Baulig, et al. 2001. Organic compounds from diesel exhaust particles elicit a proinflammatory response in human airway epithelial cells and induce cytochrome p450 1A1 expression. *American Journal of Respiratory Cell and Molecular Biology* 25:515–521.

Bridgwater, A. V. and G. V. C. Peacocke. 2000. Fast pyrolysis processes for biomass. *Renewable & Sustainable Energy Reviews* 4:1–73.

Chisti, Y. 2007. Biodiesel from microalgae. *Biotechnology Advances* 25:294–306.

Czernik, S. and A. V. Bridgwater. 2004. Overview of applications of biomass fast pyrolysis oil. *Energy & Fuels* 18:590–598.

Diaz-Sanchez, D 1997.The role of diesel exhaust particles and their associated polyaromatic hydrocarbons in the induction of allergic airway disease. *Allergy* 52:52–56.

Diaz-Sanchez, D., A. R. Dotson, H. Takenaka, and A. Saxon. 1994. Diesel exhaust particles induce local IgE production *in vivo* and alter the pattern of IgE messenger-RNA isoforms. *Journal of Clinical Investigation* 94:1417–1425.

Diaz-Sanchez, D., M. P. Garcia, M. Wang, M. Jyrala, and A. Saxon. 1999. Nasal challenge with diesel exhaust particles can induce sensitization to a neoallergen in the human mucosa. *Journal of Allergy and Clinical Immunology* 104:1183–1188.

Diaz-Sanchez, D., A. Tsien, A. Casillas, A. R. Dotson, and A. Saxon. 1996. Enhanced nasal cytokine production in human beings after *in vivo* challenge with diesel exhaust particles. *Journal of Allergy and Clinical Immunology* 98:114–123.

Diaz-Sanchez, D., A. Tsien, J. Fleming, J, and A. Saxon. 1997. Combined diesel exhaust particulate and ragweed allergen challenge markedly enhances human *in vivo* nasal ragweed-specific IgE and skews cytokine production to a T helper cell 2-type pattern. *Journal of Immunology* 158:2406–2413.

Eder, W., M. J. Ege, and E. von Mutius. 2006. The asthma epidemic. *New England Journal of Medicine* 355:2226–2235.

Force, A. D. T., V. M. Ranieri, and G. D. Rubenfeld, et al. 2012. Acute respiratory distress syndrome. *JAMA* 307:2526–2533.

Galli, S. J., M. Tsai, and A. M. Piliponsky. 2008. The development of allergic inflammation. *Nature* 454:445–454.

Hill, J., E. Nelson, D. Tilman, S. Polasky, and D. Tiffany. 2006. Environmental, economic, and energetic costs and benefits of biodiesel and ethanol biofuels. *Proceedings of the National Academy of Sciences of the United States of America* 103:11206–11210.

Kay, A. B. 2001. Allergy and allergic diseases. *New England Journal of Medicine* 344:30–37.

Kilian, L. 2009. Not all oil price shocks are alike: Disentangling demand and supply shocks in the crude oil market. *American Economic Review* 99:1053–1069.

Knox, R. B., C. Suphioglu, and P. Taylor, et al. 1997. Major grass pollen allergen Lol p 1 binds to diesel exhaust particles: Implications for asthma and air pollution. *Clinical and Experimental Allergy* 27:246–251.

Konur, O. 2000. Creating enforceable civil rights for disabled students in higher education: An institutional theory perspective. *Disability & Society* 15:1041–1063.

Konur, O. 2002a. Access to Nursing Education by disabled students: Rights and duties of nursing programs. *Nurse Education Today* 22:364–374.

Konur, O. 2002b. Assessment of disabled students in higher education: Current public policy issues. *Assessment and Evaluation in Higher Education* 27:131–152.

Konur, O. 2002c. Access to employment by disabled people in the UK: Is the Disability Discrimination Act working? *International Journal of Discrimination and the Law* 5:247–279.

Konur, O. 2006a. Participation of children with dyslexia in compulsory education: Current public policy issues. *Dyslexia* 12:51–67.

Konur, O. 2006b. Teaching disabled students in Higher Education. *Teaching in Higher Education* 11:351–363.

Konur, O. 2007a. A judicial outcome analysis of the Disability Discrimination Act: A windfall for the employers? *Disability & Society* 22:187–204.

Konur, O. 2007b. Computer-assisted teaching and assessment of disabled students in higher education: The interface between academic standards and disability rights. *Journal of Computer Assisted Learning* 23:207–219.

Konur, O., ed. 2021a. *Handbook of Biodiesel and Petrodiesel Fuels: Science, Technology, Health, and Environment.* Boca Raton, FL: CRC Press.

Konur, O., ed. 2021b. *Handbook of Biodiesel and Petrodiesel Fuels: Science, Technology, Health, and Environment. Volume 1. Biodiesel Fuels: Science, Technology, Health, and Environment.* Boca Raton, FL: CRC Press.

Konur, O., ed. 2021c. *Handbook of Biodiesel and Petrodiesel Fuels: Science, Technology, Health, and Environment. Volume 2. Biodiesel Fuels based on the Edible and Nonedible Feedstocks, Wastes, and Algae: Science, Technology, Health, and Environment.* Boca Raton, FL: CRC Press.

Konur, O., ed. 2021d. *Handbook of Biodiesel and Petrodiesel Fuels: Science, Technology, Health, and Environment. Volume 3. Petrodiesel Fuels: Science, Technology, Health, and Environment.* Boca Raton, FL: CRC Press.

Konur, O. 2021e. Biodiesel and petrodiesel fuels: Science, technology, health, and environment. In *Handbook of Biodiesel and Petrodiesel Fuels: Science, Technology, Health, and Environment. Volume 1. Biodiesel Fuels: Science, Technology, Health, and Environment*, ed. O. Konur. Boca Raton, FL: CRC Press.

Konur, O. 2021f. Biodiesel and petrodiesel fuels: A scientometric review of the research. In *Handbook of Biodiesel and Petrodiesel Fuels: Science, Technology, Health, and Environment. Volume 1. Biodiesel Fuels: Science, Technology, Health, and Environment*, ed. O. Konur. Boca Raton, FL: CRC Press.

Konur, O. 2021g. Biodiesel and petrodiesel fuels: A review of the research. In *Handbook of Biodiesel and Petrodiesel Fuels: Science, Technology, Health, and Environment. Volume 1. Biodiesel Fuels: Science, Technology, Health, and Environment*, ed. O. Konur. Boca Raton, FL: CRC Press.

Konur, O. 2021h. Nanotechnology applications in the diesel fuels and the related research fields: A review of the research. In *Handbook of Biodiesel and Petrodiesel Fuels: Science, Technology, Health, and Environment. Volume 1. Biodiesel Fuels: Science, Technology, Health, and Environment*, ed. O. Konur. Boca Raton, FL: CRC Press.

Konur, O. 2021i. Biooils: A scientometric review of the research. In *Handbook of Biodiesel and Petrodiesel Fuels: Science, Technology, Health, and Environment. Volume 1. Biodiesel Fuels: Science, Technology, Health, and Environment*, ed. O. Konur. Boca Raton, FL: CRC Press.

Konur, O. 2021j. Characterization and properties of biooils: A review of the research. In *Handbook of Biodiesel and Petrodiesel Fuels: Science, Technology, Health, and Environment. Volume 1. Biodiesel Fuels: Science, Technology, Health, and Environment*, ed. O. Konur. Boca Raton, FL: CRC Press.

Konur, O. 2021k. Biomass pyrolysis and pyrolysis oils: A review of the research. In *Handbook of Biodiesel and Petrodiesel Fuels: Science, Technology, Health, and Environment. Volume 1. Biodiesel Fuels: Science, Technology, Health, and Environment*, ed. O. Konur. Boca Raton, FL: CRC Press.

Konur, O. 2021l. Biodiesel fuels: A scientometric review of the research. In *Handbook of Biodiesel and Petrodiesel Fuels: Science, Technology, Health, and Environment. Volume 1. Biodiesel Fuels: Science, Technology, Health, and Environment*, ed. O. Konur. Boca Raton, FL: CRC Press.

Konur, O. 2021m. Glycerol: A scientometric review of the research. In *Handbook of Biodiesel and Petrodiesel Fuels: Science, Technology, Health, and Environment. Volume 1. Biodiesel Fuels: Science, Technology, Health, and Environment*, ed. O. Konur. Boca Raton, FL: CRC Press.

Konur, O. 2021n. Propanediol production from glycerol: A review of the research. In *Handbook of Biodiesel and Petrodiesel Fuels: Science, Technology, Health, and Environment. Volume 1. Biodiesel Fuels: Science, Technology, Health, and Environment*, ed. O. Konur. Boca Raton, FL: CRC Press.

Konur, O. 2021o. Edible oil-based biodiesel fuels: A scientometric review of the research. In *Handbook of Biodiesel and Petrodiesel Fuels: Science, Technology, Health, and Environment. Volume 2. Biodiesel Fuels based on the Edible and Nonedible Feedstocks, Wastes, and Algae: Science, Technology, Health, and Environment*, ed. O. Konur. Boca Raton, FL: CRC Press.

Konur, O. 2021p. Palm oil-based biodiesel fuels: A review of the research. In *Handbook of Biodiesel and Petrodiesel Fuels: Science, Technology, Health, and Environment. Volume 2. Biodiesel Fuels based on the Edible and Nonedible Feedstocks, Wastes, and Algae*, ed. O. Konur. Boca Raton, FL: CRC Press.

Konur, O. 2021q. Rapeseed oil-based biodiesel fuels: A review of the research. In *Handbook of Biodiesel and Petrodiesel Fuels: Science, Technology, Health, and Environment. Volume 2. Biodiesel Fuels based on the Edible and Nonedible Feedstocks, Wastes, and Algae*, ed. O. Konur. Boca Raton, FL: CRC Press.

Konur, O. 2021r. Nonedible oil-based biodiesel fuels: A scientometric review of the research. In *Handbook of Biodiesel and Petrodiesel Fuels: Science, Technology, Health, and Environment. Volume 2. Biodiesel Fuels based on the Edible and Nonedible Feedstocks, Wastes, and Algae: Science, Technology, Health, and Environment*, ed. O. Konur. Boca Raton, FL: CRC Press.

Konur, O. 2021s. Waste oil-based biodiesel fuels: A scientometric review of the research. In *Handbook of Biodiesel and Petrodiesel Fuels: Science, Technology, Health, and Environment. Volume 2. Biodiesel Fuels based on the Edible and Nonedible Feedstocks, Wastes, and Algae: Science, Technology, Health, and Environment*, ed. O. Konur. Boca Raton, FL: CRC Press.

Konur, O. 2021t. Algal biodiesel fuels: A scientometric review of the research. In *Handbook of Biodiesel and Petrodiesel Fuels: Science, Technology, Health, and Environment. Volume 2. Biodiesel Fuels based on the Edible and Nonedible Feedstocks, Wastes, and Algae: Science, Technology, Health, and Environment*, ed. O. Konur. Boca Raton, FL: CRC Press.

Konur, O. 2021u. Algal biomass production for biodiesel production: A review of the research. In *Handbook of Biodiesel and Petrodiesel Fuels: Science, Technology, Health, and Environment. Volume 2. Biodiesel Fuels based on the Edible and Nonedible Feedstocks, Wastes, and Algae*, ed. O. Konur. Boca Raton, FL: CRC Press.

Konur, O. 2021v. Algal biomass production in wastewaters for biodiesel production: A review of the research. In *Handbook of Biodiesel and Petrodiesel Fuels: Science, Technology, Health, and Environment. Volume 2. Biodiesel Fuels based on the Edible and Nonedible Feedstocks, Wastes, and Algae*, ed. O. Konur. Boca Raton, FL: CRC Press.

Konur, O. 2021x. Algal lipid production for biodiesel production: A review of the research. In *Handbook of Biodiesel and Petrodiesel Fuels: Science, Technology, Health, and Environment. Volume 2. Biodiesel Fuels based on the Edible and Nonedible Feedstocks, Wastes, and Algae*, ed. O. Konur. Boca Raton, FL: CRC Press.

Konur, O. 2021y. Crude oils: A scientometric review of the research. In *Handbook of Biodiesel and Petrodiesel Fuels: Science, Technology, Health, and Environment. Volume 3. Petrodiesel Fuels: Science, Technology, Health, and Environment*, ed. O. Konur. Boca Raton, FL: CRC Press.

Konur, O. 2021z. Petrodiesel fuels: A scientometric review of the research. In *Handbook of Biodiesel and Petrodiesel Fuels: Science, Technology, Health, and Environment. Volume 3. Petrodiesel Fuels: Science, Technology, Health, and Environment*, ed. O. Konur. Boca Raton, FL: CRC Press.

Konur, O. 2021aa. Bioremediation of petroleum hydrocarbons in the contaminated soils: A review of the research. In *Handbook of Biodiesel and Petrodiesel Fuels: Science, Technology, Health, and Environment. Volume 3. Petrodiesel Fuels: Science, Technology, Health, and Environment*, ed. O. Konur. Boca Raton, FL: CRC Press.

Konur, O. 2021ab. Desulfurization of diesel fuels: A review of the research. In *Handbook of Biodiesel and Petrodiesel Fuels: Science, Technology, Health, and Environment. Volume 3. Petrodiesel Fuels: Science, Technology, Health, and Environment*, ed. O. Konur. Boca Raton, FL: CRC Press.

Konur, O. 2021ac. Diesel fuel exhaust emissions: A scientometric review of the research. In *Handbook of Biodiesel and Petrodiesel Fuels: Science, Technology, Health, and Environment. Volume 3. Petrodiesel Fuels: Science, Technology, Health, and Environment*, ed. O. Konur. Boca Raton, FL: CRC Press.

Konur, O. 2021ad. The adverse health and safety impact of diesel fuels: A scientometric review of the research. In *Handbook of Biodiesel and Petrodiesel Fuels: Science, Technology, Health, and Environment. Volume 3. Petrodiesel Fuels: Science, Technology, Health, and Environment*, ed. O. Konur. Boca Raton, FL: CRC Press.

Konur, O. 2021ae. Respiratory illnesses caused by the diesel fuel exhaust emissions: A review of the research. In *Handbook of Biodiesel and Petrodiesel Fuels: Science, Technology, Health, and Environment. Volume 3. Petrodiesel Fuels: Science, Technology, Health, and Environment*, ed. O. Konur. Boca Raton, FL: CRC Press.

Konur, O. 2021af. Cancer caused by the diesel fuel exhaust emissions: A review of the research. In *Handbook of Biodiesel and Petrodiesel Fuels: Science, Technology, Health, and Environment. Volume 3. Petrodiesel Fuels: Science, Technology, Health, and Environment*, ed. O. Konur. Boca Raton, FL: CRC Press.

Konur, O. 2021ag. Cardiovascular and other illnesses caused by the diesel fuel exhaust emissions: A review of the research. In *Handbook of Biodiesel and Petrodiesel Fuels: Science, Technology, Health, and Environment. Volume 3. Petrodiesel Fuels: Science, Technology, Health, and Environment*, ed. O. Konur. Boca Raton, FL: CRC Press.

Li, N., J. Alam, M. I. Venkatesan, et al. 2004. Nrf2 is a key transcription factor that regulates antioxidant defense in macrophages and epithelial cells: Protecting against the proinflammatory and oxidizing effects of diesel exhaust chemicals. *Journal of Immunology* 173:3467–3481.

Li, N., S. Kim, M. Wang, J. Froines, C. Sioutas, and A. Nel. 2002b. Use of a stratified oxidative stress model to study the biological effects of ambient concentrated and diesel exhaust particulate matter. *Inhalation Toxicology* 14:459–486.

Li, N., M. I. Venkatesan, and A. Miguel, et al. 2000. Induction of heme oxygenase-1 expression in macrophages by diesel exhaust particle chemicals and quinones via the antioxidant-responsive element. *Journal of Immunology* 165:3393–3401.

Li, N., M. Y. Wang, T. D. Oberley, J. M. Sempf, and A. E. Nel. 2002a. Comparison of the prooxidative and proinflammatory effects of organic diesel exhaust particle chemicals in bronchial epithelial cells and macrophages. *Journal of Immunology* 169:4531–4541.

McClellan, R. O. 1987. Health-effects of exposure to diesel exhaust particles. *Annual Review of Pharmacology and Toxicology* 27:279–300.

McCreanor, J., P. Cullinan, and M. J. Nieuwenhuijsen, et al. 2007. Respiratory effects of exposure to diesel traffic in persons with asthma. *New England Journal of Medicine* 357:2348–2358.

Muranaka, M., S. Suzuki, and K. Koizumi, et al. 1986. Adjuvant activity of diesel exhaust particulates for the production of IgE antibody in mice. *Journal of Allergy and Clinical Immunology* 77:616–623.

Nel, A. E., D. Diaz-Sanchez, D. Ng, T. Hiura, and A. Saxon. 1998. Enhancement of allergic inflammation by the interaction between diesel exhaust particles and the immune system. *Journal of Allergy and Clinical Immunology* 102:539–554.

Nightingale, J. A., R. Maggs, and P. Cullinan, et al. 2000. Airway inflammation after controlled exposure to diesel exhaust particulates. *American Journal of Respiratory and Critical Care Medicine* 162:161–166.

North, D. C. 1991a. *Institutions, Institutional Change and Economic Performance*. Cambridge, Mass.: Cambridge University Press.

North, D. C. 1991b. Institutions. *Journal of Economic Perspectives* 5:97–112.

Pandya, R. J., G. Solomon, A. Kinner, and J. R. Balmes. 2002. Diesel exhaust and asthma: Hypotheses and molecular mechanisms of action. *Environmental Health Perspectives* 110:103–112.

Perron, P. 1989. The great crash, the oil price shock, and the unit root hypothesis. *Econometrica: Journal of the Econometric Society* 57:1361–1401.

Riedl, M. and D. Diaz-Sanchez. 2005. Biology of diesel exhaust effects on respiratory function. *Journal of Allergy and Clinical Immunology* 115:221–228.

Sagai, M., A. Furuyama, and T. Ichinose. 1996. Biological effects of diesel exhaust particles (DEP). 3. Pathogenesis of asthma like symptoms in mice. *Free Radical Biology and Medicine* 21:199–209.

Sagai, M., H. Saito, T. Ichinose, M. Kodama, and Y. Mori. 1993. Biological effects of diesel exhaust particles .1. *In vitro* production of superoxide and *in vivo* toxicity in mouse. *Free Radical Biology and Medicine* 14:37–47.

Salvi, S., A. Blomberg, and B. Rudell, et al. 1999. Acute inflammatory responses in the airways and peripheral blood after short-term exposure to diesel exhaust in healthy human volunteers. *American Journal of Respiratory and Critical Care Medicine* 159:702–709.

Salvi, S. S., C. Nordenhall, and A. Blomberg, et al. 2000. Acute exposure to diesel exhaust increases IL-8 and GRO-α production in healthy human airways. *American Journal of Respiratory and Critical Care Medicine* 161:550–557.

Stenfors, N., C. Nordenhall, and S. S. Salvi, et al. 2004. Different airway inflammatory responses in asthmatic and healthy humans exposed to diesel. *European Respiratory Journal* 23:82–86.

Takafuji, S., S. Suzuki, and K. Koizumi, et al. 1987. Diesel-exhaust particulates inoculated by the intranasal route have an adjuvant activity for IgE production in mice. *Journal of Allergy and Clinical Immunology* 79:639–645.

Takano, H., T. Yoshikawa, and T. Ichinose, et al. 1997. Diesel exhaust particles enhance antigen-induced airway inflammation and local cytokine expression in mice. *American Journal of Respiratory and Critical Care Medicine* 156:36–42.

Takenaka, H., K. Zhang, and D. Diaz-Sanchez, et al. 1995. Enhanced human IgE production results from exposure to the aromatic hydrocarbons from diesel exhaust: Direct effects on b-cell IgE production. *Journal of Allergy and Clinical Immunology* 95:103–115.

Wills-Karp, M., J. Luyimbazi, and X. Xu, et al., 1998. Interleukin-13: Central mediator of allergic asthma. *Science* 282:2258–2261.

Xiao, G. G., M. Y. Wang, N. Li, J. A. Loo, and A. E. Nel. 2003. Use of proteomics to demonstrate a hierarchical oxidative stress response to diesel exhaust particle chemicals in a macrophage cell line. *Journal of Biological Chemistry* 278:50781–50790.

57 Cancer Caused by Diesel Fuel Exhaust Emissions

A Review of the Research

Ozcan Konur

CONTENTS

57.1 INTRODUCTION

Crude oils have been primary sources of energy and fuels, such as petrodiesel. However, significant public concerns about the sustainability, price fluctuations, and adverse environmental impact of crude oils have emerged since the 1970s (Ahmadun et al., 2009; Atlas, 1981; Babich and Moulijn, 2003; Kilian, 2009; Perron, 1989). Thus, biooils (Bridgwater and Peacocke, 2000; Czernik and Bridgwater, 2004) and biooil-based biodiesel fuels (Chisti, 2007; Hill et al., 2006) have emerged as alternatives to crude oils and crude oil-based petrodiesel fuels in recent decades. Nowadays, although biodiesel fuels are being used increasingly in the transportation and power sectors, petrodiesel fuels are still used extensively (Konur, 2021a–ag). Therefore, the adverse health impact of 'diesel fuel exhaust particles' (DEPs) has been a great source of public concern (Konur 2021r). In this context, the causation of cancer (Ford et al., 1994; Jemal et al., 2007, 2009; Siegel et al., 2012, 2015) by DEPs in humans has been studied more extensively (Konur, 2021ad).

However, for the efficient progression of the research in this field, it is necessary to develop efficient incentive structures for the primary stakeholders and to inform these stakeholders about the research (Konur, 2000, 2002a–c, 2006a–b, 2007a–b; North, 1991a–b).

Although there have been a number of reviews and book chapters in this field (Bhatia et al., 1998; Kagawa, 2002; Lipsett and Campleman, 1999; Schuetzle et al., 1981), there has been no review of the 25-most-cited articles. Thus, this chapter reviews these articles by highlighting the key findings of these most-prolific studies on the causation of cancer by DEPs. Then, it discusses these key findings.

57.2 MATERIALS AND METHODOLOGY

The search for the literature was carried out in the 'Web of Science' (WOS) database in February 2020. It contains the 'Science Citation Index-Expanded' (SCI-E), the 'Social Sciences Citation Index' (SSCI), the 'Book Citation Index-Science' (BCI-S), the 'Conference Proceedings Citation Index-Science' (CPCI-S), the 'Emerging Sources Citation Index' (ESCI), the 'Book Citation Index-Social Sciences and Humanities' (BCI-SSH), the 'Conference Proceedings Citation Index-Social Sciences and Humanities' (CPCI-SSH), and the 'Arts and Humanities Citation Index' (A&HCI).

The keywords for the search of the literature were collated from the screening of abstract pages for the first 500 highly cited papers on the adverse health impact of DEPs. These keywords sets are provided in the Appendix of the related chapter (Konur, 2021ad).

The 25-most-cited articles are selected for this review and the key findings are presented and discussed briefly.

57.3 RESULTS

57.3.1 HUMAN STUDIES

Silverman et al. (2012) present the 'Diesel Exhaust in Miners Study', focusing on the relationship between 'diesel exhaust' (DE) and lung cancer in a paper with 201 citations. They performed a nested case-control study in a cohort of 12,315 workers in eight non-metal mining facilities, which included 198 lung cancer deaths and 562 incidence density-sampled control subjects. They found statistically significant increasing trends in lung cancer risk with increasing cumulative 'respirable elemental carbon' (REC) and average REC intensity. Cumulative REC, lagged by 15 years, yielded a statistically significant positive gradient in lung cancer risk overall among heavily exposed workers – the risk was approximately three times greater than that among workers in the lowest quartile of exposure. Among never smokers, odd ratios were 1.0, 1.47, and 7.30 for workers with 15-year lagged cumulative REC tertiles of less than 8, 8 to less than 304, and 304 $\mu g/m^3/y$ or more, respectively. They also found an interaction between smoking and 15-year lagged cumulative REC such that the effect of each of these exposures was attenuated in the presence of high levels of the other. They conclude that DE exposure may cause lung cancer in humans and may represent a potential public health burden.

Attfield et al. (2012) study the association between DE exposure and lung cancer in a paper with 157 citations. They performed a cohort mortality study of 12,315 workers exposed to DE at eight US non-metal mining facilities. They found that standardized mortality ratios for lung cancer (1.26), esophageal cancer (1.83), and pneumoconiosis (12.20) were elevated in the complete cohort compared with state-based mortality rates, but all-cause bladder cancer, heart disease, and chronic obstructive pulmonary disease mortality were not. Differences in risk by worker location (ever-underground vs surface only) initially obscured a positive DE exposure–response relationship with lung cancer in the complete cohort, although it became apparent after adjustment for worker location. The 'hazard ratios' (HRs) for lung cancer mortality increased with increasing 15-year lagged cumulative REC exposure for ever-underground workers with five or more years of tenure to a maximum in the 640 to less than 1,280 $\mu g/m^3/y$ category compared with the reference category (0 to < 20 $\mu g/m^3/y$) but declined at higher exposures. Average REC intensity HRs rose to a plateau around 32 $\mu g/m^3$. Elevated HRs and evidence of exposure response were also seen for surface workers. The association between DE exposure and lung cancer risk remained after inclusion of other work-related potentially confounding exposures in the models, which were robust to alternative approaches to exposure derivation. They conclude that exposure to DE increases risk of mortality from lung cancer and has important public health implications.

Garshick et al. (1988) assess the risk of lung cancer as a result of exposure to DE from railroad workers in locomotives in a paper with 139 citations. The sample included a cohort of 55,407 white male railroad workers, aged 40 to 64 in 1959, who had started service 10 to 20 years earlier; 1,694 lung cancer cases were identified. They obtained a relative risk of 1.45 for lung cancer in the group of workers 40 to 44 years of age in 1959, the group with the longest possible duration of diesel exposure. The cohort was selected to minimize the effect of past railroad asbestos exposure, and analysis with workers with possible asbestos exposure was excluded, which resulted in a similarly elevated risk. Workers with 20 years or more elapsed since 1959, the effective start of diesel exposure for the cohort, had the highest relative risk. They conclude that occupational exposure to DE results in a small but significantly elevated risk of lung cancer.

Garshcik et al. (2004) study lung cancer mortality in 54,973 US railroad workers between 1959 and 1996 in a paper with 130 citations. There were 43,593 total deaths, including 4,351 lung cancer deaths. Adjusting for a healthy worker survivor effect and age, railroad workers in jobs associated with operating trains had a relative risk of lung cancer mortality of 1.40. This did not increase with increasing years of work in these jobs. It was elevated in jobs associated with work on trains powered by diesel locomotives. Although a contribution from exposure to coal combustion products before 1959 cannot be excluded, they conclude that the exposure to DE contributed to lung cancer mortality in this cohort.

Garshick et al. (1987) perform a case-control study of deaths among US railroad workers to test the hypothesis that lung cancer is associated with exposure to DE in a paper with 127 citations. Employed and retired male workers with more than nine years of service who were born on or after January 1, 1,900 and who died between March 1981 and March 1982 were eligible. Workers 64 years of age or younger at the time of death with work in a DE-exposed job for 20 years had a significantly increased relative odds (odds ratio = 1.41) for lung cancer. No effect of DE exposure was seen in workers 65 years of age or older because many of these men retired shortly after the transition to diesel-powered locomotives. They conclude that occupational exposure to DE increases lung cancer risk.

Steenland et al. (1998) study DE and lung cancer in the trucking industry in a paper with 113 citations. They conducted exposure–response analyses among workers in the trucking industry, adjusted for smoking. They found that regardless of assumptions about past exposure, all analyses resulted in significant positive trends in lung cancer risk with increasing cumulative exposure. A male truck driver exposed to 5 µg/m3 of elemental carbon would have a lifetime excess risk of lung cancer of 1–2%, above a background risk of 5%. They found a lifetime excess risk ten times higher than the 1 per 1,000 excess risk allowed by the 'Occupational Safety and Health and Administration' (OSHA) in setting regulations. They conclude that occupational exposure to DE increases lung cancer risk.

57.3.2 Animal Studies

Heinrich et al. (1995) study the chronic inhalation exposure of rats and mice to DE, 'carbon black' (CB), and titanium dioxide in a paper with 317 citations. They exposed them for two years and keep them in clean air for six months. They found that the average particle exposure concentrations for diesel soot, CB, and TiO_2 were 7, 11.6, and 10 mg/m^3, respectively. Lung tumor rates in these rats increased with increasing cumulative particle exposure, independent of the type of particle employed. The exposure to 2.5 mg/m^1 diesel soot also induced a significantly increased lung tumor rate. The carbon core of diesel soot is mainly responsible for the occurrence of diesel engine exhaust-related lung tumors; the role of diesel soot-attached 'polycyclic aromatic hydrocarbon' (PAH) and NO_2-PAH is probably of minor importance in the rats' lungs. NMRI mice that were kept in the same exposure atmospheres as the rats did not show an increased lung tumor rate. Furthermore, there was no treatment-related tumor response in NMRI nor in C57BL/6N mice exposed to DE containing 4.5 mg/m^3 diesel soot or to the same exhaust dilution but devoid of soot particles. C57BU6N mice were exposed for 24 months and were subsequently kept in clean air for another six months. Not only the average survival time but also the particle load per gram lung wet weight of the C57BU6N mice was very similar to rats exposed to 7 mg/m^3 diesel soot. They conclude that lung tumor rates in these rats increased with increasing cumulative particle exposure, independent of the type of particle employed, and that the carbon core of diesel soot is mainly responsible for the occurrence of diesel engine exhaust-related lung tumors.

Aoki et al. (2001) study the accelerated DNA adduct formation in the lung of the 'nuclear factor erythroid 2–related factor 2' (Nrf2) knockout mouse exposed to DE in a paper with 231 citations. They hypothesized that the Nrf2 gene knockout mouse might serve as an excellent model system for analyzing DE toxicity. They examined lungs from Nrf2(–/–) and Nrf2(+/–) mice for the production of xenobiotic–DNA adducts after exposure to DE (3 mg/m^3 suspended particulate matter) for four weeks. They found that whereas the 'relative adduct levels' (RAL) were significantly increased in the lungs of both Nrf2(+/–) and Nrf2(–/–) mice upon exposure to DE, the increase of RAL in the lungs of Nrf2(–/–) mice exposed to DE were approximately 2.3-fold higher than that of Nrf2(+/–) mice exposed to DE. In contrast, cytochrome P4501A1 mRNA levels in the Nrf2(–/–) mouse lungs were similar to those in the Nrf2(+/–) mouse lungs even after exposure to DE, suggesting that suppressed activity of phase II drug-metabolizing enzymes is important in giving rise to the increased level of DNA adducts in the Nrf2-null mutant mouse subjected to DE. Importantly, they observed severe hyperplasia and accumulation of the oxidative DNA adduct '8-hydroxydeoxyguanosine' (8-OHdG) in the bronchial epidermis of Nrf2(–/–) mice following DE exposure. They showed the increased susceptibility of the Nrf2 germ line mutant mouse to DE exposure and conclude that the Nrf2 gene knockout mouse may represent a valuable model for the assessment of respiratory DE toxicity.

Nikula et al. (1995) study the pulmonary toxicities and carcinogenicities of chronically inhaled DE and CB in F344 rats to explore the importance of the DE soot-associated organic compounds in the lung tumor response of rats in a paper with 223 citations. They exposed male and female F344 rats chronically to diluted whole DE or aerosolized CB 16 hours/day, five days/week at target particle concentrations of 2.5 mg/m^3 (LDE, LCB) or 6.5 mg/m^3 (HDE, HCB) or to filtered air. The CB served as a surrogate for the elemental carbon matrix of DE soot. Considering both the mass fraction of solvent-extractable matter and its mutagenicity in the Ames *Salmonella* assay, the mutagenicity in revertants per unit particle mass of the CB was three orders of magnitude less than that of the DE soot. They found that both DE soot and CB particles accumulated progressively in the lungs of exposed rats, but that the rate of accumulation was higher for DE soot. In general, DE and CB caused similar, dose-related, no neoplastic lesions. CB and DE caused significant, exposure-concentration-related increases, of similar magnitudes, in the incidences and prevalence of the same types of malignant and benign lung neoplasms in female rats. The incidences of neoplasms were much lower in males than females, and the incidences were slightly higher among DE- than CB-exposed males. Survival was shortened in the CB-exposed males, and this shortened survival may have suppressed the expression of carcinogenicity as measured by crude incidence. They conclude that the organic fraction of DE may not play an important role in the carcinogenicity of DE in rats.

Mauderly et al. (1987) study DE as a pulmonary carcinogen in rats exposed chronically by inhalation in a paper with 207 citations. They exposed male and female F344 rats seven hours/day, five days/week for up to 30 months to automotive diesel engine exhaust at soot concentrations of 0.35, 3.5, or 7.0 mg/m^3 or to clean air. They found that survival and body weight were unaffected by exposure. Focal fibrotic and proliferative lung disease accompanied a progressive accumulation of soot in the lung. The prevalence of lung tumors was significantly increased at the high (13%) and medium (4%) dose levels above the control prevalence (1%). They observed four tumor types, all of epithelial origin: adenoma, adenocarcinoma, squamous cyst, and squamous cell carcinoma. There was a significant relationship between tumor prevalence and both exposure concentration and soot lung burden. They conclude that DE, inhaled chronically at a high concentration, is a pulmonary carcinogen in the rat.

Heinrich et al. (1986) study the chronic effects on the respiratory tract of hamsters, mice, and rats after long-term inhalation of high concentrations of filtered and unfiltered DE to investigate the effects of chronic toxicity and, predominantly, carcinogenicity in the respiratory tract in a paper with 168 citations. In hamsters and rats, they found significant changes only after exposure to unfiltered DE and, predominantly, in rats. No lung tumors were found in hamsters. Spontaneous tumor rates occurred in mice, and both types of DE increased the incidence of adenocarcinomas in the lungs. In rats, only the unfiltered DE caused a lung tumor incidence. This amounted to 16% with no tumors in the controls. The heavy load of particulate matter in the lungs of rats was caused by an exposure-related impairment of the alveolar lung clearance and may have been instrumental in the induction of squamous cell tumors. However, an effect of particle-associated PAH cannot be excluded. They found the carcinogenic effects of DE after initial carcinogen treatment only in the respiratory tract of rats. They conclude that DE has carcinogenic effects.

Hiura et al. (2000) study the role of a mitochondrial pathway in the induction of apoptosis by chemicals extracted from DEPs in a paper with 158 citations. They showed that methanol extracts made from DEPs induce apoptosis and 'reactive oxygen species' (ROS) in pulmonary alveolar macrophages and RAW 264.7 cells. The toxicity of these organic extracts mimics the cytotoxicity of the intact particles and could be suppressed by the synthetic sulfhydryl compounds, 'N-acetylcysteine' (NAC) and bucillamine. Because DEP-induced apoptosis follows cytochrome c release, they examined the effect of DEP chemicals on mitochondrially regulated death mechanisms. They found that crude DEP extracts induced ROS production and perturbed mitochondrial function before and at the onset of apoptosis. This mitochondrial perturbation follows an orderly sequence of events, which commence with a change in mitochondrial membrane potential, followed by cytochrome c release, development of membrane asymmetry, and propidium iodide uptake. Structural damage to the mitochondrial inner membrane, evidenced by a decrease in cardiolipin mass, leads to O_2 generation and uncoupling of oxidative phosphorylation. NAC reversed these mitochondrial effects and ROS production. Overexpression of the mitochondrial apoptosis regulator, 'B-cell lymphoma 2' (Bcl-2), delayed but did not suppress apoptosis. They conclude that DEP chemicals induce apoptosis in macrophages via a toxic effect on mitochondria.

Ichinose et al. (1997) study the relationship between lung tumor response and the formation of 8-OHdG in lung DNA to clarify the involvement of oxygen radicals in lung carcinogenesis induced by DEPs in a paper with 136 citations. They further studied the role of high dietary fat and β-carotene on these responses. They injected mice intratracheally with 0.05, 0.1, and 0.2 mg of DEP per animal once weekly for ten weeks. After 12 months, they found that the lung tumor incidence in mice treated with 0.05 and 0.1 mg showed similar increases (30 and 31%), but which decreased to 24% at 0.2 mg. High dietary fat enhanced the incidence of both benign and malignant tumors. β-carotene partially prevented the tumor development. After the ten weekly treatments of DEP, they observed inflammatory reaction in the respiratory tract and alveoli. The formation of 8-OHdG in lung DNA from mice treated with DEP showed a dose dependent increase. 8-OHdG formation was enhanced by high dietary fat and partially reduced by β-carotene. Formation of 8-OHdG was significantly correlated with the lung tumor incidence except at 0.2 mg. They conclude that the induction of oxidative DNA damage may be an important factor in the initiation of DEP-induced lung carcinogenesis, and that β-carotene and high dietary fat may play a role in the regulation of tumor development via modulation of the formation of 8-OHdG.

Tsurudome et al. (1999) study the levels of '8-hydroxyguanine' (8-OH-Gua), its total repair, and the repair enzyme OGG1 mRNA in female Fischer 344 rat lungs, as markers of the response to ROS, after DEP was intratracheally instilled in a paper with 127 citations. They find that the 8-OH-Gua levels in both DEP-treated groups (2 and 4 mg) were increased during the two to eight hours following exposure to DEP. The 8-OH-Gua repair activities in the DEP-treated groups decreased during the period from two hours to two days following DEP exposure and then recovered to the level of the control group at five days after exposure. OGG1 mRNA was induced in rats treated with 4 mg DEP for five to seven days after administration. They conclude that the 8-OH-Gua level in rat lung DNA increases markedly at an early phase after DEP exposure, by the generation of ROS and the inhibition of 8-OH-Gua repair activity, and induction of OGG1 mRNA is also a good marker of cellular oxidative stress during carcinogenesis.

57.3.3 *In vitro* Studies

Hiura et al. (1999) study the generation of 'reactive oxygen radicals' (RORs) and induction of apoptosis in macrophages by DEP in a paper with 252 citations. They showed that the phagocytosis of DEP by primary alveolar macrophages or macrophage cell lines leads to the induction of apoptosis through generation of RORs. This oxidative stress initiates two caspase cascades and a series of cellular events, including loss of surface membrane asymmetry and DNA damage. The apoptotic effect on macrophages is cell specific, because DEP did not induce similar effects in no phagocytic cells. DEP that had their organic constituents extracted were no longer able to induce apoptosis or generate ROR. The organic extracts were, however, able to induce apoptosis. DEP chemicals also induced the activation of stress-activated protein kinases, which play a role in cellular apoptotic pathways. The injurious effects of native particles or DEP extracts on macrophages could be reversed by the antioxidant, NAC. They conclude that organic compounds contained in DEP may exert acute toxic effects via the generation of ROR in macrophages.

Enya et al. (1997) study the mutagenicity of 3-nitrobenzanthrone from DE in a paper with 230 citations. They isolated the '3-Nitrobenzanthrone (3-nitro-7H-benz[d,e]anthracen-7-one)' from the organic extracts of both DE and airborne particles and identified it as a new class of powerful direct mutagen. Its mutagenicity by Ames *Salmonella* assay is very high (208,000 revertants/nmol in *Salmonella typhimurium* TA98 and 6,290,000 revertants/nmol in YG1024) and compares with that of 1,8-dinitropyrene. They show the new mutagen induces micronuclei in mouse peripheral blood reticulocytes after intraperitoneal administration (micronucleated reticulocytes, 0.64% against 25 mg/kg dose after 48 h), suggesting its potential genotoxicity to mammalians. They conclude that 3-nitrobenzanthrone from DE is a powerful mutagen.

Pitts et al. (1982) study the mutagenic activities of '6-nitrobenzo[a]pyrene', '9-nitroanthracene', '1-nitropyrene', and '5H-phenanthro[4,5-bcd]pyran-5-one' in DEPs in a paper with 212 citations.

Kumagai et al. (1997) study the generation of ROS during interaction of DEP components with 'nicotinamide adenine dinucleotide phosphate' (NADPH)-cytochrome P450 reductase in a paper with 191 citations. NADPH reduced oxidation was stimulated during interaction of a methanol extract of DEP with the 'Triton N-101' treated microsomal preparation of mouse lung whereas the cytosolic fraction was less active, suggesting that DEP contains substrates for NADPH-cytochrome P450 reductase rather than 'dithiol' (DT)-diaphorase. When purified P450 reductase was used as the enzyme source and the turnover value was enhanced by approximately 260-fold. Quinones served as a substrate for P450 reductase because reaction was inhibited by the addition of 'glutathione' (GSH) to form GSH adduct or pretreatment with NaBH4 to reduce them to the hydroxy compounds, although a possibility of nitroarenes as the alternative substrates cannot be excluded. A methanol extract of DEP (37.5 micrograms) caused a significant formation of superoxide (3,240 nmol/min/mg protein) in the presence of P450 reductase. The reactive species generated by DEP in the presence of P450 reductase caused DNA scission which was reduced in the presence of superoxide dismutase (SOD), catalase, or hydroxyl radical scavenging agents. They conclude that DEP components, probably quinoid or nitroaromatic structures, promote DNA damage through the redox-cycling-based generation of superoxide.

Kumagai et al. (2002) study the oxidation of proximal protein sulfhydryls by phenanthraquinone, a component of DEPs, using thiol compounds and protein preparation in a paper with 193 citations. They found that phenanthraquinone reacted readily with DT compounds such as 'dithiothreitol' (DTT), '2,3-dimercapto-1-propanol' (BAL), and '2,3-dimercapto-1-propanesulfonic acid' (DMPS), resulting in modification of the thiol groups, whereas minimal reactivities of this quinone with monothiol compounds such as GSH, 2-mercaptoethanol, and NAC were seen. The modification of DTT dithiol caused by phenanthraquinone proceeded under anaerobic conditions but was accelerated by molecular oxygen. Phenanthraquinone was also capable of modifying thiol groups in pulmonary microsomes from rats and total membrane preparation isolated from 'bovine aortic endothelial cells' (BAEC), but not 'bovine serum albumin' (BSA), which has a Cys34 as a reactive monothiol group. A comparison of the thiol alkylating agent 'N-ethylmaleimide' (NEM) with that of phenanthraquinone indicates that the two mechanisms of thiol modification are distinct. Thiyl radical intermediates and ROS were generated during the interaction of phenanthraquinone with DTT. They conclude that phenanthraquinone-mediated destruction of protein sulfhydryls involves the oxidation of presumably proximal thiols and the reduction of molecular oxygen.

Rosenkranz (1982) studies the direct-acting mutagens in DEs in a paper with 167 citations.

Bagley et al. (1998) study the effects of an 'oxidation catalytic converter' (OCC) and a soy-based biodiesel fuel on the chemical, mutagenic, and particle size characteristics of emissions from a diesel engine in a paper with 158 citations. They found that compared to emissions with the diesel fuel without the OCC, use of the diesel

(D2) and biodiesel fuel with the OCC had similar reductions (50–80%) in 'total particulate matter' (TPM). The solid portion of the TPM was lowered with the biodiesel fuel. Particle-associated polynuclear aromatic hydrocarbon and 1-nitropyrene emissions were lower with use of the biodiesel fuel as compared to the D2 fuel, with or without the OCC. Vapor-phase PAH emissions were reduced (up to 90%) when the OCC was used with either fuel. Use of the OCC resulted in over 50% reductions in both particle and vapor-phase-associated mutagenic activity with both fuels. No vapor-phase-associated mutagenic activity was detected with the biodiesel fuel; only very low levels were detected with the D2 fuel and the OCC. Use of the OCC caused a moderate shift in the particle size/volume distribution of the accumulation mode particles to smaller particles for the diesel fuel and a reduction of particle volume concentrations at some of the tested conditions for both fuels. The nuclei mode did not contribute significantly to total particle volume concentrations within the measured particle size range (~0.01–1.0 μm). The biodiesel fuel reduced total particle volume concentrations. They conclude that use of this OCC for the engine conditions tested with the biodiesel fuel, in particular, resulted in generally similar or greater reductions in emissions than for use of the D2 fuel. Use of the biodiesel fuel should not increase any of the potentially toxic, health-related emissions.

Salmeen et al. (1982) study the contribution of 1-nitropyrene to direct-acting Ames assay mutagenicities of diesel particulate extracts in a paper with 132 citations.

Jardim et al. (2009) study the pollutant-induced changes in miRNA expression in airway epithelial cells using DEPs in a paper with 122 citations. They hypothesized that DEP exposure can lead to disruption of normal miRNA expression patterns, representing a plausible novel mechanism through which DEP can mediate disease initiation. Human bronchial epithelial cells were grown at an air–liquid interface until they reached mucociliary differentiation and they treated these cells with 10 mu g/cm(2) DEP for 24 hours. They found that DEP exposure changed the miRNA expression profile in human airway epithelial cells. Specifically, 197 of 313 detectable miRNAs (62.9%) were either up-regulated or down-regulated by 1.5-fold. Molecular network analysis of putative targets of the 12 most altered miRNAs indicated that DEP exposure is associated with inflammatory responses pathways and a strong tumorigenic disease signature. They conclude that the alteration of miRNA expression profiles by environmental pollutants such as DEP can modify cellular processes by regulation of gene expression, which may lead to disease pathogenesis.

Salmeen et al. (1984) study the Ames assay chromatograms to identify mutagens in DEP extracts in a paper with 117 citations.

Xu et al. (1982) isolate and identify mutagenic nitro-PAH in DEPs in a paper with 122 citations. They identified more than 50 nitro-PAH in an extract of DEPs. Identifications were based on high-resolution mass spectrometry of directly mutagenic fractions derived from sequential fractionation of the extract by both low and high-resolution liquid chromatography. The diversity of nitro-PAH tentatively identified suggests that large numbers of such compounds are formed during or following the combustion process. They conclude that DEPs contain mutagenic fractions.

57.4 DISCUSSION

Table 57.1 provides information on the research fronts in this field. As this table shows the research fronts of 'human studies', 'animal studies', and '*in vitro* studies' comprise 24, 32, and 44% of these papers, respectively.

57.4.1 Human Studies

In the first group of papers, Silverman et al. (2012) present the 'Diesel Exhaust in Miners Study' focusing on the relationship between DE and lung cancer in a paper with 201 citations. They conclude that DE exposure may cause lung cancer in humans and may represent a potential public health burden.

Attfield et al. (2012) study the association between DE exposure and lung cancer in a paper with 157 citations. They conclude that exposure to DE increases risk of mortality from lung cancer and has important public health implications.

Garshick et al. (1988) assess the risk of lung cancer as a result of exposure to DE from railroad locomotives in workers in a paper with 139 citations. They conclude that occupational exposure to DE results in a small but significantly elevated risk of lung cancer.

Garshick et al. (2004) study lung cancer mortality in 54,973 US railroad workers between 1959 and 1996 in a paper with 130 citations. They conclude that the exposure to DE contributed to lung cancer mortality in this cohort.

Garshick et al. (1987) perform a case-control study of deaths among US railroad workers to test the hypothesis that lung cancer is associated with exposure to DE in a paper with 127 citations. They conclude that occupational exposure to DE increases lung cancer risk.

Steenland et al. (1998) study DE and lung cancer in the trucking industry in a paper with 113 citations. They conclude that occupational exposure to DE increases lung cancer risk.

These prolific studies show that occupational exposure to DE increases lung cancer risk with strong public policy implications.

57.4.2 Animal Studies

Heinrich et al. (1995) study the chronic inhalation exposure of rats and mice to DE, CB, and titanium dioxide in a paper with 317 citations. They conclude that lung tumor rates in these rats increased with increasing cumulative particle exposure,

TABLE 57.1
Research Fronts

	Research Front	Papers (%)
1	Human studies	24
2	Animal studies	32
3	*In vitro* studies	44

independent of the type of particle employed, and that the carbon core of diesel soot is mainly responsible for the occurrence of diesel engine exhaust-related lung tumors.

Aoki et al. (2001) study the accelerated DNA adduct formation in the lung of the Nrf2 knockout mouse exposed to DE in a paper with 231 citations. They show the increased susceptibility of the Nrf2 germ line mutant mouse to DE exposure and conclude that the Nrf2 gene knockout mouse may represent a valuable model for the assessment of respiratory DE toxicity.

Nikula et al. (1995) study the pulmonary toxicities and carcinogenicities of chronically inhaled DE and CB in F344 rats to explore the importance of the DE soot-associated organic compounds in the lung tumor response of rats in a paper with 223 citations. They conclude that the organic fraction of DE may not play an important role in the carcinogenicity of DE in rats.

Mauderly et al. (1987) study the DE as a pulmonary carcinogen in rats exposed chronically by inhalation in a paper with 207 citations. They conclude that DE, inhaled chronically at a high concentration, is a pulmonary carcinogen in the rat.

Heinrich et al. (1986) study the chronic effects on the respiratory tract of hamsters, mice, and rats after long-term inhalation of high concentrations of filtered and unfiltered DE to investigate effects of chronic toxicity and, predominantly, carcinogenicity in the respiratory tract in a paper with 168 citations. They conclude that DE has carcinogenic effects.

Hiura et al. (2000) study the role of a mitochondrial pathway in the induction of apoptosis by chemicals extracted from DEPs in a paper with 158 citations. They conclude that DEP chemicals induce apoptosis in macrophages via a toxic effect on mitochondria.

Ichinose et al. (1997) study the relationship between lung tumor response and formation of 8-OHdG in lung DNA to clarify the involvement of oxygen radicals in lung carcinogenesis induced by DEPs in a paper with 136 citations. They conclude that the induction of oxidative DNA damage may be an important factor in the initiation of DEP-induced lung carcinogenesis, and that β-carotene and high dietary fat may play a role in the regulation of tumor development via modulation of the formation of 8-OHdG.

Tsurudome et al. (1999) study the levels of 8-OH-Gua, its total repair, and the repair enzyme OGG1 mRNA in female Fischer 344 rat lungs, as markers of the response to ROS, after DEP was intratracheally instilled in a paper with 127 citations. They conclude that the 8-OH-Gua level in rat lung DNA increases markedly at an early phase after DEP exposure, by the generation of ROS and the inhibition of 8-OH-Gua repair activity, and that induction of OGG1 mRNA is also a good marker of cellular oxidative stress during carcinogenesis.

These prolific studies show that DEPs can cause cancer in animals, which has strong public policy implications.

57.4.3 *In vitro* Studies

Hiura et al. (1999) study the generation of RORs and induction of apoptosis in macrophages by DEP in a paper with 252 citations. They conclude that organic compounds contained in DEP may exert acute toxic effects via the generation of ROR in macrophages.

Enya et al. (1997) study the mutagenicity of 3-nitrobenzanthrone from DE in a paper with 230 citations. They conclude that 3-nitrobenzanthrone from DE is a powerful mutagen.

Pitts et al. (1982) study the mutagenic activities of ‘6-nitrobenzo[a]pyrene’, ‘9-nitroanthracene’, ‘1-nitropyrene’, and ‘5H-phenanthro[4,5-bcd]pyran-5-one’ in DEPs in a paper with 212 citations.

Kumagai et al. (1997) study the generation of ROS during interaction of DEP components with NADPH-cytochrome P450 reductase in a paper with 191 citations. They conclude that DEP components, probably quinoid or nitroaromatic structures, promote DNA damage through the redox-cycling-based generation of superoxide.

Kumagai et al. (2002) study the oxidation of proximal protein sulfhydryls by phenanthraquinone, a component of DEPs, using thiol compounds and protein preparation in a paper with 193 citations. They conclude that phenanthraquinone-mediated destruction of protein sulfhydryls involves the oxidation of presumably proximal thiols and the reduction of molecular oxygen.

Rosenkranz (1982) studies the direct-acting mutagens in DEs in a paper with 167 citations.

Bagley et al. (1998) study the effects of an OCC and a soy-based biodiesel fuel on the chemical, mutagenic, and particle size characteristics of emissions from a diesel engine in a paper with 158 citations. They conclude that use of this OCC for the engine conditions tested with biodiesel fuel, in particular, resulted in generally similar or greater reductions in emissions than for use of D2 fuel. Use of biodiesel fuel should not increase any of the potentially toxic, health-related emissions.

Salmeen et al. (1982) study the contribution of 1-nitropyrene to direct-acting Ames assay mutagenicities of diesel particulate extracts in a paper with 132 citations.

Jardim et al. (2009) study the pollutant-induced changes in miRNA expression in airway epithelial cells using DEPs in a paper with 122 citations. They conclude that the alteration of miRNA expression profiles by environmental pollutants such as DEP can modify cellular processes by regulation of gene expression, which may lead to disease pathogenesis.

Salmeen et al. (1984) study the Ames assay chromatograms to identify mutagens in DEP extracts in a paper with 117 citations.

Xu et al. (1982) isolate and identify mutagenic nitro-PAH in DEPs in a paper with 122 citations. They conclude that DEPs contain mutagenic fractions.

These *in vitro* prolific studies show that DEPs contain mutagenic fractions which can cause cancer with strong public policy implications.

57.5 CONCLUSION

This chapter has presented the key findings of the 25-most-cited article papers in this field. Table 57.1 provides information on the research fronts in this field.

As this table shows the research fronts of ‘human studies’, ‘animal studies’, and ‘*in vitro* studies’ comprise 24, 32, and 44% of these papers, respectively.

These prolific studies in three different research fronts provide valuable evidence on the causation of cancer by the mutagenic fractions of DEPs with strong public policy implications.

It is recommended that similar studies are carried out for each research front as well.

ACKNOWLEDGMENTS

The contribution of the highly cited researchers in this field is greatly acknowledged.

REFERENCES

Ahmadun, F. R., A. Pendashteh, and L. C. Abdullah, et al. 2009. Review of technologies for oil and gas produced water treatment. *Journal of Hazardous Materials* 170:530–551.

Aoki, Y., H. Sato, and N. Nishimura, et al. 2001. Accelerated DNA adduct formation in the lung of the *Nrf2* knockout mouse exposed to diesel exhaust. *Toxicology and Applied Pharmacology* 173:154–160.

Atlas, R. M. 1981. Microbial degradation of petroleum hydrocarbons: An environmental perspective. *Microbiological Reviews* 45:180–209.

Attfield, M. D., P. L. Schleiff, and J. H. Lubin, et al. 2012. The Diesel Exhaust in Miners Study: A cohort mortality study with emphasis on lung cancer. *JNCI-Journal of the National Cancer Institute* 104:869–883.

Babich, I. V. and J. A. Moulijn. 2003. Science and technology of novel processes for deep desulfurization of oil refinery streams: A review. *Fuel* 82:607–631.

Bagley, S. T., L. D. Gratz, J. H. Johnson, and J. F. McDonald. 1998. Effects of an oxidation catalytic converter and a biodiesel fuel on the chemical, mutagenic, and particle size characteristics of emissions from a diesel engine. *Environmental Science & Technology* 32:1183–1191.

Bhatia, R., P. Lopipero, and A. H. Smith. 1998. Diesel exhaust exposure and lung cancer. *Epidemiology* 9:84–91.

Bridgwater, A. V. and G. V. C. Peacocke. 2000. Fast pyrolysis processes for biomass. *Renewable and Sustainable Energy Reviews* 4:1–73.

Chisti, Y. 2007. Biodiesel from microalgae. *Biotechnology Advances* 25:294–306.

Czernik, S. and A. V. Bridgwater. 2004. Overview of applications of biomass fast pyrolysis oil. *Energy & Fuels* 18:590–598.

Enya, T., H. Suzuki, T. Watanabe, T. Hirayama, and Y. Hisamatsu. 1997. 3-nitrobenzanthrone, a powerful bacterial mutagen and suspected human carcinogen found in diesel exhaust and airborne particulates. *Environmental Science & Technology* 31:2772–2776.

Ford, D., D. F. Easton, D. T. Bishop, S. A. Narod, and D. E. Goldgar. 1994. Risks of cancer in BRCA1-mutation carriers. *Lancet* 343:692–695.

Galli, S. J., M. Tsai, and A. M. Piliponsky. 2008. The development of allergic inflammation. *Nature* 454:445–454.

Garshick, E., F. Laden, and J. E. Hart, et al. 2004. Lung cancer in railroad workers exposed to diesel exhaust. *Environmental Health Perspectives* 112:1539–1543.

Garshick, E., M. B. Schenker, and A. Munoz, et al. 1987. A case-control study of lung cancer and diesel exhaust exposure in railroad workers. *American Review of Respiratory Disease* 135:1242–1248.

Garshick, E., M. B. Schenker, and A. Munoz, et al. 1988. A retrospective cohort study of lung-cancer and diesel exhaust exposure in railroad workers. *American Review of Respiratory Disease* 137:820–825.

Heinrich, U., R. Fuhst, and S. Rittinghausen, et al. 1995. Chronic inhalation exposure of Wistar rats and two different strains of mice to diesel engine exhaust, carbon black, and titanium dioxide. *Inhalation Toxicology* 7:533–556.

Heinrich, U., H. Muhle, and S. Takenaka, et al.1986. Chronic effects on the respiratory tract of hamsters, mice and rats after long-term inhalation of high concentrations of filtered and unfiltered diesel engine emissions. *Journal of Applied Toxicology* 6:383–395.

Hill, J., E. Nelson, D. Tilman, S. Polasky, and D. Tiffany. 2006. Environmental, economic, and energetic costs and benefits of biodiesel and ethanol biofuels. *Proceedings of the National Academy of Sciences of the United States of America* 103:11206–11210.

Hiura, T. S., M. P. Kaszubowski, N. Li, and A. E. Nel. 1999. Chemicals in diesel exhaust particles generate reactive oxygen radicals and induce apoptosis in macrophages. *Journal of Immunology* 163:5582–5591.

Hiura, T. S., N. Li, and R. Kaplan, et al. 2000. The role of a mitochondrial pathway in the induction of apoptosis by chemicals extracted from diesel exhaust particles. *Journal of Immunology* 165:2703–2711.

Ichinose, T., Y. Yajima, and M. Nagashima, et al. 1997. Lung carcinogenesis and formation of 8-hydroxy-deoxyguanosine in mice by diesel exhaust particles. *Carcinogenesis* 18:185–192.

Jardim, M. J., R. C. Fry, I. Jaspers, L. Dailey, and D. Diaz-Sanchez. 2009. Disruption of microRNA expression in human airway cells by diesel exhaust particles is linked to tumorigenesis-associated pathways. *Environmental Health Perspectives* 117:1745–1751.

Jemal, A., R. Siegel, and E. Ward, et al. 2007. Cancer statistics, 2007. *CA: A Cancer Journal for Clinicians* 57:43–66.

Jemal, A., R. Siegel, and E. Ward, et al. 2009. Cancer statistics, 2009. *CA: A Cancer Journal for Clinicians* 59:225–249.

Kagawa, J. 2002. Health effects of diesel exhaust emissions: A mixture of air pollutants of worldwide concern. *Toxicology* 181:349–353.

Kilian, L. 2009. Not all oil price shocks are alike: Disentangling demand and supply shocks in the crude oil market. *American Economic Review* 99:1053–1069.

Konur, O. 2000. Creating enforceable civil rights for disabled students in higher education: An institutional theory perspective. *Disability & Society* 15:1041–1063.

Konur, O. 2002a. Access to Nursing Education by disabled students: Rights and duties of nursing programs. *Nurse Education Today* 22:364–374.

Konur, O. 2002b. Assessment of disabled students in higher education: Current public policy issues. *Assessment and Evaluation in Higher Education* 27:131–152.

Konur, O. 2002c. Access to employment by disabled people in the UK: Is the Disability Discrimination Act working? *International Journal of Discrimination and the Law* 5:247–279.

Konur, O. 2006a. Participation of children with dyslexia in compulsory education: Current public policy issues. *Dyslexia* 12:51–67.

Konur, O. 2006b. Teaching disabled students in Higher Education. *Teaching in Higher Education* 11:351–363.

Konur, O. 2007a. A judicial outcome analysis of the Disability Discrimination Act: A windfall for the employers? *Disability & Society* 22:187–204.

Konur, O. 2007b. Computer-assisted teaching and assessment of disabled students in higher education: The interface between academic standards and disability rights. *Journal of Computer Assisted Learning* 23:207–219.

Konur, O., ed. 2021a. *Handbook of Biodiesel and Petrodiesel Fuels: Science, Technology, Health, and Environment*. Boca Raton, FL: CRC Press.

Konur, O., ed. 2021b. *Handbook of Biodiesel and Petrodiesel Fuels: Science, Technology, Health, and Environment. Volume 1. Biodiesel Fuels: Science, Technology, Health, and Environment.* Boca Raton, FL: CRC Press.

Konur, O., ed. 2021c. *Handbook of Biodiesel and Petrodiesel Fuels: Science, Technology, Health, and Environment. Volume 2. Biodiesel Fuels based on the Edible and Nonedible Feedstocks, Wastes, and Algae: Science, Technology, Health, and Environment.* Boca Raton, FL: CRC Press.

Konur, O., ed. 2021d. *Handbook of Biodiesel and Petrodiesel Fuels: Science, Technology, Health, and Environment. Volume 3. Petrodiesel Fuels: Science, Technology, Health, and Environment.* Boca Raton, FL: CRC Press.

Konur, O. 2021e. Biodiesel and petrodiesel fuels: Science, technology, health, and environment. In *Handbook of Biodiesel and Petrodiesel Fuels: Science, Technology, Health, and Environment. Volume 1. Biodiesel Fuels: Science, Technology, Health, and Environment*, ed. O. Konur. Boca Raton, FL: CRC Press.

Konur, O. 2021f. Biodiesel and petrodiesel fuels: A scientometric review of the research. In *Handbook of Biodiesel and Petrodiesel Fuels: Science, Technology, Health, and Environment. Volume 1. Biodiesel Fuels: Science, Technology, Health, and Environment*, ed. O. Konur. Boca Raton, FL: CRC Press.

Konur, O. 2021g. Biodiesel and petrodiesel fuels: A review of the research. In *Handbook of Biodiesel and Petrodiesel Fuels: Science, Technology, Health, and Environment. Volume 1. Biodiesel Fuels: Science, Technology, Health, and Environment*, ed. O. Konur. Boca Raton, FL: CRC Press.

Konur, O. 2021h Nanotechnology applications in the diesel fuels and the related research fields: A review of the research. In *Handbook of Biodiesel and Petrodiesel Fuels: Science, Technology, Health, and Environment. Volume 1. Biodiesel Fuels: Science, Technology, Health, and Environment*, ed. O. Konur. Boca Raton, FL: CRC Press.

Konur, O. 2021i. Biooils: A scientometric review of the research. In *Handbook of Biodiesel and Petrodiesel Fuels: Science, Technology, Health, and Environment. Volume 1. Biodiesel Fuels: Science, Technology, Health, and Environment*, ed. O. Konur. Boca Raton, FL: CRC Press.

Konur, O. 2021j. Characterization and properties of biooils: A review of the research. In *Handbook of Biodiesel and Petrodiesel Fuels: Science, Technology, Health, and Environment. Volume 1. Biodiesel Fuels: Science, Technology, Health, and Environment*, ed. O. Konur. Boca Raton, FL: CRC Press.

Konur, O. 2021k. Biomass pyrolysis and pyrolysis oils: A review of the research. In *Handbook of Biodiesel and Petrodiesel Fuels: Science, Technology, Health, and Environment. Volume 1. Biodiesel Fuels: Science, Technology, Health, and Environment*, ed. O. Konur. Boca Raton, FL: CRC Press.

Konur, O. 2021l. Biodiesel fuels: A scientometric review of the research. In *Handbook of Biodiesel and Petrodiesel Fuels: Science, Technology, Health, and Environment. Volume 1. Biodiesel Fuels: Science, Technology, Health, and Environment*, ed. O. Konur. Boca Raton, FL: CRC Press.

Konur, O. 2021m. Glycerol: A scientometric review of the research. In *Handbook of Biodiesel and Petrodiesel Fuels: Science, Technology, Health, and Environment. Volume 1. Biodiesel Fuels: Science, Technology, Health, and Environment*, ed. O. Konur. Boca Raton, FL: CRC Press.

Konur, O. 2021n. Propanediol production from glycerol: A review of the research. In *Handbook of Biodiesel and Petrodiesel Fuels: Science, Technology, Health, and Environment. Volume 1. Biodiesel Fuels: Science, Technology, Health, and Environment*, ed. O. Konur. Boca Raton, FL: CRC Press.

Konur, O. 2021o. Edible oil-based biodiesel fuels: A scientometric review of the research. In *Handbook of Biodiesel and Petrodiesel Fuels: Science, Technology, Health, and Environment. Volume 2. Biodiesel Fuels based on the Edible and Nonedible Feedstocks, Wastes, and Algae: Science, Technology, Health, and Environment*, ed. O. Konur. Boca Raton, FL: CRC Press.

Konur, O. 2021p. Palm oil-based biodiesel fuels: A review of the research. In *Handbook of Biodiesel and Petrodiesel Fuels: Science, Technology, Health, and Environment. Volume 2. Biodiesel Fuels based on the Edible and Nonedible Feedstocks, Wastes, and Algae*, ed. O. Konur. Boca Raton, FL: CRC Press.

Konur, O. 2021q. Rapeseed oil-based biodiesel fuels: A review of the research. In *Handbook of Biodiesel and Petrodiesel Fuels: Science, Technology, Health, and Environment. Volume 2. Biodiesel Fuels based on the Edible and Nonedible Feedstocks, Wastes, and Algae*, ed. O. Konur. Boca Raton, FL: CRC Press.

Konur, O. 2021r. Nonedible oil-based biodiesel fuels: A scientometric review of the research. In *Handbook of Biodiesel and Petrodiesel Fuels: Science, Technology, Health, and Environment. Volume 2. Biodiesel Fuels based on the Edible and Nonedible Feedstocks, Wastes, and Algae: Science, Technology, Health, and Environment*, ed. O. Konur. Boca Raton, FL: CRC Press.

Konur, O. 2021s. Waste oil-based biodiesel fuels: A scientometric review of the research. In *Handbook of Biodiesel and Petrodiesel Fuels: Science, Technology, Health, and Environment. Volume 2. Biodiesel Fuels based on the Edible and Nonedible Feedstocks, Wastes, and Algae: Science, Technology, Health, and Environment*, ed. O. Konur. Boca Raton, FL: CRC Press.

Konur, O. 2021t. Algal biodiesel fuels: A scientometric review of the research. In *Handbook of Biodiesel and Petrodiesel Fuels: Science, Technology, Health, and Environment. Volume 2. Biodiesel Fuels based on the Edible and Nonedible Feedstocks, Wastes, and Algae: Science, Technology, Health, and Environment*, ed. O. Konur. Boca Raton, FL: CRC Press.

Konur, O. 2021u. Algal biomass production for biodiesel production: A review of the research. In *Handbook of Biodiesel and Petrodiesel Fuels: Science, Technology, Health, and Environment. Volume 2. Biodiesel Fuels based on the Edible and Nonedible Feedstocks, Wastes, and Algae*, ed. O. Konur. Boca Raton, FL: CRC Press.

Konur, O. 2021v. Algal biomass production in wastewaters for biodiesel production: A review of the research. In *Handbook of Biodiesel and Petrodiesel Fuels: Science, Technology, Health, and Environment. Volume 2. Biodiesel Fuels based on the Edible and Nonedible Feedstocks, Wastes, and Algae*, ed. O. Konur. Boca Raton, FL: CRC Press.

Konur, O. 2021x. Algal lipid production for biodiesel production: A review of the research. In *Handbook of Biodiesel and Petrodiesel Fuels: Science, Technology, Health, and Environment. Volume 2. Biodiesel Fuels based on the Edible and Nonedible Feedstocks, Wastes, and Algae*, ed. O. Konur. Boca Raton, FL: CRC Press.

Konur, O. 2021y. Crude oils: A scientometric review of the research. In *Handbook of Biodiesel and Petrodiesel Fuels: Science, Technology, Health, and Environment. Volume 3. Petrodiesel Fuels: Science, Technology, Health, and Environment*, ed. O. Konur. Boca Raton, FL: CRC Press.

Konur, O. 2021z. Petrodiesel fuels: A scientometric review of the research. In *Handbook of Biodiesel and Petrodiesel Fuels: Science, Technology, Health, and Environment. Volume 3. Petrodiesel Fuels: Science, Technology, Health, and Environment*, ed. O. Konur. Boca Raton, FL: CRC Press.

Konur, O. 2021aa. Bioremediation of petroleum hydrocarbons in the contaminated soils: A review of the research. In *Handbook of Biodiesel and Petrodiesel Fuels: Science,*

Technology, Health, and Environment. Volume 3. Petrodiesel Fuels: Science, Technology, Health, and Environment, ed. O. Konur. Boca Raton, FL: CRC Press.

Konur, O. 2021ab. Desulfurization of diesel fuels: A review of the research. In *Handbook of Biodiesel and Petrodiesel Fuels: Science, Technology, Health, and Environment. Volume 3. Petrodiesel Fuels: Science, Technology, Health, and Environment*, ed. O. Konur. Boca Raton, FL: CRC Press.

Konur, O. 2021ac. Diesel fuel exhaust emissions: A scientometric review of the research. In *Handbook of Biodiesel and Petrodiesel Fuels: Science, Technology, Health, and Environment. Volume 3. Petrodiesel Fuels: Science, Technology, Health, and Environment*, ed. O. Konur. Boca Raton, FL: CRC Press.

Konur, O. 2021ad. The adverse health and safety impact of diesel fuels: A scientometric review of the research. In *Handbook of Biodiesel and Petrodiesel Fuels: Science, Technology, Health, and Environment. Volume 3. Petrodiesel Fuels: Science, Technology, Health, and Environment*, ed. O. Konur. Boca Raton, FL: CRC Press.

Konur, O. 2021ae. Respiratory illnesses caused by the diesel fuel exhaust emissions: A review of the research. In *Handbook of Biodiesel and Petrodiesel Fuels: Science, Technology, Health, and Environment. Volume 3. Petrodiesel Fuels: Science, Technology, Health, and Environment*, ed. O. Konur. Boca Raton, FL: CRC Press.

Konur, O. 2021af. Cancer caused by the diesel fuel exhaust emissions: A review of the research. In *Handbook of Biodiesel and Petrodiesel Fuels: Science, Technology, Health, and Environment. Volume 3. Petrodiesel Fuels: Science, Technology, Health, and Environment*, ed. O. Konur. Boca Raton, FL: CRC Press.

Konur, O. 2021ag. Cardiovascular and other illnesses caused by the diesel fuel exhaust emissions: A review of the research. In *Handbook of Biodiesel and Petrodiesel Fuels: Science, Technology, Health, and Environment. Volume 3. Petrodiesel Fuels: Science, Technology, Health, and Environment*, ed. O. Konur. Boca Raton, FL: CRC Press.

Kumagai, Y., T. Arimoto, and M. Shinyashiki, et al. 1997. Generation of reactive oxygen species during interaction of diesel exhaust particle components with NADPH-cytochrome P450 reductase and involvement of the bioactivation in the DNA damage. *Free Radical Biology and Medicine* 22:479–487.

Kumagai, Y, S. Koide, and K. Taguchi, et al. 2002. Oxidation of proximal protein sulfhydryls by phenanthraquinone, a component of diesel exhaust particles. *Chemical Research in Toxicology* 15:483–489.

Lipsett, M. and S. Campleman. 1999. Occupational exposure to diesel exhaust and lung cancer: A meta-analysis. *American Journal of Public Health* 89:1009–1017.

Mauderly, J. L., R. K. Jones, W. C. Griffith, R. F. Henderson, and R. O. McClellan. 1987. Diesel exhaust is a pulmonary carcinogen in rats exposed chronically by inhalation. *Fundamental and Applied Toxicology* 9:208–221.

Nikula, K. J., M. B. Snipes, and E. B. Barr, et al. 1995. Comparative pulmonary toxicities and carcinogenicities of chronically inhaled diesel exhaust and carbon-black in F344 rats. *Fundamental and Applied Toxicology* 25:80–94.

North, D. C. 1991a. *Institutions, Institutional Change and Economic Performance*. Cambridge, Mass: Cambridge University Press.

North, D. C. 1991b. Institutions. *Journal of Economic Perspectives* 5:97–112.

Perron, P. 1989. The great crash, the oil price shock, and the unit root hypothesis. *Econometrica: Journal of the Econometric Society* 57:1361–1401.

Pitts, J. N., D. M. Lokensgard, and W. Harger, et al. 1982. Mutagens in diesel exhaust particulate identification and direct activities of 6-nitrobenzo[a]pyrene, 9-nitroanthracene, 1-nitropyrene and 5h-phenanthro[4,5-bcd]pyran-5-one. *Mutation Research* 103:241–249.

Rosenkranz, H. S. 1982. Direct-acting mutagens in diesel exhausts: Magnitude of the problem. *Mutation Research* 101:1–10.

Salmeen, I., A. M. Durisin, and T. J. Prater, et al. 1982. Contribution of 1-nitropyrene to direct-acting Ames assay mutagenicities of diesel particulate extracts. *Mutation Research* 104:17–23.

Salmeen, I. T., A. M. Pero, R. Zator, D. Schuetzle, and T. I. Riley. 1984. Ames assay chromatograms and the identification of mutagens in diesel particle extracts. *Environmental Science & Technology* 18:375–382.

Schuetzle, D., F. S. C. Lee, T. J. Prater, and S. B. Tejada. 1981. The identification of polynuclear aromatic hydrocarbon (PAH) derivatives in mutagenic fractions of diesel particulate extracts. *International Journal of Environmental Analytical Chemistry* 9:93–144.

Siegel, R., D. Naishadham, and A. Jemal. 2012. Cancer statistics, 2012. *CA: A Cancer Journal for Clinicians* 62:10–29.

Siegel, R. L., K. D. Miller, and A. Jemal. 2015. Cancer statistics, 2015. *CA: A Cancer Journal for Clinicians* 65:5–29.

Silverman, D. T., C. M. Samanic, and J. H. Lubin, et al. 2012. The diesel exhaust in miners study: A nested case-control study of lung cancer and diesel exhaust. *Journal of the National Cancer Institute* 104:855–868.

Steenland, K., J. Deddens, and L. Stayner. 1998. Diesel exhaust and lung cancer in the trucking industry: Exposure-response analyses and risk assessment. *American Journal of Industrial Medicine* 34:220–228.

Tsurudome, Y., T. Hirano, and H. Yamato, et al. 1999. Changes in levels of 8-hydroxyguanine in DNA, its repair and OGG1 mRNA in rat lungs after intratracheal administration of diesel exhaust particles. *Carcinogenesis* 20:1573–1576.

Xu, X. B., J. P. Nachtman, and Z. L. Jin, et al. 1982. Isolation and identification of mutagenic nitro-PAH in diesel-exhaust particulates. *Analytica Chimica Acta* 136:163–174.

58 Cardiovascular and Other Illnesses Caused by Diesel Fuel Exhaust Emissions

A Review of the Research

Ozcan Konur

CONTENTS

58.1 INTRODUCTION

Crude oils have been primary sources of energy and fuels, such as petrodiesel. However, significant public concerns about the sustainability, price fluctuations, and adverse environmental impact of crude oils have emerged since the 1970s (Ahmadun et al., 2009; Atlas, 1981; Babich and Moulijn, 2003; Kilian, 2009; Perron, 1989). Thus, biooils (Bridgwater and Peacocke, 2000; Czernik and Bridgwater, 2004) and biooil-based biodiesel fuels (Chisti, 2007; Hill et al., 2006) have emerged as alternatives to crude oils and crude oil-based petrodiesel fuels in recent decades. Nowadays, although biodiesel fuels are being used increasingly in the transportation and power sectors, petrodiesel fuels are still used extensively (Konur, 2021a–ag). Therefore, the adverse health impact of 'diesel fuel exhaust particles' (DEPs) has been a great source of public concern (Konur 2021ac–ad). In this context, the causation of cardiovascular illnesses (Anderson et al., 1991; Fried et al., 1991; Kannel et al. 1987), brain illnesses (Abbott et al., 2006; Gelb et al., 1999; Heneka et al., 2015), and reproductive system illnesses (Barash et al., 1996; Hess et al., 1997; Toppari et al., 1996) by DEPs in humans has been studied more extensively (Konur, 2021ac–ad).

However, for the efficient progression of the research in this field, it is necessary to develop efficient incentive structures for the primary stakeholders and to inform these stakeholders about the research (Konur, 2000, 2002a–c, 2006a–b, 2007a–b; North, 1991a–b).

Although there have been a number of reviews and book chapters in this field (Delfino et al., 2005; Ema et al., 2013; Lawal et al., 2016; Wilson et al., 2018; Yoshida and Takeda, 2004), there has been no review of the 25-most-cited articles in this field. Thus, this chapter reviews these articles by highlighting the key findings of these most-prolific studies on the causation of these illnesses by DEPs. Then, it discusses these key findings.

58.2 MATERIALS AND METHODOLOGY

The search for the literature was carried out in the 'Web of Science' (WOS) database in February 2020. It contains the 'Science Citation Index-Expanded' (SCI-E), the 'Social Sciences Citation Index' (SSCI), the 'Book Citation Index-Science' (BCI-S), the 'Conference Proceedings Citation Index-Science' (CPCI-S), the 'Emerging Sources Citation Index' (ESCI), the 'Book Citation Index-Social Sciences and Humanities' (BCI-SSH), the 'Conference Proceedings Citation Index-Social Sciences and Humanities' (CPCI-SSH), and the 'Arts and Humanities Citation Index' (A&HCI).

The keywords for the search of the literature were collated from the screening of abstract pages for the first 500 highly cited papers on the adverse health impact of DEPs. These keywords sets are provided in the Appendix of the related chapter (Konur, 2021ad).

The 25-most-cited articles are selected for this review and the key findings are presented and discussed briefly.

58.3 RESULTS

58.3.1 Cardiovascular Diseases

58.3.1.1 Human Studies

Mills et al. (2007) study ischemic and thrombotic effects of dilute 'diesel exhaust' (DE) inhalation using a sample of 20 men with coronary heart disease to determine the direct effect of air pollution on myocardial, vascular, and fibrinolytic function in a paper with 422 citations. They exposed these men to dilute DE or 'filtered air' (FA) for one hour during periods of rest and moderate exercise in a controlled-exposure facility. They found that during both exposure sessions, the heart rate increased with exercise at similar rates. Exercise-induced ST-segment depression was present in all patients, but there was a greater increase in the ischemic burden during exposure to DE. Exposure to DE did not aggravate preexisting vasomotor dysfunction, but it did reduce the acute release of endothelial tissue plasminogen activator. They conclude that brief exposure to dilute DE promotes myocardial ischemia and inhibits endogenous fibrinolytic capacity in men with stable coronary heart disease. They point to ischemic and thrombotic mechanisms for these effects.

Mills et al. (2005) study the effects of DE inhalation on vascular and endothelial function using a sample of 30 healthy men in a paper with 381 citations. They exposed these men to diluted DE or air for one hour during intermittent exercise. They found that there were no differences in resting-forearm blood flow or inflammatory markers after exposure to DE or air. Although there was a dose-dependent increase in blood flow with each vasodilator, this response was attenuated with bradykinin, acetylcholine, and sodium nitroprusside infusions two hours after exposure to DE, which persisted at six hours. Bradykinin caused a dose-dependent increase in plasma tissue plasminogen activator that was suppressed six hours after exposure to DE. They conclude that at levels encountered in an urban environment, inhalation of dilute DE impairs the regulation of vascular tone and endogenous fibrinolysis. These findings provide a potential mechanism that links DE exposure to the pathogenesis of atherothrombosis and acute myocardial infarction.

Tornqvist et al. (2007) study the vascular and systemic effects of DE using a sample of 15 healthy men 24 hours after inhalation in a paper with 235 citations. They exposed these men to DE or FA for one hour. They found that resting-forearm blood flow, blood pressure, and basal fibrinolytic markers were similar 24 hours after either exposure. DE increased plasma cytokine concentrations ('tumor necrosis factor-α' (TNF-α) and 'interleukin-6' (IL-6)) but reduced acetylcholine, and bradykinin induced forearm vasodilatation. In contrast, there were no differences in either endothelium-independent vasodilatation or bradykinin-induced acute plasma tissue plasminogen activator release. They conclude that there is a selective and persistent impairment of endothelium-dependent vasodilatation that occurs in the presence of mild systemic inflammation.

Lucking et al. (2008) study the effect of DE inhalation on platelet activation and thrombus formation in men in a paper with 172 citations. They exposed 20 healthy volunteers to dilute DE (350 μg/m^3) and FA. They found that compared with FA, DE inhalation increased thrombus formation under low and high-shear conditions by 24 and 19%, respectively. This increased thrombogenicity was seen at 2 and 6 h, using two different diesel engines and fuels. DE also increased platelet-neutrophil and platelet-monocyte aggregates by 52 and 30%, respectively, at 2 h following exposure, compared with FA. They conclude that inhalation of DE increases *ex vivo* thrombus formation and causes *in vivo* platelet activation in men.

Peretz et al. (2008) study the effects of short-term exposure to DE on vascular reactivity and on mediators of vascular tone in a paper with 145 citations. They exposed 27 adult volunteers (10 healthy and 17 with 'metabolic syndrome') to FA and each of two levels of diluted DE (100 or 200 microg/m^3 of fine particulate matter) in two-hour sessions. They found that compared with FA, DE at 200 microg/m^3 elicited a decrease in BAd (0.11 mm), and the effect appeared linearly dose related with a smaller effect at 100 microg/m^3. Plasma levels of ET-1 increased after 200 microg/m^3 DE but not after FA. There was no consistent impact of DE on plasma catecholamines or FMD. They conclude that short-term exposure to DE is associated with acute endothelial response and vasoconstriction of a conductance artery.

Bai et al. (2001) study the cytotoxic mechanism of DEP on human pulmonary artery endothelial cells, focusing on the role of active oxygen species *in vitro* in a paper with 128 citations. They assessed endothelial cell viability by WST-8, a novel tetrazolium salt. They found that DEP extracts damaged endothelial cells under both subconfluent and confluent conditions. The DEP-extract-induced cytotoxicity was markedly reduced by treatment with 'superoxide dismutase' (SOD), catalase, 'N-(2-mercaptopropionyl)-glycine' (MPG), or ebselen (a selenium-containing compound with glutathione peroxidase-like activity). Thus superoxide, hydrogen peroxide, and other oxygen-derived free radicals are likely to be implicated in DEP-extract-induced endothelial cell damage. Moreover, L-NAME and L-NMA, inhibitors of 'nitrogen oxide' (NO) synthase, also attenuated DEP-extract-induced cytotoxicity, while sepiapterin, the precursor of tetrahydrobiopterin (BH4, a NO synthase cofactor), interestingly enhanced DEP-extract-induced cell damage. They conclude that NO is also involved in DEP-extract-mediated cytotoxicity, which was confirmed by the direct measurement of NO production. These active oxygen species, including peroxynitrite, may explain the mechanism of endothelial cell damage upon DEP exposure during the early stage.

Lucking et al. (2011) study the adverse vascular and prothrombotic effects of DE inhalation in men in a paper with 103 citations. They exposed 19 healthy volunteers to FA and DE in the presence or absence of a particle trap for one hour. They determined bilateral forearm blood flow and plasma fibrinolytic factors with venous occlusion plethysmography and blood sampling during intraarterial infusion of acetylcholine, bradykinin, sodium nitroprusside, and verapamil. They found that compared with FA, DE inhalation was associated with reduced vasodilatation and increased *ex vivo* thrombus formation under both low and high-shear conditions. The particle trap markedly reduced DE particulate number and mass, and was associated with increased vasodilatation, reduced thrombus formation, and an increase in tissue-type plasminogen activator release. They conclude that exhaust particle traps are a highly efficient method of reducing particle emissions from diesel engines. With a range of surrogate measures, the use of a particle trap prevents several adverse cardiovascular effects of exhaust inhalation in men. Given these beneficial effects on biomarkers of cardiovascular health, they recommend that the widespread use of particle traps on diesel-powered vehicles may have substantial public health benefits and reduce the burden of cardiovascular disease. They conclude that DE inhalation was associated with reduced vasodilatation and increased *ex vivo* thrombus formation under both low and high-shear conditions.

Mills et al. (2011) study the adverse *in vivo* vascular effects of DE inhalation in a paper with 99 citations. They focused on the role of combustion-derived nanoparticles in mediating the adverse cardiovascular effects of air pollution. They exposed 16 healthy volunteers to dilute DE, pure carbon nanoparticulate, filtered DE, or FA. Following each exposure, they measured forearm blood flow during intrabrachial bradykinin, acetylcholine, sodium nitroprusside, and verapamil infusions. They found that compared with FA, inhalation of DE increased systolic blood pressure and attenuated vasodilatation to bradykinin, acetylcholine, and sodium nitroprusside. Exposure to pure carbon nanoparticulate or filtered exhaust had no effect on endothelium-dependent or independent vasodilatation. To determine the direct vascular effects of nanoparticulates, they assessed isolated rat aortic rings ($n = 6–9$ per group) *in vitro* by wire myography and exposed to DE particulate, pure carbon nanoparticulates. Compared with vehicle, DE particulate (but not pure carbon nanoparticulate) attenuated both acetylcholine and sodium-nitroprusside-induced vasorelaxation. These effects were partially attributable to both soluble and insoluble components of the particulate. They conclude that combustion-derived nanoparticulate predominately mediate the adverse vascular effects of DE inhalation.

Lundback et al. (2009) determine the effect of DE exposure on arterial compliance using a validated non-invasive measure of arterial stiffness in humans in a paper with 81 citations. In a double-blind randomized fashion, they exposed 12 healthy volunteers to DE (approximately 350 μg/m^3) or FA for one hour during moderate exercise. They found that blood pressure, 'augmentation pressure' (AP), and the 'augmentation index' (AIx) were generally low, reflecting compliant arteries. In comparison to FA, DE exposure induced an increase in AP of 2.5 mmHg and in AIx of 7.8%, along with a 16 ms reduction in Tr, 10 minutes post-exposure. They conclude that acute exposure to DE is associated with an immediate and transient increase in arterial stiffness. This may, in part, explain the increased risk of cardiovascular disease associated with air pollution exposure.

58.3.1.2 Animal Studies

Nemmar et al. (2003b) study the increase in cardiovascular morbidity and mortality in hamsters exposed to DEPs in a paper with 197 citations. They instilled DEPs into the trachea of hamsters and examined blood platelet activation, experimental thrombosis, and lung inflammation. They found that doses of 5 to 500 microg. of DEPs per animal induced neutrophil influx into the 'bronchoalveolar lavage' (BAL) fluid with elevation of protein and histamine but without lactate dehydrogenase release. The same doses enhanced experimental arterial and venous platelet rich-thrombus formation *in vivo*. Blood samples taken from hamsters 30 and 60 minutes after instillation of 50 microg. of DEPs yielded accelerated aperture closure (i.e. platelet activation) *ex vivo*. The direct addition of as little as 0.5 microg./mL DEPs to untreated hamster blood significantly shortened closure time *in vitro*. They conclude that the intratracheal instillation of DEPs leads to lung inflammation as well as a rapid activation of circulating blood platelets. The kinetics of platelet activation is consistent with the reported clinical occurrence of thrombotic complications after exposure to pollutants. They provide a plausible explanation for the increase in cardiovascular morbidity and mortality accompanying urban air pollution.

Nemmar et al. (2004a) study the mechanisms of the cardiovascular effects of DEPs focusing on the systemic translocation and prothrombotic effects in a paper with 195 citations. They evaluated the acute effect (1 h) of DEPs in a hamster model of peripheral vascular thrombosis induced by free-radical-mediated endothelial injury. They found that intratracheal doses of 5–500 microg. of DEP per animal induced inflammation with elevation of neutrophils, total proteins, and histamine in BAL. DEP enhanced experimental arterial and venous platelet rich-thrombus formation *in vivo*. Blood samples taken from hamsters 30 and 60 min after instillation of DEP caused platelet activation. The direct addition of DEP to untreated hamster blood also caused platelet aggregation. These effects persisted up to 24 h after instillation. They provide plausible mechanistic explanations for the epidemiologically established link between DEP pollution and acute cardiovascular effects.

Nemmar et al. (2003a) study the pulmonary inflammation and thrombogenicity caused by diesel particles in hamsters in a paper with 92 citations. At 1, 6, and 24 hours after instillation of 50 μg DEPs per hamster, the mean size of *in vivo*-induced and quantified venous thrombosis was increased by 480, 770, and 460%, respectively. In BAL, neutrophils and histamine levels were increased at all time points. In plasma, histamine was increased at 6 and 24 hours but not at 1 and 3 hours. Pretreatment with a histamine H1-receptor antagonist (diphenhydramine, 30 mg/kg intraperitoneally) abolished the DEP-induced neutrophil influx in BAL at all-time points. However, diphenhydramine pretreatment did not affect DEP-induced thrombosis or platelet activation at 1 hour, whereas both were markedly reduced at 6 and 24 hours. They conclude that pulmonary inflammation and peripheral thrombosis are correlated at 6 and 24 hours, but at 1 hour, the prothrombotic effects do not result from pulmonary inflammation but possibly from the blood penetration of DEP-associated components or by DEP particles themselves.

Nemmar et al. (2004b) study the relationship between airway inflammation and thrombosis 24 hours after intratracheal instillation of DEPs (50 μg/hamster) in a paper with 89 citations. They induced mild thrombosis in the femoral vein by endothelial injury and studied the consequences of airway inflammation on thrombogenicity via online video microscopy. They performed lung inflammation and histamine analysis in BAL and plasma after pretreatment with 'dexamethasone' (DEX) or 'sodium cromoglycate' (SC). They found that DEP-induced airway inflammation and histamine release in BAL and in plasma, and increased thrombosis, without elevating plasma von 'Willebrand factor' (vWF) levels. The IT instillation of 400 nm positively charged polystyrene particles (500 μg/hamster), serving as particles that do not penetrate into the circulation, equally produced airway inflammation, histamine release, and enhanced thrombosis. Histamine in plasma resulted from basophil activation. 'Intraperitoneal' (IP) pretreatment with DEX (5 mg/kg) abolished the

DEP-induced histamine increase in BAL and plasma and abrogated airway inflammation and thrombogenicity. The IT pretreatment with DEX (0.5 mg/kg) showed a partial but parallel inhibition of all of these parameters. Pretreatment with SC (40 mg/kg, IP) strongly inhibited airway inflammation, thrombogenicity, and histamine release. Their results are compatible with the triggering of mast cell degranulation and histamine release by DEP. They conclude that histamine plays an initial central role in airway inflammation, further release of histamine by circulating basophils, and peripheral thrombotic events. Antiinflammatory pretreatment can abrogate the peripheral thrombogenicity by preventing histamine release from mast cells.

Hirano et al. (2003) study the cytotoxicity and oxidative stress potency of 'organic extracts of DEPs' (OE-DEP) and 'urban fine particles' (OE-UFP) in rat heart microvessel endothelial (RHMVE) cells in a paper with 83 citations. They found that the LC_{50} values of OE-DEP and OE-UFP were 17 and 34 microg./ml, respectively, suggesting that OE-DEP was more cytotoxic than OE-UFP. The viability of OE-DEP- and OE-UFP-exposed cells was ameliorated by 'N-acetylcysteine' (NAC). They exposed the cell monolayer to 0 (control), 1, 3, and 10 microg./ml OE-DEP for 6 h and mRNA levels of antioxidant enzymes such as they quantify 'heme oxygenase-1' (HO-1), 'thioredoxin peroxidase 2' (TRPO), 'glutathione S-transferase P subunit' (GST-P), and 'NADPH dehydrogenase' (NADPHD) by northern analysis. All those mRNA levels increased dose-dependently with OE-DEP and HO-1 mRNA showing the most marked response to OE-DEP. mRNA levels of those antioxidant enzymes and the 'heat shock protein 72' (HSP72) in OE-DEP-exposed cells were higher than those of OE-UFP-exposed cells as compared at the same concentration. The transcription levels of HO-1 and HSP72 in OE-DEP- and OE-UFP-exposed cells were also reduced by NAC. They conclude that the OE-DEPs in urban air has a potency to cause oxidative stress to endothelial cells and may be implicated in cardiovascular diseases through functional changes of endothelial cells.

Hazari et al. (2011) study the increased risk of triggered cardiac arrhythmias in hypertensive rats exposed to DE in a paper with 80 citations. They hypothesized that increased risk of triggered arrhythmias one day after DE exposure is mediated by airway sensory nerves bearing 'transient receptor potential' (TRP) channels (e.g. 'TRP cation channel, member A1' (TRPA1)) that, when activated by noxious chemicals, can cause a centrally mediated autonomic imbalance and heightened risk of arrhythmia. They exposed spontaneously hypertensive rats implanted with radiotelemeters to either 500 $\mu g/m^3$ (high) or 150 $\mu g/m^3$ (low) 'whole DE' (wDE) or 'filtered DE' (fDE), or to FA (controls), for four hours. They found that rats exposed to wDE or fDE had a slightly higher heart rate and increased low-frequency: high-frequency ratios (sympathetic modulation) than did controls; 'electrocardiogram' (ECG) showed prolonged ventricular depolarization and shortened repolarization periods. Rats exposed to wDE developed arrhythmia at lower doses of aconitine than did controls; the dose was even lower in rats exposed to fDE. Pretreatment of low wDE-exposed rats with a TRPA1 antagonist or sympathetic blockade prevented the heightened sensitivity to arrhythmia. They conclude that a single exposure to DE increases the sensitivity of the heart to triggered arrhythmias. The gaseous components play an important role in the proarrhythmic response, which may be mediated by activation of TRPA1, and subsequent sympathetic modulation. As such, toxic inhalants may partly exhibit their toxicity by lowering the threshold for secondary triggers, complicating assessment of their risk.

Campen et al. (2005) study the effects of nonparticulate DE constituents on electrocardiographic traces from $ApoE^{-/-}$ mice exposed whole-body and isolated, pressurized septal coronary arteries from $ApoE^{-/-}$ mice in a paper with 73 citations. They hypothesized that air pollutants, primarily vapor phase organic compounds, cause an enhancement of coronary vascular constriction. They implanted $ApoE^{-/-}$ mice with radiotelemetry devices to assess ECG waveforms continuously throughout exposures (6 h/day × 3 days) to DE (0.5 and 3.6 mg/m^3) in whole-body inhalation chambers with or without particulates filtered. They observed significant bradycardia and T-wave depression, regardless of the presence of particulates. Pulmonary inflammation was present only in the whole DE-exposed animals at the highest concentration. Fresh DE or air was bubbled through the physiologic saline tissue bath prior to experiments to enable isolated tissue exposure; exposed saline contained elevated levels of several volatile carbonyls and alkanes, but low to absent levels of polycyclic aromatic hydrocarbons. They then assay vessels for constrictive and dilatory function. Diesel components enhanced the vasoconstrictive effects of endothelin-1 and reduced the dilatory response to sodium nitroprusside. They conclude that nonparticulate compounds in whole DE elicit ECG changes consistent with myocardial ischemia. Furthermore, the volatile organic compounds in the vapor phase caused enhanced constriction and reduced dilatation in isolated coronary arteries caused by nonparticulate components of DE.

58.3.2 Brain Diseases

58.3.2.1 Human Studies

Cruts et al. (2008) study the functional effect of DE exposure in the human brain in a paper with 81 citations. They exposed ten human volunteers to dilute DE (300 μg/m^3) as a model for ambient PM exposure and FA for one hour using a double blind randomized crossover design. They monitored brain activity during and for one hour following each exposure using quantitative electroencephalography at eight different sites on the scalp. They found a significant increase in 'median power frequency' (MPF) in response to DE in the frontal cortex 30 minutes into exposure. The increase in MPF was primarily caused by an increase in fast wave activity (β2) and continued to rise during the one hour post-exposure interval. They conclude that there is a functional effect of DE exposure in the human brain, indicating a general cortical stress response.

58.3.2.2 Animal Studies

Block et al. (2004) study the role of microglia, phagocytosis, and NADPH oxidase to the development of Parkinson's disease in mice exposed to DEPs in a paper with 201 citations. They found that mesencephalic neuron-glia cultures treated with DE particles (DEP; 0.22 microM) (5–50 microg./ml) resulted in a dose-dependent decrease in 'dopaminergic' (DA) neurons, as determined by DA-uptake assay and tyrosine-hydroxylase 'immunocytochemistry' (ICC). They show the selective toxicity of DEP for DA neurons by the lack of DEP effect on both GABA uptake and Neu-N immunoreactive cell number. They show the critical role of microglia by the failure of neuron-enriched cultures to exhibit DEP-induced DA neurotoxicity, where DEP-induced DA neuron death was reinstated with the addition of microglia to neuron-enriched cultures. OX-42 ICC staining of DEP treated neuron-glia cultures revealed changes in microglia morphology, indicative of activation. Intracellular reactive oxygen species and superoxide were produced from enriched-microglia cultures in response to DEP. Neuron-glia cultures from NADPH oxidase deficient ($PHOX^{-/-}$) mice were insensitive to DEP neurotoxicity when compared with control mice ($PHOX^{+/+}$). Cytochalasin D inhibited DEP-induced superoxide production in enriched-microglia cultures, implying that DEP must be phagocytized by microglia to produce superoxide. They conclude that DEP selectively damages DA neurons through the phagocytic activation of microglial NADPH oxidase and consequent oxidative insult.

Hartz et al. (2008) study the impact of DEPs on blood-brain barrier function in a paper with 145 citations. They reported that DEPs affect blood-brain barrier function at the tissue, cellular, and molecular levels. Isolated rat brain capillaries exposed to DEPs showed increased expression and transport activity of the key drug efflux transporter, P-glycoprotein. Up-regulation of P-glycoprotein was abolished by blocking transcription or protein synthesis. Inhibition of NADPH oxidase or pretreatment of capillaries with radical scavengers ameliorated DEP-induced P-glycoprotein up-regulation, indicating a role for reactive oxygen species in signaling. DEP exposure also increased brain capillary TNF-α levels. DEP-induced P-glycoprotein up-regulation was abolished when 'TNF-receptor 1' (TNF-R1) was blocked and was not evident in experiments with capillaries from TNF-R1 knockout mice. Inhibition of JNK, but not NF-κB, blocked DEP-induced P-glycoprotein up-regulation, indicating a role for 'activator protein 1' (AP-1) in the signaling pathway. Consistent with this, DEPs increased phosphorylation of c-jun. They conclude that DEPs alter blood-brain barrier function through oxidative stress and proinflammatory cytokine production.

Levesque et al. (2011b) study the central nervous system consequences of subchronic exposure to DE and address the minimum levels necessary to elicit neuroinflammation and markers of early neuropathology in a paper with 136 citations. They exposed male Fischer 344 rats to DE (992, 311, 100, 35, and 0 μg PM/m^3) by inhalation over six months. They found that DE exposure resulted in elevated levels of TNF-α at high concentrations in all regions tested, with the exception of the cerebellum. The midbrain region was the most sensitive, where exposures as low as 100 μg PM/m^3 significantly increased brain TNF-α levels. However, this sensitivity to DE was not conferred to all markers of neuroinflammation, as the midbrain showed no increase in IL-6 expression at any concentration tested, an increase in IL-1β at only high concentrations, and a decrease in 'macrophage inflammatory protein-1α' (MIP-1α) expression, supporting the hypothesis that compensatory mechanisms may occur with subchronic exposure. Aβ42 levels were the highest in the frontal lobe of mice exposed to 992 μg PM/m^3 and tau [pS199] levels were elevated at the higher DE concentrations (992 and 311 μg PM/m^3) in both the temporal lobe and frontal lobe, indicating that proteins linked to preclinical Alzheimer's disease were affected. α Synuclein levels were elevated in the midbrain in response to the 992 μg PM/m^3 exposure, supporting the hypothesis that air pollution may be associated with early

Parkinson's disease-like pathology. They conclude that the midbrain may be more sensitive to the neuroinflammatory effects of subchronic air pollution exposure. However, the DE-induced elevation of proteins associated with neurodegenerative diseases was limited to only the higher exposures, suggesting that DE-induced neuroinflammation may precede preclinical markers of neurodegenerative disease in the midbrain.

Levesque et al. (2011a) study the brain-region-specific effects of DE and key cellular mechanisms underlying DE-induced microglia activation, neuroinflammation, and DA neurotoxicity in a paper with 136 citations. They exposed rats to DE (2.0, 0.5, and 0.0 mg/m^3) by inhalation over four weeks or as a single intratracheal administration of DE particles (DEP; 20 mg/kg). They find that rats exposed to DE by inhalation demonstrated elevated levels of whole-brain IL-6 protein, nitrated proteins, and IBA-1 (ionized calcium-binding adaptor molecule 1) protein (microglial marker), indicating generalized neuroinflammation. DE increased TNF-α , IL-1β, IL-6, MIP-1α RAGE (receptor for advanced glycation end products), fractalkine, and the IBA-1 microglial marker in most regions tested, with the midbrain showing the greatest DE response. Intratracheal administration of DEP increased microglial IBA-1 staining in the substantia nigra and elevated both serum and whole-brain TNF-α at six hour post-treatment. Although DEP alone failed to cause the production of cytokines and chemokines, DEP (5 μg/mL) pretreatment followed by lipopolysaccharide (2.5 ng/mL) *in vitro* synergistically amplified nitric oxide production, TNF-α release, and DA neurotoxicity. Pretreatment with fractalkine (50 pg/mL) *in vitro* ameliorated DEP (50 μg/mL)-induced microglial hydrogen peroxide production and DA neurotoxicity. They conclude that complex, interacting mechanisms are responsible for how DE may cause neuroinflammation and DA neurotoxicity.

Gerlofs-Nijland et al. (2010) study the association between exposure to DE and neuroinflammation in a paper with 94 citations. To elucidate whether specific regions of the brain are more susceptible to exposure to diesel-derived AP, they analyzed various loci of the brain separately. They exposed rats for six hours a day, five days a week, for four weeks to DE using a nose-only exposure chamber. They found that baseline levels of the proinflammatory cytokines TNF-α and 'interleukin-1 α' (IL-1α) were dependent on the region analyzed and increased in the striatum after exposure to DE. In addition, the baseline level of activation of the transcription factors (NF-ƙB) and (AP-1) was also region dependent, but the levels were not significantly altered after exposure to DE. A similar, though not significant, trend was seen with the mRNA expression levels of TNF-α and 'TNF receptor-subtype I' (TNF-RI). They conclude that different brain regions may be uniquely responsive to changes induced by exposure to DE.

58.3.3 Reproductive System

Yoshida et al. (1999) study the effect of the exposure to DE in the male reproductive system of mice in a paper with 104 citations. They hypothesized that DE is an environmental pollutant with the potential to influence male reproductive function. They observed ultrastructural changes in Leydig cells of mice exposed to DE (0.3 mg DE particles (DEP)/m^3 through the airway, twelve hours daily, up to six months) and they observed reduction in LH receptor mRNA expression in Leydig cells at a concentration of 1 mg DEP/m^3. Daily sperm production per gram of testis dose-dependently decreased with exposure to DE for six months; 29, 36, and 53% reductions were observed at 0.3, 1.0, and 3.0 mg DEP/m^3, respectively. They finally observed a 'no-observed-adverse-effect level' (NOAEL) with approximately 30 micrograms DEP/m^3, which is lower than the WHO-recommended limit. They conclude that DE adversely influences male reproductive function.

Watanabe and Oonuki (1999) study the effect of DE on the reproductive endocrine function in growing male rats in a paper with 98 citations. They assigned the rats to three groups: (i) a group exposed to total DE containing 5.63 mg/m^3 particulate matter, 4.10 ppm nitrogen dioxide, and 8.10 ppm nitrogen oxide; (ii) a group exposed to filtered exhaust without particulate matter; and (iii) a group exposed to clean air. They performed dosing experiments for three months beginning at birth (six hours/day for five days/week). They find that serum levels of testosterone and estradiol were significantly higher in animals exposed to total DE and filtered exhaust as compared to the controls. Follicle-stimulating hormone was significantly decreased in the two groups exposed to DE as compared to the control group. Luteinizing hormone was significantly decreased in the total exhaust-exposed group as compared to the control and filtered groups. Although testis weight did not show any significant difference among the groups, sperm production and activity of testicular hyaluronidase were significantly reduced in both exhaust-exposed groups as compared to the control group. Histological examination showed decreased numbers of step 18 and 19 spermatids in stage VI, VII, and VIII tubules in the testes of both DE-exposed groups. They conclude that DE stimulates hormonal secretion of the adrenal cortex, depresses gonadotropin-releasing-hormone, and inhibits spermatogenesis in rats. Because these effects were not inhibited by filtration, the gaseous phase of the exhaust is more responsible than particulate matter for disrupting the endocrine system.

Hougaard et al. (2008) study whether exposure to DEPs could affect gestation, postnatal development, activity, learning and memory, and biomarkers of transplacental toxicity in a paper with 78 citations. Pregnant mice (C57BL/6; BomTac) were exposed to 19 mg/m^3 DEP (~1.106 particles/cm^3; mass median diameter congruent with 240 nm) on gestational days 9–19, for 1 h/day. They found that gestational parameters were similar in control and diesel groups. Shortly after birth, the body weights of DEP offspring were slightly lower than in controls. This difference increased during lactation, so, by weaning, the DEP-exposed offspring weighed significantly less than the control progeny. They observed only slight effects of exposure on cognitive function in female DEP offspring and on biomarkers of exposure to particles or genotoxic substances. They conclude that *in utero* exposure to DEP decreased weight gain during lactation. Cognitive function and levels of biomarkers of exposure to particles or to genotoxic substances were generally similar in exposed and control offspring. The particle size and chemical composition of DEP and differences in exposure methods (fresh, whole exhaust versus aged, resuspended DEP) may play a significant role on the biological effects observed in this, compared to other studies.

58.4 DISCUSSION

Table 58.1 provides information on the research fronts in this field. As this table shows the research fronts of 'cardiovascular illnesses', 'brain illnesses', and 'reproductive system illnesses' comprise 64, 24, and 12% of these papers, respectively. 'Human studies', 'animal studies', and *in vitro* studies comprise 32, 28, and 4% of the papers in the first group of papers, respectively. Similarly, 'human studies' and 'animal studies' comprise 4 and 20% of the papers in the second group of papers, respectively. Finally, 'animal studies' comprise 12% of all the papers in the final group of papers.

58.4.1 Cardiovascular Diseases

58.4.1.1 Human Studies

Mills et al. (2007) study ischemic and thrombotic effects of dilute DE inhalation using a sample of 20 men with coronary heart disease to determine the direct effect of air pollution on myocardial, vascular, and fibrinolytic function in a paper with 422 citations. They conclude that brief exposure to dilute DE promotes myocardial ischemia and inhibits endogenous fibrinolytic capacity in men with stable coronary heart disease. They point to ischemic and thrombotic mechanisms for these effects.

Mills et al. (2005) study the effects of DE inhalation on vascular and endothelial function using a sample of 30 healthy men in a paper with 381 citations. They conclude that at levels encountered in an urban environment, inhalation of dilute DE impairs the regulation of vascular tone and endogenous fibrinolysis. These findings provide a potential mechanism that links DE to the pathogenesis of atherothrombosis and acute myocardial infarction.

Tornqvist et al. (2007) study the vascular and systemic effects of DE using a sample of 15 healthy men 24 hours after inhalation in a paper with 235 citations. They conclude that there is a selective and persistent impairment of endothelium-dependent vasodilatation that occurs in the presence of mild systemic inflammation.

Lucking et al. (2008) study the effect of DE inhalation on platelet activation and thrombus formation in men in a paper with 172 citations. They conclude that

TABLE 58.1
Research Fronts

	Research Front	Papers (%)
1	**Cardiovascular illnesses**	**64**
1.1	Human studies	32
1.2	Animal studies	28
1.3	*In vitro* studies	4
2	**Brain illnesses**	**24**
2.1	Human studies	4
2.2	Animal studies	20
3	**Reproductive system illnesses**	**12**
3.1	Animal studies	12

inhalation of DE increases *ex vivo* thrombus formation and causes *in vivo* platelet activation in humans.

Peretz et al. (2008) study the effects of short-term exposure to DE on vascular reactivity and on mediators of vascular tone in a paper with 145 citations. They conclude that short-term exposure to DE is associated with acute endothelial response and vasoconstriction of a conductance artery.

Lucking et al. (2011) study the adverse vascular and prothrombotic effects of DE inhalation in men in a paper with 103 citations. They conclude that DE inhalation was associated with reduced vasodilatation and increased *ex vivo* thrombus formation under both low and high-shear conditions.

Mills et al. (2011) study the adverse *in vivo* vascular effects of DE inhalation in a paper with 99 citations. They conclude that combustion-derived nanoparticulates predominately mediate the adverse vascular effects of DE inhalation.

Lundback et al. (2009) determine the effect of DE exposure on arterial compliance using a validated non-invasive measure of arterial stiffness in humans in a paper with 81 citations. They conclude that acute exposure to DE is associated with an immediate and transient increase in arterial stiffness. This may, in part, explain the increased risk of cardiovascular disease associated with air pollution exposure.

These prolific studies show that DEPs can cause cardiovascular illnesses in humans, which has strong public policy implications.

58.4.1.2 Animal Studies

Nemmar et al. (2003b) study the increase in cardiovascular morbidity and mortality in hamsters exposed to DEPs in a paper with 197 citations. They conclude that the intratracheal instillation of DEPs leads to lung inflammation as well as a rapid activation of circulating blood platelets. The kinetics of platelet activation is consistent with the reported clinical occurrence of thrombotic complications after exposure to pollutants. They provide a plausible explanation for the increase in cardiovascular morbidity and mortality accompanying urban air pollution.

Nemmar et al. (2004a) study mechanisms of the cardiovascular effects of DEPs focusing on the systemic translocation and prothrombotic effects in a paper with 195 citations. They provide plausible mechanistic explanations for the epidemiologically established link between DEP pollution and acute cardiovascular effects.

Nemmar et al. (2003a) study the pulmonary inflammation and thrombogenicity caused by diesel particles in hamsters in a paper with 92 citations. They conclude that pulmonary inflammation and peripheral thrombosis are correlated at 6 and 24 hours, but at 1 hour, the prothrombotic effects do not result from pulmonary inflammation but possibly from the blood penetration of DEP-associated components or by DEP particles themselves.

Nemmar et al. (2004b) study the relationship between airway inflammation and thrombosis 24 hours after intratracheal instillation of DEPs (50 μg/hamster) in a paper with 89 citations. They conclude that histamine plays an initial central role in airway inflammation, further release of histamine by circulating basophils, and peripheral thrombotic events. Antiinflammatory pretreatment can abrogate the peripheral thrombogenicity by preventing histamine release from mast cells.

Hirano et al. (2003) study the cytotoxicity and oxidative stress potency of OE-DEPs and OE-UFP in rat heart microvessel endothelial cells in a paper with 83 citations. They conclude that the organic fraction of particulate materials in the urban air has a potency to cause oxidative stress to endothelial cells and may be implicated in cardiovascular diseases through functional changes of endothelial cells.

Hazari et al. (2011) study the increased risk of triggered cardiac arrhythmias in hypertensive rats exposed to DE in a paper with 80 citations. They conclude that a single exposure to DE increases the sensitivity of the heart to triggered arrhythmias. The gaseous components play an important role in the proarrhythmic response, which may be mediated by activation of TRPA1, and subsequent sympathetic modulation. As such, toxic inhalants may partly exhibit their toxicity by lowering the threshold for secondary triggers, complicating assessment of their risk.

Campen et al. (2005) study the effects of nonparticulate DE constituents on electrocardiographic traces from $ApoE^{-/-}$ mice exposed whole-body and isolated, pressurized septal coronary arteries from $ApoE^{-/-}$ mice in a paper with 73 citations. They conclude that nonparticulate compounds in whole DE elicit ECG changes consistent with myocardial ischemia. Furthermore, the volatile organic compounds in the vapor phase caused enhanced constriction and reduced dilatation in isolated coronary arteries caused by nonparticulate components of DE.

These prolific studies show that DEPs can cause cardiovascular illnesses in animals, which has strong public policy implications.

58.4.1.3 *In vitro* Studies

Bai et al. (2001) study the cytotoxic mechanism of DEP on human pulmonary artery endothelial cells, focusing on the role of active oxygen species *in vitro* in a paper with 128 citations. They conclude that NO is also involved in DEP-extract-mediated cytotoxicity, which was confirmed by the direct measurement of NO production. These active oxygen species, including peroxynitrite, may explain the mechanism of endothelial cell damage upon DEP exposure during the early stage.

This single prolific study shows that DEPs can cause cardiovascular illnesses in *in vitro* studies, which has strong public policy implications.

58.4.2 Brain Diseases

58.4.2.1 Human Studies

Cruts et al. (2008) study the functional effect of DE exposure in the human brain in a paper with 81 citations. They conclude that there is a functional effect of DE exposure in the human brain, indicating a general cortical stress response.

This single prolific study shows that DEPs can cause brain illnesses in humans, which has strong public policy implications.

58.4.2.2 Animal Studies

Block et al. (2004) study the role of microglia, phagocytosis, and NADPH oxidase to the development of Parkinson's disease in mice exposed to DEPs in a paper with 201 citations. They conclude that DEP selectively damages DA neurons through the phagocytic activation of microglial NADPH oxidase and consequent oxidative insult.

Hartz et al. (2008) study the impact of DEPs on the blood-brain-barrier function in a paper with 145 citations. They conclude that DEPs alter this function through oxidative stress and proinflammatory cytokine production.

Levesque et al. (2011b) study the central nervous system consequences of subchronic exposure to DE and address the minimum levels necessary to elicit neuroinflammation and markers of early neuropathology in a paper with 136 citations. They conclude that the midbrain may be more sensitive to the neuroinflammatory effects of subchronic air pollution exposure. However, the DE-induced elevation of proteins associated with neurodegenerative diseases was limited to only the higher exposures, suggesting that air pollution-induced neuroinflammation may precede preclinical markers of neurodegenerative disease in the midbrain.

Levesque et al. (2011a) study the brain-region-specific effects of DE and key cellular mechanisms underlying DE-induced microglia activation, neuroinflammation, and DA neurotoxicity in a paper with 136 citations. They conclude that complex, interacting mechanisms are responsible for how DE may cause neuroinflammation and DA neurotoxicity.

Gerlofs-Nijland et al. (2010) study the association between exposure to DE and neuroinflammation in a paper with 94 citations. They conclude that different brain regions may be uniquely responsive to changes induced by exposure to DE.

These prolific studies show that DEPs can cause brain illnesses in animals, which has strong public policy implications.

58.4.3 Reproductive System

Yoshida et al. (1999) study e effect of the exposure to DE in the male reproductive system of mice in a paper with 104 citations. They conclude that DE adversely influences the male reproductive function.

Watanabe and Oonuki (1999) study the effect of DE on the reproductive endocrine function in growing male rats in a paper with 98 citations. They conclude that DE stimulates hormonal secretion of the adrenal cortex, depresses gonadotropin-releasing-hormone, and inhibits spermatogenesis in rats. Because these effects were not inhibited by filtration, the gaseous phase of the exhaust is more responsible than particulate matter for disrupting the endocrine system.

Hougaard et al. (2008) study whether exposure to DEPs would affect gestation, postnatal development, activity, learning and memory, and biomarkers of transplacental toxicity in a paper with 78 citations. They conclude that *in utero* exposure to DEP decreased weight gain during lactation. Cognitive function and levels of biomarkers of exposure to particles or to genotoxic substances were generally similar in exposed and control offspring. The particle size and chemical composition of the DEP and differences in exposure methods (fresh, whole exhaust versus aged, resuspended DEP) may play a significant role on the biological effects observed in this, compared to other studies.

These prolific studies show that DEPs can cause illnesses in the reproductive system of animals, which has strong public policy implications.

58.5 CONCLUSION

This chapter has presented the key findings of the 25-most-cited article papers in this field. Table 58.1 provides information on the research fronts in this field. As this table shows the research fronts of 'cardiovascular illnesses', 'brain illnesses', and 'reproductive system illnesses' comprise 64, 24, and 12% of these papers, respectively. 'Human studies', 'animal studies', and *in vitro* studies comprise 32, 28, and 4% of the papers in the first group, respectively. Similarly, 'human studies' and 'animal studies' comprise 4 and 20% of the papers in the second group, respectively. Finally, 'animal studies' comprise 12% of all the papers in the final group.

These prolific studies in three different research fronts provide valuable evidence on the causation of cardiovascular illnesses, brain illnesses, and reproductive system illnesses by DEPs, which has strong public policy implications.

It is recommended that similar studies are carried out for each research front as well.

ACKNOWLEDGMENTS

The contribution of the highly cited researchers in this field is greatly acknowledged.

REFERENCES

Abbott, N. J., L. Ronnback, and E. Hansson. 2006. Astrocyte-endothelial interactions at the blood–brain barrier. *Nature Reviews Neuroscience* 7:41.

Ahmadun, F. R., A. Pendashteh, and L. C. Abdullah, et al. 2009. Review of technologies for oil and gas produced water treatment. *Journal of Hazardous Materials* 170:530–551.

Anderson, K. M., P. M. Odell, P. W. Wilson, and W. B. Kannel. 1991. Cardiovascular disease risk profiles. *American Heart Journal* 121:293–298.

Atlas, R. M.. 1981. Microbial degradation of petroleum hydrocarbons: An environmental perspective. *Microbiological Reviews* 45:180–209.

Babich, I. V. and J. A. Moulijn. 2003. Science and technology of novel processes for deep desulfurization of oil refinery streams: A review. *Fuel* 82:607–631.

Bai, Y. S., A. K. Suzuki, and M. Sagai. 2001. The cytotoxic effects of diesel exhaust particles on human pulmonary artery endothelial cells in vitro: Role of active oxygen species. *Free Radical Biology and Medicine* 30:555–562.

Barash, I. A., C. C. Cheung, and D. S. Weigle, et al. 1996. Leptin is a metabolic signal to the reproductive system. *Endocrinology* 137:3144–3147.

Block, M. L., X. Wu, and Z. Pei, et al. 2004. Nanometer size diesel exhaust particles are selectively toxic to dopaminergic neurons: The role of microglia, phagocytosis, and NADPH oxidase. *FASEB Journal* 18:1618–1620.

Bridgwater, A. V. and G. V. C. Peacocke. 2000. Fast pyrolysis processes for biomass. *Renewable & Sustainable Energy Reviews* 4:1–73.

Campen, M. J., N. S. Babu, and G. A. Helms, et al. 2005. Nonparticulate components of diesel exhaust promote constriction in coronary arteries from $ApoE^{-/-}$ mice. *Toxicological Sciences* 88:95–102.

Chisti, Y. 2007. Biodiesel from microalgae. *Biotechnology Advances* 25:294–306.
Cruts, B., L. van Etten, and H. Tornqvist, et al. 2008. Exposure to diesel exhaust induces changes in EEG in human volunteers. *Particle and Fibre Toxicology* 5:4.
Czernik, S. and A. V. Bridgwater. 2004. Overview of applications of biomass fast pyrolysis oil. *Energy & Fuels* 18:590–598.
Delfino, R. J., C. Sioutas, and S. Malik. 2005. Potential role of ultrafine particles in associations between airborne particle mass and cardiovascular health. *Environmental Health Perspectives* 113:934–936.
Ema, M., M. Naya, M. Horimoto, and H. Kato. 2013. Developmental toxicity of diesel exhaust: A review of studies in experimental animals. *Reproductive Toxicology* 42:1–17.
Fried, L. P., N. O. Borhani, and P. Enright, et al. 1991. The cardiovascular health study: Design and rationale. *Annals of Epidemiology* 1:263–276.
Gelb, D. J., E. Oliver, and S. Gilman. 1999. Diagnostic criteria for Parkinson disease. *Archives of Neurology* 56:33–39.
Gerlofs-Nijland, M. E., D. van Berlo, and F. R. Cassee, et al. 2010. Effect of prolonged exposure to diesel engine exhaust on proinflammatory markers in different regions of the rat brain. *Particle and Fibre Toxicology* 7:12.
Hartz, A. M. S., B. Bauer, M. L. Block, J. S. Hong, and D. S. Miller. 2008. Diesel exhaust particles induce oxidative stress, proinflammatory signaling, and P-glycoprotein upregulation at the blood-brain barrier. *FASEB Journal* 22:2723–2733.
Hazari, M. S., N. Haykal-Coates, and D. W. Winsett, et al. 2011. TRPA1 and sympathetic activation contribute to increased risk of triggered cardiac arrhythmias in hypertensive rats exposed to diesel exhaust. *Environmental Health Perspectives* 19:951–957.
Heneka, M. T., M. J. Carson, and J. El Khoury, et al. 2015. Neuroinflammation in Alzheimer's disease. *Lancet Neurology* 14:388–405.
Hess, R. A., D. Bunick, and K. H. Lee, et al. 1997. A role for oestrogens in the male reproductive system. *Nature* 390:509–512.
Hill, J., E. Nelson, D. Tilman, S. Polasky, and D. Tiffany. 2006. Environmental, economic, and energetic costs and benefits of biodiesel and ethanol biofuels. *Proceedings of the National Academy of Sciences of the United States of America* 103:11206–11210.
Hirano, S., A. Furuyama, E. Koike, and T. Kobayashi. 2003. Oxidative-stress potency of organic extracts of diesel exhaust and urban fine particles in rat heart microvessel endothelial cells. *Toxicology* 187:161–170.
Hougaard, K. S., K. A. Jensen, and P. Nordly, et al. 2008. Effects of prenatal exposure to diesel exhaust particles on postnatal development, behavior, genotoxicity and inflammation in mice. *Particle and Fibre Toxicology* 5:3.
Kannel, W. B., C. Kannel, R. S. Paffenbarger, and L. A. Cupples. 1987. Heart rate and cardiovascular mortality: The Framingham Study. *American Heart Journal* 113:1489–1494.
Kilian, L. 2009. Not all oil price shocks are alike: Disentangling demand and supply shocks in the crude oil market. *American Economic Review* 99:1053–1069.
Konur, O. 2000. Creating enforceable civil rights for disabled students in higher education: An institutional theory perspective. *Disability & Society* 15:1041–1063.
Konur, O. 2002a. Access to Nursing Education by disabled students: Rights and duties of nursing programs. *Nurse Education Today* 22:364–374.
Konur, O. 2002b. Assessment of disabled students in higher education: Current public policy issues. *Assessment and Evaluation in Higher Education* 27:131–152.
Konur, O. 2002c. Access to employment by disabled people in the UK: Is the Disability Discrimination Act working? *International Journal of Discrimination and the Law* 5:247–279.
Konur, O. 2006a. Participation of children with dyslexia in compulsory education: Current public policy issues. *Dyslexia* 12:51–67.

Konur, O. 2006b. Teaching disabled students in Higher Education. *Teaching in Higher Education* 11:351–363.
Konur, O. 2007a. A judicial outcome analysis of the Disability Discrimination Act: A windfall for the employers? *Disability & Society* 22:187–204.
Konur, O. 2007b. Computer-assisted teaching and assessment of disabled students in higher education: The interface between academic standards and disability rights. *Journal of Computer Assisted Learning* 23:207–219.
Konur, O., ed. 2021a. *Handbook of Biodiesel and Petrodiesel Fuels: Science, Technology, Health, and Environment.* Boca Raton, FL: CRC Press.
Konur, O., ed. 2021b. *Handbook of Biodiesel and Petrodiesel Fuels: Science, Technology, Health, and Environment. Volume 1. Biodiesel Fuels: Science, Technology, Health, and Environment.* Boca Raton, FL: CRC Press.
Konur, O., ed. 2021c. *Handbook of Biodiesel and Petrodiesel Fuels: Science, Technology, Health, and Environment. Volume 2. Biodiesel Fuels based on the Edible and Nonedible Feedstocks, Wastes, and Algae: Science, Technology, Health, and Environment.* Boca Raton, FL: CRC Press.
Konur, O., ed. 2021d. *Handbook of Biodiesel and Petrodiesel Fuels: Science, Technology, Health, and Environment. Volume 3. Petrodiesel Fuels: Science, Technology, Health, and Environment.* Boca Raton, FL: CRC Press.
Konur, O. 2021e. Biodiesel and petrodiesel fuels: Science, technology, health, and environment. In *Handbook of Biodiesel and Petrodiesel Fuels: Science, Technology, Health, and Environment. Volume 1. Biodiesel Fuels: Science, Technology, Health, and Environment*, ed. O. Konur. Boca Raton, FL: CRC Press.
Konur, O. 2021f. Biodiesel and petrodiesel fuels: A scientometric review of the research. In *Handbook of Biodiesel and Petrodiesel Fuels: Science, Technology, Health, and Environment. Volume 1. Biodiesel Fuels: Science, Technology, Health, and Environment*, ed. O. Konur. Boca Raton, FL: CRC Press.
Konur, O. 2021g. Biodiesel and petrodiesel fuels: A review of the research. In *Handbook of Biodiesel and Petrodiesel Fuels: Science, Technology, Health, and Environment. Volume 1. Biodiesel Fuels: Science, Technology, Health, and Environment*, ed. O. Konur. Boca Raton, FL: CRC Press.
Konur, O. 2021h Nanotechnology applications in the diesel fuels and the related research fields: A review of the research. In *Handbook of Biodiesel and Petrodiesel Fuels: Science, Technology, Health, and Environment. Volume 1. Biodiesel Fuels: Science, Technology, Health, and Environment*, ed. O. Konur. Boca Raton, FL: CRC Press.
Konur, O. 2021i. Biooils: A scientometric review of the research. In *Handbook of Biodiesel and Petrodiesel Fuels: Science, Technology, Health, and Environment. Volume 1. Biodiesel Fuels: Science, Technology, Health, and Environment*, ed. O. Konur. Boca Raton, FL: CRC Press.
Konur, O. 2021j. Characterization and properties of biooils: A review of the research. In *Handbook of Biodiesel and Petrodiesel Fuels: Science, Technology, Health, and Environment. Volume 1. Biodiesel Fuels: Science, Technology, Health, and Environment*, ed. O. Konur. Boca Raton, FL: CRC Press.
Konur, O. 2021k. Biomass pyrolysis and pyrolysis oils: A review of the research. In *Handbook of Biodiesel and Petrodiesel Fuels: Science, Technology, Health, and Environment. Volume 1. Biodiesel Fuels: Science, Technology, Health, and Environment*, ed. O. Konur. Boca Raton, FL: CRC Press.
Konur, O. 2021l. Biodiesel fuels: A scientometric review of the research. In *Handbook of Biodiesel and Petrodiesel Fuels: Science, Technology, Health, and Environment. Volume 1. Biodiesel Fuels: Science, Technology, Health, and Environment*, ed. O. Konur. Boca Raton, FL: CRC Press.

Konur, O. 2021m. Glycerol: A scientometric review of the research. In *Handbook of Biodiesel and Petrodiesel Fuels: Science, Technology, Health, and Environment. Volume 1. Biodiesel Fuels: Science, Technology, Health, and Environment*, ed. O. Konur. Boca Raton, FL: CRC Press.

Konur, O. 2021n. Propanediol production from glycerol: A review of the research. In *Handbook of Biodiesel and Petrodiesel Fuels: Science, Technology, Health, and Environment. Volume 1. Biodiesel Fuels: Science, Technology, Health, and Environment*, ed. O. Konur. Boca Raton, FL: CRC Press.

Konur, O. 2021o. Edible oil-based biodiesel fuels: A scientometric review of the research. In *Handbook of Biodiesel and Petrodiesel Fuels: Science, Technology, Health, and Environment. Volume 2. Biodiesel Fuels based on the Edible and Nonedible Feedstocks, Wastes, and Algae: Science, Technology, Health, and Environment*, ed. O. Konur. Boca Raton, FL: CRC Press.

Konur, O. 2021p. Palm oil-based biodiesel fuels: A review of the research. In *Handbook of Biodiesel and Petrodiesel Fuels: Science, Technology, Health, and Environment. Volume 2. Biodiesel Fuels based on the Edible and Nonedible Feedstocks, Wastes, and Algae*, ed. O. Konur. Boca Raton, FL: CRC Press.

Konur, O. 2021q. Rapeseed oil-based biodiesel fuels: A review of the research. In *Handbook of Biodiesel and Petrodiesel Fuels: Science, Technology, Health, and Environment. Volume 2. Biodiesel Fuels based on the Edible and Nonedible Feedstocks, Wastes, and Algae*, ed. O. Konur. Boca Raton, FL: CRC Press.

Konur, O. 2021r. Nonedible oil-based biodiesel fuels: A scientometric review of the research. In *Handbook of Biodiesel and Petrodiesel Fuels: Science, Technology, Health, and Environment. Volume 2. Biodiesel Fuels based on the Edible and Nonedible Feedstocks, Wastes, and Algae: Science, Technology, Health, and Environment*, ed. O. Konur. Boca Raton, FL: CRC Press.

Konur, O. 2021s. Waste oil-based biodiesel fuels: A scientometric review of the research. In *Handbook of Biodiesel and Petrodiesel Fuels: Science, Technology, Health, and Environment. Volume 2. Biodiesel Fuels based on the Edible and Nonedible Feedstocks, Wastes, and Algae: Science, Technology, Health, and Environment*, ed. O. Konur. Boca Raton, FL: CRC Press.

Konur, O. 2021t. Algal biodiesel fuels: A scientometric review of the research. In *Handbook of Biodiesel and Petrodiesel Fuels: Science, Technology, Health, and Environment. Volume 2. Biodiesel Fuels based on the Edible and Nonedible Feedstocks, Wastes, and Algae: Science, Technology, Health, and Environment*, ed. O. Konur. Boca Raton, FL: CRC Press.

Konur, O. 2021u. Algal biomass production for biodiesel production: A review of the research. In *Handbook of Biodiesel and Petrodiesel Fuels: Science, Technology, Health, and Environment. Volume 2. Biodiesel Fuels based on the Edible and Nonedible Feedstocks, Wastes, and Algae*, ed. O. Konur. Boca Raton, FL: CRC Press.

Konur, O. 2021v. Algal biomass production in wastewaters for biodiesel production: A review of the research. In *Handbook of Biodiesel and Petrodiesel Fuels: Science, Technology, Health, and Environment. Volume 2. Biodiesel Fuels based on the Edible and Nonedible Feedstocks, Wastes, and Algae*, ed. O. Konur. Boca Raton, FL: CRC Press.

Konur, O. 2021x. Algal lipid production for biodiesel production: A review of the research. In *Handbook of Biodiesel and Petrodiesel Fuels: Science, Technology, Health, and Environment. Volume 2. Biodiesel Fuels based on the Edible and Nonedible Feedstocks, Wastes, and Algae*, ed. O. Konur Boca Raton, FL: CRC Press.

Konur, O. 2021y. Crude oils: A scientometric review of the research. In *Handbook of Biodiesel and Petrodiesel Fuels: Science, Technology, Health, and Environment. Volume 3. Petrodiesel Fuels: Science, Technology, Health, and Environment*, ed. O. Konur. Boca Raton, FL: CRC Press.

Konur, O. 2021z. Petrodiesel fuels: A scientometric review of the research. In *Handbook of Biodiesel and Petrodiesel Fuels: Science, Technology, Health, and Environment. Volume 3. Petrodiesel Fuels: Science, Technology, Health, and Environment*, ed. O. Konur. Boca Raton, FL: CRC Press.

Konur, O. 2021aa. Bioremediation of petroleum hydrocarbons in the contaminated soils: A review of the research. In *Handbook of Biodiesel and Petrodiesel Fuels: Science, Technology, Health, and Environment. Volume 3. Petrodiesel Fuels: Science, Technology, Health, and Environment*, ed. O. Konur. Boca Raton, FL: CRC Press.

Konur, O. 2021ab. Desulfurization of diesel fuels: A review of the research. In *Handbook of Biodiesel and Petrodiesel Fuels: Science, Technology, Health, and Environment. Volume 3. Petrodiesel Fuels: Science, Technology, Health, and Environment*, ed. O. Konur. Boca Raton, FL: CRC Press.

Konur, O. 2021ac. Diesel fuel exhaust emissions: A scientometric review of the research. In *Handbook of Biodiesel and Petrodiesel Fuels: Science, Technology, Health, and Environment. Volume 3. Petrodiesel Fuels: Science, Technology, Health, and Environment*, ed. O. Konur. Boca Raton, FL: CRC Press.

Konur, O. 2021ad. The adverse health and safety impact of diesel fuels: A scientometric review of the research. In *Handbook of Biodiesel and Petrodiesel Fuels: Science, Technology, Health, and Environment. Volume 3. Petrodiesel Fuels: Science, Technology, Health, and Environment*, ed. O. Konur. Boca Raton, FL: CRC Press.

Konur, O. 2021ae. Respiratory illnesses caused by the diesel fuel exhaust emissions: A review of the research. In *Handbook of Biodiesel and Petrodiesel Fuels: Science, Technology, Health, and Environment. Volume 3. Petrodiesel Fuels: Science, Technology, Health, and Environment*, ed. O. Konur. Boca Raton, FL: CRC Press.

Konur, O. 2021af. Cancer caused by the diesel fuel exhaust emissions: A review of the research. In *Handbook of Biodiesel and Petrodiesel Fuels: Science, Technology, Health, and Environment. Volume 3. Petrodiesel Fuels: Science, Technology, Health, and Environment*, ed. O. Konur. Boca Raton, FL: CRC Press.

Konur, O. 2021ag. Cardiovascular and other illnesses caused by the diesel fuel exhaust emissions: A review of the research. In *Handbook of Biodiesel and Petrodiesel Fuels: Science, Technology, Health, and Environment. Volume 3. Petrodiesel Fuels: Science, Technology, Health, and Environment*, ed. O. Konur. Boca Raton, FL: CRC Press.

Lawal, A. O., L. M. Davids, and J. L. Marnewick. 2016. Diesel exhaust particles and endothelial cells dysfunction: An update. *Toxicology in Vitro* 32:92–104.

Levesque, S., T. Taetzsch, and M. E. Lull, et al. 2011a. Diesel exhaust activates and primes microglia: Air pollution, neuroinflammation, and regulation of dopaminergic neurotoxicity. *Environmental Health Perspectives* 119:1149–1155.

Levesque, S., M. J. Surace, J. McDonald, and M. L. Block. 2011b. Air pollution & the brain: Subchronic diesel exhaust exposure causes neuroinflammation and elevates early markers of neurodegenerative disease. *Journal of Neuroinflammation* 8:105.

Lucking, A. J., M. Lundback, and S. L. Barath, et al. 2011. Particle traps prevent adverse vascular and prothrombotic effects of diesel engine exhaust inhalation in men. *Circulation* 123(16), 1721–1728.

Lucking, A. J., M. Lundback, and N. L. Mills, et al. 2008. Diesel exhaust inhalation increases thrombus formation in man. *European Heart Journal* 29:3043–3051.

Lundback, M., N. L. Mills, and A. Lucking, 2009. Experimental exposure to diesel exhaust increases arterial stiffness in man. *Particle and Fibre Toxicology* 6:7.

Mills, N. L., M. R. Miller, and A. J. Lucking, et al. 2011. Combustion-derived nanoparticulate induces the adverse vascular effects of diesel exhaust inhalation. *European Heart Journal* 32:2660–2671.

Mills, N. L., H. Tornqvist, and M. C. Gonzalez, et al. 2007. Ischemic and thrombotic effects of dilute diesel-exhaust inhalation in men with coronary heart disease. *New England Journal of Medicine* 357:1075–1082.

Mills, N. L., H. Tornqvist, and S. D. Robinson, et al. 2005. Diesel exhaust inhalation causes vascular dysfunction and impaired endogenous fibrinolysis. *Circulation* 112:3930–3936.

Nemmar, A., B. Nemery, P. H. M. Hoet, J. Vermylen, and M. F. Hoylaerts. 2003a. Pulmonary inflammation and thrombogenicity caused by diesel particles in hamsters: Role of histamine. *American Journal of Respiratory and Critical Care Medicine* 168:1366–1372.

Nemmar, A., P. H. M. Hoet, and D. Dinsdale, et al. 2003b. Diesel exhaust particles in lung acutely enhance experimental peripheral thrombosis. *Circulation* 107:1202–1208.

Nemmar, A., M. F. Hoylaerts, P. H. M. Hoet, and B. Nemery 2004a. Possible mechanisms of the cardiovascular effects of inhaled particles: Systemic translocation and prothrombotic effects. *Toxicology Letters* 149:243–253.

Nemmar, A., P. H. M. Hoet, J. Vermylen, B. Nemery, and M. F. Hoylaerts. 2004b. Pharmacological stabilization of mast cells abrogates late thrombotic events induced by diesel exhaust particles in hamsters. *Circulation* 110:1670–1677.

North, D. C. 1991a. *Institutions, Institutional Change and Economic Performance*. Cambridge, Mass.: Cambridge University Press.

North, D. C. 1991b. Institutions. *Journal of Economic Perspectives* 5:97–112.

Peretz, A., J. H. Sullivan, and D. F. Leotta, et al. 2008. Diesel exhaust inhalation elicits acute vasoconstriction *in vivo*. *Environmental Health Perspectives* 116:937–942.

Perron, P. 1989. The great crash, the oil price shock, and the unit root hypothesis. *Econometrica: Journal of the Econometric Society* 57:1361–1401.

Toppari, J., J. C. Larsen, and P. Christiansen, et al. 1996. Male reproductive health and environmental xenoestrogens. *Environmental Health Perspectives* 104:741–803.

Tornqvist, H., N. L. Mills, and M. Gonzalez, et al. 2007. Persistent endothelial dysfunction in humans after diesel exhaust inhalation. *American Journal of Respiratory and Critical Care Medicine* 176:395–400.

Watanabe, N. and Y. Oonuki. 1999. Inhalation of diesel engine exhaust affects spermatogenesis in growing male rats. *Environmental Health Perspectives* 107:539–544.

Wilson, S. J., M. R. Miller, and D. E. Newby. 2018. Effects of diesel exhaust on cardiovascular function and oxidative stress. *Antioxidants & Redox Signaling* 28:819–836.

Yoshida, S., M. Sagai, and S. Oshio, et al. 1999. Exposure to diesel exhaust affects the male reproductive system of mice. *International Journal of Andrology* 22:307–315.

Yoshida, S. and K. Takeda. 2004. The effects of diesel exhaust on murine male reproductive function. *Journal of Health Science* 50:210–214.

Index

C

D

E

F

G

H

N

O

P

Q

R

S

T

Printed in the United States
by Baker & Taylor Publisher Services